Polymer Electrolytes

Polymer Electrolytes

Characterization Techniques and Energy Applications

Edited by

Tan Winie, Abdul K. Arof, and Sabu Thomas

WILEY-VCH

Editors

Dr. Tan Winie
School of Physics and Material Science
Faculty of Applied Sciences
Universiti Teknologi MARA
40450 Shah Alam, Selangor
Malaysia

Dr. Abdul K. Arof
Centre for Ionics University of Malaya
Department of Physics
University of Malaya
50603 Kuala Lumpur
Malaysia

Dr. Sabu Thomas
School of Chemical Sciences
Mahatma Gandhi University
Priyadarsini Hills
Kottayam 686560
Kerala, India

Library of Congress Card No.:
applied for

British Library Cataloguing-in-Publication Data
A catalogue record for this book is available from the British Library.

Bibliographic information published by the Deutsche Nationalbibliothek
The Deutsche Nationalbibliothek lists this publication in the Deutsche Nationalbibliografie; detailed bibliographic data are available on the Internet at <http://dnb.d-nb.de>.

Print ISBN: 978-3-527-34200-6
ePDF ISBN: 978-3-527-80543-3
ePub ISBN: 978-3-527-80546-4
oBook ISBN: 978-3-527-80545-7

Typesetting SPi Global, Chennai, India
Printing and Binding Markono Print Media Pte Ltd, Singapore

Printed on acid-free paper

10 9 8 7 6 5 4 3 2 1

Contents

Preface

Polymer electrolytes resulting from the complexation of polymer with organic or inorganic salts were first introduced by P.V. Wright in 1975 and proposed as potential material for electrochemical devices by M. Armand in 1978. The polymeric property of polymer electrolytes gives it advantages over liquid electrolytes in terms of leak-proof, size and shape flexibility. Polymer electrolytes are substituted for the liquid electrolyte in new generation electrochemical devices such as batteries, supercapacitors and solar cells.

Chapter 1 of this book provides an important review of the development of polymer electrolytes. Chapters 2 and 3 present the electrical and thermal properties of polymer electrolytes characterized by impedance spectroscopy, TGA and DSC, respectively. These chapters include theoretical considerations, the know-how needed to set up the experiments and how to analyze their results. There are also discussions on sources of errors along with suggestions for improvement. Chapters 4 to 12 focus on applications of polymer electrolytes, which cover the batteries, supercapacitors, solar cells and electrochromic windows. Challenges in fabrication and performance improvement of these devices were indentified along with suggestions for improvement.

This book brings together prestigious international authors. Novice reader will find an outline of basic theory, experimental set up and a discussion of experimental methods and data analysis, with examples and appropriate references. Leading researchers and faculty members will find this book very valuable as an excellent review and a comprehensive summary of the literature on the subject with a discussion of current theoretical and experimental issues.

Finally, we wish to express our gratitude and appreciation to the chapter contributors. Comments from reviewers are gratefully appreciated.

October 8, 2019

Tan Winie
Abdul K. Arof
Sabu Thomas

1

Polymer Electrolytes: State of the Art

Masashi Kotobuki

National University of Singapore, Department of Mechanical Engineering, 21 Lower Kent Ridge Rd., Singapore 119077, Singapore

Polymers are defined as large molecules or macromolecules, which consist of repeated subunits. The polymers may be synthetic plastics or natural biopolymers such as protein, DNA, and so on. In the past 20 years, polymers have been tailored as electron or ion conductors. When appropriate salt is added into some polymers, their ionic conductivity can be improved to the value that can be used as electrolyte. In the past three decades, many researchers have endeavored to develop new polymer electrolytes (PEs) due to their potential application in electrochemical/electrical power generation, storage, and conversion systems. As a result, a lot of new PEs have been found, characterized, and tried to be applied in electrochemical/electrical devices. Particularly, Li ion-conductive PEs have been of interest for application in Li batteries due to their high energy density. In this chapter, the state-of-the-art development of Li ion-conductive PEs is described.

1.1 Introduction

PE was first introduced in 1973 [1]. Since then, the research on PE has been eagerly performed by many researchers, especially in the early 1980s, due to the recognition of PEs in industrial applications.

PE is a membrane composed of salts dissolved in a polymer [2]. Some polymer matrixes such as polyethylene oxide (PEO) and poly(methyl methacrylate) (PMMA) can dissolve salts and form salt–polymer complexes due to the interaction between oxygen atom in the polymer chain and cation in the salt. This solvent-free and ion-conductive system has been expected to be widely applied in electrochemical devices like rechargeable solid-state batteries, especially rechargeable Li ion batteries. In recent years, PEs have other prospective applications in advanced electrochemical, electrochromic, and electronic devices such as fuel cells, supercapacitors, electrochemical sensors, analog memory devices, and electrochromic windows [3–7]. Figure 1.1 shows the structure of commercial Li ion batteries using graphite and $LiCoO_2$ as an anode and cathode, respectively. Li ions only exist in the cathode side when the batteries are constructed. The Li

Polymer Electrolytes: Characterization Techniques and Energy Applications,
First Edition. Edited by Tan Winie, Abdul K. Arof, and Sabu Thomas.

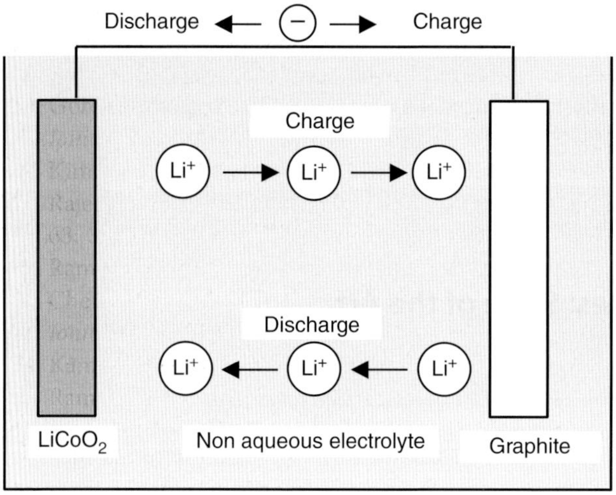

Figure 1.1 Li battery.

ions move from cathode to anode in a charge process. In a discharge process, Li ions migrate to the opposite direction. The electrolyte does not get involved in battery reactions in the Li batteries and just acts as Li ion-conductive media. In general, the electrodes of Li batteries are prepared by mixing three components, i.e. active material, binder, and conductive material. Poly(vinylidene fluoride) (PVdF) has been used as a binder thus far. As a conductive material, acetylene black and Ketjen black have been normally employed. These three components are mixed and added into a solvent *N*-methyl-2-pyrrolidone (NMP) to make a slurry. The slurry is painted onto Cu (for anode) or Al (for cathode) foil, which is used as a current collector. As for the electrolytes, nonaqueous electrolytes have been used so far due to narrow electrochemical window of aqueous electrolytes. The nonaqueous electrolytes are composed of Li salt (usually $LiPF_6$) dissolved into organic solvents, which are a mixture of acyclic solvent with low viscosity like dimethyl carbonate (DMC) and ethyl methyl carbonate (EMC) and cyclic solvent with high dielectric constant like ethylene carbonate (EC). However, flammability of these nonaqueous organic solvents sometimes has caused serious safety issues such as fire hazard and leakage of electrolyte [8–10]. Contrary to the nonaqueous electrolytes, solid polymer electrolytes (SPEs) can solve the issue of leakage. Additionally, PEs possess much lower flammability than the organic electrolytes due to low vapor pressure. The PEs should have the following physical, chemical, and electrochemical properties [11]:

(1) High ionic conductivity at operating temperature (normally room temperature), while electronic conductivity can be negligible.
(2) Sufficient mechanical strength at the operating temperature for self-supported cell.
(3) High electrochemical decomposition voltage (wide electrochemical window).
(4) High cationic or anionic transference number.

(5) Environmental benign, non-hygroscopic, low cost, and ease of preparation.
(6) Stability against chemical and electrochemical reactions with both electrodes (cathode and anode) during preparation and operation of the battery.
(7) Thermal expansion coefficient matches with that of the electrodes to ensure good contacts between PE and electrodes.

Table 1.1 summarizes a comparison of properties of conventional nonaqueous electrolyte and PE. Ceramic electrolyte, which is another type of solid electrolyte, is also compared. The ceramic electrolytes possess excellent safety due to their nonflammable nature although structural flexibility is very low and conductivity is low because of high grain boundary resistance. PEs normally possess high conductivity and structural flexibility as well as relatively wide electrochemical window.

PEs can be categorized into three groups based on their physical state and composition: (i) Solid polymer electrolyte (SPE), (ii) gel polymer electrolyte (GPE), and (iii) composite polymer electrolyte (CPE) [12, 13]. Also, PEs can be divided into two groups by polymer source, i.e. synthetic polymer and natural polymer (Figure 1.2). Synthetic polymers such as PEO and polycarbonate have been used as polymer matrix. Natural polymers like chitosan are usually low cost, eco-friendly, biodegradable, and abundant. Therefore, researches on natural polymer for PEs have been eagerly performed in recent years.

Table 1.1 Comparison of various properties among nonaqueous, polymer, and ceramics electrolytes.

	Nonaqueous electrolyte	Polymer electrolyte	Ceramics electrolyte
Conductivity	High	Middle	Low
Safety	Low	High	High
Structural flexibility	Middle	High	Low
Electrochemical window	Middle	Middle	Wide

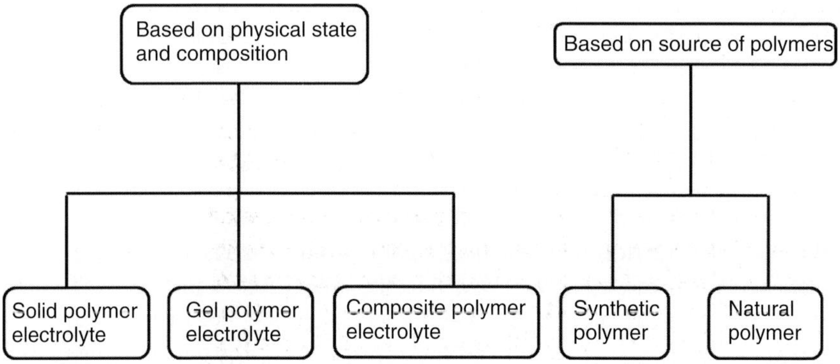

Figure 1.2 Classification of polymer electrolytes.

1.2 Solid Polymer Electrolyte

SPE is composed of host polymer matrix(es) and Li salts. The salts are dissolved into the polymer matrix(es) and provide ionic conduction. The research on SPEs commenced three decades ago [14]. The PEO-based SPE was investigated first and has been most widely researched so far [1]. The ether oxygen atoms in the PEO matrix complex with Li ion [15] and dissolve the Li salts. It is widely believed that the cation transport is related to the complexing segmental motion of PEO chain [16, 17] (Figure 1.3). Other polymer hosts such as PVdF, poly(vinylidene fluoride–hexafluoropropylene) (PVdF–HFP), PMMA, poly(vinyl chloride) (PVC), poly(acrylonitrile) (PAN), poly(acrylic acid) (PAA), poly(ethyl methacrylate) (PEMA), and so on also contain oxygen, nitrogen, chlorine, or fluorine atom, which can form a complex with Li ion, and the segmental motion would cause ion conduction. Table 1.2 summarizes repeat unit of polymer

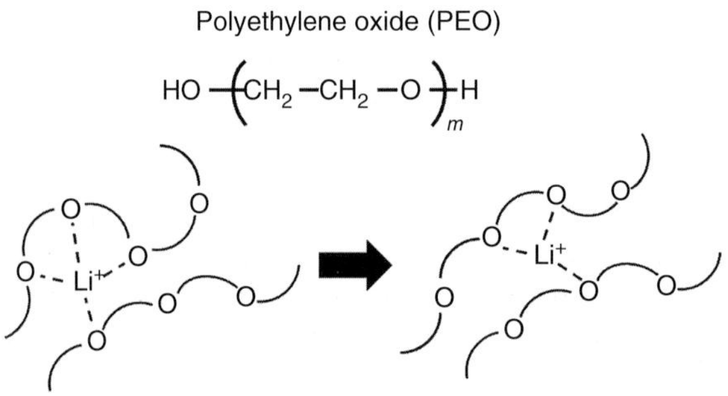

Figure 1.3 Segmental motion-assisted Li ion conduction in PEO-based polymer.

Table 1.2 Polymer host generally used in polymer electrolytes.

Polymer host	Repeat unit	Example of polymer electrolyte	Ionic conductivity at room temperature ($S\,cm^{-1}$)	References
PEO	$-(CH_2CH_2O)_n-$	(PEO-HBP)–LiTFSI–$BaTiO_3$	2.6×10^{-4}	[18]
PVdF	$-(CH_2CF_2)_n-$	PVdF–PEO–LiTFSI	5.4×10^{-4}	[19]
PVdF-HFP	$-[(CH_2CF_2)-(CF_2CFCF_3)]_n-$	P(VdF-HFP–SiO_2–LiTFSI	4.3×10^{-3}	[20]
PMMA	$-[(CH_2C(-CH_3)-COOCH_3)]_n-$	PMMA–$LiClO_4$–DMP–CeO_2	7.3×10^{-6}	[21]
PVC	$-(CH_2CHCl)_n-$	PVC–$Li_2B_4O_7$–DBP	2.83×10^{-6}	[22]
PAN	$-[CH_2CH(-CN)]_n-$	PAN–$LiClO_4$–Al_2O_3	5.7×10^{-4}	[23]
PAA	$-[CH_2CH(-COOH)]_n-$	PAA–$LiClO_4$–Sb_2O_3	2.15×10^{-4}	[24]
PEMA	$-[(CH_2C(-CH_3)-COOCH_2CH_3)]_n-$	PEMA–LiTf-IL	1.17×10^{-4}	[25]

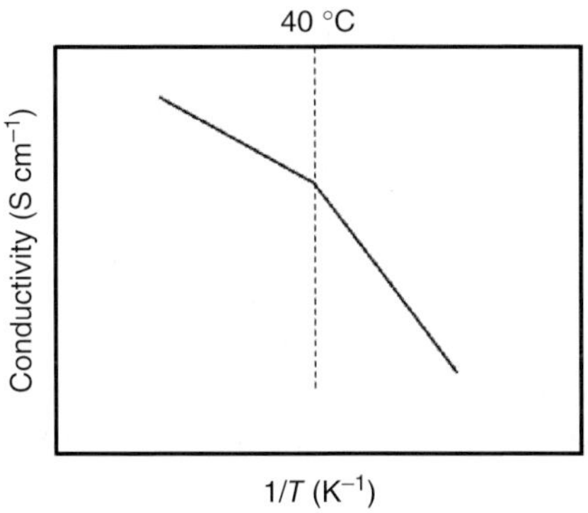

Figure 1.4 Schematic Arrhenius plot of PEO–$LiClO_4$ polymer electrolyte.

hosts and examples of PEs. The structural flexibility directly relates to the ionic conductivity. In other words, the amorphous phase in the SPEs supports ionic conduction. The continuous segmental motion of the amorphous chain occurs above the glass transition temperature (T_g) [26]. T_g of the PEs should be lower than room temperature. Moreover, increase in the amorphous regions with raised temperature improves the ionic conductivity. Figure 1.4 depicts a schematic Arrhenius plot of PEO–polystyrene (PSt)-$LiClO_4$ PE [27]. An inflection point around 40 °C can be observed in this plot. This corresponds to a phase transition from crystalline to amorphous phase of the polymer host. Due to the phase transition, the activation energy of ionic conduction is lowered considerably. However, mechanical strength of the SPE system is related to the movement of polymer chain. Increase in the amorphous regions implies that the polymer chains move actively, which enhances the ionic conductivity, but decreases the mechanical strength. This adverse effect leads to difficulty in construction of self-supported polymer batteries.

In order to solve this problem, many efforts have been devoted to develop novel polymer matrixes containing amorphous state, including blending of polymers, cross-linking, copolymerization, and so on. These approaches can lower crystallinity or T_g of the SPE systems and increase the ionic conductivity and mechanical strength [28–30].

Blend PEs are prepared by mixing two or more kinds of molecular chains (Figure 1.5). The chains are mixed with/without any chemical bonding between them. This blended polymer chain destroys the regularity of one polymer chain and prevents its rearrangement, resulting in formation of amorphous structure. Tanaka et al. prepared a polymer blend comprising of PEO and polyethylene imine (PEI). The ionic conductivity of [(8 : 2) PEO/PEI]–$LiClO_4$ was ~10^{-4} S cm^{-1} at 30 °C [31]. This high ionic conductivity is considered to

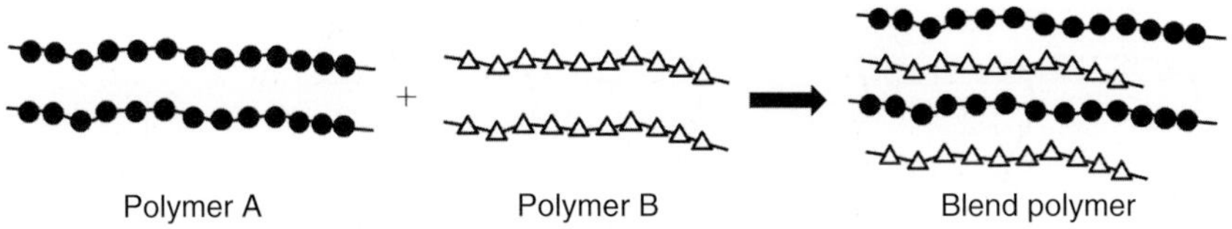

Figure 1.5 Schematic image of blend polymer.

be due to mixing of PEO with PEI that hindered their crystallization and led to more useful matrices.

Block copolymer electrolytes (BCEs) have been proposed as a novel SPE to satisfy both the ionic conductivity and the mechanical strength [32]. The most common structures researched are the diblock and triblock copolymers. A lithium salt-solvating polymer is used to ensure continuous ionic conduction pathways, and another polymer host, which forms a polymer framework of the SPEs, is chosen to provide the mechanical strength to the SPEs [33]. Polymer films with good mechanical strength can be obtained without sacrificing the ionic conductivity. Thus, a balance of the salt-solvating polymer and the framework polymer is important. Niitani et al. reported novel PSt–poly(ethylene glycol) (PEG) methyl ether methacrylate (PME)–PSt–$LiClO_4$ triblock copolymers (Figure 1.6) [34, 35]. The triblock SPE exhibited high ionic conductivity of 2.0×10^{-4} S cm^{-1} without plasticizer. The PSt block was used to improve mechanical properties, while the PEO moiety increased the ionic conductivity. An all-solid-state battery of $LiCoO_2$/SPE/Li cell demonstrated excellent charge–discharge property at room temperature. This concept can also increase Li ion transference number. Bouchet et al. reported a single-ion PE based on poly[lithium 4-styrenesulfonyl(trifluoromethylsulfonyl) imide] [P(STFSILi)–PEO–P(STFSILi)] polyanionic triblock copolymers [36]. This material demonstrated high Li ion transference number (>0.85), excellent mechanical strength, and good ionic conductivity (1.3×10^{-5} S cm^{-1} at 60 °C). The battery tests exhibited good power and cycling performances at 60 °C.

Cross-linking PEs show good ionic conductivity at ambient temperature and fully amorphous feature (Figure 1.7) [37]. However, the cross-linking polymer generally exhibits low elasticity and brittleness as well as low processability [38]. A cross-linked high molecular weight poly(oxyethylene) has been reported as a

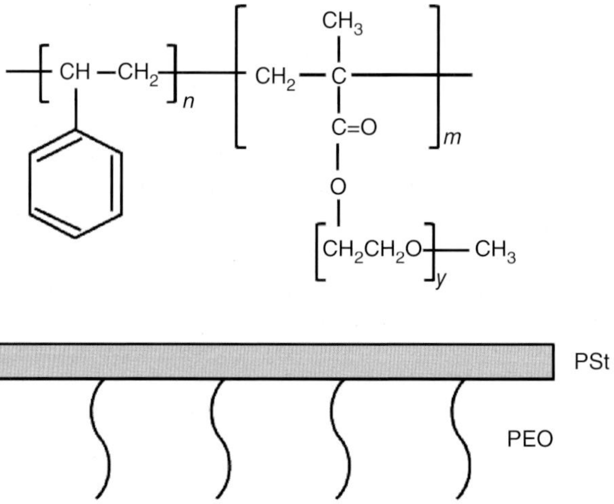

Figure 1.6 Schematic image of novel nanostructure-controlled SPE and its synthetic scheme.

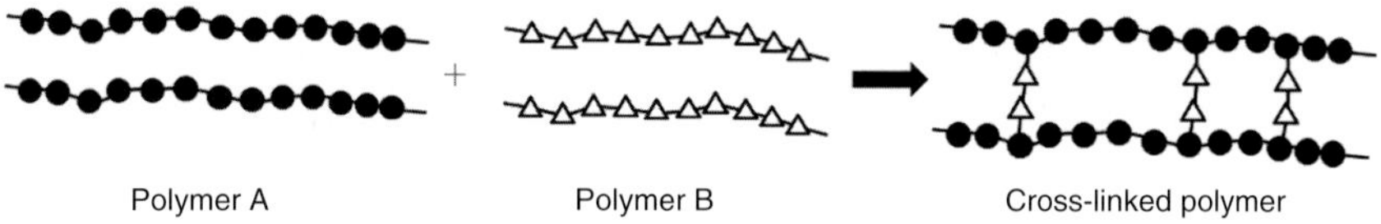

Figure 1.7 Schematic image of the cross-linked polymer.

new SPE, which demonstrated favorable ionic conductivity and good mechanical strength [39].

Copolymer is a polymer prepared from at least two different types of monomers. PVdF-HFP is the most common copolymer researched as PE. The PVdF-HFP copolymer is prepared by copolymerization of PVdF and HFP (Figure 1.8). The copolymer exhibits better features compared with the mono-polymers alone, which could be attributed to the synergistic effects in the combined structure. Also, PVdF has received much attention due to good electrochemical stability and high dielectric constant [40, 41]. The presence of strong electron-withdrawing fluorine atoms (C–F) promotes dissociation of salts and increases the concentration of charge carriers, leading to high ionic conductivity. Jiang et al. reported ionic conductivity of PVdF-based SPEs of above $10^{-4}\ S\ cm^{-1}$ at room temperature [41].

Other polymer hosts also have been researched. The potential of PMMA as a polymer host was reported by Iijima et al [42]. The PMMA-based SPEs showed low mechanical integrity and high brittleness [43]. Blending of PMMA with PVC has also been researched [44, 45]. PVC is also an attractive polymer host due to its low cost and easy processing. GPEs based on PVC with plasticizers have been widely researched. Poly(vinyl alcohol) (PVA) is nontoxic and cost effective and possess good tensile strength, good mechanical strength, good optical properties, high temperature resistance, high abrasion resistance, good flexibility, biocompatibilities, high hydrophilicity, and excellent chemical and thermal stabilities [46–50]. PVA contains a large amount of polar hydroxyl group, leading to high hydrophilicity. This provides other advantages like ease in preparation and high dielectric constant. Due to these superior properties, PVA has received considerable research interest as electrolytes for fuel cells and electrical double layer capacitors [51]. A PAN-based PE has some outstanding characteristics such as high thermal stability and high ionic conductivity [52]. PAN is superior over PVdF with respect to mechanical stability [53]. The –CN groups in PAN can interact with cations. Structures of these polymer hosts are shown in Figure 1.9.

Additionally, natural polymer such as chitosan [54], rice starch [55], and corn starch [56] has also been studied. These have an advantage as novel polymer hosts due to being low cost, biodegradable, eco-friendly, and abundant.

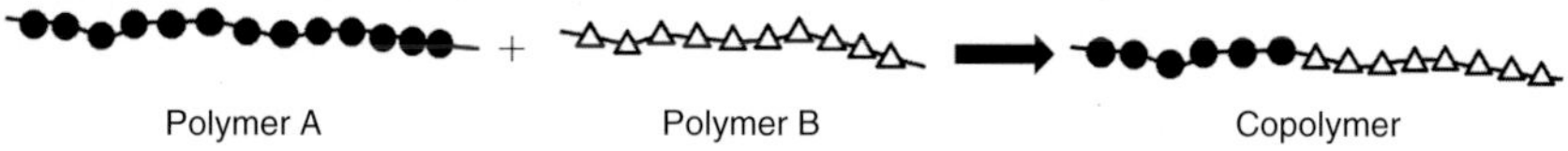

Figure 1.8 Schematic image of the copolymer.

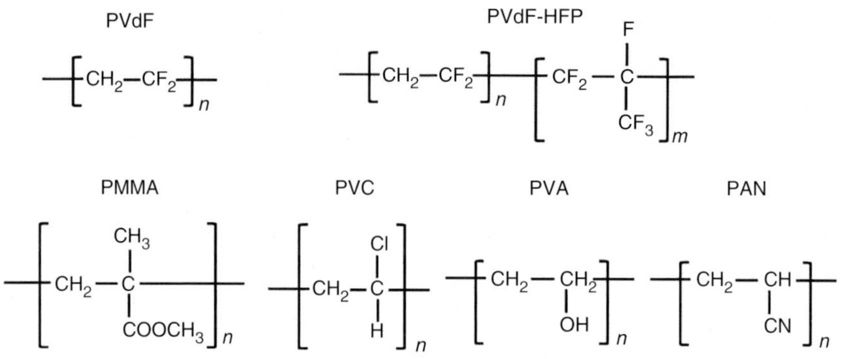

Figure 1.9 Structure of polymer hosts.

1.3 Gel Polymer Electrolyte

GPE is also known as plasticized PE, which was first introduced by Feuillade and Perche in 1975 [57]. GPE contains a plasticizer or gelled polymer matrix, which is swollen by addition of the plasticizers [58], and can be prepared by simply heating a mixture of polymer and Li salt with solvent. By introducing a plasticizer and/or solvent, the ion transport is not dominated by the segmental motion of polymer chains but occurs in the swollen gelled phase or liquid phase. In general, when the polymer is composed of interconnected micropores, the ionic conductivity of GPEs mainly depends on the properties of trapped liquid electrolyte. On the contrary, ion transport mainly occurs in the swollen gelled phase if the polymer does not contain many interconnected pores. The GPEs should possess good mechanical strength, capability of holding a liquid electrolyte, high ionic conductivity, and electrochemical stability toward both cathode and anode. Many kinds of polymer matrix such as PEO [59], PMMA [60], PAN [61], PVC [62], PVdF [63], and P(VdF-HFP) copolymer [64] have been widely studied as a framework for GPEs.

Plasticizers, which are usually low molar-mass organics, organic solvents, or ionic liquids (ILs), largely affect the properties of GPEs. A plasticizer can increase the content of the amorphous phase in a PE and promote segmental motion [65]. In addition, it can also promote dissociation of ion pairs. As a result, the number of charge carriers is increased, leading to enhanced ionic conductivity [59]. PEG has been widely used as a low molar-mass plasticizer. It was reported that the ionic conductivity of PEO–$LiCF_3SO_3$ complex increased with the decrease of molecular weight of PEG and with the increase of PEG content [66]. However, the hydroxyl end groups in PEG react with electrode materials such as lithium metal. Therefore, various modified forms of PEG were synthesized by replacing active oxygen atoms in PEG with monomethoxy (MMPEG), dimethoxy (DMPEG) groups, or lithium (LPEG) ions [67]. The LPEG plasticizer can improve the compatibility of the PE with lithium metal anode [13]. On the contrary, in some polymer systems such as PEO-PMMA, no significant improvement of ionic conductivity by addition of PEG was reported. The plasticizer must be chosen carefully depending on the polymer host used. Other low molar-mass plasticizers such as polyethylene glycol dimethyl ether (PEGDME) [68], borate ester such as PEG borate ester [69], tris(2-(2-methoxyethoxy)ethyl) borate (B_2), and tris(2-(2-(2-methoxyethoxy)ethyl borate (B_3) [70], phthalates such as dibutyl phthalate (DBP) [71], dimethyl phthalate (DMP) [65], dioctyl phthalate (DOP) [72], succinonitrile (SN) [73], and so on have also been studied. The ionic conductivity of the PEs containing these plasticizers is summarized in Table 1.3.

The organic solvents usually used as plasticizer are polar and nonvolatile solvents such as EC, propylene carbonate (PC), diethyl carbonate (DEC), and DMC. The solvents help to solvate ions and facilitate their transportation. Therefore, high dielectric constant and low viscosity are required for the solvents. Individual solvent is difficult to meet all the requirements; thus, a mixture of the solvents usually has been employed. The mixture of solvents is more efficient to enhance the ionic conductivity compared with a single solvent, which is due to the combined action of dielectric constant and viscosity [75]. The ionic conductivity of some GPEs using the plasticizers is tabulated in Table 1.4. The ionic conductivity can be increased to $\sim 10^{-3}$ S cm^{-1} by the addition of suitable solvents. Choi et al. studied the ionic conductivity of PAN polymer swollen by 1 M $LiPF_4$ in EC/DMC (1 : 2 wt%), EC/DMC (1 : 1), EC/EMC (1 : 1), EC/DEC (1 : 1), and EC/DMC/DEC (1 : 1 : 1) [61]. The order of ionic conductivity was EC/DMC/DEC (1 : 1 : 1) > EC/DMC (1 : 1) > EC/EMC (1 : 1) > EC/DEC (1 : 1) > EC/DMC (2 : 1).

ILs have attracted considerable interest as a novel plasticizer. They are room temperature molten salts, which are composed of a bulky organic cation and a large delocalized inorganic anion. The ILs possess some unique properties such as high chemical and thermal stabilities, nonflammability, negligible volatility, and high electrochemical stability [83–85]. Due to these unique properties, the incorporation of room temperature ionic liquids (RTILs) into PEs can overcome inherent limitations of the ionic conductivity in SPEs as proposed by Passerini

Table 1.3 Ionic conductivity of various polymer electrolytes with plasticizers.

Polymer	Ionic conductivity (S cm^{-1})	Temperature (°C)	References
$(PEO)_{15}$/LiTFSI/10 wt% PEGDME	$>10^{-3}$	50	[59]
$(PEO)_{20}$/$LiCF_3SO_3$/50 wt% PEGDME	1.2×10^{-4}	30	[68]
30 wt% PVA/10 wt% $LiClO_4$/60 wt% DMP	1.5×10^{-4}	29	[65]
17.5 wt% PVA+7.5 wt% PMMA/8 wt% $LiClO_4$/67 wt% DMP	6×10^{-5}	30	[74]
PEO/15 wt% $LiCF_3SO_3$/15 wt% PEG	1.7×10^{-5}	RT	[72]
PEO/15 wt% $LiCF_3SO_3$/20 wt% DOP	7.6×10^{-4}	RT	[72]
$(PEO)_{20}$/LiBOB/24 mol% SN	$>6\times10^{-4}$	60	[73]
PEO/13 wt% $LiCF_3SO_3$/10 wt% DBP	1.6×10^{-4}	27	[71]
$(PEO)_{20}$/LiTFSI/100 wt% PEG borate ester	1.36×10^{-5}	30	[69]
PSi/15 wt% $LiCF_3SO_3$/40 wt% borate ester B_2	3.70×10^{-5}	RT	[70]
PSi/15 wt% $LiCF_3SO_3$/40 wt% borate ester B_3	1.60×10^{-4}	RT	[70]

Table 1.4 Ionic conductivity of various polymer electrolytes with organic solvents.

Polymer	Ionic conductivity (S cm^{-1})	Temperature (°C)	References
P(VdF-HFP)/1.0 M $LiPF_6$/EC + DEC	1.0×10^{-3}	RT	[76]
P(VdF-HFP)/1.0 M $LiPF_6$/EC + DMC + DEC	1.43×10^{-3}	RT	[64]
PVdF/1.0 M $LiPF_6$/EC + DMC + DEC	1.0×10^{-3}	RT	[77]
30 wt% PVC/8 wt% $LiClO_4$/62 wt% PC	6.70×10^{-6}	30	[62]
PEO/15 wt% $LiCF_3SO_3$/20 wt% EC	8.12×10^{-5}	RT	[78]
$(PEO)_{16}$/$LiClO_4$/40 wt% EC	2.67×10^{-4}	RT	[79]
4.5 wt% PMMA/46.5 wt% $LiClO_4$/30 wt% PC + 19 wt% EC	5.0×10^{-4}	RT	[80]
PAN/1 M $LiPF_6$/EC + DMC + DEC	$>1.0 \times 10^{-3}$	RT	[61]
PAN/1 M $LiBF_4$/EC + DEC	2.80×10^{-3}	RT	[81]
PVdF/1.0 M LiPF6/EC + DMC	1.00×10^{-3}	RT	[63]
7.5 wt% PVC/5 wt% $LiBF_4$/42 wt% EC + 28 wt% PC	8.60×10^{-5}	RT	[82]

et al. in 2003 [86]. Since then, many groups have devoted much effort to study GPEs containing ILs. Many types of IL comprising cations based on pyridium, imidazolium, piperidinium, quaternary ammonium, and so on and anions based on $[BF_4]^-$, $[PF_6]^-$, $[N(CF_3SO_2)_2]^-$, $[CF_3SO_3]^-$, $[C_4F_9SO_3]^-$, $[N(CN)_2]^-$, $[CF_3CO_2]^-$, $[CF_3CONCF_3SO_2]^-$, and so on have been investigated. In most cases, the ILs contain the same anion as salts such as IL containing $[PF_6]^-$ anion in $LiPF_6$. This is because the solubility of the salt into the IL incorporating the same anion is much higher than in a system of different anions. Passerini et al. studied a series of ILs containing pyrrolidinium-based cations and TFSI anions [87–94]. The structure of *N*-alkyl-*N*-methyl-pyrrolidium bis(trifluoromethanesulfonyl) imide ($PYR_{1A}TFSI$, A = C_nH_{2n+1}, $1 < n < 10$) is depicted in Figure 1.10. The commonly used ILs for PEs are $PYR_{13}TFSI$ (1-propyl-1-methylpyrrolidinium bis(fluorosulfonyl) imide) and $PYR_{14}TFSI$ (1-butyl-1-methylpyrrolidinium bis(trifluorosulfonyl) imide). The ionic conductivity of PEO/LiTFSI/$PYR_{13}TFSI$ PE is $\sim 10^{-4}$ S cm^{-1} at 20 °C, which is about two orders of magnitude higher than that without the IL [86]. It was reported that the $PYR_{13}TFSI$ also improved the ionic conductivity of PVdF-based polymer. The GPE composed of P(VdF-HFP)/LiTFSI/$PYR_{13}TFSI$ showed high ionic conductivity of 2.7×10^{-4} S cm^{-1} [95]. Also, incorporation of $PYR_{14}TFSI$ plasticizer into P(VdF-HFP)/LiTFSI PE showed a good ionic conductivity of 4.0×10^{-4} S cm^{-1} and high thermal stability [96]. An interesting study was performed by Winter et al. [83, 84]. They conducted in situ UV photoradiation of a complex of PRO/LiTFSI/$PYR_{14}TFSI$ using benzophenone as a cross-linking agent to obtain high conductive PEs with high mechanical properties.

ILs containing 1-alkyl-3-methylimidazolium cation have also been used as a plasticizer in GPEs. The structure of 1-alkyl-3-methylimidazolium

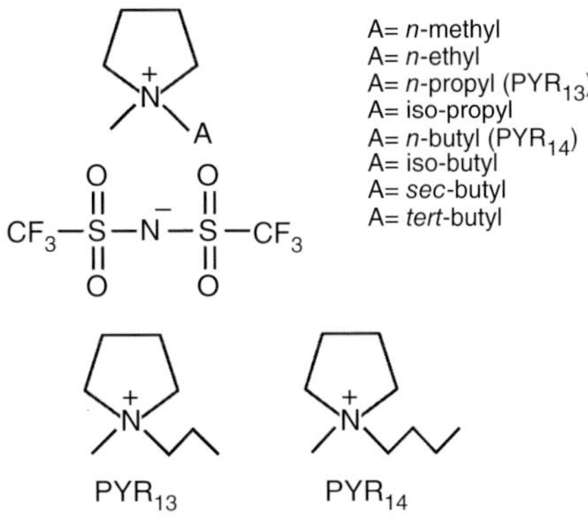

Figure 1.10 Structure of *N*-alkyl-*N*-methyl-pyrrolidium bis(trifluoromethanesulfonyl) imide.

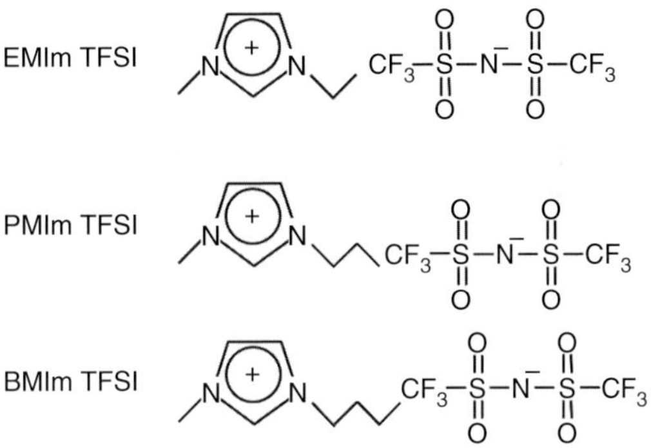

Figure 1.11 Structure of 1-alkyl-3-methylimidazolium bis(trifluoromethylsulfonyl) imide.

bis(trifluoromethylsulfonyl) imide IL is depicted in Figure 1.11. The alkyl group in the cation affects the ionic conductivity of GPEs. EMImTFSI, PMImTFSI, and BMImTFSI were employed as plasticizers to form GPEs based on P(VdF-HFP) matrix [97]. All these GPEs revealed a high ionic conductivity in the range of 2.4×10^{-3} to $4.5\times10^{-3}\,S\,cm^{-1}$. The GPE containing EMImTFSI showed the highest conductivity. Also, effects of anion on conductivity of PEO-based GPE were studied using ILs containing BMIm cation [98]. In the study, three different anions ($TFSI^-$, BF_4^-, and $CF_3SO_3^-$) and BMIm cation were employed. The PEO/LiTFSI/BMImTFSI electrolyte showed the highest ionic conductivity and good electrochemical stability [99].

As expected, compatibility between ILs and polymer is critical to determine the properties of GPEs. It was reported that quaternary ammonium-based ILs were more compatible than imidazolium-based ILs with PEO-PMMA copolymer [100]. This is thought to be due to preferable interaction between IL and polymer matrix.

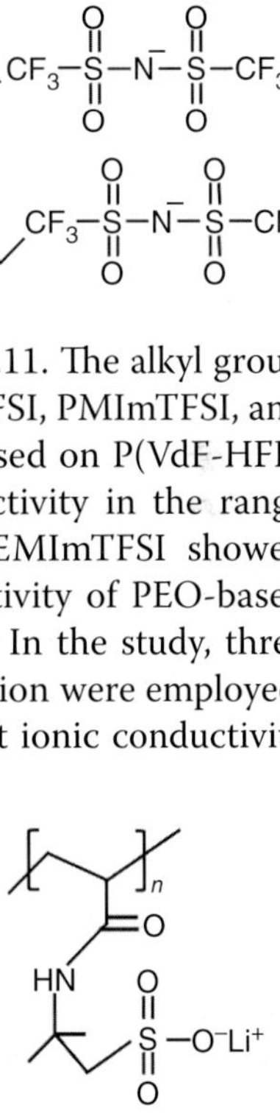

Figure 1.12 Structure of PAMPSLi.

The GPEs comprising polymeric lithium salts and ILs have been proposed as novel GPEs. Lithium poly(2-acrylamido-2-methyl propanesulfonate) (PAMPSLi, Figure 1.12) is used as the polymeric lithium salt. The PAMPSLi-PVdF copolymer was combined with 1-ethyl-3-methylimidazolium tricyanomethanide (EMImTCM). The ionic conductivity of PAMPSLi-PVdF

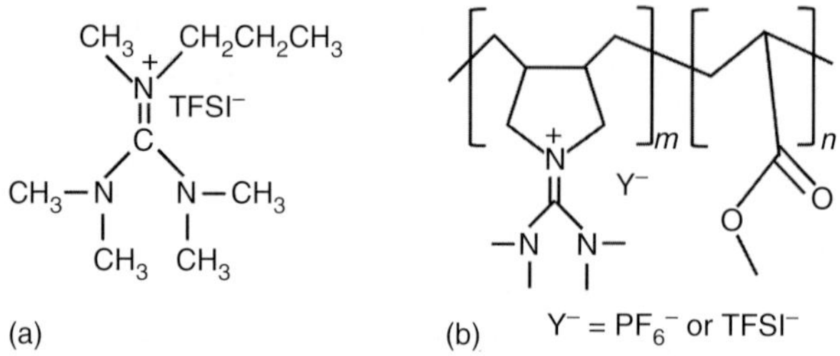

Figure 1.13 Structure of (a) guanidinium IL and (b) guanidinium-based PIL.

copolymer with EMImTCM IL was 5.43×10^{-3} S cm^{-1}, which was four times higher than that of homopolymer system of PAMPSLi-PVdF [101].

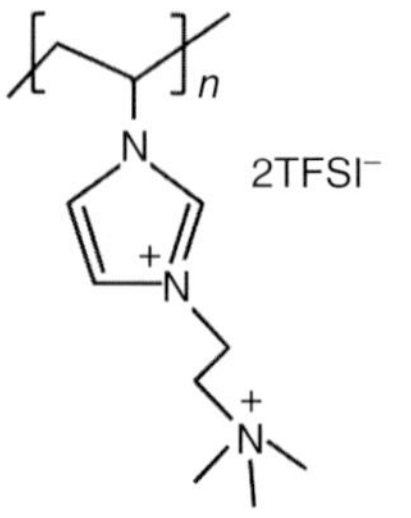

Figure 1.14 Structure of imidazolium–tetraalkylammonium-based PIL.

Also, polymeric ionic liquids (PILs) obtained by the polymerization of an IL monomer have attracted much attention as the polymer matrix in GPEs. The main advantages of the PIL-based GPEs are low flammability and high anodic stability. Guanidinium-based PILs containing different anions (PF_6^- and $TFSI^-$) were prepared by the polymerization of a guanidinium IL monomer with methyl acrylate (Figure 1.13). A quaternary GPE with a guanidinium PIL (matrix), a guanidinium IL, LiTFSI salt, and nano-SiO_2 particles showed high conductivity and wide electrochemical stability window [102]. Yin et al. synthesized a novel dicationic PIL and employed it for the GPE matrix (Figure 1.14) [103]. The novel GPE with dicationic PIL, poly(*N*,*N*,*N*-trimethyl-*N*-(1-vinylimidazolium-3-ethyl) ammonium bis(trifluoromethanesulfonyl)imide), 1,2-dimethyl-3-ethoxyethylimidazolium TFSI (IM(2o2)11TFSI) IL, and LiTFSI salt exhibited a low T_g (−54 °C), high thermal stability (330 °C), good ionic conductivity (about 10^{-4} S cm^{-1} at 25–40 °C), and high electrochemical stability. In another research using PIL, a less flammable GPE was prepared by the in situ polymerization of an IL monomer, 1-methyl-2-(2-acryloyloxyethyl)imidazolium tetrafluoroborate (MAHI-BF_4) [104].

1.4 Composite Polymer Electrolyte

CPEs were discovered in an attempt to overcome disadvantages and limitations of PEs. Generally, the CPEs are prepared by dispersing inorganic fillers such as insulative ceramic fillers (Al_2O_3, SiO_2, TiO_2, etc.), ferroelectric ceramic fillers ($BaTiO_3$, $PbTiO_3$, $LiNbO_3$, etc.), clay, carbon nanotubes (CNTs), and powder of fast ionic conductors into the polymer matrix [80, 105, 106], which can improve the mechanical properties and ionic conductivity of PEs [107–109]. Also, the fillers can increase the ionic conductivity by percolating interfacial effect, i.e. anions adsorb on the surface of fillers due to Lewis acid–base character and then promote dissociation of ion pairs, leading to increase interfacial ionic conductivity. Another reason for improvement of ionic conductivity by the fillers is thought to be due to a reduction in the crystallinity of the polymer–salt system because the nanoparticles can reduce crystallinity [110–112]. Accordingly, particle size, content, and surface functional groups of fillers are critical factors to determine the properties of the CPEs.

A novel PE that was PEO-based incorporated with inert ceramic fillers such as Al_2O_3, TiO_2, and SiO_2 was reported as a successful approach to enhance ionic conductivity [113, 114]. Figure 1.15 depicts an image of the Arrhenius plots of $P(EO)_8LiClO_4 + 10\,wt\%$ ceramic nanoparticle electrolyte during cooling. In the filler-free electrolyte, an inflection point was observed around 84 °C. This was attributed to the crystallization of PEO polymer. On the contrary, the inflection point did not appear in the electrolytes containing TiO_2 or SiO_2 filler. The fillers prevented the crystallization of PEO polymer and helped to keep high conductive amorphous phase. As a result, the filler containing PEO polymer exhibited high ionic conductivity.

The amorphous state of PEO polymer was stable over 60 days at 30 °C by the addition of filler. The fillers stabilized the high conductive amorphous state at ambient temperature for a long time. Similar behavior was also observed in Al_2O_3-added PEO-$LiCF_3SO_3$ composite electrolyte [72, 115].

The addition of fillers can also enhance the iconic conductivity of PEO-PVdF blend PE [116]. Yoon et al. fabricated a composite electrolyte based on PEO-PVdF–$LiClO_4$–2 wt% silica aerogel. Highest conductivity of $1.7 \times 10^{-4}\,S\,cm^{-1}$ was obtained at 30 °C in the composite composed of PEO:PVdF = 3 : 1 and polymer:salt = 6 : 1. The ionic conductivity decreased with an increase of silica aerogel content owing to degradation of mobility caused by the aggregation of silica aerogel particles. The surface functional groups on the fillers affect properties of the CPEs. It was reported that addition of acid-modified nano-SiO_2 reduced the interfacial resistance and prevented dendrite formation of Li metal significantly [117].

Among the nanosized ceramic fillers, TiO_2 was thought to be the best candidate for the CPEs. Lin et al. studied the effects of particle size of TiO_2 on ion transport properties of PEO-based composite electrolytes [118]. It was

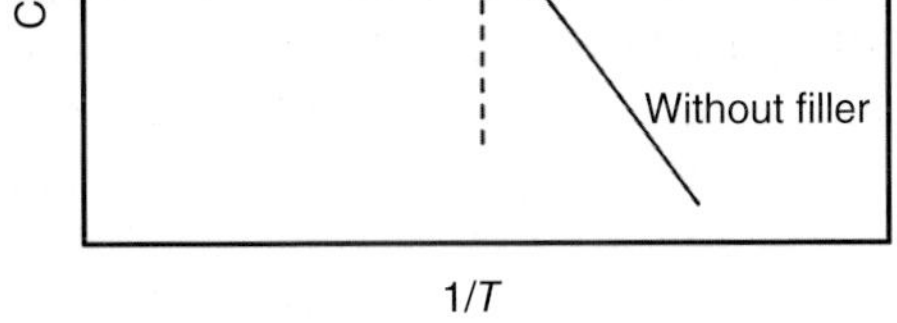

Figure 1.15 Arrhenius plots of the ionic conductivity of $P(EO)_8LiClO_4$.

found that the ionic conductivity increased with a decrease in the particle size of TiO_2. The conductivity of 1.40×10^{-4} S cm^{-1} was obtained for PEO–10 wt% $LiClO_4$–5 wt% TiO_2 (3.7 nm) electrolyte. The transference number also increased from 0.21 to 0.51 by the addition of TiO_2. This result indicated that nanosized TiO_2 particles in the PEO-$LiClO_4$ matrix formed a new pathway for Li ion transport. It is thought that TiO_2–Li^+ interaction changed the Li ion environment and provided a fast Li ion-conducting pathway at the interface between the fillers and the polymer [119].

A highly conductive layer on the surface of filler particles is created by an interaction between the filler and cations. Therefore, ferroelectric ceramic fillers are expected to interact with the cations easily owing to the permanent dipole of the ferroelectric ceramic materials, resulting in enhancement in ionic conductivity [120]. Sun et al. studied electrochemical properties of composite electrolytes based on PEO, Li salts ($LiClO_4$, $LiBF_4$, $LiCF_3SO_3$, and $LiN(CF_3SO_2)_2$), and ferroelectric ceramic materials ($BaTiO_3$, $PbTiO_3$, and $LiNbO_3$) [121]. The results showed that the ionic conductivity was greatly enhanced by only a small amount of ferroelectric filler addition. The highest ionic conductivity was obtained for 2 wt% ferroelectric $BaTiO_3$ addition. Also, it was found that the enhancement in ionic conductivity was affected by anions of the Li salt. The enhancement in conductivity was highest in ClO_4^- anion. On the contrary, no enhancement was observed in the composite with $LiCF_3SO_3$ salt. This enhancement of conductivity can be explained by the association tendency of anions with Li ions and the spontaneous polarization of the ferroelectric ceramics due to their particular crystal structure.

The usage of layered clays such as montmorillonite (MMT) and hecorite as fillers has also been reported [122, 123]. The layered clay can contain cations due to intercalation reaction. It was reported that the ionic conductivity of PEO–LiTFSI composite containing 10 wt% MMT was 3.22×10^{-4} and 2.75×10^{-5} S cm^{-1} at 60 and 25 °C, respectively [124]. Li-intercalated clay (bentonite) also increased the ionic conductivity about one order of magnitude when it was incorporated into PEO-based PE [125]. CNT has also been researched as fillers to enhance the ionic conductivity of PEs. The strong affinity between CNT's electron cloud and cation promotes salt dissociation, resulting in high conductivity. However, compositing CNTs with the PE may enhance the electronic conductivity due to high electronic conductivity of CNTs. Therefore, Tang et al. prepared CNT packaged in insulating clay for nanofiller. This hybrid nanofiller can block electron conduction in CNT and eliminate the risk of increased electronic conductivity of CPEs. This hybrid filler could increase the ionic conductivity of PEO-$LiClO_4$ electrolyte (Figure 1.16) [126]. The highest conductivity was obtained for PEO-$LiClO_4$ containing 10 wt% clay–CNT hybrid filler.

Inorganic ceramic solid electrolytes, namely, fast ceramic ionic conductors, are also considered as promising fillers because it can provide another pathway for ionic transport (Figure 1.17).

The CPEs with many Na superionic conductor (NASICON) type ionic conductors with the general formula of $LiM_2(PO_4)_3$ (M = Ti, Ge, Sr, Zr, Sn, etc.) (Figure 1.18) have been investigated in the past decade [127]. Among the

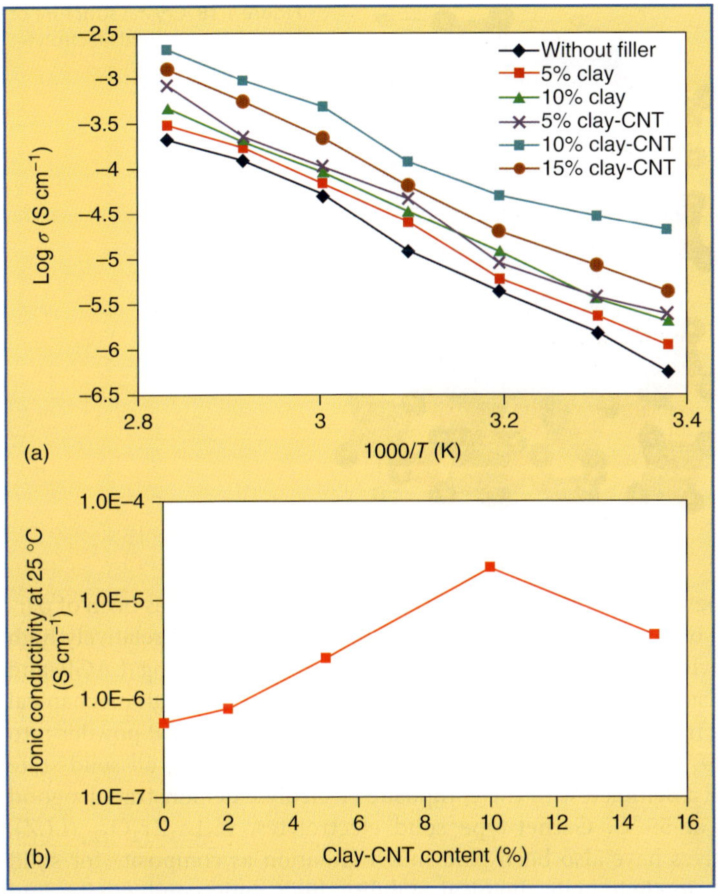

Figure 1.16 (a) Ionic conductivity of PEO-based composite electrolytes and (b) ionic conductivity of PEO-based composite electrolyte with various contents of the hybrid filler.

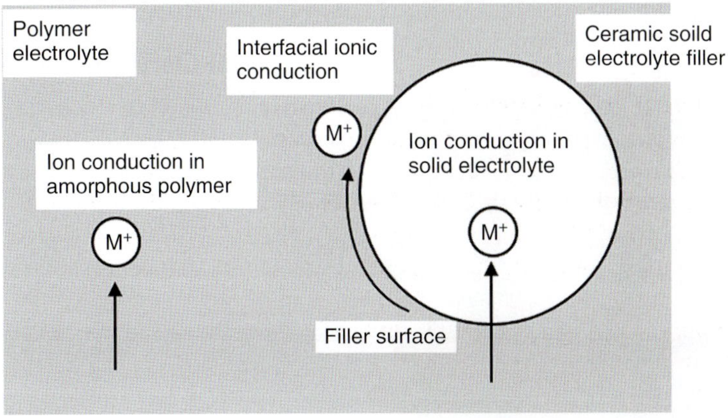

Figure 1.17 Three conduction pathway in composite electrolyte with inorganic solid electrolyte filler.

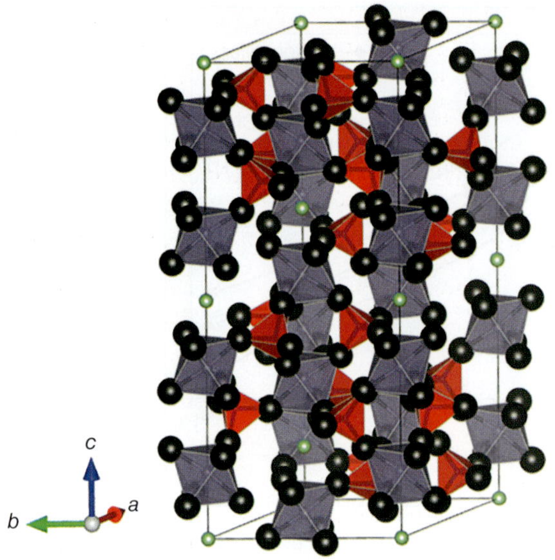

Figure 1.18 Crystal structure of NASICON-type solid electrolyte.

NASICON-type ionic conductors, $Li_{1.5}Al_{0.5}Ge_{1.5}(PO_4)_3$ (LAGP) has been of particular interest because of its wide electrochemical window and relatively high ionic conductivity. Solvent-free composite electrolyte comprising LAGP and PEO–$LiClO_4$ has been studied [128]. The electrolyte possessed high mechanical strength and is free standing. By addition of ion-conductive LAGP powder into the PEO-$LiClO_4$ PE, the ionic conductivity was improved. The all-solid-state $LiFePO_4$/Li cell assembled with the composite electrolyte demonstrated a good cycle stability at 55 °C. Garnet-type solid electrolytes, $Li_7La_3Zr_2O_{12}$ (LLZ), and its derivatives have also been paid much attention as composite for solid electrolyte due to their high chemical stability, high ionic conductivity, and wide electrochemical window [129–131]. The structure of LLZ is depicted in Figure 1.19. The framework of LLZ garnet is composed of dodecahedral LaO_8 and octahedral ZrO_6. The CPEs composed of LLZ and PEO–$LiClO_4$

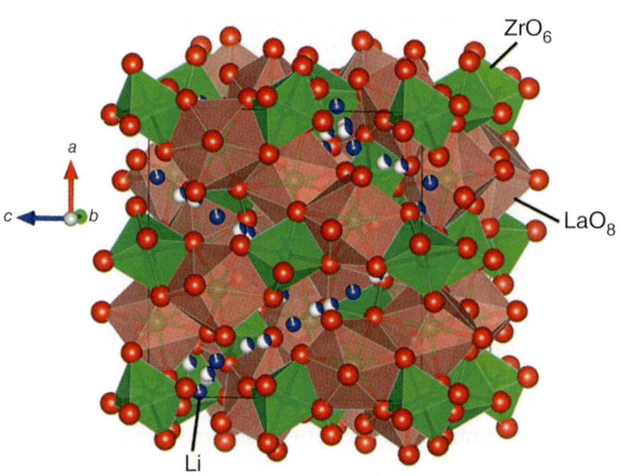

Figure 1.19 Structure of LLZ.

matrix showed high ionic conductivity of 4.42×10^{-4} S cm^{-1} [132]. Furthermore, the charge and discharge properties of Li/$LiNi_{0.6}Co_{0.2}Mn_{0.2}O_2$ cell using the composite electrolyte were better than those of pure PEO membrane.

The sulfide electrolytes have extra high ionic conductivity and wide electrochemical window. The addition of sulfide electrolytes into the polymer matrix has also been studied. $Li_{10}GeP_2S_{12}$ (LGPS) was incorporated into PEO-LiTFSI matrix [133]. The composite electrolyte with 1 wt% LGPS exhibited the highest conductivity of 1.21×10^{-3} and 1.18×10^{-5} S cm^{-1} at 80 and 25 °C, respectively.

1.5 Summary

Recent researches on PEs for Li battery are described. The battery using PEs has a lot of advantages such as their structural flexibility and high safety. The researches on PEs have focused on the increase in ionic conductivity and improving mechanical strength, but a compatibility of the PEs with electrode materials, which is another critical property, has not been extensively studied. The compatibility, particularly the interface between electrode materials and the PEs during electrochemical reactions, should be studied in nanoscale. The characterization of the interface and a development of the characterization technique for this type of study will promote the development of Li batteries using PEs.

References

1 Fenton, D.E., Parker, J.M., and Wright, P.V. (1973). *Polymer* 14: 589.

2 Ramesh, S. and Lu, S.C. (2012). *J. Appl. Polym. Sci.* 126: 484–492.

3 Bruce, P.G. (1995). *Solid State Electrochemistry*. Cambridge University Press.

4 Kim, J.H., Kang, M.S., Kim, Y.J. et al. (2004). *Chem. Commun.* 14: 1662–1663.

5 MacCallum, J.R. and Vincent, C.A. *Polymer Electrolytes Reviews*, vol. 2. Elsevier.

6 Scrosati, B. *Applications of Electroactive Polymers*. London: Chapman Hall.

7 Vincent, C.A. (1987). *Prog. Solid State Chem.* 17: 145–261.

8 Kotobuki, M., Suzuki, Y., Sato, Y. et al. (2011). *J. Power Sources* 196: 9815–9819.

9 Hoshina, K., Yoshima, K., Kotobuki, M., and Kanamura, K. (2012). *Solid State Ionics* 209–210: 30–35.

10 Kotobuki, M. and Koishi, M. (2015). *Ceram. Int.* 41: 8562–8567.

11 Thangadurai, V. and Weppner, W. (2005). *J. Power Sources* 142: 339–344.

12 Bloise, A.C., Tambelli, C.C., Franco, R.W.A. et al. (2001). *Electrochim. Acta* 46: 1571–1579.

13 Stephan, A.M. (2006). *Eur. Polym.* 42: 21–42.

14 Wright, P.V. (1975). *Br. Polym. J.* 7: 319–327.

15 Golodnitsky, D., Strauss, E., Peled, E., and Greenbaum, S. (2015). *Electrochim. Acta* 162: A2551–A2566.

16 Meyer, W.H. (1998). *Adv. Mater.* 10: 439–448.

17 Wright, P.V. *Electrochim. Acta* 43: 1998 1137–1998 1143.

18 Itoh, T., Ichikawa, Y., Uno, T. et al. (2003). *Solid State Ionics* 156: 393–399.

19 Goren, A., Costa, C.M., Tamano Machiavello, M.N. et al. (2015). *Solid State Ionics* 280: 1–9.

20 Kim, J.K., Cheruvally, G., Li, X. et al. (2008). *J. Power Sources* 178: 815–820.

21 Rajendran, S., Mahendran, O., and Kannan, R. (2002). *J. Phys. Chem. Solids* 63: 303–307.

22 Ramesh, S., Yin, T.S., and Liew, C.W. (2011). *Ionics* 17: 705–713.

23 Chen-Yang, Y.W., Chen, H.C., Lin, F.J., and Chen, C.C. (2002). *Solid State Ionics* 150: 327–335.

24 Kam, W., Liew, C.W., Lim, J.Y., and Ramesh, S. (2014). *Ionics* 20: 665–674.

25 Ramesh, S., Uma, O., Shanti, R. et al. (2014). *Measurement* 48: 263–273.

26 Fullertron-Shirey, S.K. and Maranas, J.K. (2009). *Macromolecules* 42: 2142–2156.

27 Sun, H.Y., Sohn, H.-J., Yamamoto, O. et al. (1999). *J. Electrochem. Soc.* 146 (5): 1672–1676.

28 Acosta Luqure, J.-L. and Bergas, E.M. (1996). *Solid State Ionics* 91: 75–80.

29 Khurana, R., Schaefer, J.L., Archer, L.A., and Coates, G.W. (2014). *J. Am. Chem. Soc.* 136: 7395–7402.

30 Porcarelli, L., Gerbaldi, C., Bella, F., and Nair, J.R. (2016). *Sci. Rep.* 6: 19892.

31 Tanaka, R., Sakurai, M., Sekiguchi, H. et al. (2001). *Electrochim. Acta* 46: 1709–1715.

32 Soo, P.P., Huang, B., Jang, Y.-I. et al. (1999). *J. Electrochem. Soc.* 146: 32–37.

33 Sadoway, D.R. (2004). *J. Power Sources* 129: 1–3.

34 Niitani, T., Shimada, M., Kawamura, K., and Kanamura, K. (2005). *J. Power Sources* 146: 386–390.

35 Niitani, T., Shimada, M., Kawamura, K. et al. (2005). *Electrochem. Solid-State Lett.* 8: A385–A388.

36 Bouchet, R., Maria, S., Meziane, R. et al. (2013). *Nat. Mater.* 12: 452–457.

37 Nishimoto, A., Agehara, K., Furuya, N. et al. (1999). *Macromolecules* 32: 1541–1548.

38 Wen, Z., Itoh, T., Uno, T. et al. (2003). *Solid State Ionics* 160: 141–148.

39 Matoba, Y., Ikeda, Y., and Kohjiya, S. (2003). *Solid State Ionics* 147: 403–409.

40 Gopalan, A.I., Santhosh, P., Manesh, K.M. et al. (2008). *J. Membr. Sci.* 325: 683–690.

41 Jiang, Z., Carroll, B., and Abraham, K.M. (1997). *Electrochim. Acta* 42: 2667–2677.

42 Iijima, T., Toyoguchi, Y., and Eda, N. (1985). *Electrochemistry* 53: 619–621.

43 Liew, C.W., Durairaj, R., and Ramesh, S. (2012). *J. Mater. Res.* 27: 2996–3004.

44 Rhoo, H.J., Kim, H.T., Park, J.K., and Hwang, T.S. (1997). *Electrochim. Acta* 42: 1571–1579.

45 Croce, F., Persi, L., Scrosati, B. et al. (2001). *Electrochim. Acta* 46: 2457–2461.

46 Dasenbropck, C.O., Ridgway, T.H., Seliskar, C.J., and Heineman, R.W. (1998). *Electrochim. Acta* 43: 3497–3502.

47 Yang, J.M., Wang, Z.W., and Yang, C.C. (2008). *J. Membr. Sci.* 322: 74–80.
48 Yang, C.C. and Wu, G.M. (2009). *Mater. Chem. Phys.* 114: 948–955.
49 Lu, Y., Wang, D., Li, T. et al. (2009). *Biomaterials* 30: 4143–4151.
50 Qiao, J., Okada, T., and Ono, H. (2009). *Solid State Ionics* 180: 1318–1323.
51 Hirankumar, G., Selvasekarapandian, S., Kuwata, N. et al. (2005). *J. Power Sources* 144: 262–267.
52 Tsutsumi, H., Matsuo, A., Takase, K. et al. (2010). *J. Power Sources* 90: 33–38.
53 Tatsuma, T., Taguchi, M., Iwaku, M. et al. (1999). *J. Electroanal. Chem.* 472: 142–146.
54 Osman, Z. and Arof, A.K. (2003). *Electrochim. Acta* 48: 993–999.
55 Khanmirzaei, M.H. and Ramesh, S. (2013). *Int. J. Electrochem. Sci.* 8: 9977–9991.
56 Teoh, K.H., Ramesh, S., and Arof, A.K. (2012). *J. Solid State Electrochem.* 16: 3165–3170.
57 Feuillade, G. and Perche, P. (1975). *J. Appl. Electrochem.* 5: 63–69.
58 Saikia, D., Chen-Yang, Y.W., Chen, Y.T. et al. (2008). *Desalination* 234: 24–32.
59 Kim, Y.T. and Smotkin, E.S. (2002). *Solid State Ionics* 149: 29–37.
60 Ramesh, S. and Ang, G.P. (2010). *Ionics* 16: 465–473.
61 Choi, S.W., Kim, J.R., Jo, S.M. et al. (2005). *J. Electrochem. Soc.* 152: A989–A995.
62 Rajendran, S., Babu, R., and Sivakumar, P. (2009). *J. Appl. Polym. Sci.* 113: 1651–1656.
63 Choi, S.W., Jo, S.M., Lee, W.S., and Kim, Y.R. (2003). *Adv. Mater.* 15: 2027–2032.
64 Li, Z.H., Cheng, C., Zhen, X.Y. et al. (2009). *Electrochim. Acta* 54: 4403–4407.
65 Rajendran, S., Sivakumar, M., and Subadevi, R. (2004). *Solid State Ionics* 167: 335–339.
66 Ito, Y., Kanehori, K., Miyauchi, K., and Kudo, T. (1987). *J. Mater. Sci.* 22: 1845–1849.
67 Yang, L., Lin, J., Wang, Z. et al. (1990). *Solid State Ionics* 40-41: 616–619.
68 Kim, H., Oh, B., and Kang, Y. (2000). *Polym. Bull.* 44: 509–515.
69 Chakrabarti, A., Filler, R., and Mandal, B.K. (2010). *Solid State Ionics* 180: 1640–1645.
70 Karatas, Y., Banhatti, R.D., Kaskhedikar, N. et al. (2009). *J. Mater. Chem. B* 113: 15473–15484.
71 Johan, M. and Fen, L. (2010). *Ionics* 16: 335–338.
72 Klongkan, S. and Pumchusak, J. (2015). *Electrochim. Acta* 161: 171–176.
73 Wu, X.-L., Xin, S., Seo, H.-H. et al. (2011). *Solid State Ionics* 186: 1–6.
74 Rajendran, S. and Mahendran, O. (2001). *Ionics* 7: 463–468.
75 Bandara, L.R.A.K., Dissanayake, M.A.K.L., and Mellander, B.E. (1998). *Electrochim. Acta* 43: 1447–1451.
76 Yan, P., Huang, Z., Lin, Y. et al. (2015). *Ionics* 21: 593–599.
77 Kim, J.R., Choi, S.W., Jo, S.M. et al. (2004). *Electrochim. Acta* 50: 69–75.
78 Johan, M.R., Shy, O.H., Ibrahim, S. et al. (2011). *Solid State Ionics* 196: 41–47.

79 Qian, X., Gu, N., Cheng, Z. et al. (2002). *Mater. Chem. Phys.* 74: 98–103.
80 Meneghetti, P., Qutubuddin, S., and Webber, A. (2004). *Electrochim. Acta* 49: 4923–4931.
81 Min, H.-S., Ko, J.-M., Kim, D.-W., and Power, J. (2003). *Sources* 119–121: 469–472.
82 Vivkraman, P. and Ramamurthy, S. (2006). *Mater. Lett.* 60: 3431–3436.
83 Wetjen, M., Kim, G.-T., Joost, M. et al. (2013). *Electrochim. Acta* 87: 779–787.
84 Kim, G.T., Appetecchi, G.B., Carewska, M. et al. (2010). *J. Power Sources* 195: 6130–6137.
85 Rupp, B., Schmuck, M., Balducci, A. et al. (2008). *Eur. Polym. J.* 44: 2986–2990.
86 Shin, J.-H., Henderson, W.A., and Passerini, S. (2003). *Electrochem. Commun.* 5: 1016–1020.
87 Appetecchi, G.B., Montanino, M., Zane, D. et al. (2009). *Electrochim. Acta* 54: 1325–1332.
88 Shin, J.-H., Henderson, W.A., and Passerini, S. (2005). *J. Electrochem. Soc.* 152: A978–A983.
89 Shin, J.-H., Henderson, W.A., and Passerini, S. (2005). *Electrochem. Solid-State Lett.* 8: A125–A127.
90 Shi, J.-H., Henderson, W.A., Scaccia, S. et al. (2006). *J. Power Sources* 156: 560–566.
91 Tizzani, C., Appetecchi, G.B., Carewska, M. et al. (2007). *Aust. J. Chem.* 60: 47–50.
92 Shi, J.-H., Henderson, W.A., Tizzani, C. et al. (2006). *J. Electrochem. Soc.* 153: A1649–A1654.
93 Appetecchi, G.B., Kim, G.T., Montanino, M. et al. (2011). *J. Power Sources* 196: 6703–6709.
94 Kim, G.T., Jeong, S.S., Xue, M.Z. et al. (2012). *J. Power Sources* 199: 239–246.
95 Ye, H., Huang, J., Xu, J.J. et al. (2007). *J. Electrochem. Soc.* 154: A1048–A1057.
96 Li, L.B., Wang, J.J., Yang, P.X. et al. (2013). *Electrochim. Acta* 88: 147–156.
97 Raghavan, P., Zhao, X., Choi, H. et al. (2014). *Solid State Ionics* 262: 77–82.
98 Kim, Y.H., Cheruvally, G., Choi, J.W. et al. (2007). *Macromol. Symp.* 249–250: 183–189.
99 Choi, J.-W., Cheruvally, G., Kim, Y.-H. et al. (2007). *Solid State Ionics* 178: 1235–1241.
100 Egashira, M., Todo, H., Yoshimoto, N., and Morita, M. (2008). *J. Power Sources* 178: 729–735.
101 Cha, E.H., Lim, S.A., Park, J.H. et al. (2008). *J. Power Sources* 178: 779–782.
102 Li, M.T., Yang, L., Fang, S.H. et al. (2012). *Polym. Int.* 61: 259–264.
103 Yin, K., Zhang, Z.X., Li, X.W. et al. (2015). *J. Mater. Chem. A* 3: 170–178.
104 Lee, Y.-S. and Kim, D.–.W. (2013). *Electrochim. Acta* 106: 460–464.
105 Kumar, B. (2004). *J. Power Sources* 135: 215–231.
106 Adebahr, J., Byrne, N., Forsyth, M. et al. (2003). *Electrochim. Acta* 48: 2099–2103.
107 Appetecchi, G.B. and Passerini, S. (2000). *Electrochim. Acta* 45: 2139–2145.

108 Michael, M.S., Jacon, M.M.E., Prabaharan, S.R.S., and Radhakrishna, S. (1997). *Solid State Ionics* 98: 167–174.
109 Slane, S. and Salomon, M. (1995). *J. Power Sources* 55: 7–10.
110 Krawiec, W., Scanlon, L.G. Jr., Fellner, P. et al. (1995). *J. Power Sources* 54: 310–315.
111 Croce, F., Appetecchi, G.B., Persi, L., and Scrosati, B. (1998). *Nature* 394: 456–458.
112 Jayathilaka, P.A.R.D., Dissanayake, M.A.K.L., Albinsson, I., and Mellander, B.-E. (2002). *Electrochim. Acta* 47: 3257–3268.
113 Croce, F., Persi, L., Ronci, F., and Scrosati, B. (2000). *Solid State Ionics* 135: 47–52.
114 Scrosati, B., Croce, F., and Persi, L. (2000). *J. Electrochem. Soc.* 147: 1718–1721.
115 Pitawala, H.M.J.C., Dissanayake, M.A.K.L., and Seneviratne, V.A. (2007). *Solid State Ionics* 178: 885–888.
116 Yoon, M.Y., Hong, S.K., and Hwang, H.J. (2013). *Ceram. Int.* 39: 9659–9663.
117 Liu, S., Imanishi, N., Zhang, T. et al. (2010). *J. Power Sources* 195: 6847–6853.
118 Lin, C.W., Hung, C.L., Venkateswarlu, M., and Hwang, B.J. (2005). *J. Power Sources* 146: 397–401.
119 Adebahr, J., Best, A.S.>., Byrne, N. et al. (2003). *Phys. Chem. Chem. Phys.* 5: 720–725.
120 Shanmukaraj, D., Wang, G.X., Murugan, R., and Liu, H.K. (2008). *J. Phys. Chem. Solids* 69: 243–248.
121 Sun, H.Y., Yakeda, Y., Imanishi, N. et al. (2000). *J. Electrochem. Soc.* 147: 2462–2467.
122 Huang, G., Li, Y., Han, L. et al. (2011). *Appl. Clay Sci.* 51: 360–365.
123 Walls, H.J., Riley, M.W., Singhal, R.R. et al. (2003). *Adv. Funct. Mater.* 13: 710–717.
124 Zhang, Y., Zhao, Y., Gosselimk, D., and Chen, P. (2015). *Ionics* 21: 381–385.
125 Moreno, M., Quijada, R., Santa Ana, M.A. et al. (2011). *Electrochim. Acta* 58: 112–118.
126 Tang, C., Hackenberg, K., Fu, Q. et al. (2012). *Nano Lett.* 12: 1152–1156.
127 Zhu, Y., Zhang, Y., and Lu, L. (2015). *J. Power Sources* 290: 123–129.
128 Jung, Y.C., Lee, S.M., Choi, J.-H. et al. (2015). *J. Electrochem. Soc.* 162: A704–A710.
129 Kotobuki, M., Munakata, H., Kanamura, K. et al. (2010). *J. Electrochem. Soc.* 157: A1076–A1079.
130 Kotobuki, M., Kanamura, K., Sato, Y., and Yoshida, T. (2011). *J. Power Sources* 196: 7750–7754.
131 Kotobuki, M., Kanamura, K., Sato, Y. et al. (2012). *J. Power Sources* 199: 346–349.
132 Choi, J.-H., Lee, C.-H., Yu, J.-H. et al. (2015). *J. Power Sources* 274: 458–463.
133 Zhao, Y., Wu, C., Peng, G. et al. (2016). *J. Power Sources* 301: 47–53.

2

Impedance Spectroscopy in Polymer Electrolyte Characterization

Mohamed Abdul Careem[1], Ikhwan Syafiq Mohd Noor[2], and Abdul K. Arof[1]

[1] *Centre for Ionics University of Malaya, Department of Physics, Faculty of Science, University of Malaya, 50603 Kuala Lumpur, Malaysia*
[2] *Physics Division, Centre of Foundation Studies for Agricultural Science, Universiti Putra Malaysia, 43400 UPM Serdang, Selangor Darul Ehsan, Malaysia*

2.1 Introduction

The resistance to the passage of an alternating current (*ac*) in a material is generally called electrical impedance or simply impedance. The impedance concept was first introduced by Oliver Heaviside in the 1880s in respect of electrical circuits [1]. The concept was further developed by A.E. Kennelley with the use of vector diagrams and complex number representations [2]. Impedance is dependent on frequency of the applied alternating signal. The impedance response of a system can be measured as a function of frequency and represented in a complex plane with frequency as an implicit variable similar to Argand diagrams in mathematics. Hence, the term impedance spectroscopy (IS) can be used for analysis of impedance of a system as a function of frequency. The IS as a technique to study electrochemical behavior of materials was introduced in the early 1900s [3]. The technique is still being developed and has gained wide acceptance now. Since its introduction, IS technique has undergone tremendous progressive evolution. It has become an important tool in multidisciplinary areas ranging from materials science, agriculture, and medical fields. IS is a powerful technique in materials research and for studying electrochemical systems and processes.

2.2 IS: Principal Operation and Experimental Setup

2.2.1 Basic Principles of Impedance Spectroscopy

According to Ohm's law, electrical resistance in the circuit element is the ratio between the voltage applied and the current flow across that element. This widely known relationship is limited to only ideal resistor that is independent on frequency. In reality, the circuit elements are more complex that limits the function

Polymer Electrolytes: Characterization Techniques and Energy Applications,
First Edition. Edited by Tan Winie, Abdul K. Arof, and Sabu Thomas.

of direct current (*dc*). Therefore the use of impedance is more convenient to determine the electrical resistance of a material.

Impedance spectroscopy is based on measuring the impedance "feedback" of a material or an electrochemical system or process to a small *ac* signal. Since the *ac* signal is frequency dependent, the resulting impedance response also will be frequency dependent. Impedance is a complex quantity and therefore can be written in terms of its real and imaginary parts. The real and imaginary parts can be calculated from the phase shift between the applied signal and the current through the material and magnitude of the impedance. From the phase shift, it is possible to know how resistive, capacitive, or inductive the material, system, or process can be.

In the IS technique, a small alternating (e.g. sinusoidal) voltage of a few millivolts is applied across a sample. Current through the sample is measured. The small applied signal $v(t)$ ensures that the current response $i(t)$ to the alternating voltage is linear or pseudo-linear. Thus, the frequency of is also the same as that of the voltage applied, but is phase shifted. The voltage to current ratio, i.e. $v(t)/i(t)$, is the complex impedance of the sample and noted as Z. Since $v(t)$ and $i(t)$ differ with frequency, impedance is therefore frequency (f) or angular frequency (ω) dependent. A spectrum of Z versus $f(\omega)$ can be obtained for the frequency of choice.

Suppose a small *ac* voltage represented by Eq. (2.1) is applied to a sample sandwiched between two electrodes as shown in Figure 2.1.

$$v(t) = v_o \sin \omega t \tag{2.1}$$

Here v_o is the maximum voltage and $\omega = 2\pi f$ is the angular frequency of the *ac* signal.

The resulting *ac* current flowing through the sample is given by

$$i(t) = i_o \sin(\omega t + \theta) \tag{2.2}$$

where θ is a phase difference between $i(t)$ and $v(t)$. Here the current is ahead of voltage by θ as shown in Figure 2.2.

The impedance of the sample is given by

$$Z = \frac{v(t)}{i(t)} = \frac{v_o \sin \omega t}{i_o \sin(\omega t + \theta)} \tag{2.3}$$

Z is a function of frequency and has a magnitude of

$$|Z| = \frac{v_o}{i_o} = Z_o \tag{2.4}$$

and a phase angle θ. Both Z_o and θ are frequency-dependent quantities.

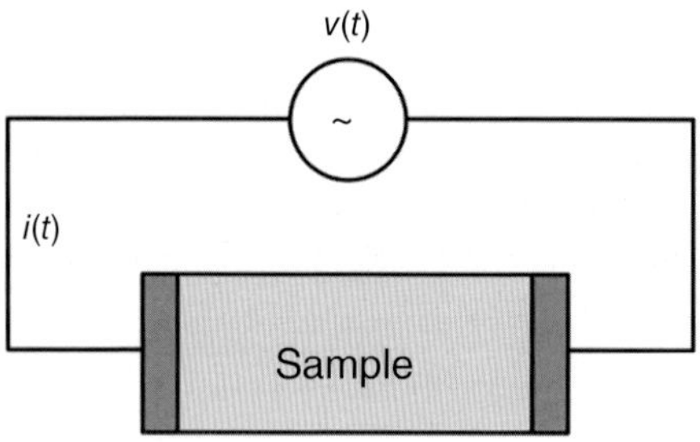

Figure 2.1 A sample with an *ac* signal applied across.

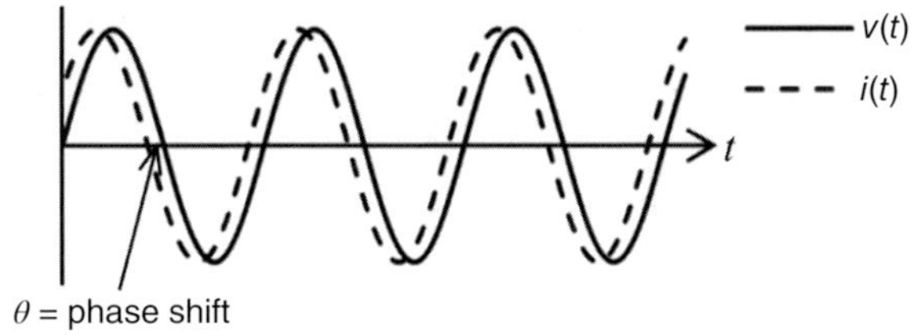

Figure 2.2 Pictorial representations of an *ac* voltage and an *ac* current. Here the current is ahead of the voltage with a phase difference of θ.

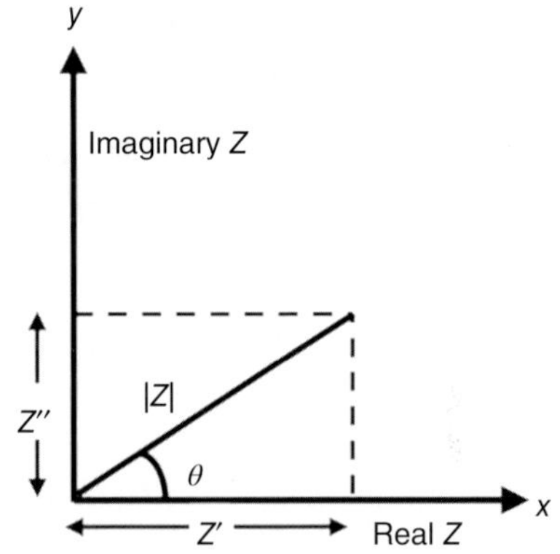

Figure 2.3 Complex impedance plot.

Generally *ac* impedances are represented by $Z(\omega)$, which is a complex quantity. Hence, impedance has real $Z'(\omega)$ and an imaginary $Z''(\omega)$ component. Then

$$Z(\omega) = Z'(\omega) + jZ''(\omega), \quad j = \sqrt{-1} \tag{2.5}$$

Here the real part, $Z'(\omega)$, and imaginary part, $Z''(\omega)$ of $Z(\omega)$, are respectively given by

$$Z'(\omega) = |Z| \cos\theta \quad \text{and} \quad Z''(\omega) = |Z| \sin\theta \tag{2.6}$$

The phase angle of $Z(\omega)$ is given by

$$\theta = \tan^{-1}\left[\frac{Z''(\omega)}{Z'(\omega)}\right] \tag{2.7}$$

The magnitude of $Z(\omega)$ is given by

$$|Z| = \sqrt{[Z'(\omega)]^2 + [Z''(\omega)]^2} \tag{2.8}$$

All these quantities are pictorially illustrated in Figure 2.3. Some people prefer to use Z_r and Z_i to represent the real part Z' and imaginary part Z'', respectively.

2.2.2 Impedance Spectroscopy (IS) Technique

The conventional IS technique involves the measuring Z as a function of frequency f ($\omega = 2\pi f$) over a wide frequency range for a sample and plotting of $Z(f)$ against f in the form of $Z''(f)$ against $Z'(f)$ for various f. The resulting graph is called a complex impedance plane plot or an impedance plot. Sometimes the plot is erroneously referred as a Nyquist plot, a term strictly applicable to electronic circuits. Such a plot can be used to evaluate the electrical or electrochemical parameters/processes of the system under study.

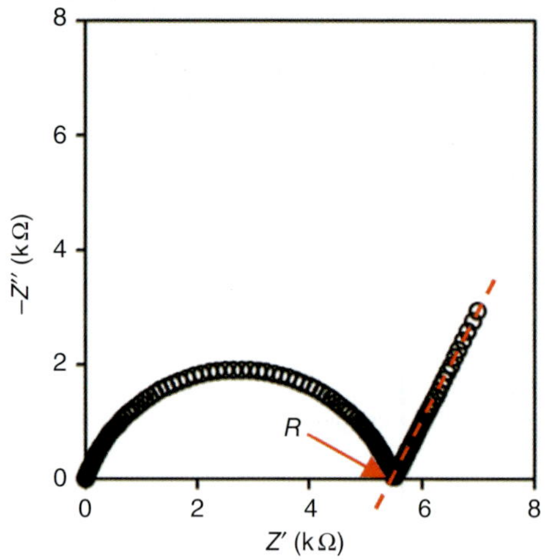

Figure 2.4 A typical impedance plot.

If $Z(\omega)$ has a magnitude $|Z|$ and phase angle $-\theta$, it can be as a complex function in the form

$$Z(\omega) = |Z|e^{-j\theta} = |Z|[\cos\theta - j\sin\theta] = Z'(\omega) - jZ''(\omega) \tag{2.9}$$

Then $-Z''(\omega)$ is plotted against $Z'(\omega)$ to obtain the impedance or Nyquist plot. An example of such a plot is shown in Figure 2.4.

In Figure 2.4, the horizontal axis represents the real impedance, $Z'(\omega)$. The vertical axis represents the imaginary impedance, $Z''(\omega)$. The intercept of the plot with the Z' axis corresponds to the resistance R of the sample. The vertical axis corresponds to the reactance X of the sample. The plot shown in Figure 2.4 corresponds to simple systems. Some systems may have more complicated impedance plots as will be discussed later on.

2.2.3 Electrical Conductivity of a Sample

For a sample in the form of a disc, cylinder, or film, the resistance can be written as

$$R = \rho\frac{L}{A} \tag{2.10}$$

where ρ represents the resistivity of the material, L is the length or thickness of the sample, and A is the cross-sectional area of the sample. The electrical conductivity, σ of the sample is given by

$$\sigma = \frac{1}{\rho} = \frac{{}^{L}/_{A}}{R} \tag{2.11}$$

where L/A is the cell constant. Since R can be obtained from IS plot and L and A are known, thus σ can be determined.

2.2.4 Conditions Necessary for IS Measurements

From the experimental point of view, three fundamental conditions must be fulfilled in order for the impedance measurement data to be valid.

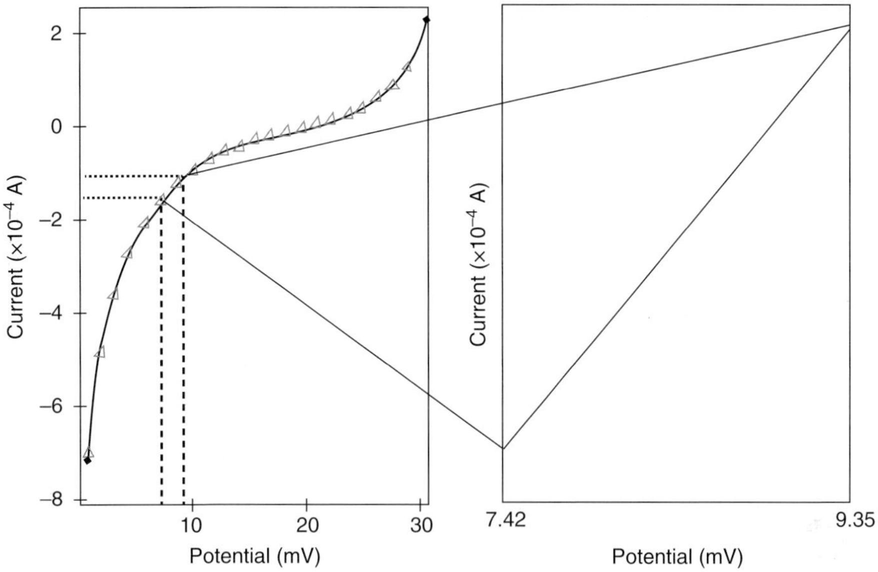

Figure 2.5 Example of a small signal that fulfills the linearity condition of an *I–V* response.

(i) Linearity:
The system under study must behave linearly under measuring conditions. For this, the applied *ac* amplitude must be **small** so that the response of the sample is linear or assumed to be linear.
The size of signal used depends on the sample under investigation. Signal must be small enough to produce a linear response and at the same time it must be big enough to measure the response. An example is illustrated in Figure 2.5 in which the solid curve represents the *I–V* response of a sample and the superimposed small signal that can be assumed to be linear is applied at a *dc* bias voltage. Usually *ac* signals of around 10–20 mV are used. For most IS measurements, signals are applied with no *dc* bias.
Whether the *ac* signal chosen for a particular measurement has satisfied the linearity condition or not can be tested practically by changing the magnitude of the signal applied and making sure that the impedance measured remains same. Theoretically the linearity can be checked using Kramers–Kronig (KK) transformation relations. On using these relations, the real and imaginary components of impedance must transform correctly into one another [4].

(ii) *Stability*:
The state of the sample or system under study should not change (significantly) during the acquisition of the data. Hence, it is essential to choose the appropriate frequency range and measurement conditions. Generally the experimental duration for impedance measurements is determined by the low frequency limit chosen. Most electrochemical systems are unstable due to corrosion of electrodes, adsorption or formation of oxides on the electrodes, discharge of the device, changes in the electrochemical interface, etc.

Therefore, precautions must be taken to avoid such changes taking place during the measuring period.

(iii) *Causality*:
The measured *ac* response of the system must be directly correlated to the applied *ac* signal only. Therefore, precautions must be taken to avoid outside interference to the sample response so that the observed response is correlated to the applied *ac* signal only. In many situations it is necessary to shield the sample cell to avoid interference from outside sources.

2.2.5 Impedance Plots of Simple Circuits

2.2.5.1 A Pure Resistance, *R*

Consider a resistor with a resistance R. Since R is a constant quantity independent of frequency, its complex impedance $Z(\omega)$ will be R for all values of f or ω. Hence $Z(\omega)$ will have a magnitude $|Z| = R$ and a phase angle $\theta = 0$. Thus for all values ω, $Z'(\omega) = R$ and $Z''(\omega) = 0$. Hence, the impedance plot is **a point** on the real axis at $Z' = R$ as shown in Figure 2.6.

2.2.5.2 A Pure Capacitance, *C*

Consider a capacitor with a capacitance C. Its complex impedance $Z(\omega)$ is given by

$$Z(\omega) = \frac{1}{j\omega C} = -\frac{j}{\omega C} \tag{2.12}$$

Hence the $Z'(\omega)$ and $Z''(\omega)$ can be written as

$$Z'(\omega) = 0 \quad \text{and} \quad Z''(\omega) = -\frac{1}{\omega C} \tag{2.13}$$

Therefore, $Z(\omega) = Z''(\omega)$, which varies with frequency. As ω increases, $Z''(\omega)$ decreases. The impedance plot will consist of points on the Z''-axis with each point corresponding to the frequency concerned. The impedance plot will be **vertical line** along the Z''-axis as depicted in Figure 2.7. The arrow indicates the

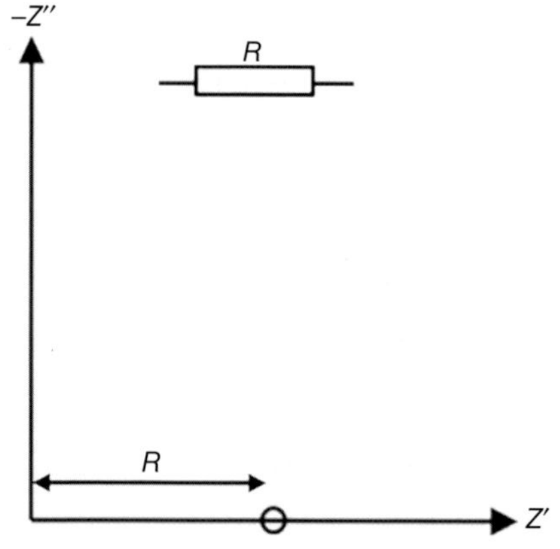

Figure 2.6 Impedance plot of a resistor.

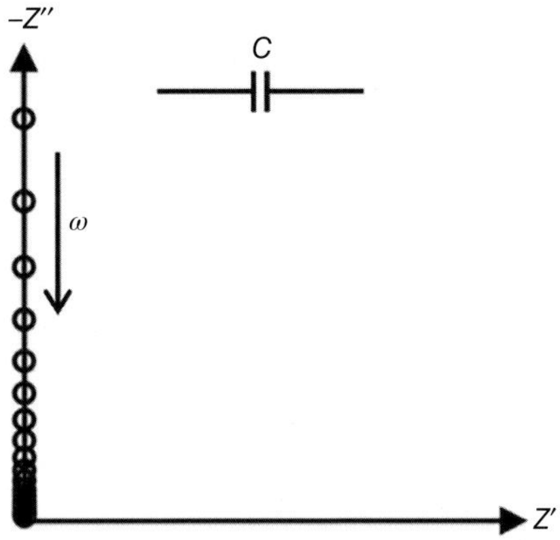

Figure 2.7 Impedance plot of a capacitor.

direction of increase in frequency values. It should be noted that the Z'' values are negative when capacitances are involved and that is why the $-Z''$ is used for the y-axis.

2.2.5.3 *R* and *C* Connected in Series

If a resistor of resistance R and a capacitor of capacitance C are connected in series, the total impedance of the system is obtained by adding the impedances of R and C as

$$Z(\omega) = R - \frac{j}{\omega C} \tag{2.14}$$

Hence,

$$Z'(\omega) = R \quad \text{and} \quad Z''(\omega) = -\frac{1}{\omega C} \tag{2.15}$$

The resulting impedance plot will have the characteristics both of a pure resistance and a pure capacitance. Hence the impedance plot becomes **a vertical line at $Z' = R$**, parallel to the Z''-axis as shown in Figure 2.8.

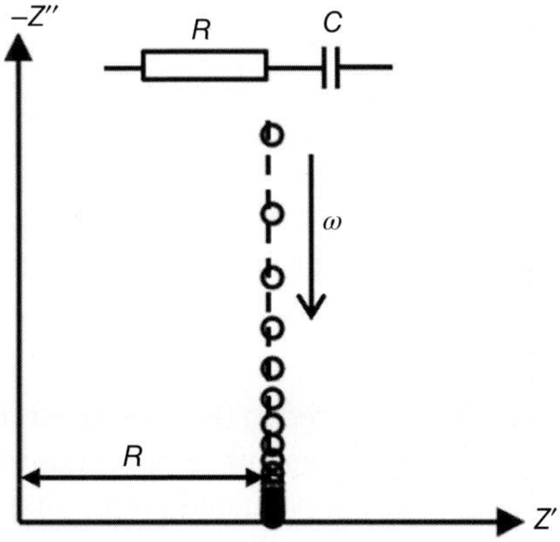

Figure 2.8 Impedance plot of a resistor and capacitor connected in series.

2.2.5.4 *R* and *C* Connected in Parallel

If a resistor of resistance R and a capacitor of capacitance C are connected in parallel, the total impedance of the combination $Z(\omega)$ is given by

$$\frac{1}{Z} = \frac{1}{R} + \frac{1}{1/j\omega C} = \frac{1}{R} + j\omega C \tag{2.16}$$

Hence,

$$\begin{aligned} Z(\omega) &= \frac{R}{1 + j\omega RC} \\ &= \frac{R(1 - j\omega RC)}{(1 + j\omega RC)(1 - j\omega RC)} \\ &= \frac{R(1 - j\omega RC)}{1 + (\omega RC)^2} \\ &= \frac{R}{1 + \omega^2 R^2 C^2} - \frac{j\omega R^2 C}{1 + \omega^2 R^2 C^2} \\ &= Z'(\omega) - Z''(\omega) \end{aligned} \tag{2.17}$$

Here,

$$Z'(\omega) = \frac{R}{1 + \omega^2 R^2 C^2} \tag{2.18}$$

and

$$Z''(\omega) = \frac{\omega R^2 C}{1 + \omega^2 R^2 C^2} \tag{2.19}$$

Then,

$$\frac{Z''(\omega)}{Z'(\omega)} = \omega RC \tag{2.20}$$

On eliminating ω in Eqs. (2.18) and (2.19), it can be shown Z' and Z'' are related by the equation

$$\left[Z'(\omega) - \left(\frac{R}{2}\right)\right]^2 + [Z''(\omega)]^2 = \left(\frac{R}{2}\right)^2 \tag{2.21}$$

The Eq. (2.21) represents a circle of radius $R/2$ with the center at the point $(R/2, 0)$. Hence the impedance plot is **a semicircle** with its center at $(R/2, 0)$ lying on the Z'-axis as shown in Figure 2.9.

Maximum point on the semicircle is given by

$$\omega_m RC = 1 \tag{2.22}$$

Hence the angular frequency at the maximum point is

$$\omega_m = \frac{1}{RC} \tag{2.23}$$

Here RC is referred to the time constant or relaxation time and is usually denoted by τ. The time constant corresponds to a single relaxation process. From the value of ω_m, C can be evaluated for an unknown circuit. Z' and Z'' axis must have the same scales to see the semicircular shape. In fact all plots of imaginary against real responses must have axis with the same scale.

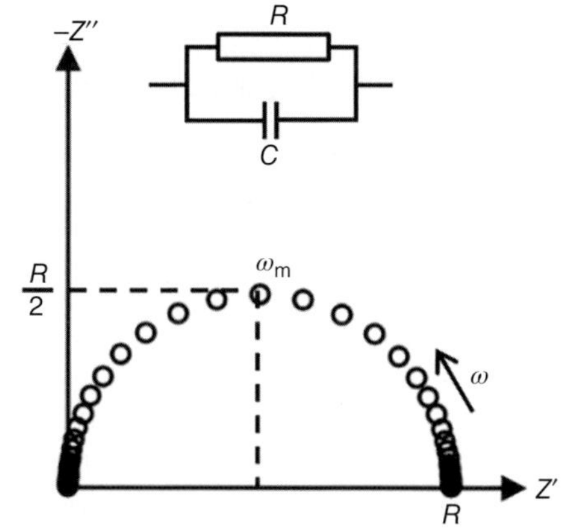

Figure 2.9 Impedance plot of a resistor and capacitor connected in parallel.

2.2.5.5 Combined Series and Parallel Circuits

Consider the combined circuits shown in Figure 2.10a,b. The corresponding impedance plots will take forms shown in Figure 2.11a,b, respectively.

For the first circuit (Figure 2.10a), the semicircle corresponding to R_1 and C_1 is displaced horizontally due to the presence of the series resistance R_s. For the second circuit (Figure 2.10b), the two parallel combinations give two semicircles. The semicircles may or may not overlap depending on the values of the components.

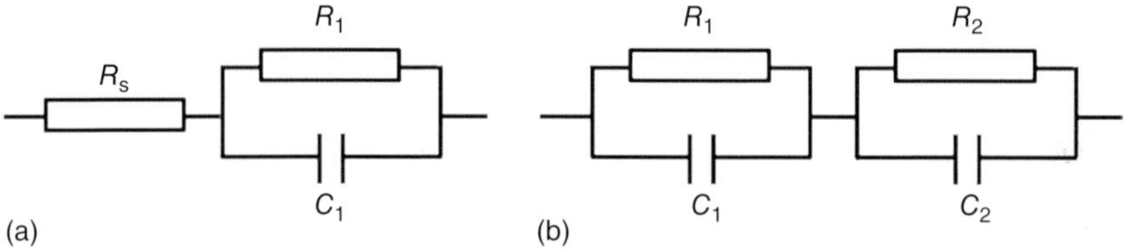

Figure 2.10 Two combined circuits in (a) series and (b) parallel.

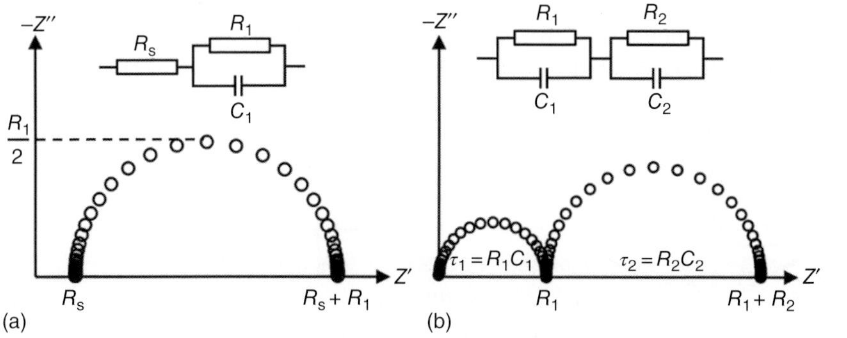

Figure 2.11 Impedance plots of (a) a parallel R_1C_1 circuit connected in series with a resistance R_s and (b) a parallel R_1C_1 circuit connected in series with a R_2C_2 parallel circuit.

2.2.5.6 Impedance Spectra of Model Electrolyte Systems

An electrolyte is a good ionic conductor in which the predominant charge carriers are ions. Electrolytes can have electronic conductivity as well. But in an ideal electrolyte the electronic contribution to the total conductivity will be zero or negligibly small. An electrolyte can be in liquid, solid and quasi-solid, or gel form. Aqueous NaCl solution is a good liquid electrolyte [5] and α-AgI is a very good solid electrolyte [6]. Ionic salts dissolved polymers can serve as polymer electrolytes. For example, $LiClO_4$ in polyethylene oxide (PEO) is a well-known polymer electrolyte [7]. Polymer electrolytes are flexible and can be prepared either in solid or quasi-solid form [8].

2.2.5.6.1 An Electrolyte Sandwiched Between Two Non-blocking Electrodes Consider an ideal electrolyte sandwiched between two electrodes, as shown in Figure 2.12a, which permit the ions to penetrate the electrodes. Such electrodes are referred to as non-blocking electrodes. Ag/AgI/Ag arrangement is a good example for such a system [9]. Since the ions are not blocked by the electrodes, under an applied field moving ions will not accumulate at the electrolyte/electrode interface and no double layer of charges will be formed at the two interfaces.

In general, since any material can show resistive and capacitive properties, the electrolyte/electrode system can be regarded as a combination of a resistor and a capacitor connected in parallel. Hence the ionic conductor can be represented by an equivalent circuit as shown in Figure 2.12b. Here it has been assumed that the electrodes do not offer any resistance to ion penetration.

In Figure 2.12, C_b represents the bulk or geometric capacitance of the cell containing ionic conductor. C_b is related to the vacuum capacitance of the cell C_o through $C_b = \varepsilon C_o$, where ε is the permittivity of ionic conductor. Here

$$C_o = \frac{\varepsilon_o A}{d} \tag{2.24}$$

where A is the area of cross section, d is the sample thickness, and ε_o is the permittivity of free space. The resistance R_b is the bulk resistance of the ionic conductor, which is given by

$$R_b = \frac{d}{A\sigma} \tag{2.25}$$

where d is the thickness of the sample, A is its cross-sectional area, and σ is its conductivity. The expected impedance plot for the cell is shown in Figure 2.13.

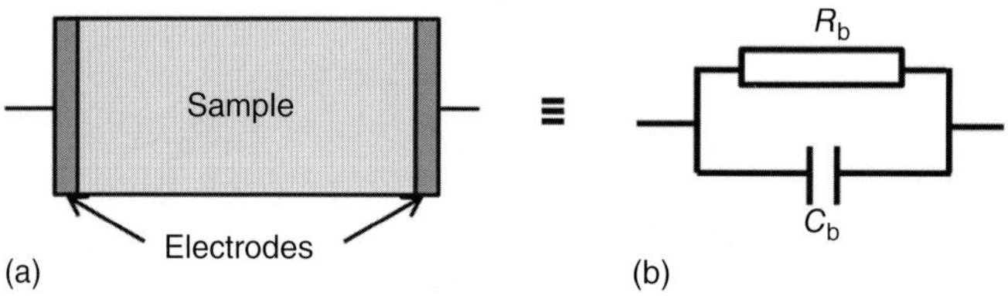

Figure 2.12 (a) An electrolyte sandwiched between two non-blocking electrodes and (b) the corresponding equivalent circuit. R_b = bulk resistance (sample) and C_b (°C_g) = bulk capacitance (sample).

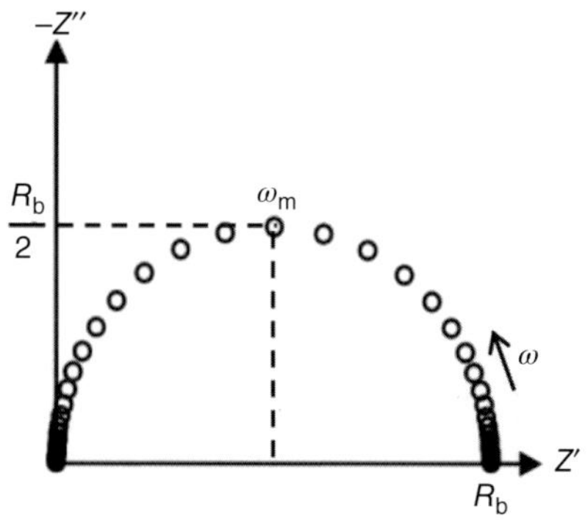

Figure 2.13 The expected impedance plot for the cell shown Figure 2.12.

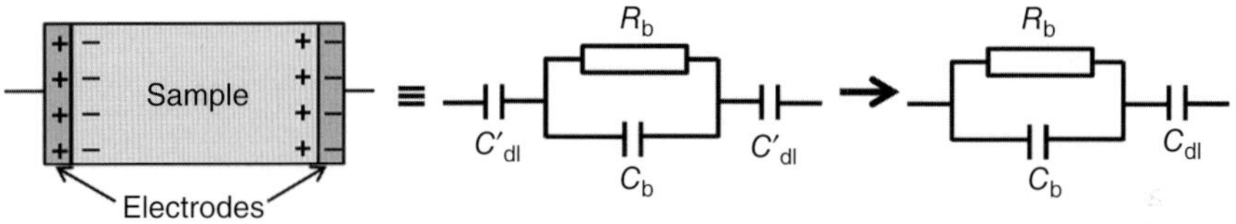

Figure 2.14 An ionic conductor sample sandwiched between two blocking electrodes and the equivalent circuit of the arrangement.

2.2.5.6.2 An Electrolyte Sandwiched Between Two Blocking Electrodes Consider an ideal electrolyte sandwiched between two blocking electrodes, i.e. the electrodes do not allow the ions to merge or react. Under an electric field, moving ions will accumulate at the electrode surface as they cannot merge or react with them. The accumulated charges will attract opposite charges to form two double layers of charge at the two ionic conductor/electrode interfaces as shown in Figure 2.14. Hence, the accumulation of charges can result in two double layer capacitances at the interfaces, each with a capacitance C'_{dl} as shown in Figure 2.14.

The two C'_{dl} capacitances can be combined to a single double layer capacitance C_{dl}. Then the equivalent circuit can be simplified to contain a single double layer of capacitance C_{dl} in series to the parallel RC circuit as shown in Figure 2.14. This effective C_{dl} will add a spike (straight line) to the semicircular arc produced in the impedance plot as in Figure 2.15.

2.2.5.6.3 A Polycrystalline Electrolyte with Two Blocking Electrodes The polycrystalline electrolyte can be regarded to consist of small crystalline regions separated by grain boundaries as shown in Figure 2.16. Conduction occurs inside the grain (intragrain or bulk conduction) and along grain boundaries (intergrain conduction).

When such a polycrystalline electrolyte is sandwiched between two blocking electrodes, for example, Pt, the cell arrangement can be regarded to be equivalent to crystalline grain + grain boundaries + electrode/electrolyte interfaces. Each of the two sample regions (grains + grain boundaries) will be equivalent to a parallel RC network, and the electrolyte/electrode interfaces will give rise to an effective double layer capacitance. Hence the polycrystalline sample–electrode system may be represented by a simple equivalent circuit of the form shown in the top of

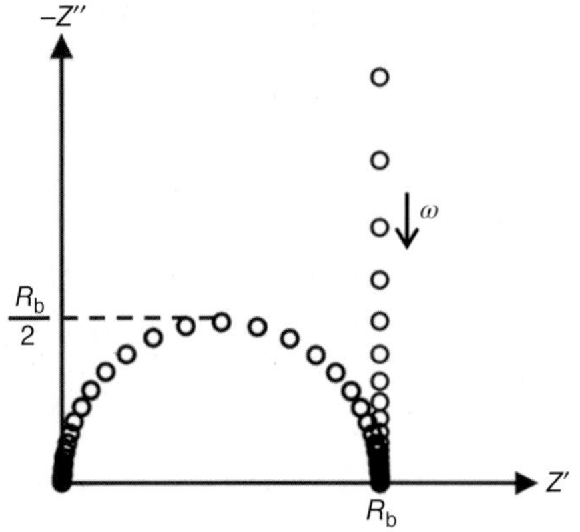

Figure 2.15 The impedance plot for a sample sandwiched with two blocking electrodes.

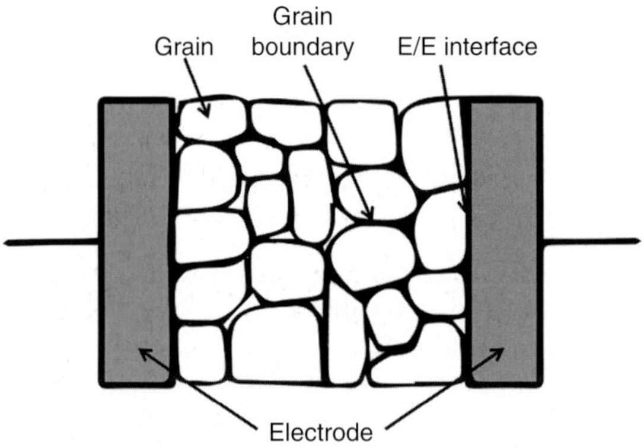

Figure 2.16 A polycrystalline sample sandwiched between two blocking electrodes.

Figure 2.17a. Here it has been assumed that grain boundaries prevent easy passage of ions from grain to grain and reduce the overall conductivity. It should be noted that the polycrystalline system can be represented by many different types of equivalent circuits depending on the electrochemical behavior of the material. In the equivalent circuit shown in Figure 2.17a, $R_b//C_b$ circuit represents the grain conduction and $R_{gb}//C_{gb}$ circuit represents the grain boundary conduction, while C_{dl} is the effective double layer capacitance of the sample/electrode interface. For the simple equivalent circuit representation assumed here, the impedance plot will have two semicircular arcs corresponding to grain and grain boundary effects, respectively, and a vertical spike corresponding to the double layer capacitance as shown in Figure 2.17b. Since the ion movements are expected to be comparatively easier in the grains and therefore the ions can respond faster to an alternating field, the high frequency semicircle can be assumed to represent the response of the ions in the grains. The other semicircle represents the grain boundary response.

The thickness of the grain boundary can be small compared with the size of the grains. Hence C_{gb} can be larger compared with C_b. Since R_{gb} can be larger compared with R_b, a larger semicircle can be expected for the grain boundary (GB) region. The total resistance, $R_b + R_{gb}$, determines the overall conductivity σ of the sample. Generally, for polycrystalline materials, the high frequency semicircle

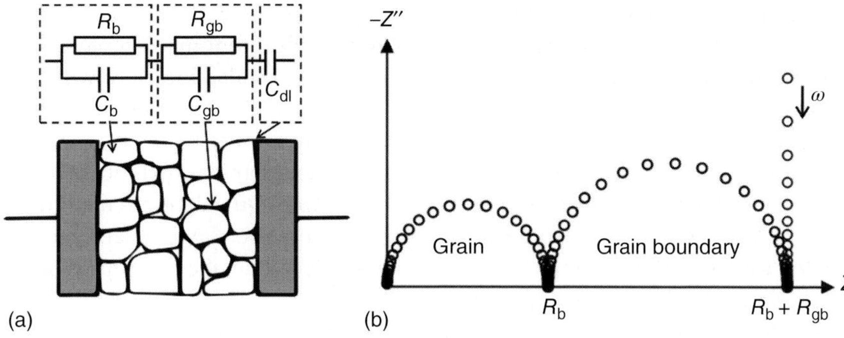

Figure 2.17 (a) A simple equivalent circuit for the polycrystalline sample cell and (b) the expected impedance plot.

(with small C) of the impedance plot represents the bulk conduction and the low frequency semicircle (with large C) corresponds to grain boundary conduction. Hence from the estimated values of the capacitances, different semicircles in an impedance plot may be associated with different conduction processes in the polycrystalline sample. Thus the shape of IS spectra can assist in understanding the microstructure of materials.

2.2.6 Possible Conduction Processes in a Solid Electrolyte

The behavior of materials is controlled by the structure of the materials. Since the structure including microstructure of a solid is easy to determine accurately, a solid electrolyte is chosen as an example to look for the possible conduction process and arrive at an equivalent circuit. Generally a number of transport processes can contribute to or affect the total conduction of an electrolyte. Some of the possible processes that can contribute to conduction in a solid electrolyte are shown in Figure 2.18.

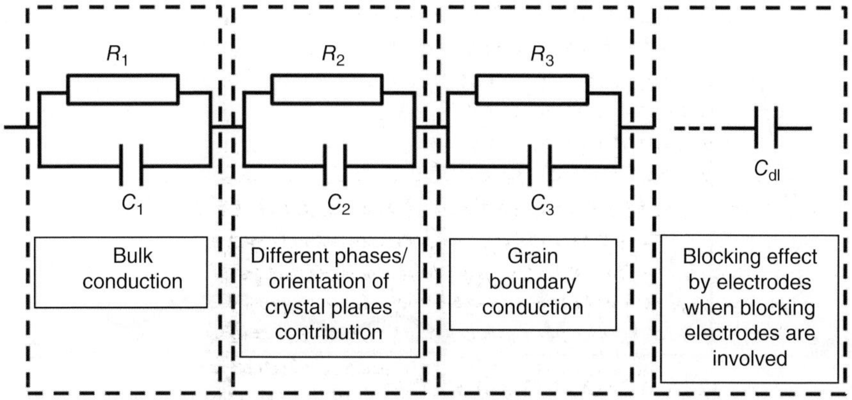

Figure 2.18 Possible conduction processes in a solid electrolyte sample and one possible simple equivalent circuit to represent the electrolyte.

The bulk process refers to conduction taking place in the crystalline regions or grains in the electrolyte. Different crystalline phases if present in a sample or different particle orientation in a pressed pellet sample can also interfere with the conduction processes. Conduction also can take place along the grain boundaries or through them if such boundaries are present in the electrolyte. If each of these processes is assumed to operate independently and contribute to conduction, the electrolyte can be represented by a simple equivalent circuit shown in Figure 2.18. Here it has also been assumed that the electrolyte is sandwiched between two perfect blocking electrodes and the electrode effect is represented by a single double layer capacitance C_{dl}.

It should be noted that the equivalent circuit model shown in Figure 2.18 is a hypothetical circuit. The equivalent circuit of an electrolyte, however, may be more complicated depending on the type of sample. The R and C values, particularly C values, differ for different transport/transfer processes [10]. The bulk capacitance C_1 is approximately $1\,\mathrm{pF\,cm^{-1}}$. The capacitance of different crystalline phases C_2 is about $10\,\mathrm{pF\,cm^{-1}}$. The grain boundary capacitance C_3 is in between $10\,\mathrm{pF\,cm^{-1}}$ and $10\,\mathrm{nF\,cm^{-1}}$, while the capacitance of effective double layer C_{dl} is in the range of $0.1–10\,\mathrm{\mu F\,cm^{-1}}$. From the approximate C values, different processes may be identified. Actual identification of different processes must be based on the dependence on temperature, pressure, etc.

2.2.7 Impedance Spectra of Real Systems

The impedance spectra of real electrolytes and devices are usually complicated. The spectra will have distorted or depressed semicircles and slanted or curved spikes. For ideal systems such as described in Section 2.2.6, we can only expect the impedance spectra to contain semicircles with their centers lying on horizontal Z'-axis and vertical spikes. For real systems, the spectra will usually contain depressed or distorted semicircles and slanted or curved spikes. Some examples of impedance plots for real systems are shown in Figure 2.19. While the spectra in (a) and (b) are for homogeneous samples, (c) corresponds to that of a composite material.

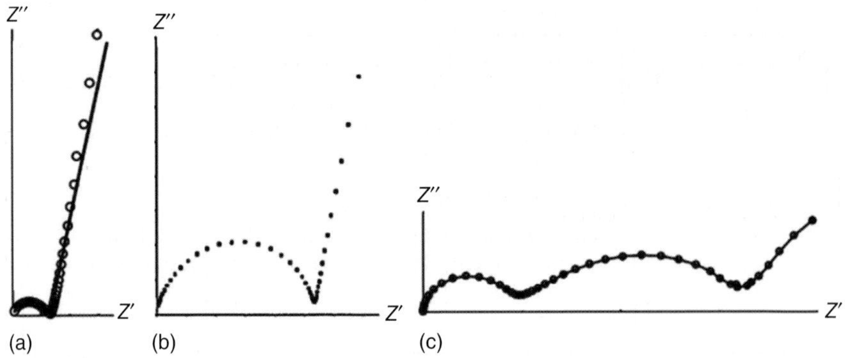

Figure 2.19 Impedance plots of three real systems. (a) A solid electrolyte, (b) a polymer electrolyte, and (c) a composite material.

The deviations of the impedance spectra from the expected ideal behavior arise due to nonideal capacitive behavior of materials (like leaky capacitors), distributed microscopic properties of the material, unevenness of the electrode/electrolyte interfaces, etc. Depressed semicircles usually arise due to the nature of the electrolyte/electrode interface, properties of the electrolyte not being homogeneous, distributed microscopic properties of the electrolyte, etc. Slanted or curved spikes may arise due to rough electrode/electrolyte interfaces, charge transfer across the electrode/electrolyte interface, diffusion of species in the electrolyte or electrode, etc. The deviations of impedance spectra from ideal behavior can be explained in terms of a new circuit parameter called **constant phase element** (**CPE**).

2.2.7.1 The Constant Phase Element (CPE)

In general the CPE has the properties of R and C (CPE is equivalent to a leaky capacitor). Mathematically impedance of a CPE can be assumed to be given by the complex quantity

$$Z_{\mathrm{CPE}} = \frac{1}{Y_o (j\omega)^n} = Z_o (j\omega)^{-n} \quad \text{with } 0 \le n \le 1 \tag{2.26}$$

where Y_o and Z_o are frequency-independent admittance and impedance quantities, respectively. When $n = 0$, $Z_{\mathrm{CPE}} = Z_o$, a frequency-independent impedance as in the case of a resistance R. Hence the CPE is equivalent to a pure resistance. When $n = 1$,

$$Z_{\mathrm{CPE}} = \frac{1}{j\omega Y_o} \tag{2.27}$$

Hence, Z_{CPE} corresponds to the impedance of a capacitance having a capacitance value of Y_o. Therefore the CPE is equivalent to a pure capacitance.

When $0 < n < 1$, the CPE acts as an intermediate between R and C. When $n = 0.5$, the CPE can be represented by a special element called Warburg element (explained in later sections). It can be shown that the circuit containing R and CPE in parallel gives a circular arc in the impedance plane as shown in Figure 2.20. Usually CPE is denoted by the circuit element Q.

In an impedance plane, the impedance plot of CPE alone will be a straight line inclined at angle $(n \times 90°)$ to the real Z'-axis. CPE//R circuit gives a tilted semicircle with its center depressed and situated below the horizontal Z'-axis so that the impedance plot appears as an arc. The diameter of the semicircle is inclined at an angle $[(1 - n) \times 90°]$ with the Z'-axis as shown in Figure 2.20.

2.2.7.2 Equivalent Circuits for Real Systems

For real systems, the capacitance involved in the equivalent circuits are usually not perfect and therefore they can be replaced with CPEs. Hence, the general equivalent circuit shown in Figure 2.18 for the hypothetical solid ion conductor between two blocking electrodes will take the form as shown in Figure 2.21. Here it has been assumed that electrodes are non-perfect as well. All capacitances in the circuit in Figure 2.18 have been replaced with corresponding CPEs to reflect the general behavior of the real sample and electrodes.

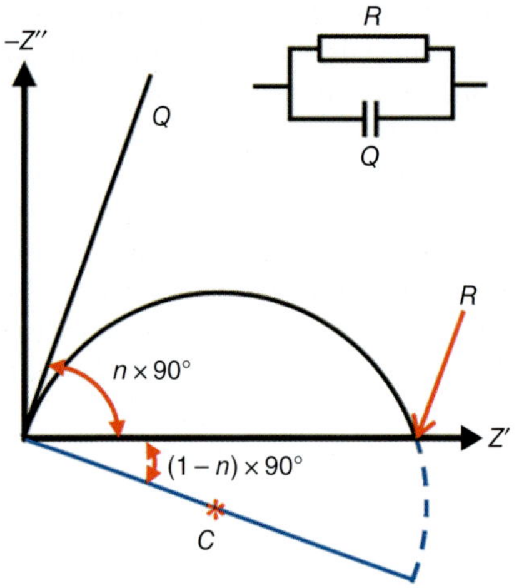

Figure 2.20 Impedance plot of a *R*//CPE circuit.

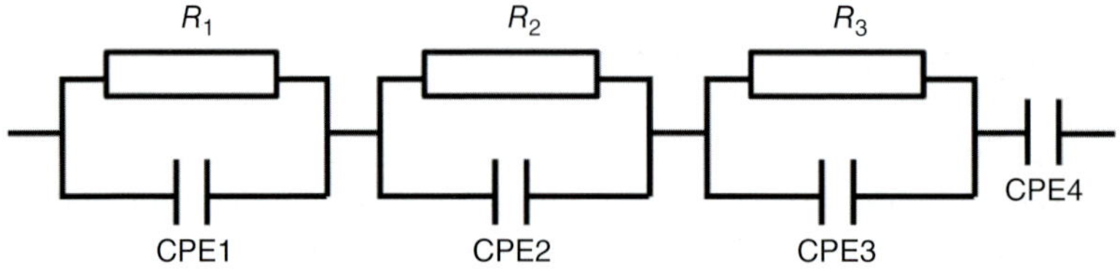

Figure 2.21 The possible equivalent circuit of ion-conducting material sandwiched between two blocking electrodes in real system.

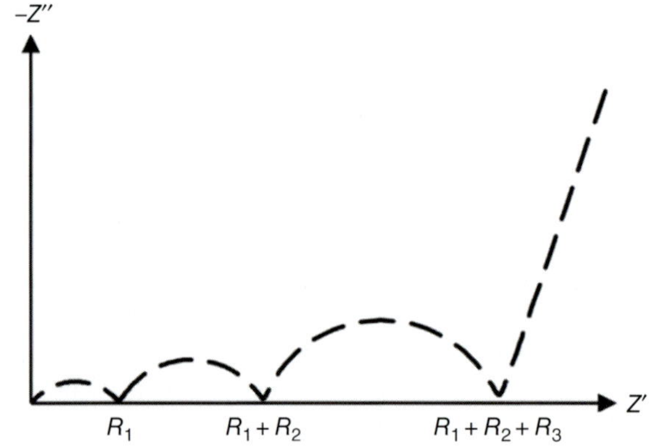

Figure 2.22 Impedance plot for the circuit shown in Figure 2.21.

The expected impedance plot will consist of depressed semicircles and a slanted spike as in Figure 2.22. Each arc corresponds to a different transport process in the electrolyte. Here it has been assumed that these processes are well separated from each other. The slanted line represents the behavior of the electrolyte/electrode interface. Each CPE will have its own exponent parameter n. The parameters n_1, n_2, and n_3 of the CPE1, CPE2, and CPE3 respectively determine the depression of the centers of the respective semicircles from the horizontal Z'-axis, and n_4 of the CPE4 determine the inclination of the spike. The spike will be inclined at an angle ($n_4 \times 90°$) to the Z'-axis. The overall resistance of this hypothetical sample that determines the effective conductivity of the sample corresponds to the point of intersection of the last depressed semicircle and the spike.

2.2.7.3 Electrolyte/Electrode (E/E) Interface

When an electrode is in contact with an electrolyte, the total impedance of the electrolyte/electrode (E/E) interface is determined by the contributions from E/E interface and electrochemical or Faraday reactions taking place at the electrode. For a perfect non-blocking electrode, charge transfer occurs across the interface freely and no buildup of charges occurs at the E/E interface and, therefore, no double layer formation occurs at the interface. The total impedance of the electrolyte/electrode system will include the contributions from the electrolyte and Faradic reactions only.

On the other hand, if the electrode is a perfectly blocking electrode, it is possible that no charge transfer takes place across the E/E interface. Hence, charges will accumulate at the E/E interface forming the double layer of charge to make the interface behave as an ideal capacitor with a double layer capacitance of C_{dl}. Hence, the equivalent circuit of the E/E system will consist of C_{dl} in series with $R_b//C_b$ circuit representing the electrolyte as shown in Figure 2.23.

If the blocking electrode is not perfect and Faradic reactions occur at the electrode, charge transfer can take place at the interface and therefore a charge transfer resistance R_{ct} has to be added in parallel with the C_{dl} to represent the behavior of the E/E interface. Hence, the equivalent circuit of such an E/E system can be represented by the circuit shown in Figure 2.24.

2.2.7.4 Diffusion Impedance or Mass Transport Impedance

If the electrode is partially blocking or non-blocking, charge transfer can take place across electrolyte/electrode interface through electrochemical reaction at the electrode, which is known as electrolysis. The charge transfer will give rise to resistance R_{ct}. As a result of the electrolysis reactions, concentrations of charged species or ions in or near the electrode/electrolyte interface will be different to those in the bulk of the electrolyte. However, all species will try to maintain

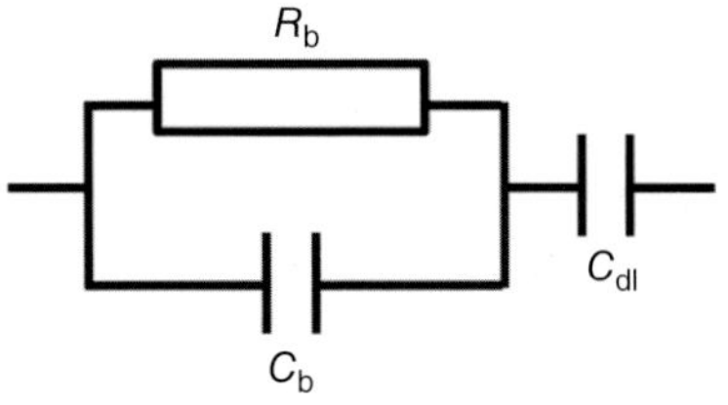

Figure 2.23 Equivalent circuit that can represent an electrolyte in contact with a perfect blocking electrode.

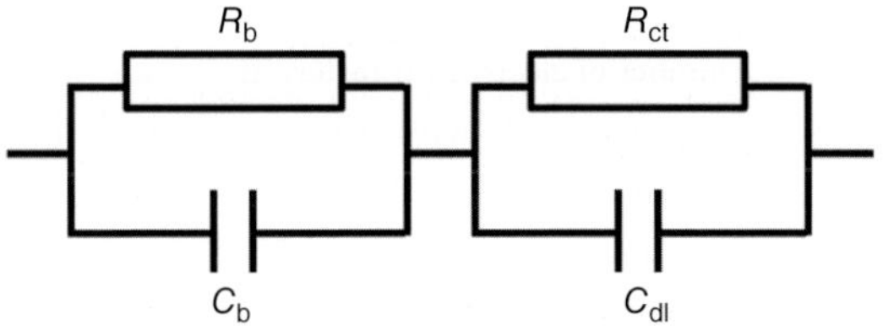

Figure 2.24 Equivalent circuit that can represent an electrolyte in contact with an electrode at which Faradaic reactions are taking place.

equal concentration distribution within a system. For example, when ions are consumed at the electrode during the electrochemical reaction, ions from the bulk of the electrolyte will move to the electrode to replenish the depleting ions. This movement to equalize concentration difference by any species is called diffusion. Similarly any new species produced at the electrode will diffuse away from the electrode. Such diffusions occur close to the E/E interface just outside double layer region. During any such diffusion, mass transport occurs as the species physically move from one place to another.

The electrical impedance produced due to diffusion of charged species is referred to as diffusion impedance or mass transport impedance and is usually denoted by Z_D. The mass transport impedance term is used as mass transport is involved in diffusion processes. Diffusion is a slow process and therefore generally observed at low frequencies only. Z_D varies with the frequency and can be considered to act in series with R_{ct} as the same ionic species are involved in charge transfer and diffusion processes. At sufficiently lower frequencies, Z_D is referred to as the Warburg impedance Z_W. In equivalent circuits Z_W is usually denoted by the Warburg element W.

2.2.7.5 Warburg Impedance

Various expressions are used to represent the diffusion impedance depending on the restrictions imposed by the diffusion environment. The expression for diffusion impedance was first derived by Warburg, assuming linear diffusion to a large planar electrode and the impedance is commonly known as the Warburg impedance and is denoted by Z_W or W. In equivalent circuits the Warburg impedance is often represented by the Warburg element W. it should be noted that, strictly speaking, the element W or Z_W actually represents the infinite Warburg impedance.

Under the assumption of infinite or semi-infinite diffusion layer, i.e. the diffusion layer has an infinite thickness, the Warburg Impedance can be written as

$$W = Z_W = \sigma\omega^{-\frac{1}{2}} - j\sigma\omega^{-\frac{1}{2}} = \sigma\omega^{-\frac{1}{2}}(1 - j) \tag{2.28}$$

where σ is the Warburg coefficient given by

$$\sigma = \frac{RT}{n^2F^2A\sqrt{2}}\left(\frac{1}{C_O^*\sqrt{D_O}} + \frac{1}{C_R^*\sqrt{D_R}}\right) \tag{2.29}$$

Here

D_O	=	Diffusion coefficient of the oxidant (O)
D_R	=	Diffusion coefficient of the reductant (R)
A	=	Surface area of the electrode
n	=	Number of electrons transferred
C^*	=	Bulk concentration of the diffusing species (mol cm^{-3})
F	=	Faraday constant
R	=	Gas constant
T	=	Absolute temperature

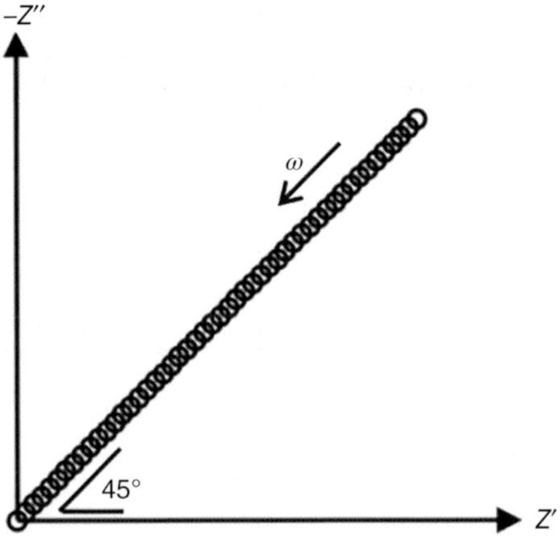

Figure 2.25 Impedance plot of infinite Warburg impedance.

Here W actually corresponds to diffusion in an infinite layer and therefore must be referred to as **infinite Warburg impedance**. However, W is generally referred to as Warburg impedance or Warburg element. For more details and derivation of Eq. (2.28), the readers are advised to refer the book Vadim F. Lvovich [11].

The magnitude of infinite Warburg impedance is given by

$$|Z_W| = \frac{\sqrt{2}\sigma}{\omega^{\frac{1}{2}}} \propto \omega^{-\frac{1}{2}} \tag{2.30}$$

Equation (2.28) shows that at all frequencies, Real Z_W = Imaginary Z_W; the real part is referred to as "diffusional resistance" and the imaginary part is known as "diffusional capacitance." Hence, on an impedance plot, the infinite Warburg impedance becomes a straight line with a slope of 0.5 as in Figure 2.25. Therefore, W can be regarded as a special CPE with $n = 0.5$.

2.2.7.6 Equivalent Circuit Representation of an E/E System

Normally not all charged species arriving at the electrode participate in the electrolysis reaction, but they accumulate at the electrode. The charge of these accumulating species are countered by the opposite charge on the electrode side (by accumulation of electrons or depletion of electrons depending on the polarity of the electrode) producing the double layer of charge at the electrolyte/electrode interface. It should be noted that the ionic species involved in the creation of the double layer are bound and different to ionic species involved in the electrolysis and diffusion. There will be two such double layers of charges if an electrolyte is bound by two electrodes and they can be assumed to have an effective double layer capacitance C_{dl}. In equivalent circuits, C_{dl} is normally placed in parallel with the R_{ct} and W ($\equiv Z_D$). Hence the equivalent circuit of an electrolyte bounded between two electrodes takes the form shown in Figure 2.26. Here C_{dl} is the effective double layer capacitance, R_{ct} is the charge transfer resistance, and W is the Warburg element. $R_b//C_b$ circuit represents the effect of the electrolyte.

The impedance plot of such a system will have a semicircle corresponding to R_b and C_b (electrolyte), another semicircle corresponding to C_{dl} and R_{ct}, and a straight part inclined at 45° to the horizontal axis corresponding to W as shown

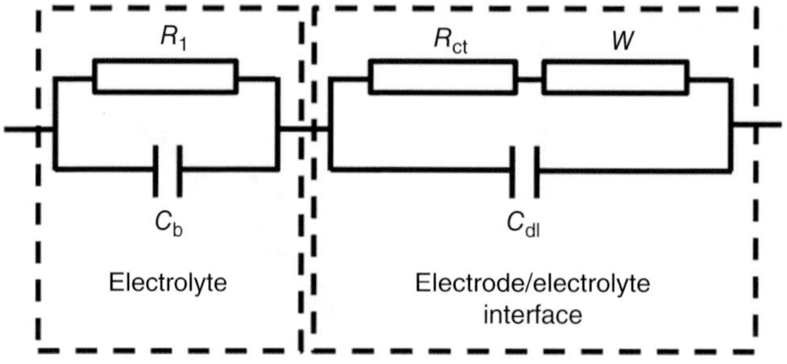

Figure 2.26 The equivalent circuit of an electrolyte/electrode system.

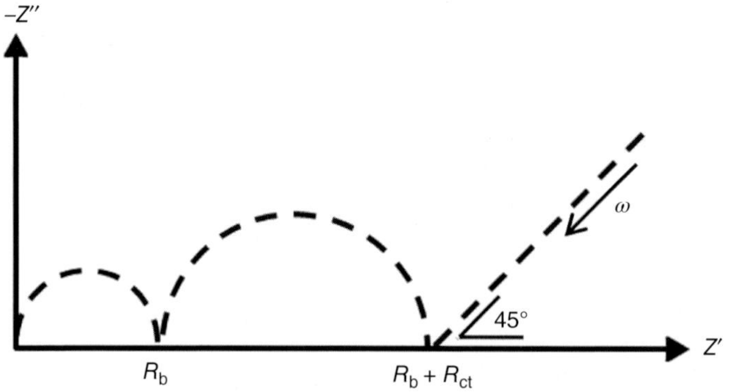

Figure 2.27 The expected ideal impedance plot for the electrolyte/electrode system shown in Figure 2.26.

in Figure 2.27. It is to be noted that, if in addition, ion adsorption at the electrode and/or any other process occurs additional circuit components have to be added to the E/E part. This will give additional features in the impedance spectra.

2.2.8 Impedance-Related Functions

By definition impedance is a complex quantity containing a real and an imaginary component. Impedance spectroscopy is useful to study the microstructure and dielectric behavior of many materials. However, in some situations impedance spectra may not be able to resolve the detailed structure of materials especially when the semicircles corresponding to different portions overlap. In some cases, only a small part of the impedance spectrum is observed and it will be difficult to estimate the impedance parameters like resistance and peak frequency. In the case of many dielectric materials, dc conductivity may prevent the identification of relaxation phenomena at low frequencies. In such cases, the features of the impedance spectra can be seen clearly when the spectra are plotted using impedance-related functions such as admittance, permittivity, and electric modulus. The relations between these functions and their applications are discussed in this section.

2.2.8.1 Immittance Functions

Several quantities are related to impedance (Z) that can play important roles in IS. Some of them are admittance (Y), permittivity (ε), and electric modulus (M). These are generally referred to as "immittances" or "immittance functions."

2.2.8.2 Relationships Between Immittance Functions

If an *ac* voltage v is applied to a sample having an impedance Z, the resulting current is given by

$$v = Zi \tag{2.31}$$

Hence,

$$i = \frac{1}{Z}v = Yv \tag{2.32}$$

Here Y is the complex admittance. The admittance is also a complex quantity and it can be written as

$$Y = Z^{-1} = Y' + jY'' \tag{2.33}$$

where Y' and Y'' are the real and imaginary components of Y. The complex electric modulus M of material is related to its complex permittivity ε through

$$M = \varepsilon^{-1} = j\omega C_o Z \tag{2.34}$$

where ω is the angular frequency of the applied signal and C_o is the capacity of the empty sample cell (vacuum capacitance), which is given by

$$C_o \equiv \frac{\varepsilon_0 A}{d} \tag{2.35}$$

Here ε_0 is the free space permittivity, A is the sample–electrode contact area, and d is the sample thickness. The complex modulus can be written in terms of the real and imaginary components as

$$M = M' + jM'' \tag{2.36}$$

By definition, the complex permittivity is related to admittance by

$$\varepsilon = \frac{Y}{j\omega C_o} \tag{2.37}$$

The complex permittivity can be represented by the real and imaginary components through

$$\varepsilon = \varepsilon' - j\varepsilon'' \tag{2.38}$$

The four immitance functions are related to each other and the relationships are shown in Table 2.1.

2.2.8.3 Immittance Plots

Plotting the impedance results of a sample in more than one formalism is practical to identify different important features. The Z, Y, M, and ε plots for simple circuits are compared in Figures 2.28–2.30. The bulk resistance R_b is usually obtained from the impedance plot (Z'' against Z' plot).

Table 2.1 Relationship between immittance functions.

	M	*Z*	*Y*	ε
M	M	μZ	μY^{-1}	ε^{-1}
Z	$\mu^{-1}M$	Z	Y^{-1}	$\mu^{-1}\varepsilon^{-1}$
Y	μM^{-1}	Z^{-1}	Y	$\mu\varepsilon$
ε	M^{-1}	$\mu^{-1}Z^{-1}$	$\mu^{-1}Y$	ε

Here $\mu = j\omega C_o$.

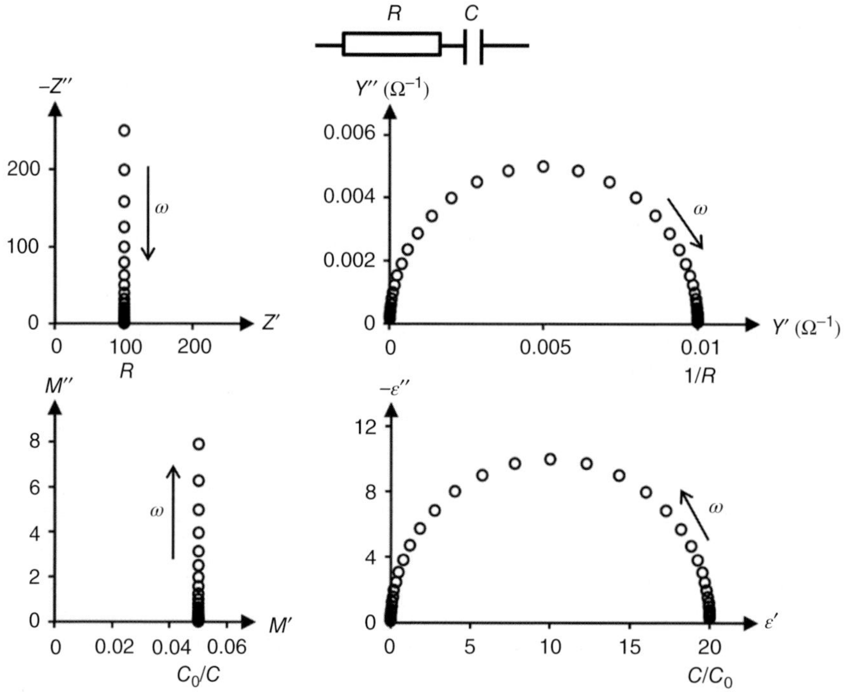

Figure 2.28 *Z*, *Y*, *M*, and ε plots for a series *RC* circuit (simulated with $R = 100\,\Omega$, $C = 20\,\mu F$, and $C_o = 1\,\mu F$).

2.2.8.3.1 ***R and C in Series*** The Z, Y, M, and ε plots expected for a circuit containing R and C in series are shown in Figure 2.28. It can be seen that the direction of increasing frequency in each plot is different.

While the impedance and electric modules plots are straight lines, the Y and ε are semicircles. However, the direction of increase of frequency is opposite in each set of plots. In Z plot, the resistance value can be obtained from the intersection of the points with the real impedance axis on the high frequency side. While the intersection point gives the resistance value directly in the impedance plot, the corresponding point in the admittance plot corresponds to the inverse of R. Modulus plot is useful to determine the value of capacitance as the intercept on

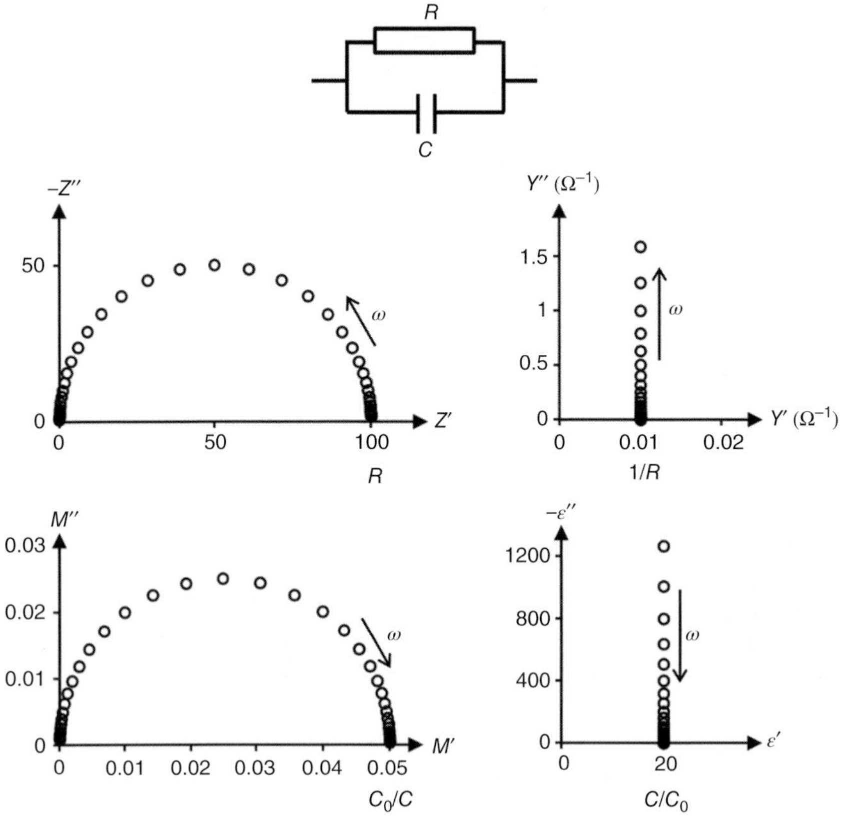

Figure 2.29 *Z*, *Y*, *M*, and *ε* plots for a parallel *RC* circuit (simulated with $R = 100\,\Omega$, $C = 20\,\mu F$, and $C_o = 1\,\mu F$).

the real axis corresponds to C_o/C where C_o is the vacuum or geometric capacitance. The value of *C* also can be obtained from the permittivity plot, which takes the semicircle shape, as the admittance plot forms the intercept with the real axis at the low frequency side.

2.2.8.3.2 ***R and C in Parallel*** The *Z*, *Y*, *M*, and *ε* plots expected for a circuit with *R* and *C* connected in parallel are shown in Figure 2.29.

When *R* and *C* are in parallel connection, the shape of the impedance and modulus plots looks the same but the direction of the increasing frequency is different. The admittance and permittivity plots however are straight lines parallel to the imaginary axis. Different sample parameters that can be extracted from the plots are indicated in the figures.

2.2.8.3.3 ***A Parallel R and C Circuit in Series with a Resistance R_s*** The *Z*, *Y*, *M*, and *ε* plots expected for the circuit with *R*//*C* connected in series with a resistance R_s are shown in Figure 2.30.

Unlike in the earlier cases, the *Z* and *Y* curves take the same shape of semicircles and both *M* and *ε* plots have a semicircular part and a straight line. However,

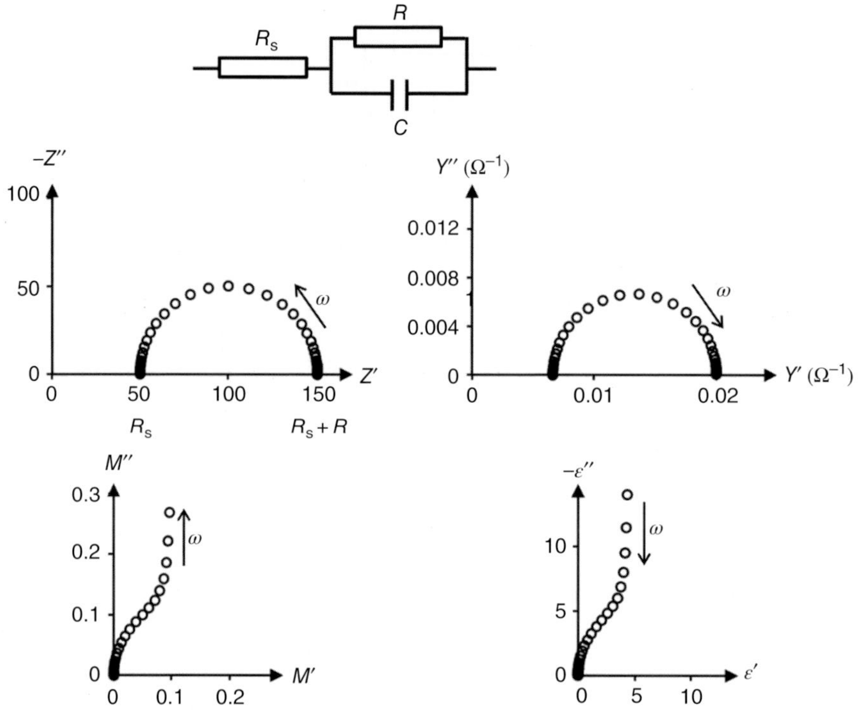

Figure 2.30 *Z*, *Y*, *M*, and *ε* plots for a parallel *RC* circuit in series with a resistance (simulated with $R = 100\,\Omega$, $C = 1\,\mu F$, $C_o = 0.1\,\mu F$, and $R_s = 50\,\Omega$).

the direction of increase in frequency is different for each set of plots. From the *Z* plot, the resistance *R* can be obtained from the diameter of semicircle and R_s is the distance between origin and the first semicircle intercepts.

2.2.8.4 Choice Between Immittance Functions

For any system, the four different formalisms correspond to giving the same information in different ways. However, each formalism highlights different features of the system. For a more complex system, it may be worthwhile to plot the data in more than one formalism in order to extract all possible information from a result. *Z* plots give prominence to most resistive elements and *M* plots give prominence to smallest capacitances. For example, *Z* plot is good to study grain boundary effects and *M* plot is good to study bulk effects.

2.2.9 Experimental Setup

Correct choice of sample cell arrangement, measuring equipment, measuring conditions, and environment are essential for the collection impedance data and analysis. This section discusses these details along with the precautions necessary for obtaining accurate data. The procedure for extracting the parameters from impedance plots is also explained here.

2.2.9.1 Sample and Cell Arrangement

2.2.9.1.1 The Sample The preferred shape for a sample is a thin cylinder with parallel faces of well-defined cross section as shown in Figure 2.31a. If the sample has a length L and a cross-sectional area A, it will have a cell constant, L/A. If the material under test is in a solid or film form, it would be easy to achieve this shape. If the material is in a liquid or gel form, suitable sample holders have to be used. Wherever possible, the surfaces of the sample must be cleaned and polished so as to have uniform and smooth surfaces. For liquid samples, suitable containers can be used to obtain the desired sample shape (e.g. coin cell). The sample is usually sandwiched between two suitable electrodes as shown in Figure 2.31b.

2.2.9.1.2 Electrodes The materials used for electrodes must be nonreactive (inert) and good conducting. Silver (Ag), gold (Au), platinum (Pt), palladium (Pa), stainless steel (SS), graphite, etc., are widely used as electrode materials. They can be attached to the solid samples using vacuum evaporation or sputtering provided the samples can withstand the temperature produced during the process. Alternatively a thin layer of conducting pastes such as silver, gold, or graphite can be used as electrodes. On using pastes, care must be taken to avoid reactions with the sample as the pastes may contain reactive chemicals. For soft materials like polymer gels or liquids, disc electrodes pressed to the samples can be used. A typical sample–electrode arrangement is shown in Figure 2.32.

2.2.9.1.3 Sample Holders A simple sample holder that is suitable for most samples is shown in Figure 2.33. It contains a small glass tube with tight-fitting cylindrical contact electrodes made of stainless steel. The top contact electrode must be removed to place the sample and pressed slowly until the two electrodes make good contact with the sample. If the sample has no attached electrodes, the stainless steel cylinders can serve as the electrodes; otherwise they can serve as current collectors. This sample holder can be constructed easily to the required size and is

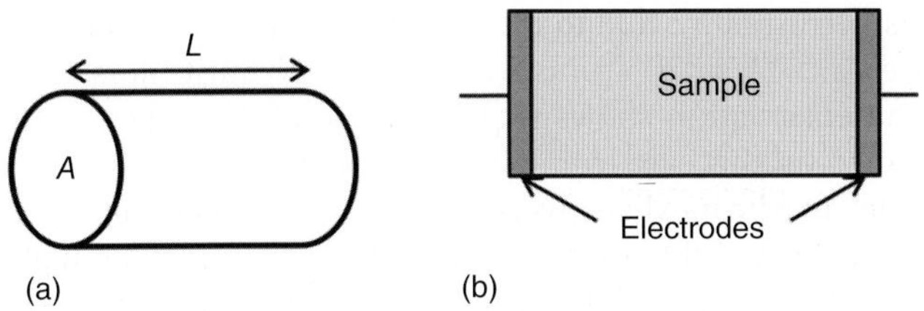

Figure 2.31 (a) A cylindrical sample and (b) a sample cell with the sample sandwiched between two electrodes.

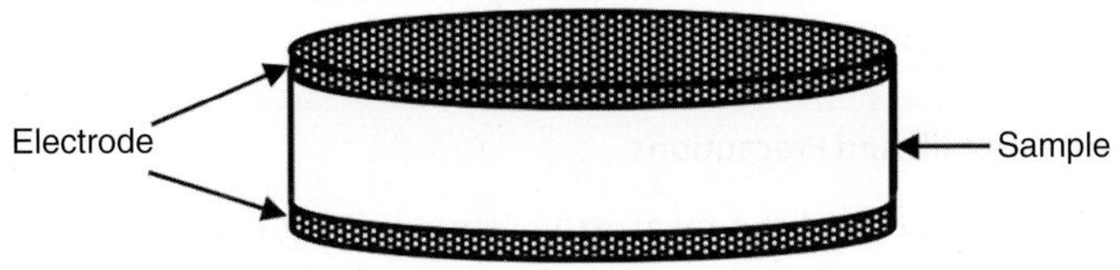

Figure 2.32 A solid sample coated with two metal electrodes.

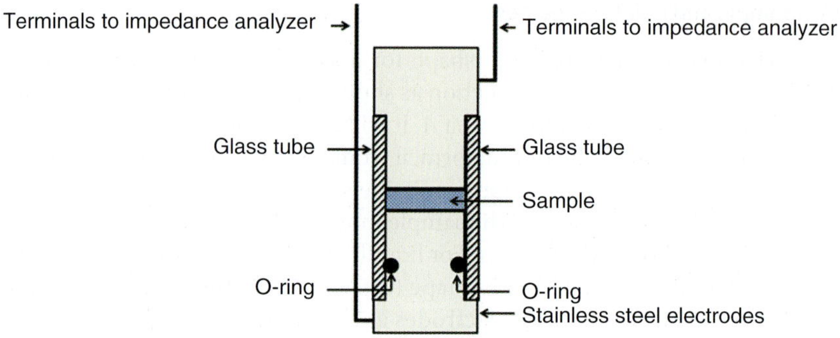

Figure 2.33 A simple sample holder arrangement.

suitable for most samples including soft solid, gel, or liquid samples. The sample holder can be placed in an oven if measurements at different temperatures have to made, provided precautions are taken to have suitable connecting arrangements to the impedance analyzer. More sophisticated sample holders with spring loading can be either constructed or bought commercially.

2.2.9.1.4 Sample–Electrode Connection to Impedance Analyzer The sample holder must be air tight if the sample is sensitive to the environment. The sample holder can be placed in a furnace if temperature variation is required. Cables connecting the sample to the impedance measuring equipment must be of low noise coaxial type. The cables must have the minimum possible lengths to avoid errors in the measurements. Two-terminal or 4-terminal configuration for the connection can be used.

In electrochemical impedance spectroscopy (EIS) measurements, the instrument will capture the magnitude of the complex impedance and phase angle (θ). Normally, the phase angle has a negative value. As such the real and imaginary parts of impedance will be given as ($Z\cos\theta$) and ($Z\sin\theta$), respectively. The imaginary impedances are usually negative due to the negative θ value. The complex impedance plot is plotted as $-(Z\sin\theta)$ against ($Z\cos\theta$). In Nyquist plots with semicircle and spike, the observation shows that the value of θ is negative in the frequency range studied (in this case from 50 Hz to 5 MHz). However, again based on experience, which maybe inexhaustive, if the Nyquist plot contains only a tilted spike, then there is a possibility of θ having a positive value at the high frequency end. Positive θ values may imply an inductive effect, which may be due to the length of the cable, which is sometimes necessary to be lengthened when the sample was placed in the furnace for temperature-dependent studies. To overcome this problem, measurements are repeated within the frequency range where the θ values are all negative to check for consistency.

2.2.9.2 Other Practical Details and Precautions

2.2.9.2.1 Accuracy Check To test of the measuring setup, it is a good practice to perform impedance measurements from time to time on a dummy cell containing a simple network of resistors and capacitors of known values. If the values obtained from impedance plots do not agree with the values of the components used in making the simple network, the cause for such discrepancies must be identified and corrected before making measurements on actual samples. Many impedance analyzers are provided with dummy cells and instructions for checking accuracy and the procedure to follow to take accurate measurements.

2.2.9.2.2 Frequency Range and Number of Readings For obtaining meaningful impedance spectra of a system, frequency range as wide as possible ideally from 0 to ∞ must be used. As it is not easy to cover such wide range in one equipment, most commercial equipment have a fixed frequency range or frequency window (for example, 1 mHz to 100 kHz) for measurements. Hence, the number of readings depends on the available range. As wide ranges of frequencies are involved, it is necessary to take readings equally spaced not on a linear scale, but on a logarithmic scale. It is possible to program impedance spectrometers to take measurements at set frequencies equally spaced on logarithmic scale. For example, the impedance analyzer can be programmed to take 6–10 readings for each decade of frequency. If the starting frequency is 1 mHz and the ending frequency is 100 kHz, with 10 points per decade setting, there will be 80 reading as the frequency range covers 8 decades.

The frequency range required to explore the main features of most electrochemical system is 10 mHz to 100 kHz corresponding to time constants of 10^2 and 10^{-5} seconds, respectively. For studies on dielectric materials and fast kinetic processes, high frequencies up to 10^6 Hz are necessary. For investigating slow processes such as diffusion and intercalation in materials, very low frequencies in the range of 10^{-3} to 10^{-6} Hz may be needed. For low frequency measurements, it is necessary to maintain constant temperature and pressure of the sample cell and very stable conditions of the current lines as the time taken for measurements is long. It should be noted that measurements at frequencies lower than 10^{-3} Hz will take longer times (greater than 16 minutes), and a complete measurement at 10^{-5} Hz will require roughly 28 hours. On the other hand, for measurements at frequencies greater than 1 MHz, there will be electrical lead/cable problems arising mainly from inductance and special precautions are to be taken to avoid these problems. Shielding the sample and proper cable arrangements will reduce errors in measurements. The range of frequencies for impedance measurements has to be decided depending on the properties and phenomena to be investigated bearing in mind the limitations stated in the preceding text.

2.3 IS: Experimental Data Interpretation and Analysis

2.3.1 Determination of Bulk Resistance from the Impedance Plots

The impedance plot is useful to obtain the overall conductivity of materials. The conductivity is determined by the bulk resistance R_b, which is obtained from the intercept of the plot on the real axis. R_b can be determined graphically by drawing the best semicircle to fit the experimental points. Alternatively, the resistance value can be obtained by fitting a $R//C$ circuit with suitable values for R and C.

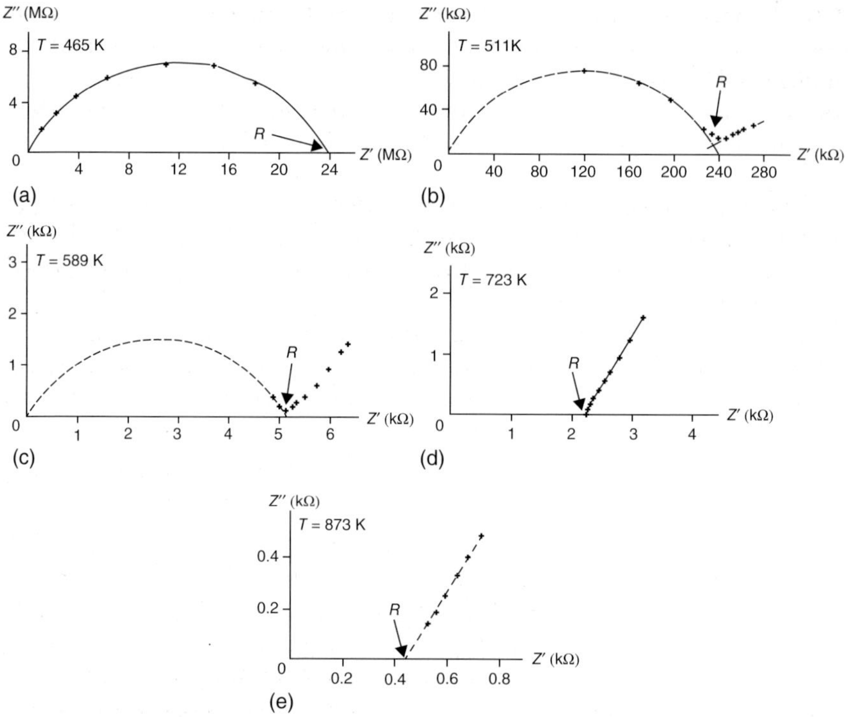

Figure 2.34 (a–e) IS spectra of a solid electrolyte system at different temperatures in the frequency range from 100 Hz to 100 kHz.

Here the value of R is the bulk resistance. It should be noted that both the Z'- and Z''-axis must have the same scales to obtain circular arcs.

For many real systems, the impedance plot takes the form of a distorted or depressed semicircle and the straight line is slanted. In most situations only parts of the semicircle or straight line will be present. This behavior can be seen in the impedance plots obtained for a solid ionic conductor sample at different temperatures shown in Figure 2.34. At the temperature 465 K (Figure 2.34a), the plot contains only a depressed semicircle and the intercept of the arc on the Z'-axis at the low frequency region gives the value of R_b. In the cases where impedance plots have a semicircle at higher frequency followed by a slanted spike at the low frequency end (see Figure 2.34b,c), R_b is obtained from the intersection of the semicircular part and the spike. If the plot consists of only a steeply rising spike as depicted in Figure 2.34d,e, R_{b} is determined from the intercept on the Z'-axis at the high frequency region as shown. Table 2.2 lists the value of R_{b} at corresponding temperature obtained in Figure 2.34.

2.3.2 Impedance Data Interpretation and Analysis

Impedance spectroscopy has brought many benefits to the field of research especially in the area of conducting materials. IS is able to carry out the experiment

Table 2.2 R_b value at corresponding temperatures obtained in Figure 2.34.

Temperature, T (K)	465	511	589	723	873
R_b (Ω)	23.7×10^{-6}	239.3×10^{-3}	5.1×10^{-3}	2.20×10^{-3}	0.4×10^{-3}

on assembled device using the device's own electrodes, which makes it easy to handle. The impedance data obtained from the IS measurement is able to

- provide the details about the electron transfer rate of reaction (refer to Section 2.3.4.3),
- yield the details on the capacitive behavior of the system,
- identify diffusion-limited reactions such as diffusion through a passive film,
- determine the number of electrochemical reaction taking part in a system.

All the features listed above for a material can be determined by interpreting and analyzing the impedance data obtained from the experiment.

2.3.2.1 Interpretation of Impedance Data

Interpretation of impedance data is usually done by identifying appropriate electrical equivalent circuits and choosing suitable parameters for the equivalent circuits to fit the impedance plots. This can be done manually using spreadsheets, but the process will take considerable time. Various computer software are now available to fit the experimental spectra and to identify the equivalent circuits. It should be noted that the equivalent circuits are merely analogs, rather than models. But in many situations, equivalent circuits can be considered as a model to describe transport processes and electrochemical reactions in electrolytes and at electrode/electrolyte interfaces. Most commercial impedance analyzers come with their own software for analyzing the impedance data. Some software has the ability to analyze multiple spectra together. Many of them support automatic pre-fit of initial guesses, which eliminates the need of manual guessing. Graphical presentation of the spectra, results and fitting parameters, and library of commonly used equivalent circuits are also included in some of the commercially available software.

In addition, the equivalent circuit model is the most recommended approach to interpret a Nyquist plot that has scattered data points. By representing the Nyquist plot with an appropriate equivalent circuit model, the impedance parameters such as bulk resistance and capacitance of the sample can be determined. The best fit plot with lowest regression value will provide the necessary information about the material under study.

2.3.2.2 Choice of Equivalent Circuits

Usually the fitting software does not give a unique equivalent circuit (model) for a particular impedance spectrum but may suggest a number of complicated circuits (multiple models). Hence, choosing the correct equivalent circuit for a system can be difficult. For example, consider an impedance spectrum with a two time constant as shown in Figure 2.35a. It can be fitted with the three equivalent circuits that are shown in Figure 2.35b. However, the values of circuit elements

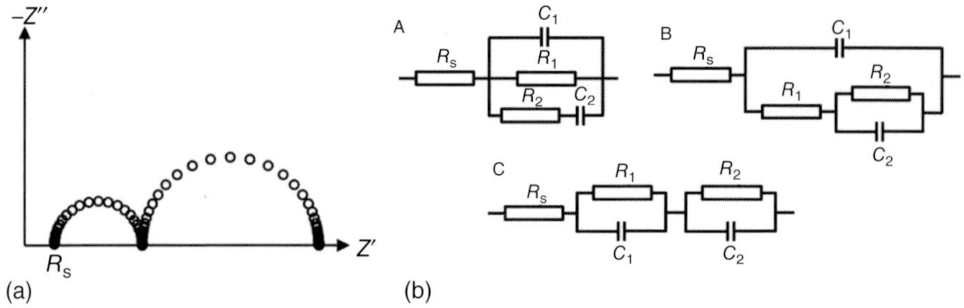

Figure 2.35 (a) Two time constant impedance spectrum and (b) some of the possible equivalent circuits (ambiguous circuits) that fit the spectrum.

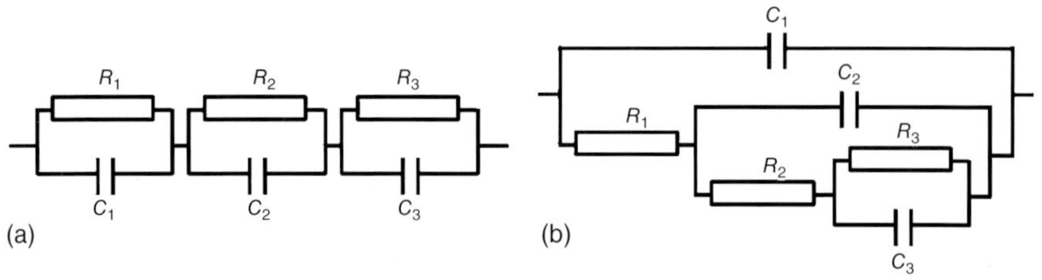

Figure 2.36 (a, b) Two ambiguous circuits.

will be different for each circuit. The fitting software will not be able distinguish between them. These circuits are referred to as ambiguous circuits.

It should be noted that the equivalent circuit model chosen should not only fit the IS data but also be verifiable through other experiments and theories and justifiable through other known facts. For example, the two circuits (a) and (b) shown in Figure 2.36 can give three distinct semicircles in the impedance spectrum if their time constants are well separated. However, the circuit in (a) is more suitable for a polycrystalline sample and (b) is more suitable for a homogeneous material. The polycrystalline sample may contain crystalline grains separated by grain boundaries. An $R//C$ circuit can represent the impedance response of the grains and the other two similar circuit can represent the response of grain boundaries and electrode/sample interface, respectively. The three $R//C$ circuits can be assumed to be connected in series if the grain boundaries and electrodes partially block the passage of ions.

On the other hand, the homogeneous material may have the same structure throughout the material, but the impedance response may be coupled to bulk effects, ion dissociation and recombination reaction, ion adsorption at electrodes, double layer at electrode, etc., and the responses can be assumed to be interconnected as in the circuit (b). If the material consists of a number of bigger homogeneous regions, then the equivalent circuit may contain more complicated circuits having a number of circuits of (b) type connected in series. In many situations it would be beneficial to choose the equivalent circuit with fewer number of circuit elements. However, equivalent circuits with more components may be necessary to agree with physical and chemical models of the system. It is to be noted that all the circuits discussed previously consist only the regular circuit elements of resistances and capacitances. However, the actual circuits may contain distributed circuit elements such as CPEs, Warburg element, etc.

Therefore, an equivalent circuit of a system must be related or can be related to the physical/chemical properties/behavior of the system. To choose the correct equivalent circuit, one must have some idea about the material under study. Temperature and/or potential dependence of the spectra will help the choice. Experience in choosing models is essential especially to drop unrealistic circuits. Electron microscopy will be very useful to identify the microstructure of materials. Therefore, additional information is necessary to decide which circuit has a physical meaning in each case.

In polymer electrolyte system, the preferred equivalent circuit model to fit the impedance spectrum is a resistor (R) in parallel with capacitor (C) and in series with another double layer capacitor, (C_{dl}) (refer to Figures 2.14 and 2.15). When the electric field is applied to the electrolyte, there is an opposition to ion flow that is represented by the resistor (R). At the same time, the *ac* at certain frequency aligns the polymer change by dipole polarization, which represents a capacitor (C). Since this process occurs at the same time, R and C are placed in parallel. It is noted that during the *ac* current flow, the ions are alternately accumulated and depleted from each electrode surface forming a double layer, which can be represented by (C_{dl}). Since the ions have to flow to the electrode surface before this happens, the C_{dl} is in series with the $R//C$.

The impedance of the R and C in parallel circuit will give an equation of circle:

$$Z'(\omega) = \frac{R}{1 + \omega^2 R^2 C^2} \tag{2.18}$$

and

$$Z''(\omega) = \frac{\omega R^2 C}{1 + \omega^2 R^2 C^2} \tag{2.19}$$

Hence, a perfect semicircle is obtained. In the case the equivalent circuit is as shown in Figure 2.14, if the capacitor is a leaky capacitor then the symbol ⊣⊢ is changes to ⊣⊢ CPE and the impedance equation is given as Eqs. (2.63) and (2.64) that will be discussed later in Section 2.3.3.2.

2.3.3 Determination of Transport Parameters from Impedance Data

In the previous sections, how impedance data can be used to determine the *dc* conductivity of electrolytes and analyze electrode/electrolyte interfaces have been discussed. In addition, the impedance data can be used to determine the transport parameters such as number density (n), mobility (μ), and diffusion coefficient (D) of charge carriers in electrolytes. Two relatively simple methods are discussed here.

2.3.3.1 Bandara–Mellander (B–M) Method

This method was proposed by Bandara and Mellander [12]. The method is discussed briefly here. According to them an electrolyte sandwiched between two blocking electrodes, can be represented by an ideal equivalent circuit as shown

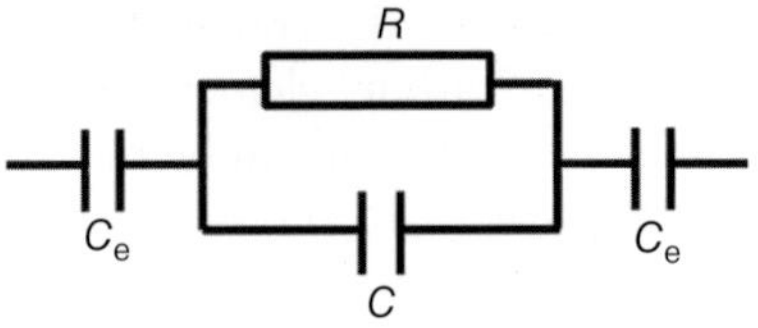

Figure 2.37 The ideal equivalent circuit for an electrolyte sandwiched between two blocking electrodes.

in Figure 2.37. Here C_e represents double layer capacitance formed at each electrode/electrolyte interface, and R and C represent the bulk resistance and geometrical capacitance of the sample, respectively.

Hence, the total impedance for an electrolyte with two blocking electrodes can be represented by

$$Z = \frac{R}{1 + j\omega RC} + \frac{2}{j\omega C_e} \tag{2.39}$$

The real (Z') and imaginary (Z'') components of this impedance respectively become

$$Z' = \frac{R}{1 + \omega^2 R^2 C^2} \tag{2.40}$$

$$Z'' = -\left(\frac{\omega R^2 C}{1 + \omega^2 R^2 C^2} + \frac{2}{\omega C_e}\right) \tag{2.41}$$

On simplifying, Z'' and Z' are related through

$$Z'' = -\left(\omega RCZ' + \frac{2}{\omega C_e}\right) \tag{2.42}$$

The impedance plot for the equivalent circuit will take the simple form of a semicircle and a vertical spike (straight line at $Z' = R$) as shown in Figure 2.38. Usually C_e is larger than C, since the electrical double layer (EDL) thickness is about that of the Debye length (λ), ~nm (very small compared to the thickness of the electrolyte which is ~1 mm). The angular frequency at the maximum and minimum of the imaginary impedance is denoted by ω_1 and ω_2. Expressions for the frequencies corresponding to the maximum and minimum of Z'' can be derived under the assumption that C is very small compared to C_e.

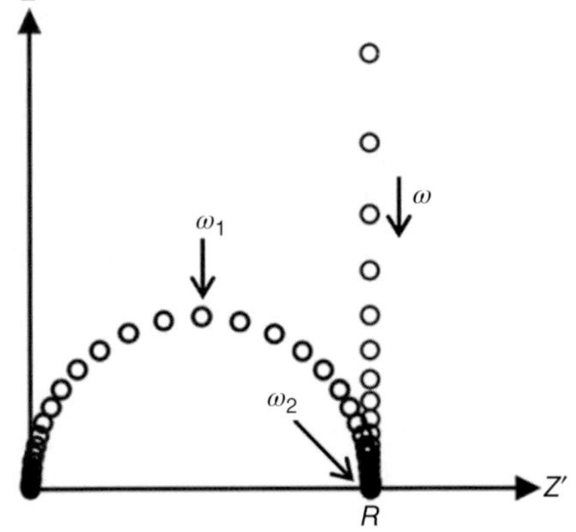

Figure 2.38 The impedance plot for the circuit shown in Figure 2.37.

The angular frequency ω_1 corresponding to the maximum in Z'' shown in Figure 2.38 can be obtained simply by substituting $Z' = R/2$ in Eq. (2.40) as at the maximum position $Z' = Z'' = R/2$. Thus

$$\omega_1 = \frac{1}{RC} \tag{2.43}$$

Hence, the characteristic time constant τ_1 corresponding to the maximum in Z'' is given by

$$\tau_1 = \frac{1}{\omega_1} = RC \tag{2.44}$$

The angular frequency ω_2 corresponding to minimum in Z'' shown in Figure 2.38 can be obtained by simply substituting $Z' = R$ and $Z'' = 0$ in Eq. (2.42). Thus

$$\omega_2 = \frac{1}{RC}\sqrt{\frac{2C}{C_e}} \tag{2.45}$$

The characteristic time constant τ_2 corresponding to the minimum in Z'' is given by

$$\tau_2 = \frac{1}{\omega_2} = RC\sqrt{\frac{C_e}{2C}} \tag{2.46}$$

For an electrolyte sample of thickness $2d$ and cross-section area A, if the mean thickness of the EDL is λ (order of Debye length), R, C, and C_e can be represented by

$$R = \frac{2d}{\sigma_{dc} A} \tag{2.47}$$

$$C = \frac{\varepsilon_r \varepsilon_o A}{2d} \tag{2.48}$$

$$C_e = \frac{\varepsilon_r \varepsilon_o A}{\lambda} \tag{2.49}$$

where σ_{dc} is the dc conductivity of the electrolyte, ε_r is the dielectric constant of the electrolyte and ε_o is the permittivity of free space. The R, C, and C_e in Eqs. (2.47), (2.48), and (2.49) can be substituted in Eqs. (2.43) and (2.45) to obtain

$$\omega_1 = \frac{1}{\tau_1} = \frac{\sigma_{dc}}{\varepsilon_o \varepsilon_r} \tag{2.50}$$

$$\omega_2 = \frac{1}{\tau_2} = \frac{\sigma_{dc}}{\varepsilon_o \varepsilon_r}\sqrt{\frac{\lambda}{d}} \tag{2.51}$$

If we substitute $\delta = \frac{d}{\lambda}$ and take $\sigma_{dc} = n\mu e$ where n is the charge carrier density, μ is the mobility of charge carriers, and e is the electronic charge (here it has been assumed that each charge carrier has a charge e), the Eqs. (2.50) and (2.51) become

$$\omega_1 = \frac{1}{\tau_1} = \frac{n\mu e}{\varepsilon_o \varepsilon_r} \tag{2.52}$$

$$\omega_2 = \frac{1}{\tau_2} = \frac{n\mu e}{\varepsilon_o \varepsilon_r \sqrt{\delta}} \tag{2.53}$$

Using Eqs. (2.52) and (2.53), the relationship between τ_1 and τ_2 can be obtained as

$$\tau_2 = \tau_1 \sqrt{\delta} \tag{2.54}$$

Usually λ is very small (~nm) in comparison to the half sample thickness d (~mm) and therefore $\delta = d/\lambda$ will be large ($\gg$1). According to Bandara and Mellander, the net diffusion coefficient of the ions in the electrolyte is D and the Debye length λ is determined by the relaxation time τ_2 through the relation

$$\lambda = (D\tau_2)^{\frac{1}{2}} \tag{2.55}$$

For a single ion conductor, $D = D_{ion}$ can be used and in case where both cations and anions are mobile, an effective D value may be used. Since $\delta = d/\lambda$, δ and D can be obtained from Eq. (2.55) as

$$\delta = \frac{d}{(D\tau_2)^{\frac{1}{2}}} \tag{2.56}$$

$$D = \frac{d^2}{\tau_2 \delta^2} \tag{2.57}$$

Now the diffusion coefficient D is related to the mobility through the well-known Nernst–Einstein relation:

$$\mu = \frac{eD}{kT} \tag{2.58}$$

Here k is the Boltzmann constant and T is the absolute temperature. Combining Eqs. (2.57) and (2.58), the mobility μ can be written as

$$\mu = \frac{ed^2}{kT\tau_2\delta^2} \tag{2.59}$$

On substituting μ from the relation $\sigma_{dc} = n\mu e$ in Eq. (2.59), the number density of charge carriers n can be obtained as

$$n = \frac{\sigma_{dc} kT\tau_2\delta^2}{e^2 d^2} \tag{2.60}$$

If d, τ_2, δ, and σ_{dc} for an electrolyte are known, the values for transport parameters D, μ, and n can be evaluated using Eqs. (2.57), (2.59), and (2.60). The quantities τ_2, δ, and σ_{dc} can be obtained from impedance plots easily. Using the frequencies corresponding to the maximum and minimum in Z'' values, ω_1 and ω_2 can be obtained (see Figure 2.38). The corresponding relaxation times τ_1 and τ_2 can be obtained using Eqs. (2.44) and (2.46). d can be obtained by measuring the thickness of the sample and σ_{dc} can be calculated by knowing the bulk resistance of the sample from the impedance plot. Using these values, the diffusion coefficient, mobility, and mobile charge carrier concentration can be calculated. Even though the circuit considered for deriving the parameter equations is an ideal one, the same equations can be used to obtain the parameters for real systems as well; only the values of ω_1 and ω_2 are needed, which can be obtained easily from the impedance plots.

Dielectric measurements too can be used to estimate τ_2 and δ using electrical and dielectric measurements. The readers are advised to refer the paper by Bandara and Mellander mentioned at the beginning of Section 2.3.3.1.

2.3.3.2 Nyquist Plot Fitting Method

Another way of representing the impedance behavior of real electrolytes sandwiched between blocking electrodes by equivalent circuit is to use CPEs instead of capacitors. The CPE is a "leaky capacitor." An equivalent circuit comprising a CPE and a resistor connected in parallel can represent a depressed semicircle. A resistor in series with a CPE can represent a tilted spike. The impedance plot that consists of a depressed semicircle with a tilted spike can be represented by a parallel combination of a resistor and CPE that is connected in series with another CPE [13]. The depressed semicircle represents the bulk material, while the tilted spike represents the EDL.

Consider an impedance plot with a depressed semicircle and tilted spike as shown in Figure 2.39a. The simple equivalent circuit that can be used to represent the impedance plot is shown in Figure 2.39b. Here it has been assumed that both double layer effects at electrolyte/electrode interface can be represented by a single CPE2.

The real (Z') and negative imaginary (Z'') parts of the complex impedance of the "leaky capacitor" or CPE1 are given by the equations

$$Z' = \frac{\cos\left(\frac{\pi p_1}{2}\right)}{k^{-1}\omega^{p_1}} \tag{2.61}$$

$$Z'' = \frac{\sin\left(\frac{\pi p_1}{2}\right)}{k^{-1}\omega^{p_1}} \tag{2.62}$$

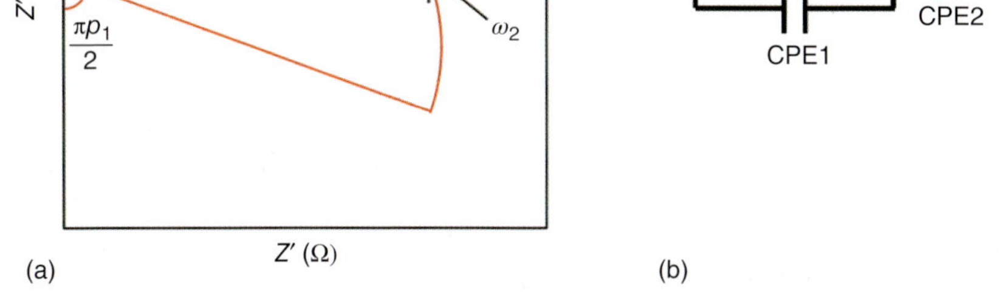

Figure 2.39 (a) A typical impedance plot of electrolyte sandwiched between two blocking electrodes and (b) the corresponding simple equivalent circuit.

In Eqs. (2.61) and (2.62), $\left(\frac{\pi p_1}{2}\right)$ is the angle between Z''-axis and the diameter of the semicircle as shown in Figure 2.39a. Similarly it can be shown that the real and imaginary parts of an impedance plot consisting of a depressed semicircle and tilted spike are given by the following equations:

$$Z' = \frac{R + R^2 k_1^{-1} \omega^{p_1} \cos\left(\frac{\pi p_1}{2}\right)}{1 + 2Rk_1^{-1}\omega^{p_1}\cos\left(\frac{\pi p_1}{2}\right) + R^2 k_1^{-2}\omega^{2p_1}} + \frac{\cos\left(\frac{\pi p_2}{2}\right)}{k_2^{-1}\omega^{p_2}} \quad (2.63)$$

$$Z'' = \frac{R^2 k_1^{-1} \omega^{p_1} \sin\left(\frac{\pi p_1}{2}\right)}{1 + 2Rk_1^{-1}\omega^{p_1}\cos\left(\frac{\pi p_1}{2}\right) + R^2 k_1^{-2}\omega^{2p_1}} + \frac{\cos\left(\frac{\pi p_2}{2}\right)}{k_2^{-1}\omega^{p_2}} \quad (2.64)$$

Here k^{-1} corresponds to the capacitance of CPE. Thus, in Eqs. (2.63) and (2.64), k_1^{-1} is the bulk geometrical capacitance of the polymer electrolyte, k_2^{-1} is the capacitance due to the effective EDL formed at the electrode/electrolyte interface during impedance measurement, which is similar to the capacitance C_e in the impedance equation used by Bandara and Mellander, but here it has been assumed the CPE2 represents both C_e. R is bulk resistance of the electrolyte. p_1 is the ratio of the angle between the diameter of the depressed semicircle and Z''-axis to the right angle subtended by the real and imaginary impedance axis. p_2 is the skew parameter. It is the ratio between the angle of inclination of the tilted spike from the Z'-axis. The values of p_1 and p_2 parameters lie between 0 and 1 as shown in Figure 2.39a.

If the impedance plot comprises a depressed semicircle and a tilted spike, Eqs. (2.63) and (2.64) are used to fit the graph. The values of R, p_1, and p_2 can be determined from the impedance plot shown in Figure 2.39a. The values of k_1^{-1} and k_2^{-1} can be obtained by trial and error until the experimental plot is fitted. The values of k_1^{-1} and k_2^{-1} can also be obtained using nonlinear least squares software.

Based on Eq. (2.55), the diffusion coefficient (D) of charge carriers in the electrolyte is given by

$$D = \frac{\lambda^2}{\tau_2} \quad (2.65)$$

where $\tau_2 = \frac{1}{\omega_2}$ with ω_2 being angular frequency corresponding to the minimum in imaginary impedance, Z''_{min} (Figure 2.39). λ is the effective thickness of the double layer capacitance, which is of the order of Debye length. Since capacitance k_2^{-1} is equivalent to effective double layer capacitance (C_e), it can be expressed as

$$k_2^{-1} = \frac{\varepsilon_r \varepsilon_o A}{\lambda} \quad (2.66)$$

Hence

$$\lambda = k_2 \varepsilon_r \varepsilon_o A \quad (2.67)$$

On substitution of λ from the Eq. (2.67) in Eq. (2.65), the diffusion coefficient (D) becomes

$$D = \frac{(k_2 \varepsilon_r \varepsilon_o A)^2}{\tau_2} \quad (2.68)$$

The mobility (μ) of charge carriers can be determined from the Nernst–Einstein relation as

$$\mu = \frac{e(k_2\varepsilon_r\varepsilon_o A)^2}{kT\tau_2} \tag{2.69}$$

Using the expression for conductivity ($\sigma_{dc} = n\mu e$), the number density of charge carriers n can be written as

$$n = \frac{\sigma_{dc} kT\tau_2}{(ek_2\varepsilon_r\varepsilon_o A)^2} \tag{2.70}$$

Thus from the fitted impedance plot, the diffusion coefficient (D), mobility (μ), and number density (n) of charge carriers in an electrolyte can be determined using Eqs. (2.68), (2.69), and (2.70). It should be noted that these equations are not dependent on the value of δ. However, the equations depend on the value of ε_r. By transforming the impedance data into the permittivity formalism and plotting $\log \varepsilon_r$ against frequency, ε_r can be obtained from the frequency-independent part in the high frequency region.

2.3.4 Some Experimental Results and Analysis

2.3.4.1 Conductivity Calculation of Impedance Plots

For very good ionic conductors, the impedance spectra take the shape of a perfect semicircle as illustrated in Figure 2.40. These plots were obtained for aqueous solutions of potassium iodide (KI) having two different concentrations of 10^{-4} and 10^{-3} M sandwiched between two blocking electrodes [14].

The impedance spectra have a semicircular part and a straight line part. The bulk resistance R_b of the solutions can be obtained by taking the intercept of

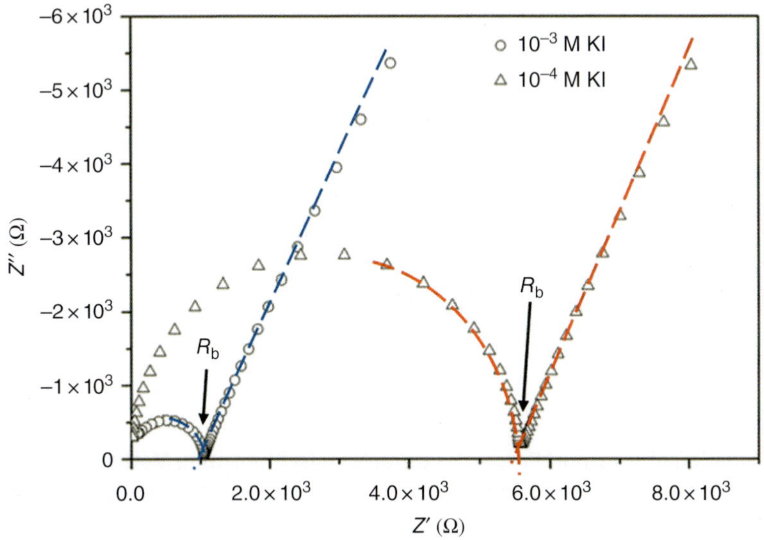

Figure 2.40 Nyquist plot of different potassium iodide liquid electrolyte [14].

Table 2.3 Conductivity and parameter values obtained in Figure 2.40 for different KI solutions.

KI concentration (M)	L/A (cm^{-1})	R_b (Ω)	σ ($S\,cm^{-1}$)
10^{-4}	1.58×10^{-1}	5.55×10^3	2.85×10^{-5}
10^{-3}	1.58×10^{-1}	1.05×10^3	1.50×10^{-4}

the semicircular part with the spike. By substituting the value of R_b in Eq. (2.11) and knowing the thickness and electrolyte/electrode contact area, the conductivity value of each sample concentration can be determined. Table 2.3 lists the conductivity and parameter values obtained from the plots. It can be seen as the concentration increases the resistance decreases, showing that the solution becomes more conducting. When a material is more conducting, the ions can move faster giving rise to more electrode polarization [14].

From Figure 2.40, one may think that the value of Z' is negative when Z'' is 0. However, it should be noted that Z' can never be negative whether θ is positive or negative. Hence the only reason that this may occur is due to limitation of frequency and may be the suitability of the instrument electrode. It is to be realized that some instruments or bridges require different electrodes at higher frequencies.

2.3.4.2 Conductivity Determination from Fitting Equivalent Circuit

Another way to determine bulk resistance of impedance plot is using equivalent circuit fitting method. Figure 2.41 shows the impedance spectra obtained for polyacronitrile (PAN) solid polymer electrolyte with different lithium bis(oxalate)borate (LiBOB) salt content. All impedance plots consist of depressed semicircle and tilted spike, which can be fitted using the simple equivalent circuit. As we mentioned earlier, the selection of correct equivalent circuit is essential to ensure that the impedance plot study is completely fitted. The depressed semicircle part can be represented by resistor in parallel with CPE, while a tilted spike can be represented by a CPE. Combination of R//CPE in series with another CPE as shown in Figure 2.42 revealed the equation of real and imaginary impedance in Eqs. (2.63) and (2.64), respectively, and both equations fitted the impedance plot of Figure 2.41. Resistor (R) in the circuit represented the bulk resistance of the sample. Using the value of R obtained from fitting method and substitute in Eq. (2.11), the conductivity of each sample can be quantified. The value of R and conductivity of each corresponding sample obtained by equivalent circuit fitting method are listed in Table 2.4.

2.3.4.3 Evaluation of Transport Properties using Nyquist Plot Fitting Method

In Section 2.3.3.2, we have discussed theoretically a method to determine diffusion coefficient (D), mobility (μ), and number density (n) of charge carries in polymer electrolytes using equivalent circuit fitting method. Arof and his co-workers have developed and used this method to analyze the

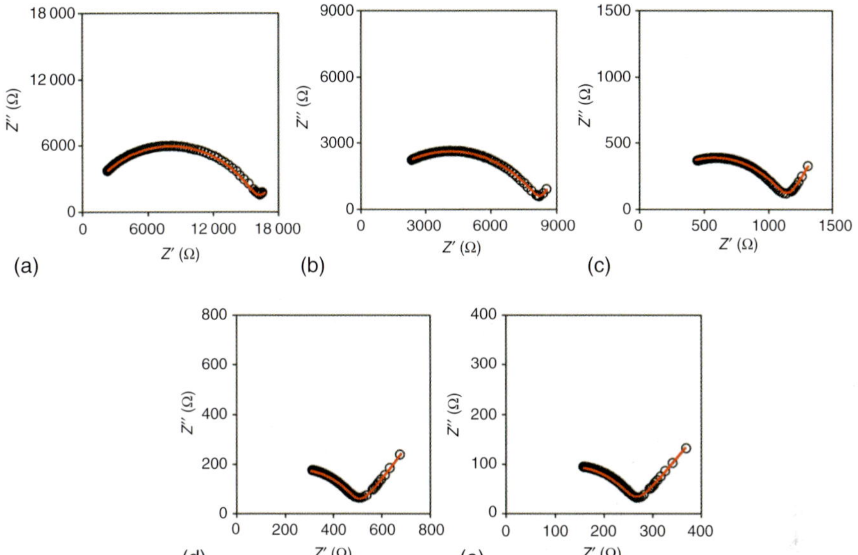

Figure 2.41 Impedance and corresponding fitted plot of PAN solid polymer electrolyte consist of (a) 10, (b) 20, (c) 30, (d) 40, and (e) 50 wt% LiBOB salt concentration [13].

Figure 2.42 Equivalent circuit fitted the Nyquist plot in Figure 2.41.

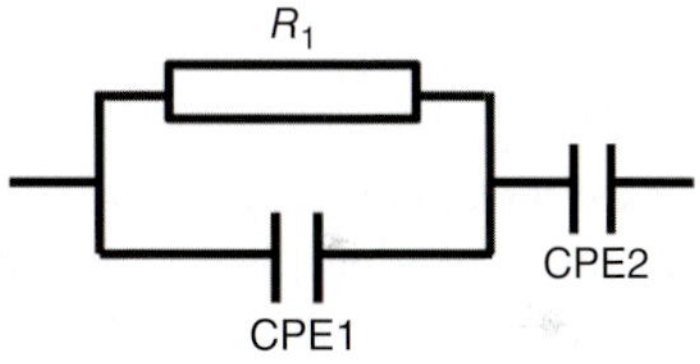

Table 2.4 Values of L/A, R_bA, and σ obtained in Figure 2.41 using equivalent circuit fitting method.

PAN:LiBOB (wt%:wt%)	L/A (cm^{-1})	R_bA (Ω cm^2)	σ (S cm^{-1})
90 : 10	2.85×10^{-3}	4.99×10^{4}	1.79×10^{-7}
80 : 20	2.75×10^{-3}	2.59×10^{4}	3.34×10^{-7}
70 : 30	1.85×10^{-3}	3.71×10^{3}	1.57×10^{-6}
60 : 40	3.18×10^{-3}	1.53×10^{3}	6.52×10^{-6}
50 : 50	5.24×10^{-3}	8.48×10^{2}	1.94×10^{-5}

transport properties of charge carriers in PAN–LiBOB solid polymer electrolyte [13]. The impedance plots obtained consists of depressed semicircle and tilted spike as shown in Figure 2.41. The real and imaginary impedance equations derived from the equivalent circuit as shown in Figure 2.42 are used to fit the Nyquist plot until all fitted matched the experimental data point at corresponding frequency. Recall the Z' and Z'' equations to fit the plots in Figure 2.41:

$$Z' = \frac{R + R^2 k_1^{-1} \omega^{p_1} \cos\left(\frac{\pi p_1}{2}\right)}{1 + 2Rk_1^{-1}\omega^{p_1}\cos\left(\frac{\pi p_1}{2}\right) + R^2 k_1^{-2}\omega^{2p_1}} + \frac{\cos\left(\frac{\pi p_2}{2}\right)}{k_2^{-1}\omega^{p_2}} \tag{2.63}$$

$$Z'' = \frac{R^2 k_1^{-1} \omega^{p_1} \sin\left(\frac{\pi p_1}{2}\right)}{1 + 2Rk_1^{-1}\omega^{p_1}\cos\left(\frac{\pi p_1}{2}\right) + R^2 k_1^{-2}\omega^{2p_1}} + \frac{\cos\left(\frac{\pi p_2}{2}\right)}{k_2^{-1}\omega^{p_2}} \tag{2.64}$$

The value of R, p_1 and p_2 were first estimated as shown in Figure 2.39a. The parameters k_1^{-1} and k_2^{-1} were determined by trial and error until the fitted points approximates the experimental points at the corresponding frequencies. The R, p_1, and p_2 values were then being adjusted slightly until the regression percentage closes to 100%. In order to calculate the transport properties parameter of charge carriers in polymer electrolyte system, Eqs. (2.68)–(2.70) were used. The value of τ_2 was taken at the frequency corresponding to a minimum in Z'' as shown in Figure 2.39a. However, these equations need dielectric constant ε_r to solve it. The ε_r value of each sample can be obtained by taking their value at high frequency region in complex permittivity ε_r against frequency, f plot. To do so, the experimental impedance data for each sample were substituted in Eq. (2.71) and the plots are depicted in Figure 2.43.

$$\varepsilon_r = \frac{Z''}{(Z'^2 + Z''^2)}\left(\frac{L}{\omega \varepsilon_o A}\right) \tag{2.71}$$

Based on Figure 2.43, the log permittivity ε_r is almost constant at high frequency region. Thus, the ε_r was taken at maximum frequency point, i.e. in this system at 10^5 Hz. Table 2.5 lists all the parameter values obtained from the fitted Nyquist plot and complex permittivity, ε_r. Inserting all the parameter values listed in Table 2.5 into Eqs. (2.68), (2.69), and (2.70), the values of diffusion coefficient (D), mobility (μ), and number density (n) of charge carriers can be evaluated.

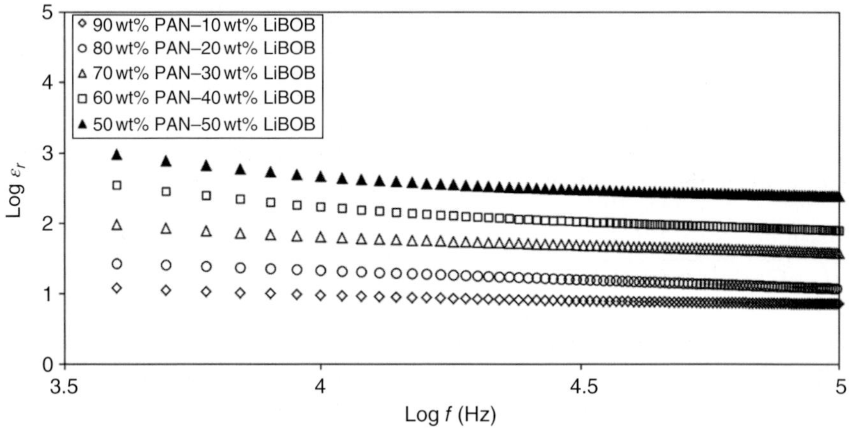

Figure 2.43 Plot of ε_r as function of frequency for PAN–LiBOB electrolyte system in the frequency range between log f of 3.5 and 5 where the log ε_r value is almost constant [13].

Table 2.5 Parameter values obtained from the fitted Nyquist plot and complex permittivity ε_r.

LiBOB content (wt.%)	R (Ω)	p_1	p_2	k_1 ($\times 10^7$ F^{-1})	k_2 (F^{-1})	ε_r	τ_2 ($\times 10^{-4}$ s^{-1})
10	15 900	0.81	0.60	25.0	150 000	5.77	1.84
20	8 240	0.72	0.73	6.5	118 333	13.13	3.04
30	1 182	0.76	0.67	2.1	35 575	34.70	0.80
40	488	0.72	0.59	0.9	15 500	83.53	0.32
50	270	0.79	0.60	1.4	8 383	208.58	0.77

Table 2.6 Transport properties of charge carrier values of PAN-LiBOB solid polymer electrolyte evaluated using Nyquist plot fitting method.

LiBOB content (wt.%)	D ($\times 10^{-10}$ cm^2 s^{-1})	μ ($\times 10^{-8}$ cm^2 V^{-1} s^{-1})	n ($\times 10^{19}$ cm^{-3})	σ (S cm^{-1})
10	3.67	1.43	7.78	1.79×10^{-7}
20	6.09	2.37	8.77	3.34×10^{-7}
30	14.92	5.81	17.30	1.57×10^{-6}
40	40.73	15.86	25.85	6.52×10^{-6}
50	94.85	36.93	32.27	1.94×10^{-5}

Table 2.6 tabulated the transport properties values of PAN-LiBOB solid polymer electrolyte evaluated using Nyquist plot fitting method based on equivalent circuit approach.

From Table 2.6, all three transport properties of charge carrier parameters increase along with the increase of LiBOB salt content in PAN polymer. The conductivity of this electrolyte system is also seen to increase as the salt concentration increases. As we have discussed in this chapter, conductivity was governed by number density and mobility of charge carriers and the elementary charge since $\sigma = n\mu e$. Thus, the conductivity of PAN-LiBOB solid polymer electrolyte in this system was dominantly influenced by both number density and mobility of charge carriers. It is important to determine D, μ, and n of charge carrier quantitatively so that we can clearly understand the parameters that modulate the conductivity behavior in our electrolyte system.

2.4 Conclusions

Impedance spectroscopy is based on measuring the impedance response of a material or an electrochemical system or process to a small *ac* signal. Plotting the impedance data in Argand diagram and apply the equivalent circuit concept to fit the impedance points from experiment have assists a lot of researchers to conclude the behavior of material under their case study. In addition, the complex information such as diffusion coefficient, mobility, and number density of charge carriers in a material especially polymer electrolyte and ionic materials can be deduced by using this basic concept. Overall, impedance spectroscopy is a powerful technique in materials research and for studying electrochemical systems and processes. It is necessary to study and understand the basic concept that is based on Ohm's law.

References

1 Heaviside, O. (1887). II. On the self-induction of wires.—Part V. *London Edinburgh Dublin Philos. Mag. J. Sci.* 23 (140): 10–29.
2 Kennelly, A.E. (1893). Impedance of mutually inductive circuits. *Electrician* 31: 699–700.
3 Delmastro, J.R. and Smith, D.E. (1966). Some considerations on the effect of alternating potential amplitude in AC polarography. *J. Electroanal. Chem. Interfacial Electrochem.* 14 (3): 261–268.
4 Urquidi-Macdonald, M., Real, S., and Macdonald, D.D. (1986). Application of Kramers-Kronig transforms in the analysis of electrochemical impedance data II. Transformations in the complex plane. *J. Electrochem. Soc.* 133 (10): 2018–2024.
5 Buchner, R., Hefter, G.T., and May, P.M. (1999). Dielectric relaxation of aqueous NaCl solutions. *J. Phys. Chem. A* 103 (1): 1–9.
6 Mellander, B.E. (1982). Electrical conductivity and activation volume of the solid electrolyte phase α-AgI and the high-pressure phase fcc AgI. *Phys. Rev. B* 26 (10): 5886.
7 Mao, G., Saboungi, M.L., Price, D.L. et al. (2001). Lithium environment in PEO-$LiClO_4$ polymer electrolyte. *Europhys. Lett.* 54 (3): 347.
8 Stephan, A.M. (2006). Review on gel polymer electrolytes for lithium batteries. *Eur. Polym. J.* 42 (1): 21–42.
9 Foley, R.T. (1969). Solid electrolyte galvanic cells. *J. Electrochem. Soc.* 116 (1): 13C–22C.
10 Sinclair, D.C. (1995). Characterisation of electro-materials using ac impedance spectroscopy. *Bol. Soc. Esp. Ceram. Vidrio* 34 (2): 55–65.
11 Lvovich, V.F. (2012). *Impedance Spectroscopy: Applications to Electrochemical and Dielectric Phenomena*. Hoboken, NJ: Wiley.
12 Bandara, T.M.W.J. and Mellander, B.E. (2011). Evaluation of mobility, diffusion coefficient and density of charge carriers in ionic liquids and novel electrolytes based on a new model for dielectric response, Chapter 17. In: *Ionic Liquids: Theory, Properties, New Approaches* (ed. A. Kokorin), 383–406. *InTech*. ISBN: 978-953-307-349-1.
13 Arof, A.K., Amirudin, S., Yusof, S.Z., and Noor, I.M. (2014). A method based on impedance spectroscopy to determine transport properties of polymer electrolytes. *Phys. Chem. Chem. Phys.* 16 (5): 1856–1867.
14 Bandara, T.M.W.J., Dissanayake, M.A.K.L., Albinsson, I., and Mellander, B.E. (2011). Mobile charge carrier concentration and mobility of a polymer electrolyte containing PEO and $Pr_4N^+I^-$ using electrical and dielectric measurements. *Solid State Ionics* 189 (1): 63–68.

3

Thermal Characterization of Polymer Electrolytes

Aparna Thankappan[1,2], Manuel Stephan[3], and Sabu Thomas[1,4]

[1] *International and Inter University Centre for Nanoscience and Nanotechnology, Mahatma Gandhi University, Priyadarsini Hills, Kottayam 686560, Kerala, India*
[2] *Department of Physics, Baselius College, Kottayam, Kerala 686001, India*
[3] *Central Electrochemical Research Institute (CSIR-CECRI), Electrochemical Power Systems Division, Karaikudi 630006, Tamil Nadu, India*
[4] *Mahatma Gandhi University, School of Chemical Science, Priyadarsini Hills, Kottayam, 686560, Kerala, India*

3.1 Introduction

Electrolyte materials play an important role in electrochemical applications, which help to transmit electrons and ions during the charge–discharge process. It consists of ionizable species, which are easily dissolved in solvents with high dielectric constants. In 1973, Fenton et al. [1] exposed that alkali salts could be dissolved in poly(ethylene oxide) (PEO), thereby opening up the scope of solvent-free polymer electrolytes (PEs). Among the various applications, the use of PEs in lithium ion batteries (LIBs) has been most generally considered and helps to improve their safety and reliability [2]. Lithium salts are combined with polysiloxanes, poly(styrenesulfonates), poly(acrylonitrile), PEO, and their derivatives to create PEs for commercialization of LIBs [3]. Solvent-free or solid polymer electrolytes (SPEs) can resolve the safety problems linked with the traditional liquid counterparts and they are currently under deep study for the use in batteries, other electro- and photochemical devices, supercapacitors, and sensors [4–12].

PE is a membrane and their charge transport is the segmental motion of polymer chains in the environs of ions. Their charge transfer properties are analogous with common liquid ionic solutions. They are inherently safer than some other classical electrolytes such as $LiPF_6$/ethylene carbonate [13]. The advance of PEs has drawn the mind of scientists in the last few decades due to their reward over liquid counterparts.

Their development has mainly three stages, depending on their physical state, SPEs [14], gel/plasticized polymer electrolytes (GPEs) [15], and microporous polymer electrolytes (MPEs) [16]. For industrial application MPEs have received major attention due to their outstanding property, for example, high chemical,

Polymer Electrolytes: Characterization Techniques and Energy Applications,
First Edition. Edited by Tan Winie, Abdul K. Arof, and Sabu Thomas.

electrochemical, and photochemical stability and ionic conductivity [17, 18]. The very first example of the SPE is the PEO. Owing to poor ionic conductivity from the thermal induced liquid–liquid separation, PEO was found unfit for the lithium batteries [1].

In the last two decades, great research pains have been devoted on electrolyte systems based on PEO. The second category of PE is called GPE, which is neither liquid nor solid, and has properties of both solids (cohesive) and liquids (diffuse). Composite PE (ceramic electrolyte) is a subset of PE with the incorporation of inert fillers (such as TiO_2, Al_2O_3, ZrO_2) and silica [19, 20]. The improvement of ionic conductivity and the stability at the interface with electrode are their major advantages [21–23].

Measurement and analysis of SPEs are complicated by various factors such as heterogeneous behavior of materials [24] and distribution of salt concentration. The anionic contribution [25] and low dielectric constant of PEO affects the analysis of SPE [26]. The intercalated polymer nanocomposite electrolytes (PNCEs) are also interesting as it creates an apt background for the intercalation of polymer chains between nanocomposites [27].

So far, more than a few types of PEs have been developed and characterized. PEO, poly(vinyl chloride) (PVC), poly(methyl methacrylate) (PMMA), poly(vinylidene fluoride) (PVdF), poly(vinyl acetate) (PVAc), poly(acrylonitrile) (PAN), and poly(ethyl methacrylate) (PEMA), poly(vinyl alcohol) (PVA) are some polymers that have been investigated as hosts for PEs. In recent years a few polymers that show hydrophilic features have received great interest for electrolyte applications [28]. PMMA was first initiated as GPE matrix [29] and its side chain-modified acrylates propose new promise for the low temperature applications.

This chapter gives general idea about the thermal analysis (TA) tools and its literature reviews. These analytical experimental techniques can be used to investigate the stiffness of electrolyte by measuring modulus and energy absorbing properties. Thermal analysis can also give an indication of thermal stability. An understanding of mechanism is important because the degree of crystallinity influences the thermal properties. The purpose of this chapter is to assist the users to interpret the transitions in TA results. Thermogravimetric analysis (TGA), differential scanning calorimetry (DSC), and dynamic mechanical analysis (DMA) are very widely explained in this chapter. Briefly, TGA is the technique in which the mass change of a substance as a function of temperature over a period of time is followed. Thermal stability and decomposition temperature of the PE can be gauged easily using this method. DSC is another fundamental tool in thermal analysis to determine their thermal transitions in which the heat capacity of the specimen is evaluated as a function of temperature. We could measure the material's deformation by using DMA. The dynamic modulus, the loss modulus, and mechanical damping or internal friction were examined from these TA measurements.

3.2 TGA: Experimental Data Interpretation and Analysis

TG curves are classified into various types are illustrated in Figure 3.1.

Type A curve (Plateau): This curve shows stable sample weight over the temperature range considered.

Type B curve: This curve exhibit a rapid initial mass loss followed by mass plateau and is characteristic of drying, desorption, or polymerization. In the absence of interaction atmosphere in the chamber, the type B curve will modify to type A curve.

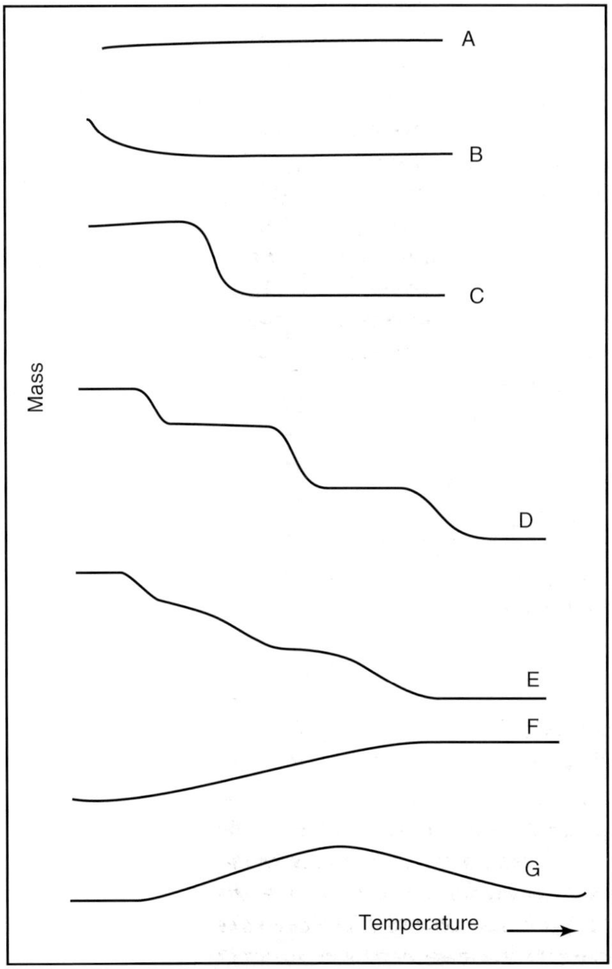

Figure 3.1 Main types of TG curves.

Type C curve: This represents a single stage decomposition of the sample (T_i and T_f).

Type D curve: This indicates multistage decomposition with relatively stable intermediates.

Type E curve: This curve also shows a multistage decomposition reaction without stable intermediates. It resembles type D at lower heating rates, while at high temperatures both type D and type E may resemble type C, and hence the detailed information for stages is lost.

Type F curve: The sample weight is increased in the presence of an interacting atmosphere, e.g. oxidation of metal sample.

Type G curve: This is a characteristic TG curve representing an oxidation reaction followed by decomposition at higher temperatures.

Plateau: The part of TG curve where the mass is basically constant (region AB in Figure 3.2).

Procedural decomposition temperature: This is the temperature at which the cumulative mass deviation reaches an extent that the thermobalance can sense. This is denoted by is T_i.

Final temperature: The temperature at which the change in cumulative mass reaches a maximum, denoted by T_f.

Reaction interval: This reaction interval is the difference in temperature between final temperature (T_f) and procedural decomposition temperature (T_i).

Isobaric mass change determination: A technique that refers to the equilibrium mass of a substance at constant partial pressure of the volatile product(s) and measured as a function of temperature, while the substance is subjected to a controlled temperature program.

As an electrolyte, the polymer membrane affords an environment for the electrochemical reactions at interfaces between electrolyte and electrode. M.M.E.

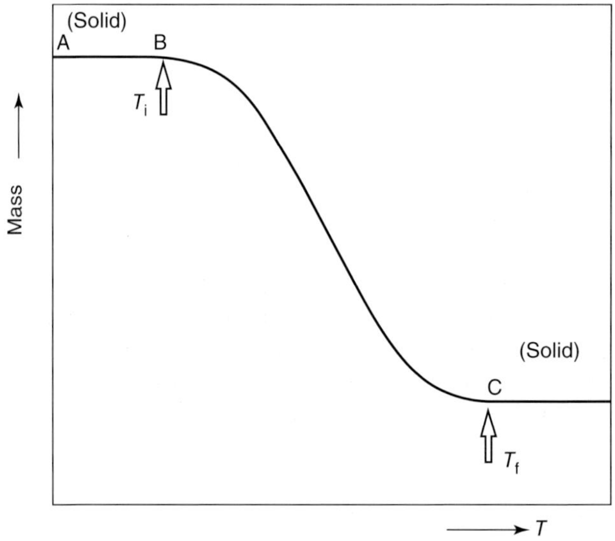

Figure 3.2 Formalized TG curve.

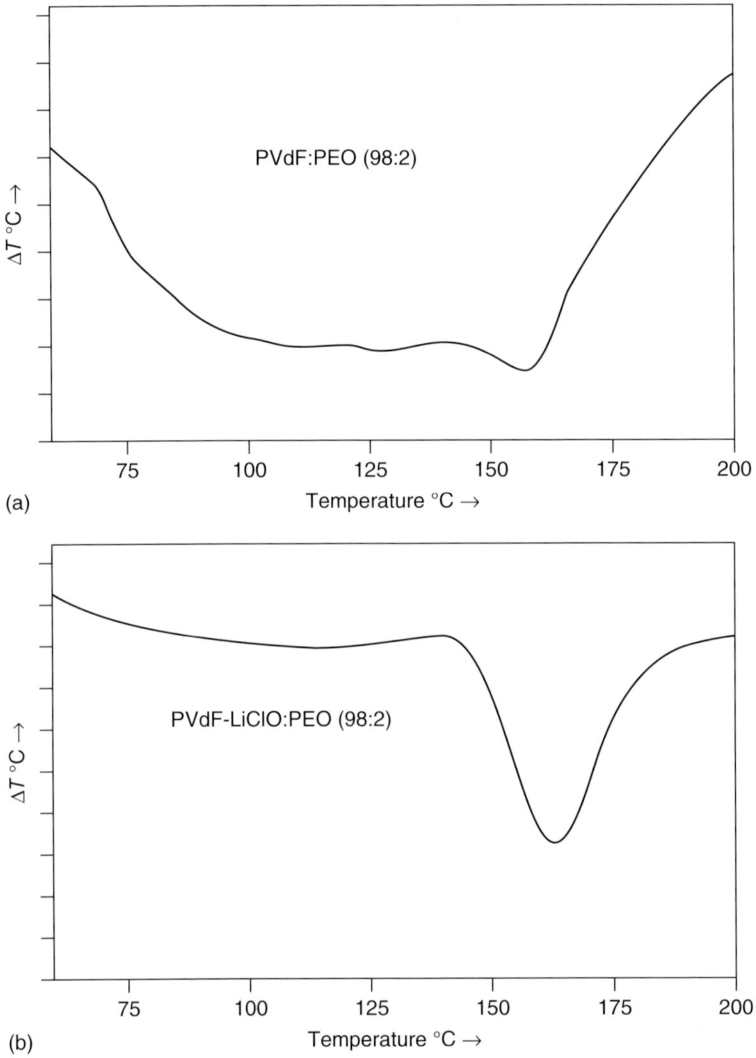

Figure 3.3 DTA thermogram of PEs of (a) PVdF:PEO (98 : 2) and (b) PVdF–$LiClO_4$:PEO (98 : 2).

Jacob et al. [30] have studied about a polymer membrane comprising PVdF, PEO, and $LiClO_4$. Their chemical compatibility was studied using differential thermal analysis (DTA) to recognize the compatibility between PVdF and PEO as shown in Figure 3.3. In their work they have highlighted the task of PVdF on the thermal, physical, and mechanical properties of the PE system. The summary is also depicted in Table 3.1.

In 2009, Manuel Stephan et al. [31] have reported chitin-incorporated PE systems with PEO prepared by dispersing appropriate amounts of chitin in PEO–$LiClO_4$ (Table 3.2) for electrochemical applications. Here the melting temperature is reduced with the chitin and lithium salt concentrations, which in

Table 3.1 Summary of ionic conductivities at 30 °C and mechanical properties of PVdF-$LiClO_4$:PEO system.

Composition of the complex (w/o)[a]	Conductivity (σ) at 30 °C S cm^{-1}	Mechanical properties
98 : 2	2.00×10^{-6}	Free standing and less stable
95 : 5	2.29×10^{-6}	Free standing and less stable
90 : 10	1.75×10^{-5}	Free standing and less stable
80 : 20	2.62×10^{-5}	Free standing and stable
70 : 30	4.24×10^{-6}	Free standing but fragile
60 : 40	—	No film

a) PVdF–$LiClO_4$:PEO.

Table 3.2 Polymer, chitin, and lithium salt compositions.

Sample	Polymer (wt%)	Chitin (wt%)	$LiClO_4$ (wt%)
S_1	95	0	5
S_2	90	5	5
S_3	85	10	5
S_4	75	17	8
S_5	70	20	10
S_6	94	5	1
S_7	93	5	2
S_8	92	5	3
S_9	91	5	4

turn broadened the endotherm with the obvious diminish in the heat of fusion (Figure 3.4).

They found that the enhanced thermal stability of chitin-added nanocomposite electrolytes (up to a temperature of 275 °C) is attributed to the intercalation of the polymer matrix with inert particles. This intercalation/exfoliation helps to prevent the thermal degradation. In order to improve the room temperature conductivity of PEO, the role of different weight percentage (wt%) of $KBrO_3$ were examined by Angesh Chandra [32] et al. They found that the room temperature conductivity is 4.36×10^{-7} S cm^{-1}, which is higher than that of pure PEO. The authors were assuming that the conductivity enhancement is due to the increase in ion mobility and mobile ion concentration. SPE exhibits a broad endothermic peak was observed at 69–70 °C and corresponds to the melting point of pure PEO, and the shifting of the peak is due to the complexation of K^+ ion to the ether oxygen of pure PEO. Their results are shown in Figure 3.5.

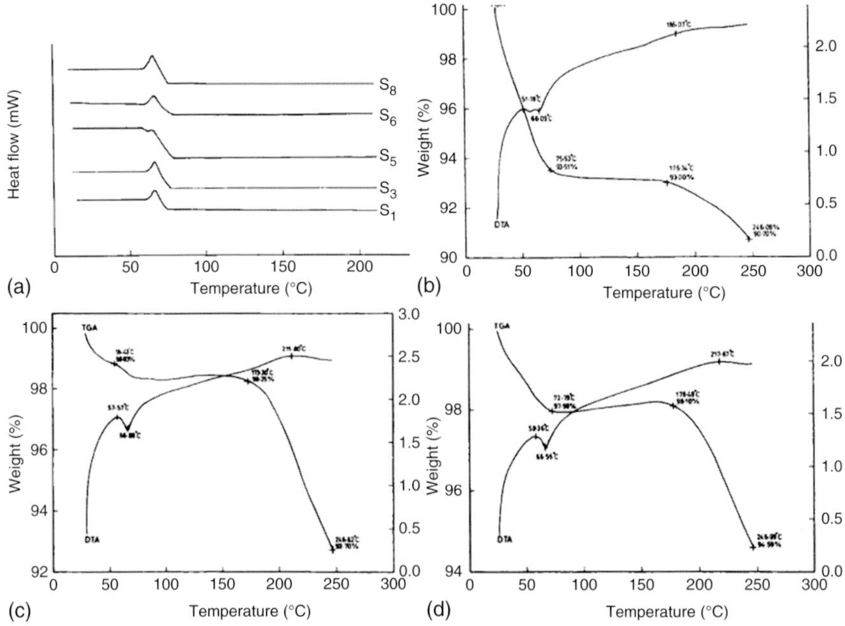

Figure 3.4 TGA–DTA curve of nanocomposite PEs with different nanochitin concentration. (a) DSC traces of NCPEs, (b) TG-DTA traces of sample S_3, (c) TG-DTA traces of sample S_1, and (d) TG-DTA traces of sample S_5.

In order to utilize the advantages of both ceramic fillers (MgO and Al_2O_3), Angulakshmi et al. [33] have synthesized and analyzed magnesium aluminate ($MgAl_2O_4$) and incorporated different percentages (see Table 3.3) into PEO comprising lithium hexafluorophosphate ($LiPF_6$). It was found that a negligible weight loss of 3% at 40 °C for PEO is associated with the removal of moisture absorbed by the sample, an endothermic peak at 65 °C attributed to the melting of PEO, and the decomposition occurred at 200 °C. They found that the thermal stability is increased to above 300 °C for the sample $LiPF_6 + MgAl_2O_4$ compared with that of $LiPF_6$ (Figure 3.6). The increased stability is due to the intercalation and exfoliation of polymer matrix with the inert particles, which hinder the thermal degradation to a certain extend.

In 2014, the same research group studied about the enhancement of physical and electrochemical properties of metal–organic framework (MOF) Mg-BTC-incorporated PE composed of different wt% of PEO and lithium bis(trifluoromethane)sulfonamide (LiTFSI) [34]; the results are shown in Figures 3.7 and 3.8. The MOF composite electrolyte is thermally stable up to a temperature of 290 °C in a nitrogen atmosphere. The incorporation of Mg-BTC could enhance the ionic conductivity of PE and also improve the thermal stability, compatibility, and elongation of the polymeric membrane. These enhanced properties could be due to the interaction between the Lewis acidic site of MOF

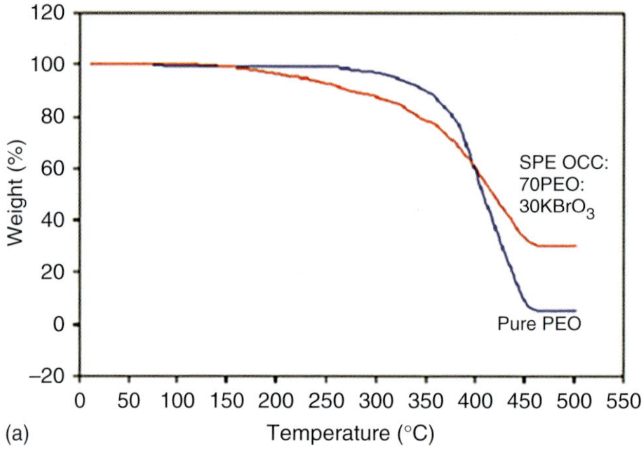

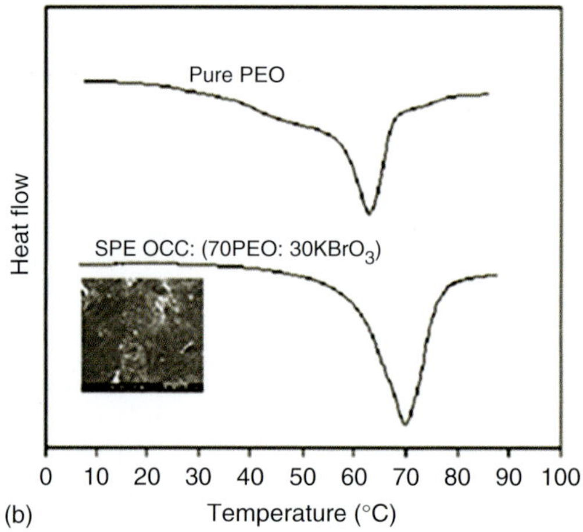

Figure 3.5 TGA and DSC curves of pure PEO and SPE modified with $KBrO_3$ (a) weight (%) and temperature (b) heat flow and temperature.

Table 3.3 Composition of PEO, $MgAl_2O_4$, and lithium salt in the different electrolyte membranes.

Sample	PEO (wt%)	$MgAl_2O_4$ (wt%)	$LiPF_6$ (wt%)
S_1	95	0	5
S_2	90	5	5
S_3	85	10	5
S_4	75	17	8
S_5	70	20	10
S_6	94	5	1
S_7	93	5	2
S_8	92	5	3
S_9	91	5	4

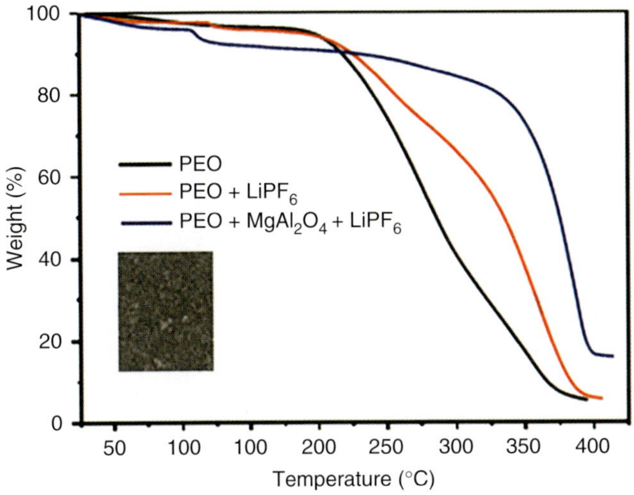

Figure 3.6 TGA of PEO, PEO + $LiPF_6$, and PEO + $MgAl_2O_4$ + $LiPF_6$. Inset is the SEM image of based composite polymer electrolyte (CPE) (PEO + $MgAl_2O_4$ + $LiPF_6$)

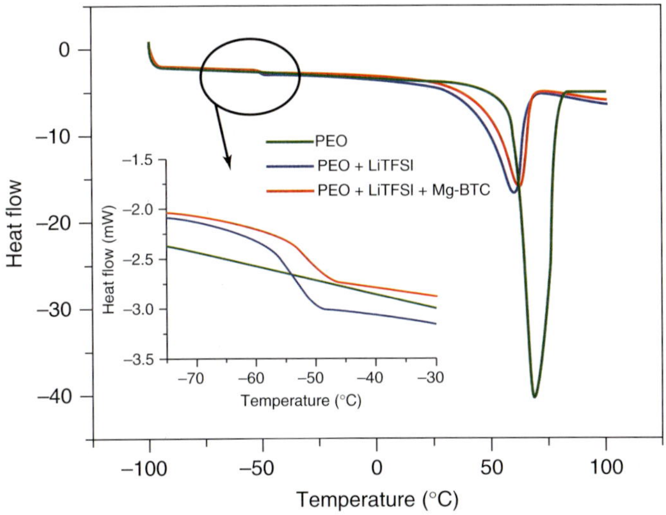

Figure 3.7 DSC thermograms of PEO, PEO + LiTFSI, and PEO + LiTFSI + Mg-BTCMOF.

and the oxygen of ethylene oxide moiety. Thus resulting intercalation restricts segmental movement of PEO chains.

Later, Savitha Thayumanasundaram et al. [35] have made a systematic study of the effect of LiTFSI on sulfonated poly(ether ether ketone) (SPEEK) membranes. They observed that that SO_3H groups crumble earlier in the SPEEK–LiTFSI membranes than in pure SPEEK, as a result of interactions between the Li^+ ions and the SO_3H groups. The plasticizing effect of the lithium salts on the polymer matrix is due to the addition of LiTFSI, which in turn diminished the crystallinity and the glass transition temperature of the polymer. The results are shown in

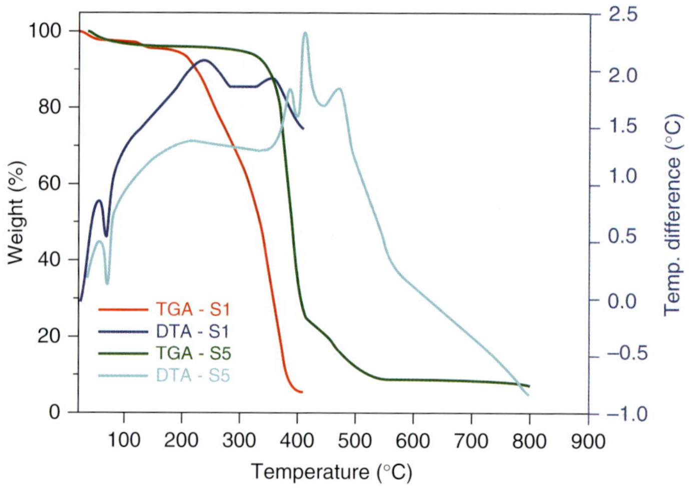

Figure 3.8 TGA–DTA curve of PEO + LiTFSI (sample S_1–5 wt% LiTFSI) and PEO + LiTFSI + Mg-BTC MOF (sample S_5–10 wt% Mg-BTC + 15 wt% LiTFSI).

Figure 3.9. They also found that the pure SPEEK membrane exhibits three main weight-loss regions as a result of the removal of adsorbed moisture, decomposition of the $-SO_3H$ ionic clusters, and decomposition of the polymer backbone as a function of temperature. The plasticization effect could be optimized for 90 SPEEK and 80 SPEEK membranes by increasing the lithium salt concentration. As in Figure 3.9 X-ray diffraction and thermogravimetric analysis confirmed the solvation of L_iTFSI by the polymer backbone, owing to the occurrence of sulfonic acid group attached to the phenyl ring. The pure SPEEK membrane is thermally stable up to 200 °C, while the lithium–SPEEK membranes are stable up to 150 °C as a result of the intercalation of salt and polymer.

In the endless hunt for the green power sources, Jijeesh R. Nair et al. in 2016 [36] have studied nanocellulose-laden composite PEs as green power source for lithium–sulfur battery application. In their work, nanoscale microfibrillated cellulose-laden polymer systems are synthesized by using a thermal induced polymerization process. This was stable up to 200 °C, which is higher than the normal operating temperature required for practical operation of lithium–sulfur batteries. The TGA curve of nanocellulose-laden composite PEs is shown in Figure 3.10.

As per their investigation, this nanocellulose-laden composite exhibited better thermal and mechanical properties under ambient conditions. The glass transition temperature was found to be −35.1 °C. The observed initial weight loss around 70 °C was credited to the removal of moisture, while the second 10% weight loss around 240 °C, referred as T_{10} value. The decomposition of 2-diethylamino ethyl methacrylate (DEAM) and bisphenol A ethoxylate (15 EO/phenol) dimethacrylate (BEA) were reflected from the weight loss at 340 °C and at 410 °C. The LiTFSI and nMFC residual content remained above 500 °C.

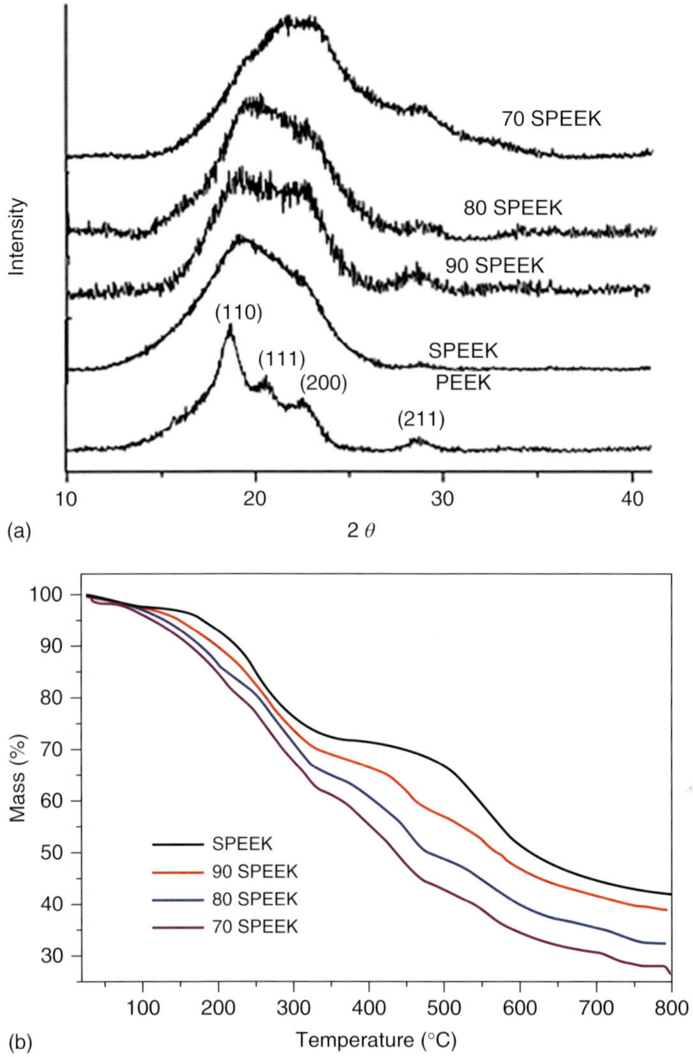

Figure 3.9 (a) X-ray diffraction (XRD) (b) TGA analysis of lithium-doped SPEEK membranes.

3.3 DSC: Experimental Data Interpretation and Analysis

In 2000, Akihiro Noda et al. [37] studied about the highly conducting PEs, 1-ethyl-3-methylimidazolium tetrafluoroborate ($EMIBF_4$) and 1-butylpyridinium tetrafluoroborate ($BPBF_4$). Below the melting points (around 15 °C) during the cooling scans, these ionic liquids form super cooled liquids and are fairly stable. Their DSC thermograms and thermal properties are shown in Figure 3.11 and their thermal properties are tabulated in Table 3.4.

C. Capiglia et al. [38] examined the thermal stability of Li-ion polymer battery based on $LiNi_{0.8}Co_{0.2}O_2$ and they found that Li-based SPE has good cycling

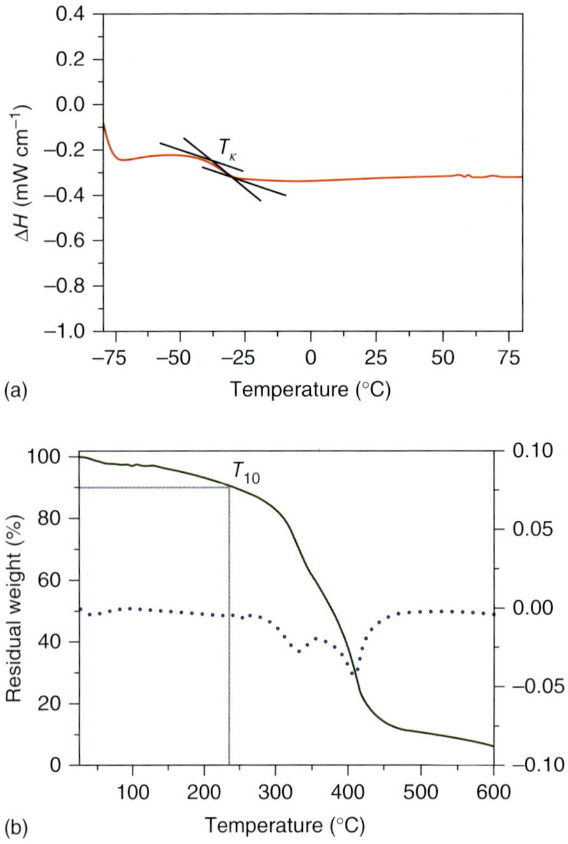

Figure 3.10 DSC trace and TGA curve of Nanocellulose-laden composite PEs (a) Δ*H* and temperature (b) residual weight and temperature.

performance with thermal stability of 300 °C. They analyzed that the DSC curves of Li-SPE that exhibits two small endothermic peaks centered at 65 and 180 °C as in Figure 3.12. They are related to the melting of PEO and lithium, and a single exothermic peak (350 °C) related to the thermal decomposition. These peaks were associated with two-step exothermic chain reactions, i.e. between PE cathode and lithium-entered gas. The heat developed during thermal decomposition was found that $\Delta H = 1284\,\text{J g}^{-1}$.

Later, L.C. Rodrigues et al. [39] have studied a cycle of electrolytes based on the poly(epichlorohydrin-*co*-ethylene oxide-*co*-allyl glycidyl ether) incorporated with lithium perchlorate and lithium bis(trifluoromethanesulfonyl)imide. The p(EEOAGE)LiTFSI (wt%) electrolyte system exhibits a higher thermal stability (293 °C) than the perchlorate electrolyte system shown in Figure 3.13. They also reported that the glass transition temperature is proportional to the salt concentration and is illustrated in Figures 3.14 and 3.15 [40].

Suriani Ibrahim and Mohd Rafie Johan [41] found that the increased room temperature conductivity for composite PEs with the complexation of $LiPF_6$. The DSC thermograms of various compositions of PEO, lithium hexafluorophosphate

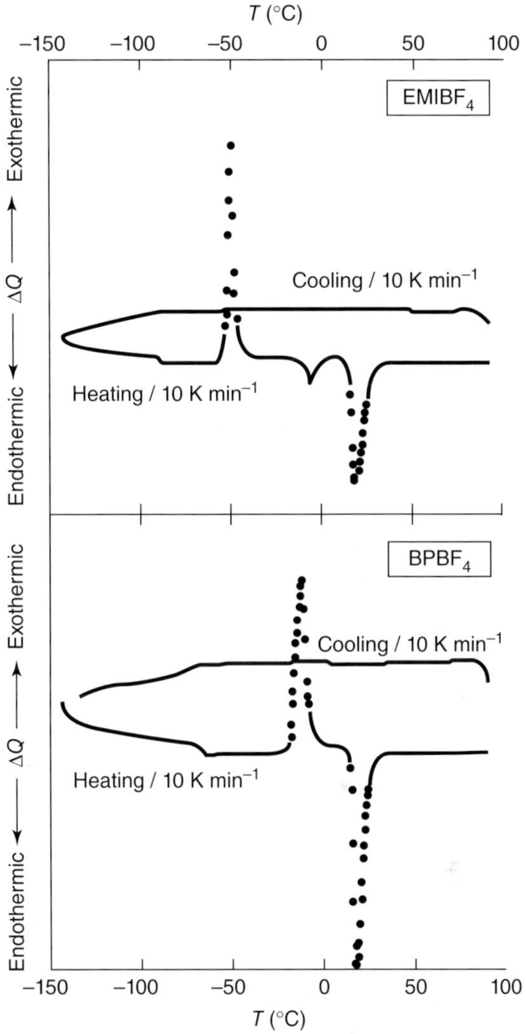

Figure 3.11 DSC thermograms of $EMIBF_4$ and $BPBF_4$.

Table 3.4 Thermal properties of $EMIBF_4$ and $BPBF_4$.

	T_m (°C)	ΔH_m ($J\,g^{-1}$)	T_c (°C)	ΔH_c ($J\,g^{-1}$)	T_g (°C)
$EMIBF_4$	14.6	63.2	−50.5	−53.4	−89.4
$BPBF_4$	15.3	45.8	−11.9	−37.5	−66.7

($LiPF_6$), ethylene carbonate (EC), and amorphous carbon nanotube (αCNTs) systems are shown in Figure 3.16. It has been found that during the heating process these exhibit a sharp endothermic peak near 69 °C for pure PEO. The decrease in T_g and X_c would increase the flexibility of PEO and it was also observed that the addition of salt into the PEO results in an enhancement in T_g, which suggests a reduction in PEO chain mobility. The authors were

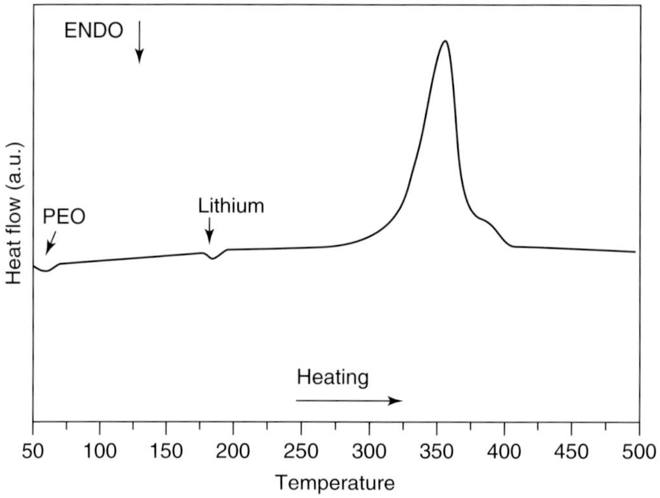

Figure 3.12 DSC curves of Li-SPE under the scan rate of 58 °C/min, N_2 flow 100 ml/min.

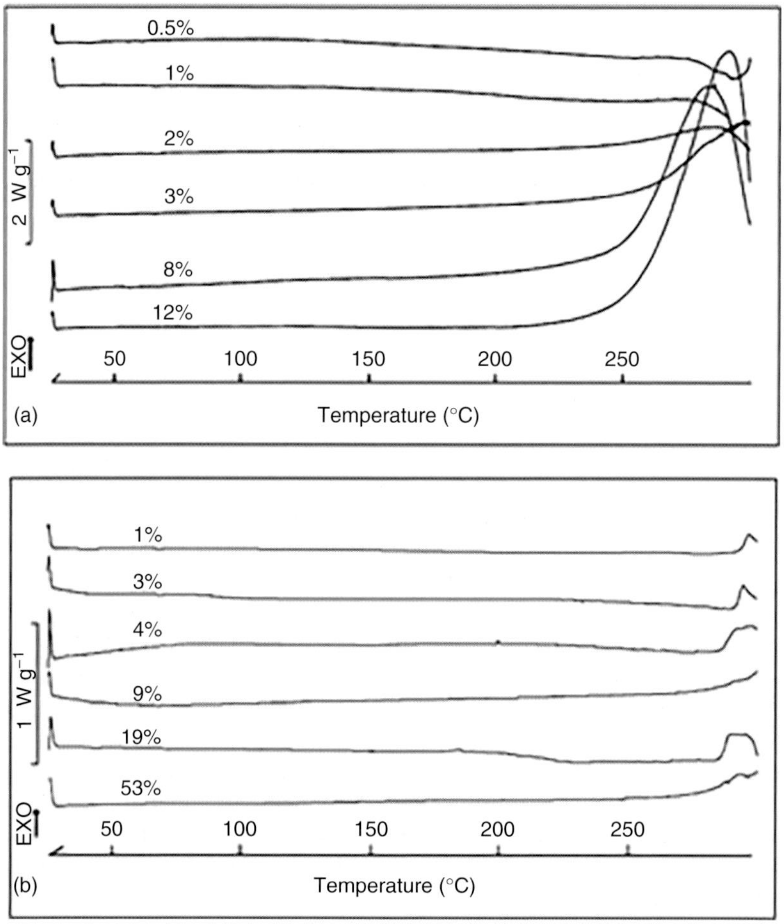

Figure 3.13 (a) Thermal analysis of the p(EEO-AGE)$LiClO_4$ (wt%) and (b) p(EEO-AGE)LiTFSI (wt%) electrolyte systems.

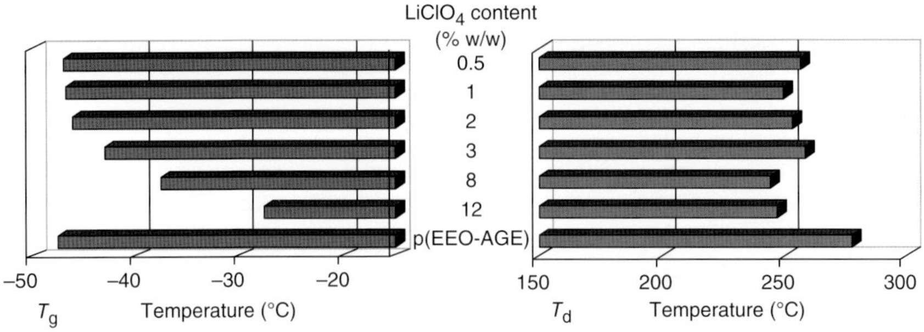

Figure 3.14 Extrapolated onset of glass transition and degradation temperatures of p(EEO-AGE)-based electrolytes: p(EEO-AGE)$LiClO_4$ (wt%).

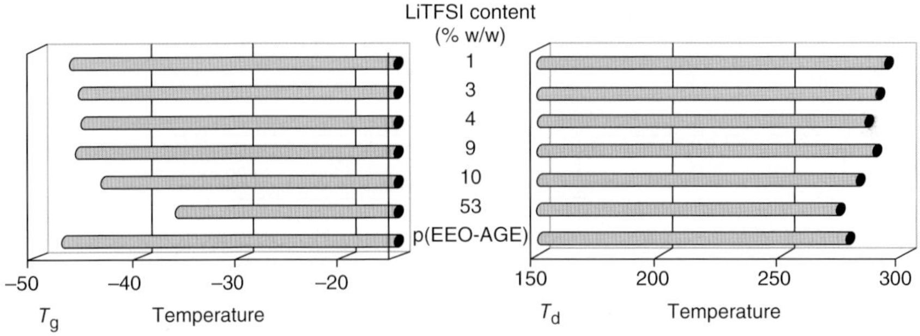

Figure 3.15 Glass transition and degradation temperatures of p(EEO-AGE)-based electrolytes: p(EEOAGE)LiTFSI (wt%).

reported that the broadening behavior of peak is due to the suppression of the crystallization of PEO, which leads to an increase in ionic conductivity, especially at temperatures lower than its melting point [42].

The thermal stability of gellan gum–$LiCF_3SO_3$-based PEs membrane have been studied by Noor et al. [43] is stable up to 234 °C (DSC curve is in Figure 3.17). From the Figure 3.17, it is clear that the samples suffer the endothermic reactions of oxidation and hydrolysis followed by polysaccharide pyrolysis exothermic reactions. They reported that the addition of salt does not change the degradation temperature of pure polymer sample, but enhanced the mechanical properties of the sample.

Figure 3.18 depicts the DSC trace of ($MgAl_2O_4$)–PEO–$LiPF_6$ composite electrolyte (see Table 3.3) [33]. The $MgAl_2O_4$-incorporated sample S_3 shows a slightly higher value of glass transition temperature of −70 °C. The increase in T_g may be attributed to the amount of dispersed particles and the confinement of the polymer chains. The marginal decrease in the melting point of PEO is due to the mild retarding effect on the crystallization.

The DSC trace of bare and doped SPEEK [35] membrane revealed the effect of LiTFSI on the T_g and the degree of crystallinity of the pure SPEEK membrane is shown in Figure 3.19 and the thermal properties are tabulated in Table 3.5. The trend of T_g exhibits the plasticization effect of LiTFSI in 90 SPEEK and 80 SPEEK. Also revealed the existence of ion pair association in these membranes. As shown in Figure 3.18, ΔH_m and T_m decreased with the addition of lithium salts.

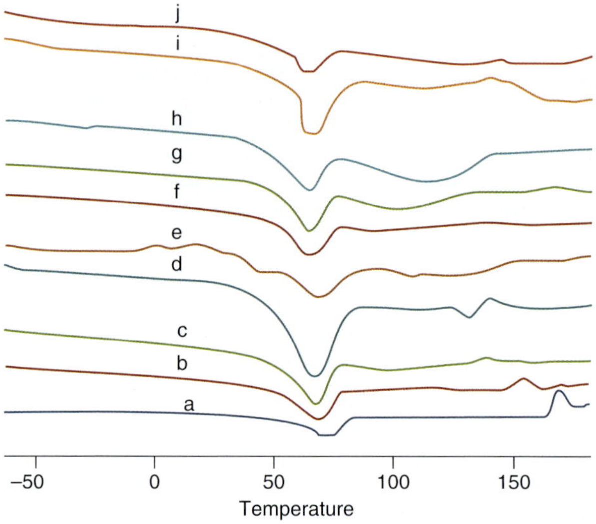

(a)

Sample	T_g (°C)	T_m (°C)	Xc (%)	σ (S cm^{-1}) at 298 K
Pure PEO	−66.01	68.8	84.87	3.25×10^{-10}
PEO–5 wt% $LiPF_6$	−67.98	67.4	67.82	1.20×10^{-6}
PEO–10 wt% $LiPF_6$	−68.00	65.1	58.46	9.03×10^{-6}
PEO–15 wt% $LiPF_6$	−70.05	64.4	55.03	1.82×10^{-5}
PEO–20 wt% $LiPF_6$	−72.00	63.5	47.52	4.10×10^{-5}
PEO–20 wt% $LiPF_6$–5 wt% EC	−72.01	63.2	59.47	5.93×10^{-5}
PEO–20 wt% $LiPF_6$–10 wt% EC	−74.01	63.1	51.27	1.43×10^{-4}
PEO–20 wt% $LiPF_6$–15 wt% EC	−76.03	62.9	47.35	2.06×10^{-4}
PEO–20 wt% $LiPF_6$–15 wt% EC –1 wt% αCNT	−78.03	62.0	46.21	2.20×10^{-4}
PEO–20 wt% $LiPF_6$–15 wt% EC –5 wt% αCNT	−80.04	61.0	27.12	1.30×10^{-3}

(b)

Figure 3.16 (A) DSC curves of (a) PEO; (b) PEO–5 wt% $LiPF_6$; (c) PEO–10 wt% $LiPF_6$; (d) PEO–15 wt% $LiPF_6$; (e) PEO–20 wt% $LiPF_6$; (f) PEO–20 wt% $LiPF_6$–5 wt% EC; (g) PEO–20 wt% $LiPF_6$–10 wt% EC; (h) PEO–20 wt% $LiPF_6$–15 wt% EC; (i) PEO–20 wt% $LiPF_6$–5 wt%EC–1 wt% αCNTs; (j) PEO–20 wt% $LiPF_6$–5 wt% EC–5 wt% αCNTs [41]. (B) Their corresponding properties are also tabulated in this figure.

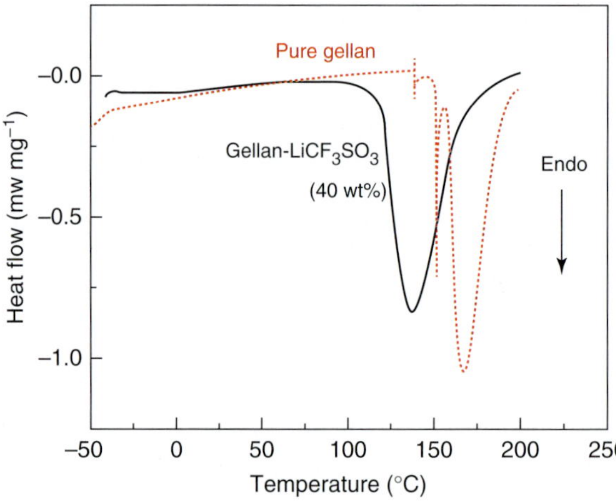

Figure 3.17 DSC curve of pure gellan and $LiCF_3SO_3$ (40 wt%) incorporated gellan.

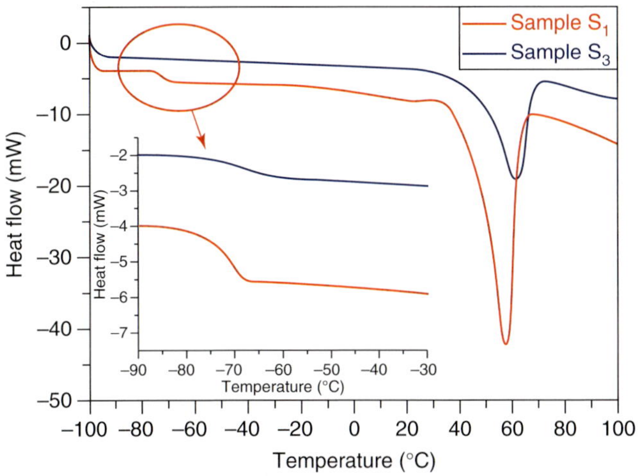

Figure 3.18 DSC trace of $MgAl_2O_4$-incorporated composite electrolytes composed of PEO and $LiPF_6$. Source: Angulakshmi et al. 2013 [33]. Reproduced with permission of Elsevier.

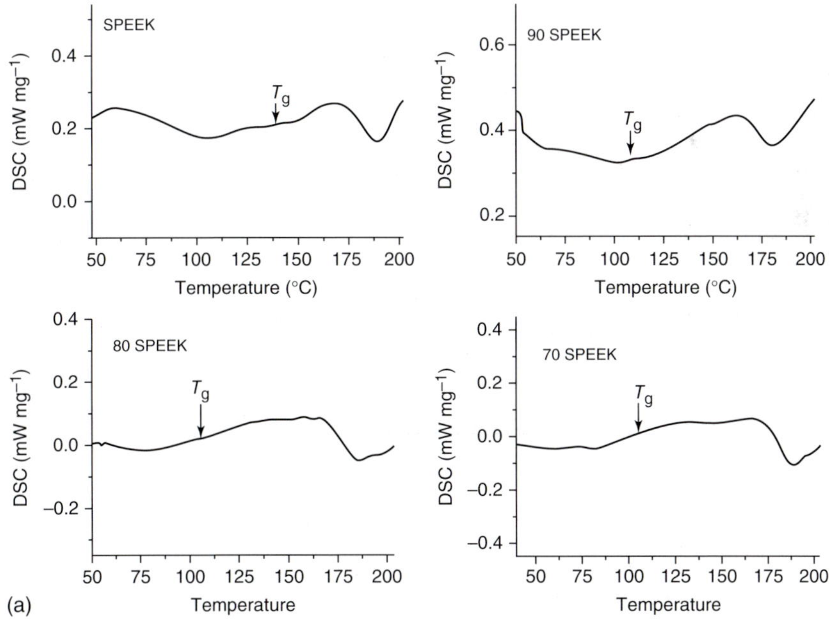

Figure 3.19 DSC traces of the pure and lithium-doped SPEEK membranes (a) in the glass transition region and (b) melting endotherm.

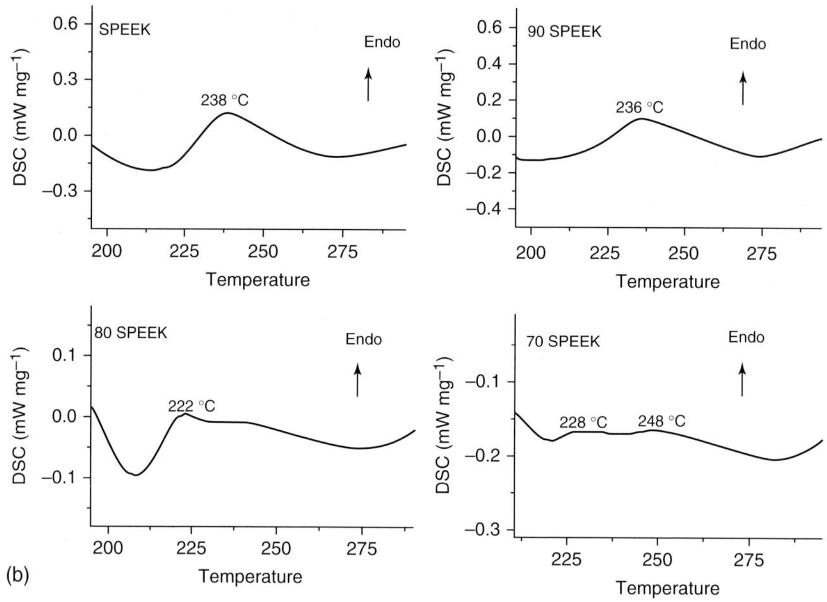

Figure 3.19 (*Continued*)

Table 3.5 Summary of pure and doped SPEEK.

Sample	T_g [a] (°C)	ΔH_m (J g^{-1})	T_g [b] (°C)	χ_C (%)
SPEEK	138	41	136	32
90 SPEEK	108	32	123	23
80 SPEEK	104	5.3	117	4
70 SPEEK	106	18	122 (broad)	14

a) Determined by DSC.
b) Determined by DMA.

3.4 DSC: Experimental Errors and Suggestion for Improvement

DSC provides quantitative and qualitative information about the physical and chemical changes of a material. Some of the specific information from DSC curve includes glass transition temperature, melting and boiling points, crystallization time and temperature, percent crystallinity, heats of fusion and reaction, oxidative stability, purity, degree of cure, specific heat, rate of cure reaction kinetics, and thermal stability. Besides providing wealth information, DSC is user-friendly and has become the most commonly used thermal analysis technique. However, some unexpected transitions in DSC data can lead to misinterpretation. Figure 3.20 is the artificial DSC curve, which was generated to illustrate these transitions.

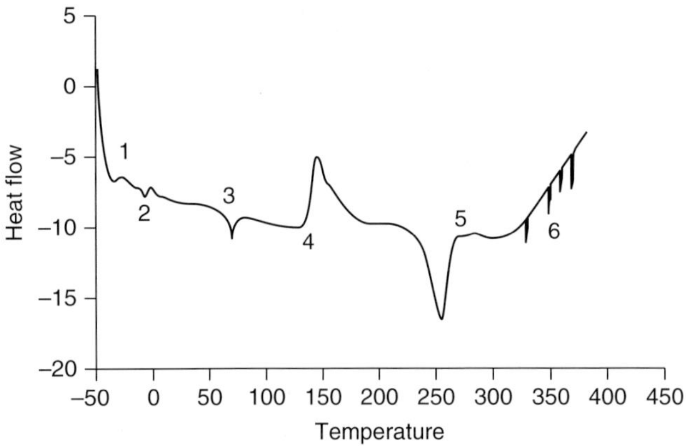

Figure 3.20 Artificial DSC curve.

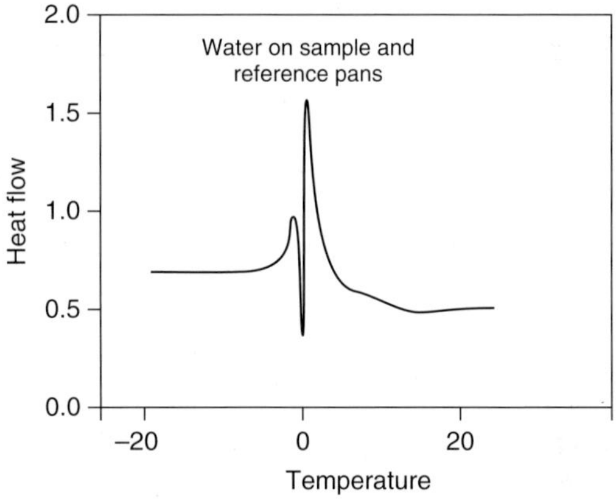

Figure 3.21 A sample transition due to purge gas.

3.4.1 Transition(s) at 0 °C

There may be weak transitions at 0 °C due to the condensation of water on both the reference and sample pans as shown in Figure 3.21. The water in the sample cell can act as a plasticizer and reduce transition temperatures. Furthermore, the peaks may appear slightly lower than 0 °C, which is due to the dissolution of impurities by the moisture from the cell and pans. The water in the purge gas can cause the perturbation in the baseline that makes a challenge for detection of real transitions around 0 °C. This can be avoided by keeping the hygroscopic samples in a desiccator.

3.4.2 Apparent Melting at T_g

When a material is heated through the glass transition, internal stresses are released caused by handling, processing, or thermal history. As a result of this shifting, the measured glass transition temperature several degrees or leading to misreading of the T_g as an endothermic melting peak. Heating it to at least 25 °C above T_g followed by quench cooling to a temperature below the T_g relieves the

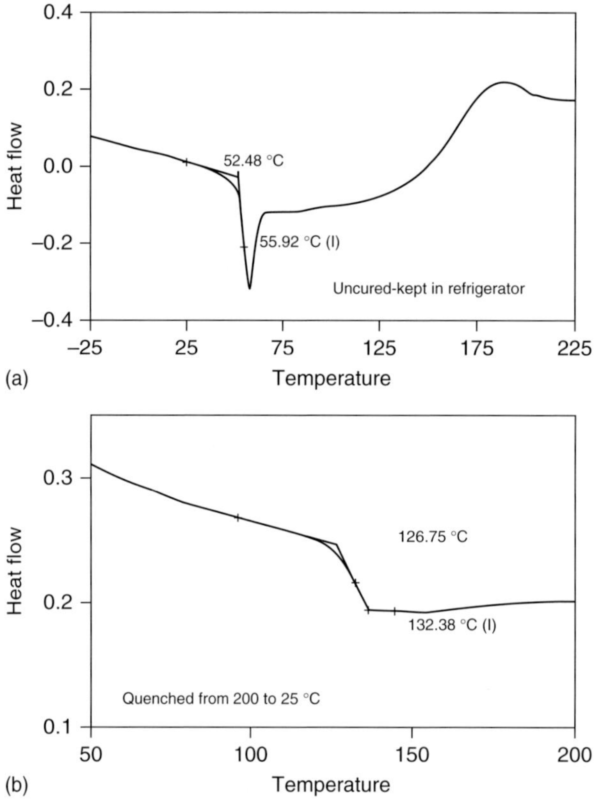

Figure 3.22 (a) Molecular relaxation can cause T_g to appear as a melt. (b) The effect of curing at 200 °C and then quench cooling to 25 °C.

internal stresses in the material. The effect of curing and quenching is clearly depicted in Figure 3.22, while the effect of cooling rate on the shape of T_g is illustrated in Figure 3.23.

The molecular relaxation becomes a weak endothermic transition that is close to the end of a glass transition. This can be pronounced to lead the misinterpretation of T_g as an endothermic melting peak.

3.4.3 Exothermic Peaks Below Decomposition Temperature While Heating

This effect occurs during the crystallization of a thermoplastic polymer or curing of a thermosetting resin. Degree of cure and percent of crystallinity can be determined from the amount of heat associated with this transition. These depend on the thermal history of the material. DSC results will not be reproducible if the thermal history is not controlled. Figure 3.24 illustrates the different results obtained for poly(ethylene terephthalate) (PET) after quench cooling and slow cooling at 10 °C/min, respectively. The well-defined T_g in the quenched PET is an indication of an amorphous nature ($DH_{crystallization} < DH_{melting}$), which rearranges

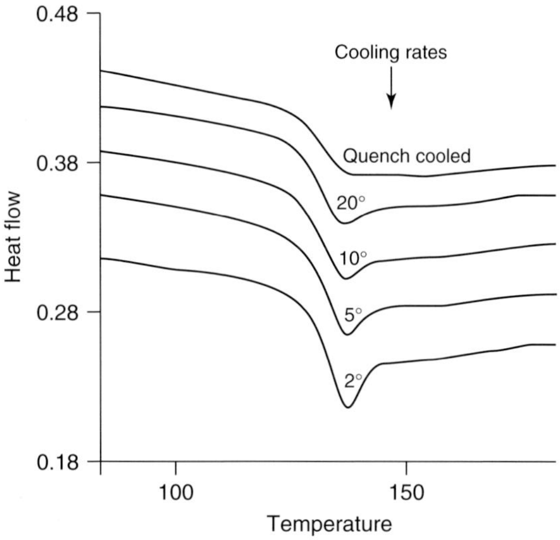

Figure 3.23 Effect of cooling rate on shape of T_g.

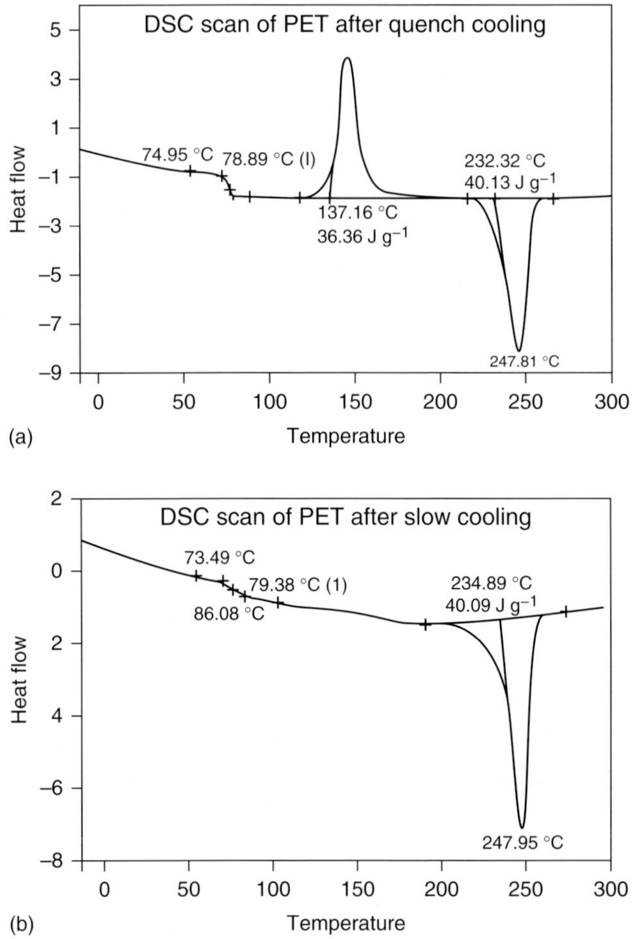

Figure 3.24 DSC scan of PET (heat flow Vs temperature).

on heating to a crystalline structure before melting at about 235 °C. A weak T_g was observed in a slowly cooled material, indicating crystalline structure, and no additional crystallization occurs prior to the melt at 235 °C. So compare materials with a recommended thermal history in the ASTM D3418-82 standard.

3.4.4 Baseline Shift after Endothermic or Exothermic Peaks

These are caused by changes in sample weight, heating rate, or the specific heat of the sample. The specific heat of sample changes subsequent to their transition and the calculation of ΔH is based on the sample weight, so any calculation of ΔH after a weight change will be in error. The sample weight is calculated before and after measurement to realize the occurrence of weight loss, and ΔH of the transitions must be compared using different limits and types of baselines. Figure 3.25 depicts an example where use of sigmoidal baseline is required.

3.4.5 Sharp Endothermic Peaks During Exothermic Reactions

They are usually the result of "experimental phenomena" rather than real material transitions. For example, rapid volatilization of gases trapped in the material can cause sharp peaks. Here also the sample weight is needed to calculate before and after a run to realize the occurrence of weight loss and reduce the temperature limit if failing to obtain useful information due to volatilization and using a pressure DSC cell is optional.

Calibration procedures are different for different calorimeters and depend on the type of the instrument and on the software used and manufacture recommendations intended to improve the measurements.

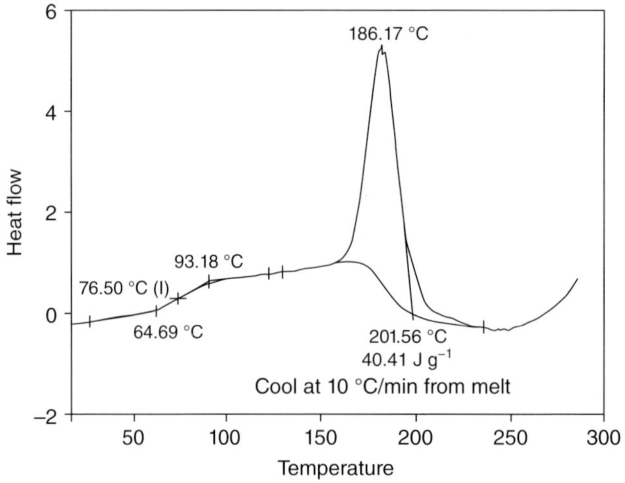

Figure 3.25 DSC scan on PET crystallization and T_g on cooling.

3.5 DMA: Experimental Data Interpretation and Analysis

Jung-Ohk Kweon and Si-Tae Noh [44] have studied about organic–inorganic hybrids, PEO–silica, and PEO–silica–$LiClO_4$ and they found that in both hybrids the peaks of loss tangent are broadened and lowered with the wt% of silica content. This may be due to the strong interactions of the PEO with silica. The impact of silica on T_g behavior and the loss tangent recommends that the reduction of T_g of the PEO with wt% of silica is caused by the weak interaction between Li^+ and PEO with increasing of silica wt% in salt-added hybrids as shown in Figure 3.26.

Later Chun-Chen Yang and Yuan-Chen [45] have examined another PE system based on PVA/montmorillonite (MMT) for fuel cell applications, scrutinizing the salt concentration in the PE system. Their detailed results are tabulated in Table 3.6 and Figure 3.27.

In 2012, Noor et al. [46] have studied about gellan gum-based PE system with a modifier of $LiCF_3SO_3$. They reported that these semicrystalline samples have thermal stability up to 234 °C. As in Figure 3.28 these samples undergo the endothermic reactions of oxidation and hydrolysis followed by polysaccharide pyrolysis exothermic reactions and also observed that the addition of salt does not change the degradation temperature of pure polymer sample but enhanced the mechanical properties of the sample.

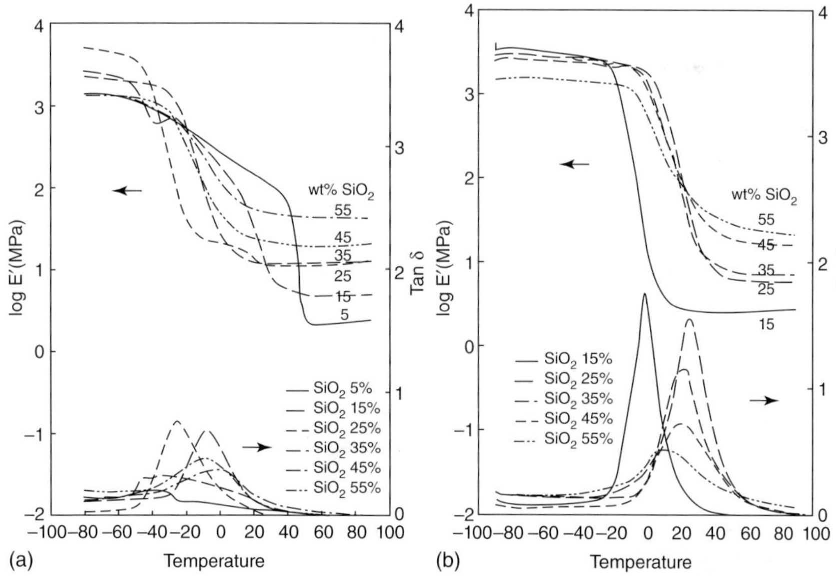

Figure 3.26 DMA spectra (storage modulus and tan δ versus temperature) of (a) PEO–silica and (b) PEO–silica–$LiClO_4$.

Table 3.6 DMA results of PVA/MMT composite for the various concentrations of MMT.

Types	E' (dyn cm^{-2})		
T (°C)	PVA film	PVA/5 wt% MMT	PVA/10 wt% MMT
30	1.36×10^{10}	5.43×10^{10}	4.01×10^{10}
60	7.85×10^{8}	2.30×10^{10}	1.78×10^{10}
100	4.36×10^{8}	2.23×10^{9}	3.88×10^{9}
120	4.08×10^{8}	1.26×10^{9}	1.95×10^{9}
150	2.32×10^{8}	7.77×10^{8}	1.08×10^{9}

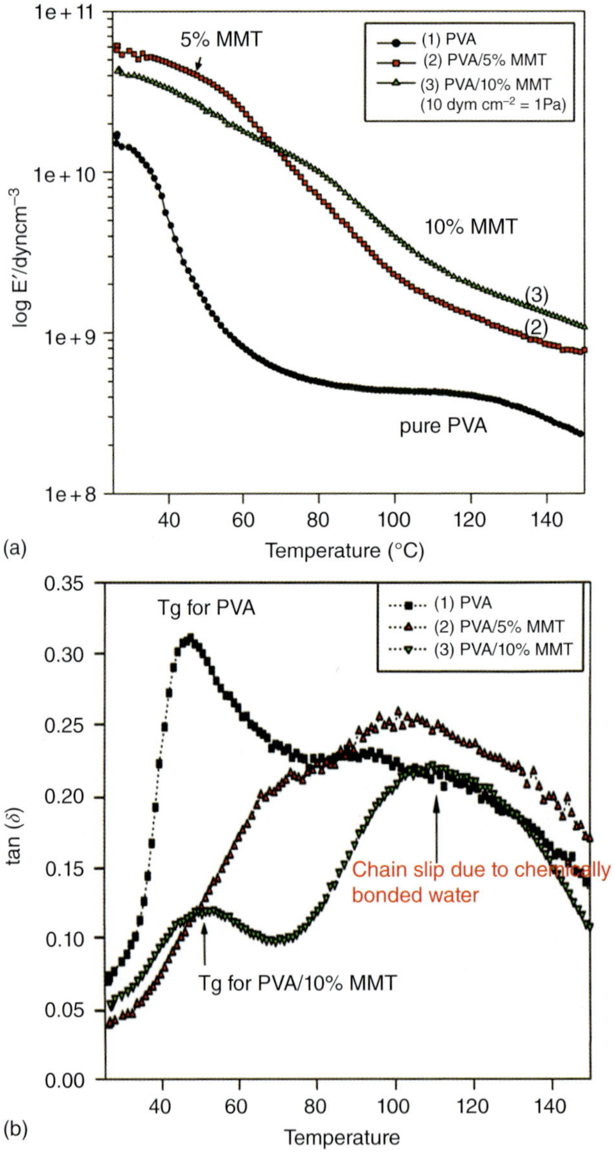

Figure 3.27 The DMA curve of PVA/MMT composite (a) Log E′ and temperature (b) tanδ and temperature.

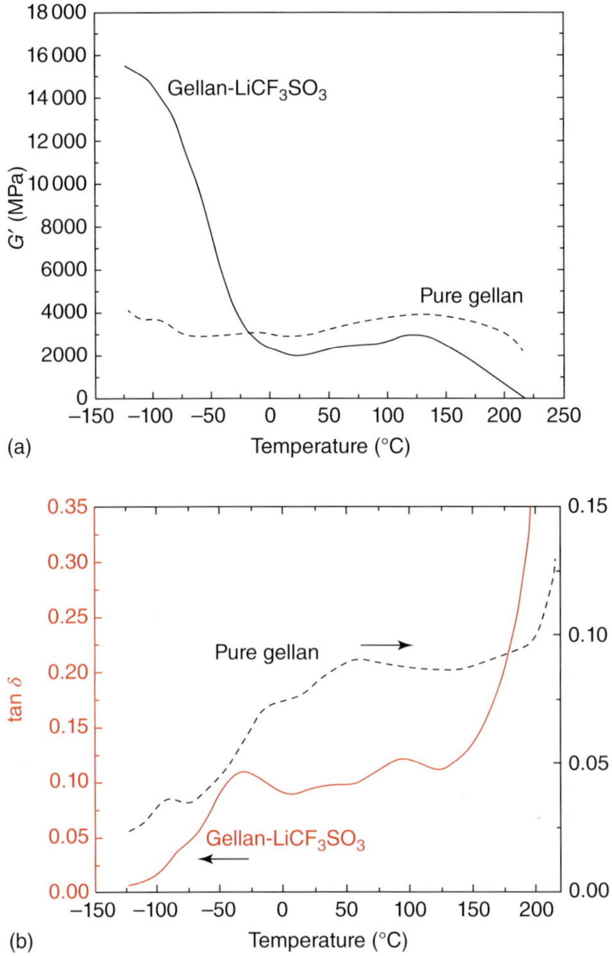

Figure 3.28 DMA analysis (a) storage modulus and (b) tan δ of pure gellan and modified gellan PE with $LiCF_3SO_3$.

In order to quantify the mechanical strength of composite electrolyte ($MgAl_2O_4$)–PEO–$LiPF_6$ (see Table 3.3), Angulakshmi et al. [33] have examined the stress–strain behavior and found that the sample with 95% PEO + 5% $LiPF_6$ exhibits brittle behavior (Figure 3.29). The tensile properties are tabulated in Table 3.7.

The changes in the macroscopic properties of the pure and doped SPEEK [35] were also examined using DMA technique (Figure 3.30). The storage modulus showed a reverse fashion with the addition of salt to 30 wt% implying a high degree of crystallinity. The trend of E" confirmed the phase transition with the increasing LiTFSI content. This may be due to the presence of crystalline domains from the ion associations.

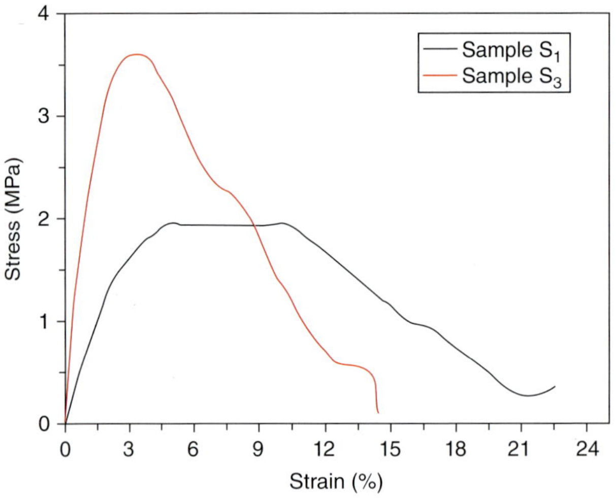

Figure 3.29 Stress versus strain behavior of sample S_1 (PEO + $LiPF_6$) and sample S_3 (PEO + $MgAl_2O_4$ + $LiPF_6$). Source: Angulakshmi et al. 2013 [33]. Reproduced with permission of Elsevier.

Table 3.7 Tensile properties of samples S_1 (PEO + $LiPF_6$) and S_3 (PEo + $MgAl_2O_4$ + $LiPF_6$).

Sample	Tensile strength (MPa)	Elongation at break (%)	Modulus of elasticity (MPa)
S_1	1.92	20.50	0.58
S_3	3.72	12.50	1.35

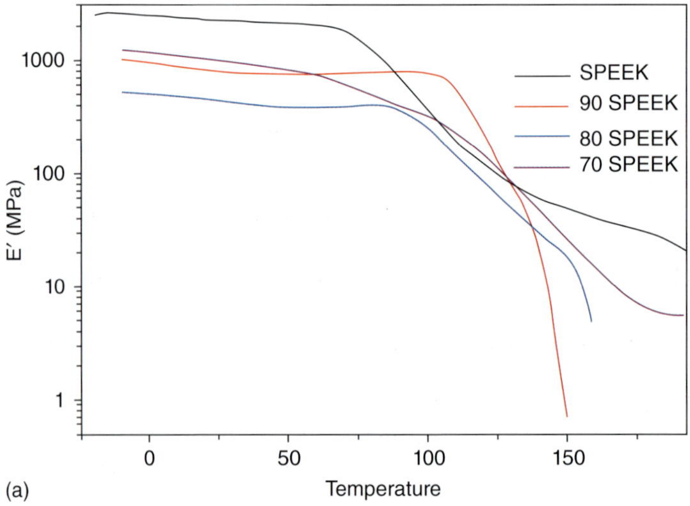

Figure 3.30 Variation of (a) the storage modulus and (b) tan δ DMA with temperature for pure SPEEK and the LiTFSI-doped SPEEK membranes. Source: Thayumanasundaram et al. 2015 [35]. Reproduced with permission of Wiley-VCH).

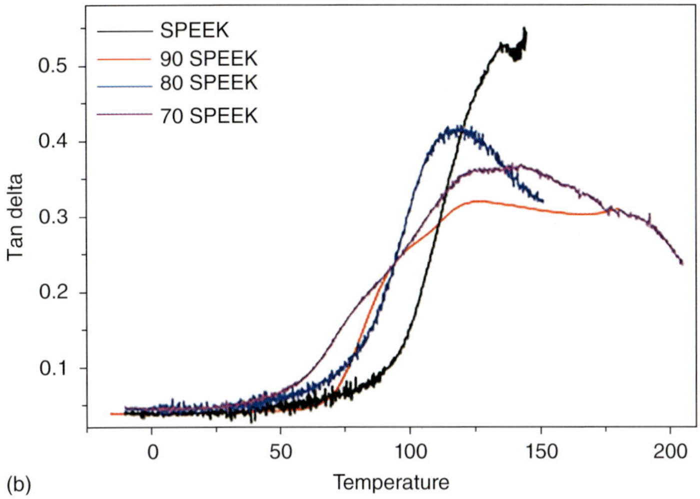

Figure 3.30 (*Continued*)

References

1 Fenton, D.E., Parker, J.M., and Wright, P.V. (1973). *Polymer* 14: 589.
2 Manuel, S.A. (2005). *Eur. Polym. J.* 42: 21–42.
3 Meyer, W.H. (1998). *Adv. Mater.* 10: 439–448.
4 Manuel, S.A., Kumar, P.T., Renganathan, N.G. et al. (2000). *J. Power Sources* 89: 80.
5 Rajendran, S., Sivakumar, P., and Babu, R.S. (2007). *J. Power Sources* 164: 815.
6 Rajendran, S., Mahendran, O., and Kannan, R. (2002). *J. Phys. Chem. Solids* 63: 303.
7 MacCallum, J.R. and Vincent, C.A. (1987). *Polymer Electrolyte Reviews—I*. London: Elsevier.
8 Xaio, L., Zhang, H., Jana, T. et al. (2005). *Fuel Cells* 5 (2): 287–295.
9 Gray, F.M. and Connor, J.A. (1997). *Polymer Electrolytes*, RSC Materials Monographs. Cambridge: The Royal Society of Chemistry.
10 Gray, F.M. (1991). *Solid Polymer Electrolytes – Fundamentals and Technological Applications*. New York, NY: VCH.
11 MacCallum, J.R. and Vincent, C.A. (1989). *Polymer Electrolyte Reviews—II*. London: Elsevier.
12 Scrosati, B. (1993). *Applications of Electroactive Polymers*. London: Chapman Hall.
13 Hallinan, D.T. and Balsara, N.P. (2013). *Annu. Rev. Mater. Res.* 43: 503–525.
14 Zhou, J. and Fedkiw, P.S. (2004). *Solid State Ionics* 166: 275.
15 Song, J.Y., Wang, Y.Y., and Wan, C.C. (1999). *J. Power Sources* 77: 183.
16 Boudin, F., Andrieu, X., Jehoulet, C., and Olsel, I.I. (1999). *J. Power Sources* 804: 81–82.
17 Choi, S.W., Jo, S.M., Lee, W.S., and Kim, Y.R. (2003). *Adv. Mater.* 15: 2027.

18 Cheng, C.L., Wan, C.C., and Wang, Y.Y. (2004). *Electrochem. Commun.* 6 (6): 531–535.
19 Golodnitsky, D., Ardel, G., Strauss, E. et al. (1997). *J. Electrochem. Soc.* 144: 3484.
20 Dai, Y., Wang, Y., Greenbaum, S.G. et al. (1998). *Electrochim. Acta* 43: 1557.
21 Borghini, M.C., Mastragostino, M., Passerini, S., and Scrosati, B. (1995). *J. Electrochem. Soc.* 142: 2118.
22 Mastragostino, M., Soavi, F., and Zanelli, A. (1999). *J. Power Sources* 729: 81–82.
23 Qian, X., Gu, N., Cheng, Z. et al. (2001). *Electrochim. Acta* 46: 1829.
24 Fauteux, D., Lupien, M.D., and Robitaille, C.D. (1987). *J. Electrochem. Soc.* 134: 2761–2767.
25 Rietman, E.A., Kaplan, M.L., and Cava, R.J. (1985). *Solid State Ionics* 17: 67–73.
26 Bruce, P.G. (1995). *Solid State Electrochemistry*. Cambridge: Cambridge University Press.
27 Sinha Ray, S. and Okamoto, M. (2003). *Prog. Polym. Sci.* 28: 1539–1641.
28 Wu, G.M., Lin, S.J., and Yang, C.C. (2006). *J. Membr. Sci.* 280: 802.
29 Feuillade, G. and Perche, P. (1975). *J. Appl. Electrochem.* 5: 63–69.
30 Jacob, M.M.E., Prabaharan, S.R.S., and Radhakrishna, S. (1997). *Solid State Ionics* 104: 267–276.
31 Stephan, A.M., Kumar, T.P., Kulandainathan, M.A., and Lakshmi, N.A. (2009). *J. Phys. Chem. B* 113: 1963–1971.
32 Chandra, A., Chandra, A., and Thakur, S. (2012). *Int. J. Chem. Res.* 3 (1): 81–85.
33 Angulakshmi, N., Nahm, K.S., Nair, J.R. et al. (2013). *Electrochim. Acta* 90: 179–185.
34 Angulakshmi, N., Kumar, R.S., Kulandainathan, M.A., and Stephan, A.M. (2014). *J. Phys. Chem. C* 118: 24240–24247.
35 Thayumanasundaram, S., Rangasamy, V.S., Seo, J.W., and Locquet, J.-P. (2015). *Eur. J. Inorg. Chem.*: 5395–5404.
36 Nair, J.R., Bella, F., Angulakshmi, N. et al. (2016). *Energy Storage Mater.* 3: 69–76.
37 Noda, A. and Watanabe, M. (2000). *Electrochim. Acta* 45: 1265–1270.
38 Capiglia, C., Yang, J., Imanishi, N. et al. (2003). *J. Power Sources* 119–121: 826–832.
39 Rodrigues, L.C., Barbosa, P.C., Silva, M.M., and Smith, M.J. (2007). *Electrochim. Acta* 53: 1427–1431.
40 Gray, F.M. (1991). *Solid Polymer Electrolytes: Fundamentals and Technological Applications*. VCH Publishers, Inc.
41 Ibrahim, S. and Johan, M.R. (2011). *Int. J. Electrochem. Sci.* 6: 5565–5587.
42 Appeteechi, G.B., Crose, F., Persi, L. et al. (2001). *Electrochim. Acta* 45: 1481.
43 Norem, S.D., O'Neill, M.J., and Gray, A.P. (1970). *Thermochim. Acta* 1: 29.
44 Kweon, J.-O. and Noh, S.-T. (2001). *J. Appl. Polym. Sci.* 81: 2471–2479.
45 Yang, C.-C. and Lee, Y.-C. (2008). *ECS Trans.* 13 (28): 1–20.
46 Noor, I.S.M., Majid, S.R., Arof, A.K. et al. (2012). *Solid State Ionics* 225: 649–653.

4

Energy in a Portable World

Noor Syuhada Zakuan[1], Woo Haw Jiunn[1], and Tan Winie[2,3]

[1] *Centre for Ionics University of Malaya, Department of Physics, Faculty of Science, University of Malaya, 50603 Kuala Lumpur, Malaysia*
[2] *Faculty of Applied Sciences, Universiti Teknologi MARA, 40450 Shah Alam, Selangor, Malaysia*
[3] *Institute of Science, Universiti Teknologi MARA, 40450 Shah Alam, Selangor, Malaysia*

4.1 Introduction

Half a century ago, not many people can imagine the scope of mobile power usage like it is today. The application of this energy in portable world nowadays starts right from our doorstep (smartphone, tablet, smart watch, Google Spec), covering the whole spectrum of life stages and even far reaches the unlimited terrestrial applications, such as satellite that has been deployed to improve the quality of human life. One must be thankful to the scientists who dare to dream the unthinkable and believe the unbelievable.

The development of mobile power was actually led by the military sector where lots of research funds were pumped in. It might not be known to everyone that the logistical cost of mobile power in military sector is huge, often taking up a big portion of the country allocated budget. Energy in portable form is essential and could be one determining winning factor in a battlefield. Therefore, portable power is one of the key technologies that can help cut down huge cost in war and minimize risk of the troops who can now operate independently. A hybrid mobile microgrid is developed to reduce and replace the noisy diesel-powered generators, saving the need to truck the heavy generators into the hostile territory.

Today, when one talks about mobile power, it is a general and broad term that refers to an energy system that integrates packs of smart battery, supercapacitor, fuel cells, solar cells, or various combinations of them. Mobile power is an accessible off-grid energy source within our reach anywhere we go. It should be clean, affordable, lightweight, of high quality, and convenient to use.

How was it been discovered? How shall we take care of it or recycle this technological waste at the end of its retirement? This chapter will briefly review the history development of mobile power, its application, and its performances. Caring for mobile power from birth to retirement will also be covered. Finally, the recycling of electronic waste will be discussed to prepare a better future for our next generation.

Polymer Electrolytes: Characterization Techniques and Energy Applications,
First Edition. Edited by Tan Winie, Abdul K. Arof, and Sabu Thomas.

4.2 History Development of Mobile Power

First wave of development came from the advancement of electrochemistry. The first battery system could be traced back to 1859 when lead-acid batteries were invented. After that, series of discoveries were reported by building the initial finding with new materials, new arrangement, and new combination. However, it did not attempt to use it as mobile devices that are lightweight, small, and thin. Second wave of development would probably be initiated from the idea of using lithium ion as charge carrier to develop super high specific energy density batteries. Further enhancement is reported to use polymeric matrix to capture the electrolyte ions, making it a lightweight, flexible thin film for various shape and fitting.

The timeline of mobile power discovery is shown in Table 4.1.

Around 1936, a German archaeologist in Baghdad uncovered an ancient mobile power, in this case, a battery composed of a clay jar holding an iron rod surrounded by copper cylinder (Figure 4.1). When this ancient battery was filled with an electrolyte, such as grape juice, the clay jar could produce 1.5–2.0 V of electricity. It was estimated to be more than 2000 years old, dated back to 150 BCE. However, some scientists believed it was not used as an energy source, but for other purposes such as electroplating, medical purpose, or even religious rituals [1].

The first battery that was based on electrochemistry was believed to come only in 1800, through the work of Alessandro Volta. He designed "voltaic cell" or "galvanic cell" by stacking zinc and copper disc separated by paper soaked in brine [2]. Volta noticed that the different potentials of metals in a battery system would eventually determine the battery voltage, and this discovery led to the invention of the first "wet battery cell" with stable potential. John Daniell had invented the Daniell cell in 1836. The cell consisted of two electrolytes that were zinc rod immersed in zinc sulfate and copper rod immersed in copper (II) sulfate. This cell managed to draw electricity evenly and opened up many potential applications including telegraphy [3].

The first efficient battery was developed by Raymond Gaston Planté in 1859 based on lead acid (Figure 4.2). Two long lead foils that were used as electrodes sandwiching a coarse cloth in between were wounded in screw shape and soaked in sulfuric acid (about 10%) liquid electrolyte [4]. It was rechargeable and able to deliver a steady current to store electrical energy for a longer time. This wet cell technology is still useful until today to start most of the car's internal combustion engine. However, there are a few shortcomings such as the bulky size, the heavy weight, and the use of corrosive liquid electrolyte that limits the mobile usage of this energy device.

Smaller size and lightweight cell was later demonstrated in 1868, by a French engineer, Georges-Lionel Leclanché. The initial invention of this "Leclanché cell" composed of amalgamated zinc bar as negative electrode (anode), manganese dioxide mixed with a bit of carbon as positive electrode (cathode), and low corrosive liquid, ammonium chloride solution, as liquid electrolyte [5]. After some years of effort, the inventor successfully substituted the liquid electrolyte into an electrolyte paste, giving birth to "dry cell" concept. Leclanché cell is a cheap 1.5 V primary cell (non-rechargeable) with long shelf life but delivers low current loading limiting its application in low-drain devices like clocks and remote controls.

Table 4.1 History of mobile power development.

Year	Activity	Inventor
150 BCE	Baghdad battery	—
1749	Linked capacitor	Franklin (American)
1791	Discovery of "animal electricity"	Galvani (Italy)
1800	Invention of voltaic cell	Volta (Italy)
1802	First electric battery capable of mass production	Cruickshank (UK)
1820	Electricity from magnetism	Ampère (France)
1833	Faraday's law	Faraday (Britain)
1836	Invention of Daniell cell	Daniell (Britain)
1839	Invention of fuel cell (H_2/O_2)	Grove (Britain)
1859	Invention of lead-acid battery	Planté (France)
1868	Invention of dry cell (carbon–zinc)	Leclanché (France)
1899	Invention of nickel–cadmium battery	Jungner (Sweden)
1900	Invention of film capacitor	Mansbridge (London)
1901	Invention of nickel–iron battery	Edison (US)
1909	Invention of mica capacitor	William Dubilier (American)
1926	Invention of ceramic capacitor	Germany
1932	Fuel cell using non corrosive alkaline electrolyte	Francis Bacon (England)
1949	Invention of alkaline–manganese battery	Lewis (Eveready battery)
1954	First generation's solar cell	Bell labs (American)
1960	Fuel cell generator for space mission	NASA
1976	Second generation's solar cell	RCA laboratories
1978	Supercapacitor	NEC
1991	Third generation's solar cell	Michael and Brian (California)
1990	Commercial nickel–metal hydride battery	Group effort
1991	Commercial lithium ion battery	Sony (Japan)
1994	Commercial of Li ion polymer	Bellcore (USA)
1996	Introduction of Li ion with manganese cathode	Moli energy (Canada)
1996	Identification of Li-phosphate ($LiFePO_4$)	University of Texas (USA)
2002	Nanomaterial batteries	Group effort
2002+	Recent developments	—

Building on the dry cell concept, little major advancements were made based on new materials to increase the power loading. In 1899, Waldemar Jungner worked on nickel–cadmium (NiCd) battery that used nickel as cathode material, while anode is composed of cadmium as active material [6]. NiCd is a secondary cell (rechargeable) that could deliver fair energy capacity with high discharge rate. However, cadmium is highly toxic and brings harm to the human body's cardiovascular, reproductive, neurological, gastrointestinal, renal, and respiratory systems. A similar system was attempted by Thomas Edison in 1901 by replacing cadmium with iron, called as nickel–iron battery (NiFe), but producing low

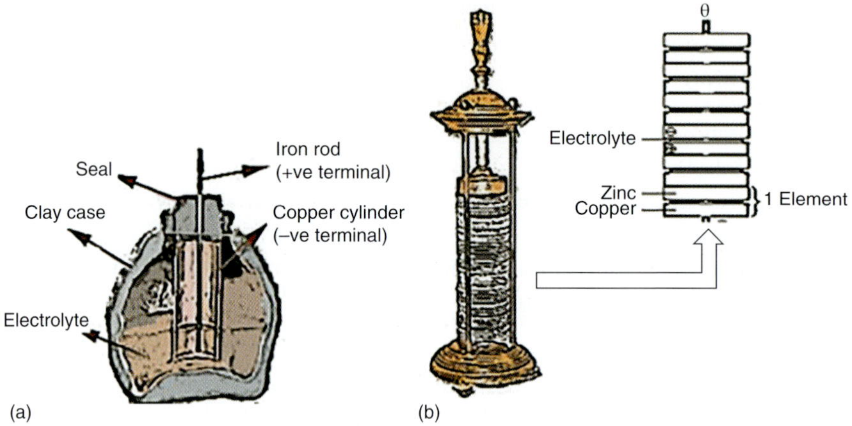

Figure 4.1 (a) Baghdad battery and (b) voltaic battery, with cross-sectional view.

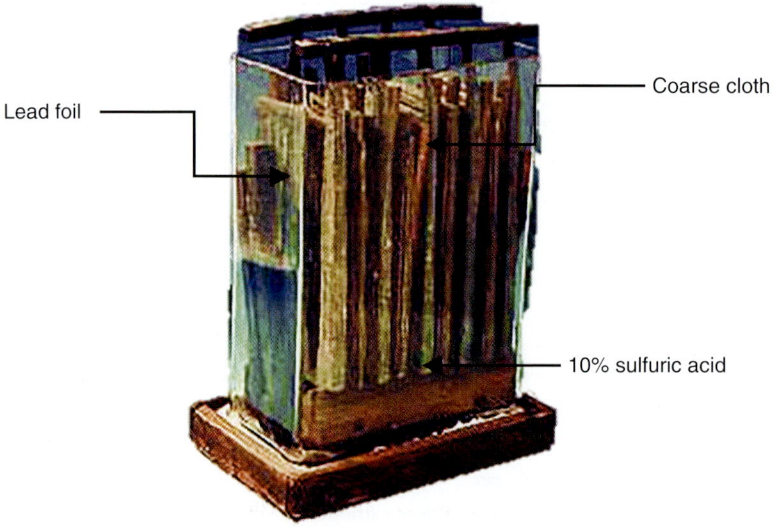

Figure 4.2 Nineteenth century illustration of Planté's original rechargeable lead-acid cell.

energy capacity, poor performance at low temperature, and high self-discharge. In 1949, Lewis Urry, an engineer at the Eveready Battery Company, discovered the high-performance version of Leclanché cell called alkaline battery. He used an alkaline potassium hydroxide to replace the acidic ammonium chloride. This primary cell could supply greater energy at higher load currents and has longer shelf life than the Leclanché cell.

A new direction of cell based on renewable solar energy was given birth in 1954 at Bell Laboratories. Three inventors, Gerald Pearson, Calvin Fuller, and Daryl Chapin, were using a silicon semiconductor-based material to draw electricity from light directly via photovoltaic effect [7]. According to them, this solar cell could achieve an energy conversion efficiency of 6% under direct sunlight. When light photon was captured and absorbed, it generated charge called electron–hole pair or excitons. These charge carriers were then separated and extracted into external circuit supplying electrical current or store up in battery for night use.

Although NiCd could provide high energy density, it has mild toxicity in nature, and this posed danger to human health and the environment upon disposal. Thus, replacement of NiCd was an ongoing effort to reduce the technological garbage danger. This safety issue has limited NiCd usefulness within aircraft industry owning to its high energy density performance. Later in 1990, nickel–metal hydride (NiMH) was invented to replace NiCd using the same nickel hydroxide (NiOOH) as the positive electrode, but the negative electrode used hydrogen-absorbing alloy, metal hydride (MH), to replace Cd [8].

The launching of lithium ion batteries (LIBs) in 1991 by Sony was a huge commercial success as LIBs beat all its rivals in terms of cost, sheet abundance, nontoxicity, long cycle life, and environmental friendliness. Using Li as anode material is a fascinating approach as Li is extremely electropositive (−3.04 V versus standard hydrogen electrode). Furthermore, it is the lightest of all metals (molar mass, $M = 6.94\,\mathrm{g\,mol^{-1}}$, and density, $\rho = 0.53\,\mathrm{g\,cm^{-3}}$) that facilitate the design of energy storage systems with extraordinary high energy density. These superior characteristics had made Li ion the number one choice to develop flying performance primary and secondary batteries in the last 20 years. The pioneering work of LIB was reported to begin in 1912 under G.N. Lewis and was sold for military applications in 1970. The breakthrough that forms the basics of LIB was laid down in 1980 via John B. Goodenough's invention of $LiCoO_2$ as cathode, combining with Rachid Yazami's (a French research assistant) invention of graphite as anode [9].

After 1991, most of the battery researches have focused on improving the performance of LIB using new active materials and electrolytes. In 1994, lithium ion polymer battery (LiPo) was reported. The liquid solvent was replaced by solid polymer composite to hold the electrolyte. The electrodes and separator were laminated, allowing the battery to be packed in a flexible wrapping [10]. The batteries can now be specifically shaped without the rigid casing to fit a device (see overview in Figure 4.3).

Solar cells gained prominence and attention when it was incorporated onto the 1958 Vanguard I satellite (Figure 4.4). The battery-based primary power source of the satellite was supplemented with solar cell-based alternative power source at the outer body. Without any major adjustments to the power system, it successfully prolonged the satellite mission time. Following the success, the United States had launched Explorer 6 in 1959, in which large wing solar arrays were displayed. These arrays consisted of 9600 Hoffman solar cells and set the standard feature of future satellite architecture. A year later, its high specific power (power-to-weight ratio) had made it the preferred power source for all satellites orbiting the Earth.

Generally, there are three types of generations of solar cells. The first-generation solar cells were reported at 1954 based on crystalline silicon wafers and currently dominate the international market (about 80% of the solar panels volume around the world) due to high efficiencies. They can be divided into two types based on their crystallization levels, the single crystalline silicon (mono-silicon) and

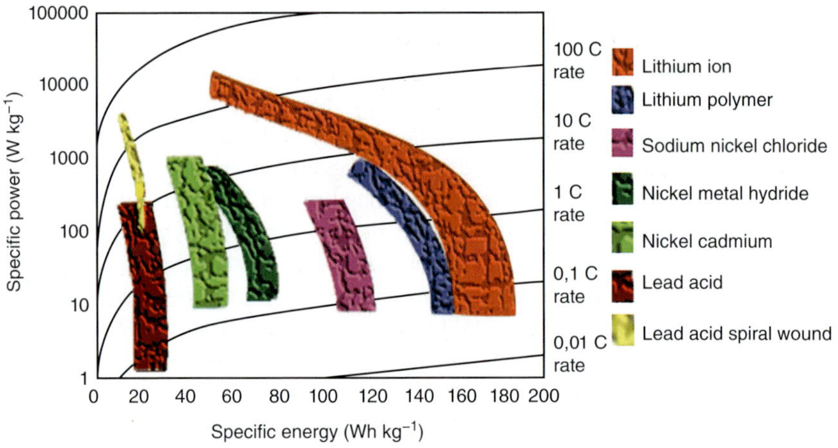

Figure 4.3 Comparison of rechargeable battery in terms of specific energy and specific power.

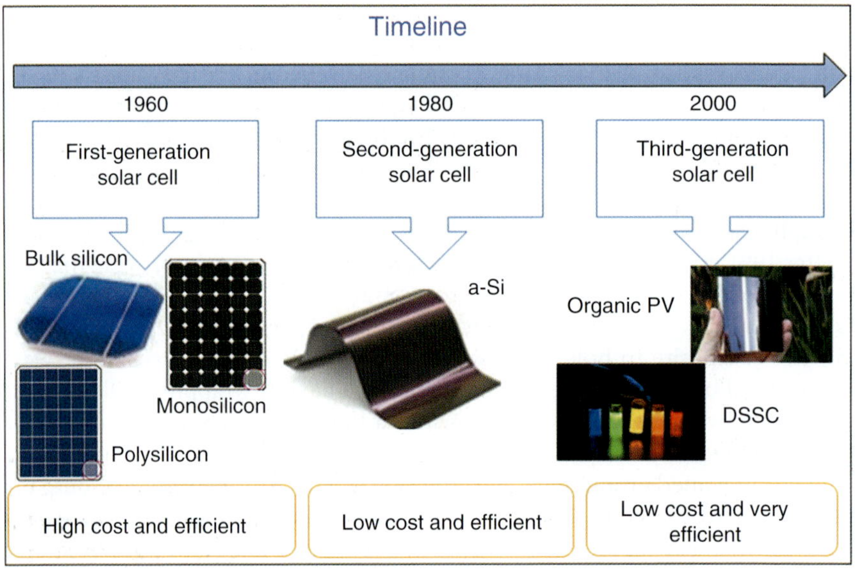

Figure 4.4 Timeline generation of solar cells.

polycrystalline silicon (polysilicon), demonstrating high efficiency in the range of 15–20% [11]. They are highly stable, but the marketability is limited by expensive price tag due to the stringent clean room manufacturing requirements.

To reduce the cost, second-generation solar cells were fabricated based on amorphous semiconductor materials in 1976 and found its usage in utility-scale photovoltaic power stations, building integrations PV, and in small stand-alone power system [12]. A few micrometers thick thin film semiconductor materials were formed on substrate other than glass. This thin film could be based on amorphous silicon (a-Si) or non-silicon materials such as cadmium telluride/cadmium sulfide (CdTe/CdS) and copper indium gallium diselenide (CIGS). The advantages of this second-generation solar cells include light weight and moderate cost with efficiency in the range of 10–15%.

Third-generation solar cells started from 1991, mainly based on organic materials, organometallic compounds, or inorganic materials, but have not been commercially proven yet. Examples of these solar cells are dye sensitized solar cell (DSSC), organic solar cell, copper zinc tin sulfide (CZTS) solar cell, polymer solar cell, quantum dot solar cell (QDSC), and perovskite solar cell. DSSCs use dye as a sensitizer to absorb the wide spectrum of sunlight, producing an excited electron that is then injected into conduction band of TiO_2 and connected to external circuit [13]. The main attractive feature of DSSCs is its simple and economical production route. On the other hand, novel thin film perovskite materials are actively investigated as light harvesting active layer and showed drastic improvement in efficiency from 3.8% in 2009 to 22.1% in early 2016 [14]. No one would underestimate the potential of third-generation solar cells although its current performance and stability are still limited compared with first- and second-generation solar cells.

In 1745, "Leyden jar" was reported separately by German cleric Ewald Georg von Kleist and Dutch scientist Pieter van Musschenbroek of Leiden. This invention was named after the city Leiden. Leyden jar is the original form of a capacitor where it stores static electricity between two electrodes of a glass jar [15]. It was later called "battery" by an American scientist Benjamin Franklin in 1749 when two jars were connected in parallel in an electricity experiment.

Capacitor can be divided into two polarized group (Figure 4.5): nonpolar capacitors such as film capacitor, mica capacitor, and ceramics capacitor and polar capacitors such as electrolytic capacitor and supercapacitor. Film capacitor using

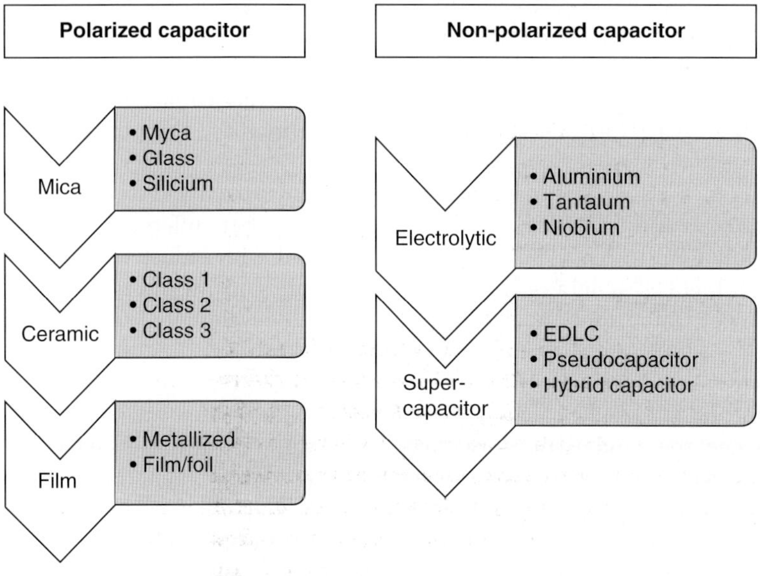

Figure 4.5 Various types of capacitors.

metallized paper was reported in 1876 and patented in 1900 by G.F. Mansbridge. A strip of impregnated paper was sandwiched between two strips of metal and rolled into a cylinder shape. They were used in the early twentieth century as decoupling capacitors in telecommunications. Small size mica capacitors were invented by William Dubilier in 1909 and used for high frequency decoupling applications. It became popular during the booming broadcasting industry where low losses and low capacitor change over time are required. In the 1920s, silver mica capacitors were developed using sheets of mica that are coated on both sides with the deposit metal. However, the scarcity of mica has led to the development of ceramic capacitor in 1926. After the Second World War, this low-cost small disc-shaped capacitor played an important role in three classes of electronic applications (Class 1, Class 2, and Class 3).

Supercapacitor is the family name for electrical double layer capacitors (EDLC), pseudocapacitors, and hybrid capacitors. They have higher capacitance value than conventional capacitors and higher power density than batteries.

In the early 1950s, H. Becker developed the high capacitance EDLC with porous carbon electrodes in which it stores charge electrostatically by separation of charge in a Helmholtz double layer [16]. Standard Oil of Ohio then rediscovered this effect in 1966 and eventually licensed it to NEC at 1978 with supercapacitor as its trademark. It had capacitances up to 1.0 F to provide backup or emergency shutdown power and is commonly used in industrial laser, medical equipment, and wind turbines.

In 1839, Sir William Robert Grove reported a galvanic device or "fuel cell" converting the chemical energy of fuel to electrical energy. Over a platinum catalyst, the electrochemical reaction between hydrogen and oxygen could generate an electric current. At that time, this fuel cell has not been able to deliver sufficient electricity for any device application yet [17].

In 1889, Charles Langer and Ludwig Mond are the first people who coined the term "fuel cell" when they are using air and industrial coal gas as fuel. The workable fuel cell was reported in the early 1900s, that conversion of electricity from coal or carbon occurred, giving birth to the internal combustion engine concept. The research in fuel cell kept growing to increase the performance of fuel cell.

In 1932, Francis Bacon invented fuel cell based on hydrogen and oxygen using nickel electrodes and noncorrosive alkaline electrolyte but suffered from high cost of production. In the 1960s, NASA collaborated with industrial partners to develop a fuel cell generator for manned space missions, while at the same time the Soviet Union spent huge budget on fuel cell development for military applications.

After many years of investigation, fuel cells begin to be used as auxiliary power units (APU) and stationary backup power in 2007, with written warranties and manage to meet the standard of the markets. Major auto vehicle makers have also started an ongoing research in fuel cell vehicles (FCVs) development (Figure 4.6). In 2008, Honda was reported to lease the FCX Clarity fuel cell in electric vehicle (EV). As this technology becomes more accessible and cheaper, coupled with the abundant hydrogen in nature, it is expected to see more fuel cell energy to work hand in hand with traditional power source in the coming years.

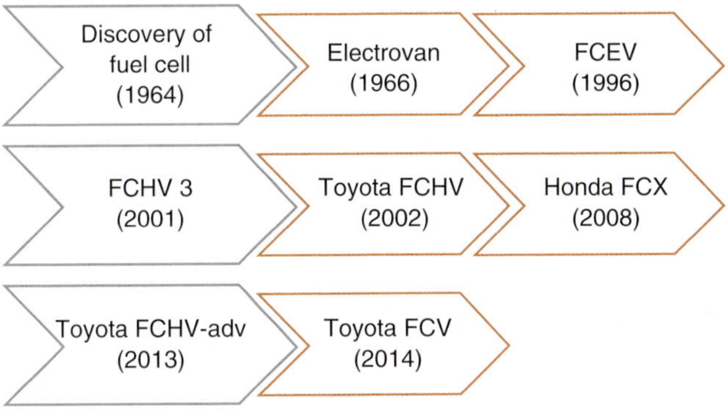

Figure 4.6 Timeline of fuel cell vehicle (FCV).

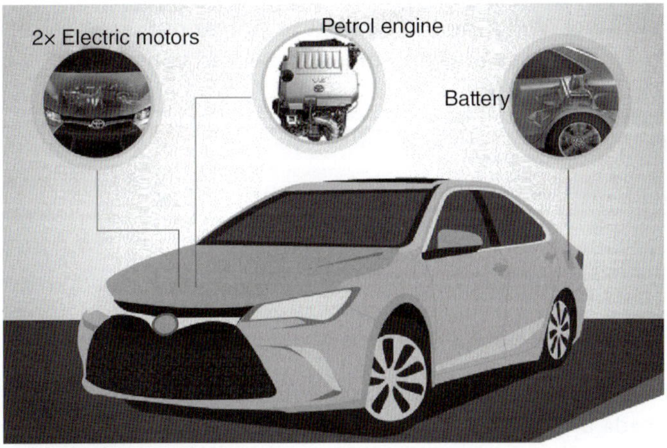

Figure 4.7 Design of a hybrid car.

The fast-growing demand for near zero emission gas electric motors presently has sped up the development of mobile power. Hybrid electric vehicles (HEVs), plug-in hybrid electric vehicles (PHEVs), and EVs are emerging automotive products using a combination of mobile power devices (Figure 4.7). Limited HEVs were first sold in 1997. Few years later, a limited production of PHEVs was introduced in 2004 and mass produced in 2011, followed by commercialization of EVs to the public in the same year [18]. Many believe EVs might be the solution to greenhouse effect as they do not produce exhaust fume. This will give better air quality in megacities and improve health.

Since 2010, many hybrid cars have been sold around the world as consumers, business, and governments are adopting the near zero emission gas approach. In 2009, the United States has invested hugely to build up PHEVs industry and was reported procuring 474 000 PHEVs. China's annual plug-in car sales have increased drastically between year 2011 and 2014 from 5579 to 45 048. It was also reported that there were 105 000 PHEVs registered in Europe in 2016, while Canada secured an annual sale of 107 000 PHEVs between years 2015 and 2020.

Daily Mail

Oh dear, Theresa!

Our fashion editor's verdict on PM's pink holiday dress

SEE PAGE THREE

Tolls on polluted roads – and ALL petrol and diesel vehicles banned from 2040

WAR ON DIESELS GETTING DIRTY

theguardian

Ban from 2040 on diesel and petrol car sales

UK plans follow French commitment to take polluting vehicles off the road

Taking a break from it all ...

60% rise in unqualified teachers, says Labour

Will America be next to phase out gasoline- and diesel-powered vehicles?

In an effort to meet the agreements of the Paris climate accord, the French government has announced the country will ban the sale of gasoline and diesel vehicles by 2040.

They are one of several European countries which have taken strides to reduce their carbon footprint through the ban of vehicles which burn fossil fuels.

Norway, which has the greatest concentration of electric cars globally, aims to allow the sale of only all-electric and plug-in hybrid cars by 2025.

Figure 4.8 Article about phasing out gasoline- and diesel-powered vehicles [20, 21].

A projection of 700 000 PHEVs running on the road was made by the National Energy Board of Canada by year 2035 within the country [19]. The "2020 Paris Agreement" has also been a big motivator for the adoption of electric and hybrid cars in Europe, according to MarketLine on 27 July 2017. France and Germany have also committed to banning all nonelectric cars from the roads (Figure 4.8).

4.3 Caring for Mobile Power from Birth to Retirement

Most of the mobile device users are concerned about their device battery level especially when they are out of office/home. However, not many users will care much about the battery eventual lifespan (probably between three and five years). In fact, most smartphone manufacturers tested their devices for 300–500 cycles. After a threshold cycle, the batteries are no longer able to hold as much electricity as the first 100 cycles. As a result, it could not deliver the rated capacity (as indicated in the manual) and would only be able to power the device for a shorter period.

Primary batteries utilize their stored energy one time in a single discharge cycle and then are discarded. On the other hand, secondary batteries or "rechargeable batteries" can be brought back to their initial (charged) state after being discharged by passing a current in the reverse direction. There are generally three types of secondary batteries, lead-acid, nickel-based, and lithium ion-based powering almost everything from portable devices to HEVs and EVs. There are ways to keep battery in tip-top condition for a long lifespan. Battery lifespan does not only depend on the type of battery and its quality but also depend on how the battery is cared for.

4.3.1 Getting the Most Out of the Primary Batteries

Single use non-rechargeable primary battery must be carefully used to ensure every volt is squeezed out. One way is to store batteries in the refrigerator, as heat is the number one battery killer. Low temperature will slow down the battery chemical reaction, lower the self-discharge rate, and eventually improve the battery's shelf life by 5%. Therefore, it is proposed to store batteries between 15 °C (59 °F) and 26.7 °C (80 °F). Under those ideal storage conditions, an alkaline battery should last from 5 to 7 years, carbon zinc battery from 3 to 5 years, and lithium cell from 10 to 15 years. Another way (practiced by many battery manufacturers) is to keep the batteries in a dry place properly, such as a sealed bag to avoid corrosion from the oxygen.

Disposable primary cells that should not be recharged as active materials will not return to their original forms via the irreversible chemical reactions. In fact, many manufacturers warn against the attempt of recharging primary cells as it will likely cause battery failure and explosion. Some internal batteries are very soft and therefore one must be careful not to squash them. The unwanted internal short circuit could be created. In addition, bending the battery might damage the case and allow air come in to contaminate the battery. Battery should be regularly checked for any swelling or damage on the surface and the correct size should be used. Old batteries should not be mixed with new batteries. The same device should use the same brands/types of batteries as mixing up batteries may cause the batteries to leak and degrade the performance of all batteries.

Batteries must be correctly connected into the circuit due to the terminals having either positive or negative polarity. Wrong connection is likely to damage both the battery and the equipment. Positive terminals should be connected to positive connections, and negative terminals should be connected to negative connections. Batteries should also be removed from device after use because many devices do not power off completely on switched off mode causing battery leakage. When battery powered equipment fails to operate and functions badly, remove the batteries immediately.

4.3.2 Getting the Most Out of the Lead-Acid Batteries

Lead-acid batteries are the most commonly used rechargeable batteries for vehicles and theoretically can retain more than five years under proper use and storage. However, there are many ways to shorten a lead-acid battery. Main maintenance is crucial to last the battery as long as possible.

Lead-acid battery tends to suffer from corrosion on connectors and terminals. This is due to the contact with hydrogen gas from the leaking acidic gasses or that has seeped from an overfilled wet cell. The electrolyte level should be regularly checked. Only distilled water should be used when topping up the electrolyte and not overfill it; otherwise, impurities will be added that will drastically shorten the life of the battery. When a white or blue powdery substance is observed to form on the battery terminals, the cell should be disconnected and washed with a mixture of water and baking soda, the battery back resecured in place, and finally the terminals tightened.

The discharge capacity of lead-acid battery declines gradually after 100–200 cycles due to a process known as sulfation. Sulfate is observed to form on the battery's lead plates as it discharges. This electrically insulating material inflates the cells internal resistance and eventually shortens the operational life of batteries. Therefore, it is recommended to recharge this battery when it is within 20–70% capacity. Charging a battery that has below 20% capacity will most likely damage the battery from the additional heat generated during charging time. A fully saturated charge lasting 14–16 hours should be applied for a battery that is being stored for too long (two years).

Lead-acid batteries should be kept in a purpose designed battery box that allows the flammable hydrogen gas to escape, but does not allow seawater to get in. The battery must be kept cool and in dark place away from direct sunlight exposure. The temperature of battery store room must be stable. Monthly equalization for 2–16 hours is recommended because extreme temperature will deteriorate batteries faster. A temperature compensation function should be built into the charge controller or inverter system to ensure that a high charge voltage limit can be delivered.

4.3.3 Getting the Most Out of the Nickel-Based Batteries

Sealed nickel batteries (NiCd) generally have low internal resistance, could provide high surge currents, and only require basic maintenance to operate. However, it is not easy to manage it running at peak performance.

One of the specialties of NiCd batteries is that it can be charged at different rates, ranging from 10 minutes to 14 hours, depending on the cell's construction. However, the internal resistance will increase with the increasing ampere-hour rating. This generates heat, slows down the electron transfer, and will cause higher chance of overcharging (the operational cell life is shortened). Near full capacity, NiCd batteries will suffer additional heat up; therefore a compatible charger with a temperature gauge should be used to automatically shut off the current flow above a certain temperature, thus preventing an overcharge.

Another issue of NiCd batteries is self-discharge, which depends on storing temperature. It is recommended first to run it down to <40% capacity before storing it in a cool place or fully expending it for lasting storage. Before using a new or a long-stored nickel-based battery, it should be charged around 16–24 hours. This will allow the cells to adjust with each other and bring them to an equal charge level.

The number of cycles required to achieve full capacity varies for different brands. Slow charging rate will redistribute the electrolyte to avoid the development of dry spots on the separator. For good quality cells, it will achieve its full capacity after five to seven cycles, while lower-cost cells will reach the acceptable capacity levels after 50 or more cycles. NiCd batteries can last up to five years in storage under proper management and this makes them a favorable choice battery. It is often used for camera flash units, cordless power tools, and remote-controlled electric model vehicles.

4.3.4 Getting the Most Out of the Lithium Ion Batteries

LIBs are so popular in home appliances and become an inseparable part of our lives due to its excellence high energy density and low self-discharge. Under normal circumstances, LIB can usually last up to five years. Like other batteries, proper management is required to keep the buildup of internal resistance with increasing battery discharges.

One good practice is to partially and fully discharge the battery at a regular time slot. This will reduce the building up of internal resistance, keep the battery meter accurate, and maximize the operational lifespan of battery. Do not put the battery in an elevated temperatures environment such as a hot dashboard, near radiator, and long duration of plug-in to power socket. This will lead to cell oxidation, was reported to reduce battery capacity, and might shrink the operating life by one year. Extreme temperatures below 0 °C can also cause the battery to crack, becoming physically damaged.

The low self-discharge makes it last very long for small electronic appliances. However, for long-term storage, it is better to keep it at partial charge rather than fully charge or fully discharge. Recommendation from manufacturers are to drain the fully charge battery to 40% or to charge up the empty battery up to 40% for storage of few months. This level of charge enables low degree of self-discharge and at the same time able to maintain the protection circuit powered. An empty battery that is stored for too long may not able to recharge again.

Every device comes with its respective charger that has been preset with a charge current rate. Therefore, consumer should use the designated charger to deliver the right amount of current rate to the battery. Inconsistent use of charger that charge too much or too little will reduce the lifespan of the battery. Some device has built in a battery protection circuit that will shut off any overload power that could cause voltage spikes and overheating.

Important basic points about the dos and don'ts on battery safety:

Dos

- Regularly servicing your battery.
- Keep track of battery life.
- Inspect the terminal clamps to see if they are tight and free from corrosion.
- Keep battery clean and dry.
- Use battery regularly to maintain the level charge in the battery.
- Disconnect the battery if you plan to keep your car in storage for a long time.
- Use the correct charger for your battery.
- Keep batteries in approved battery cases when not in use.
- Always insulate the battery.

Don'ts

- Cross connections – Explosions!
- Don't disconnect a live battery.
- Don't allow your battery to discharge completely.
- Don't place any metal object on a battery.
- Don't drain your battery.
- Excessive heat and cold are bad for batteries.
- Don't keep your battery in the pocket or contact with metal object.

4.4 Mobile Power Recycling

In modern-day world, many households use batteries regularly, and an estimate of around 345 million batteries are spent annually. These include all kinds of batteries, such as alkaline and lithium batteries as primary batteries and lead acid, NiCd, NiMH, and LIB as secondary batteries. They have become inseparable to our daily life, as they provide energy for a wide range of applications, spanning from household items such as laptop computers, mobile phones, and hearing aids to large industrial enterprise. All dead batteries are considered hazardous waste. It was reported that an average person would own about 10 household batteries per year and only two are recycled presently. Others are disposed to landfill, burned, or stored in electrical and electronic devices that are not functioning.

Batteries consist of many materials. Some heavy metals and corrosive and toxic chemicals are potentially hazardous to human health and the environment. Other materials such as lead, cadmium, manganese, nickel, lithium, zinc, and mercury are important resources of the Earth. These materials can be recycled at indefinite number of times and have a commercial value. However, improper disposal of batteries by throwing into garbage bin has resulted in soil contamination and water pollution, while burning in open space could vaporize heavy metals such as mercury into the air.

Recycling is an alternative to "conventional" waste disposal that reduces environment hazardous and aims to minimize the number of batteries as technological solid waste. It keeps them out of landfill and combustion facilities, prevents pollution, saves energy, and helps sustain the environment for future generations. Sorting process is the main challenge for battery recycling project as batteries are made from many different chemistries. Sorting by sizes, materials, and types such as lithium ion (used in laptop batteries), NiCd (used in power tool batteries), or zinc (used in AA batteries) can recover some of the materials for new production. Thus, recycling bring benefit to the community and environment (Figure 4.9).

4.4.1 Recycling Primary Batteries

The amount of used primary battery or also called as single used battery (Figure 4.10) increases rapidly over the years due to the high drain applications. Most of them are lithium, alkaline, and carbon zinc battery and are usually thrown into dustbin with the regular waste. Some major industrial players in

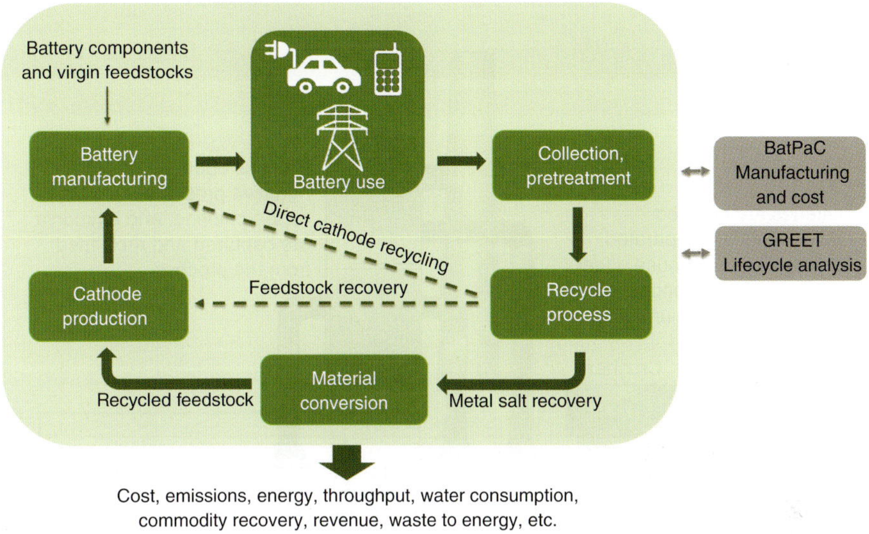

Figure 4.9 Lifecycle of a battery starting from raw material until dispose [22].

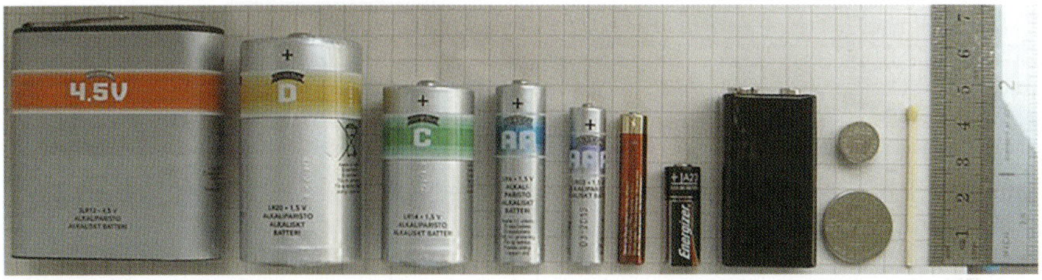

Figure 4.10 Primary battery in various size and voltage (Reproduced from [23] with permission from Yosep Bachtiar).

the world have come together to discuss the effective way to dispose batteries but could not finalize any optimum method to cover all aspects including cost of disposal, effective use of resources, amount of energy consumed, and environmental impact.

Since the early 1990s, the content of mercury in alkaline battery has been eliminated completely to make way for battery disposal with normal household waste. However, California and European countries still consider battery as hazardous waste due to the corrosive chemicals and metals that can disrupt the carefully controlled landfill chemistries. Alkaline batteries (AAA, AA, C, D, 9V, etc.) are recyclable materials that can be separated into two end products (Figure 4.11). The main products are common metal steel (battery casting), zinc (negative electrode), and manganese concentrate (positive electrode) that could be recycled or reused. Second products are paper, plastic, and brass. The steel is cleaned and separated from any impurities and sold into secondary steel markets for new production. Other components would undergo mining and high temperature refining process to extract high purity zinc and manganese oxide.

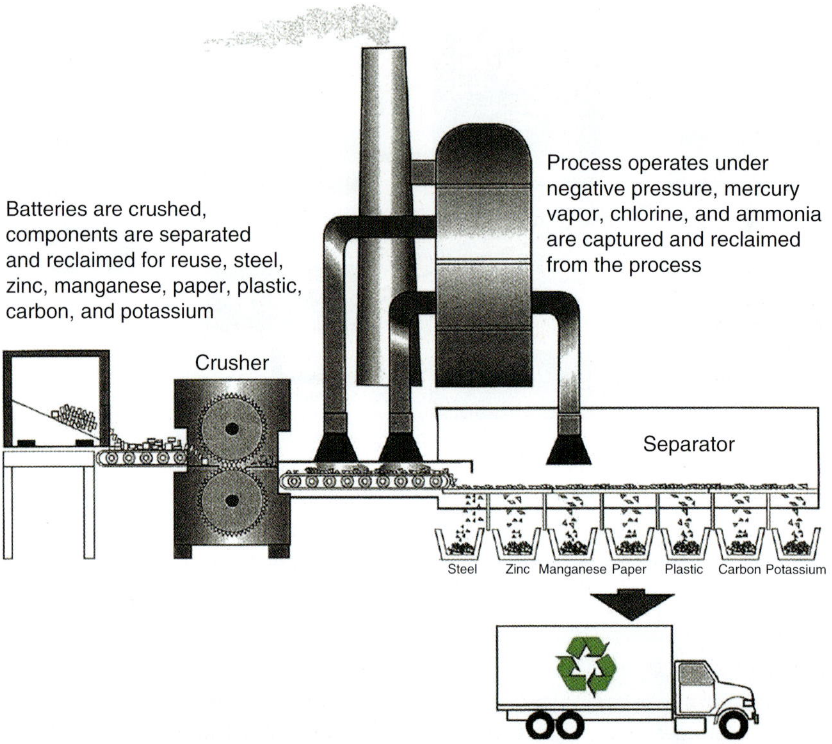

Figure 4.11 Processing of alkaline battery (Reproduced from [24] with permission from Yen-Van Tran/designlifehacks.com).

Unlike alkaline battery, there are two types of zinc air batteries. One contains mercury, while another does not contain mercury, depending on the applications. Zinc air button cells are commonly manufactured with zero mercury content and thus do not require special handling process upon disposal (Figure 4.12). For mercury-containing zinc air batteries, there is an environmental regulation to guide proper disposal or recycling option at an existing waste cost center. The recycling process would extract the zinc and other toxic metals to be reused for secondary market and send the remaining harmless material to landfills.

Among primary batteries, non-rechargeable lithium battery has the most reactive chemicals. It is used in high output applications (e.g. digital cameras), industrial and military sections. This type of battery must be handled carefully to prevent short circuit, where thermal runaway may cause fire or explosion. One important approach is to make the battery materials less reactive. A cryogenic pretreatment is carried out at very low temperature and under a liquid phase to prevent fugitive emissions.

All the recycled materials will be reused in the secondary market. This will indirectly reduce the cost of the expensive recycling process. Global industrial players are working hand in hand with the local government and other public organizations to implement the best possible disposal of used batteries.

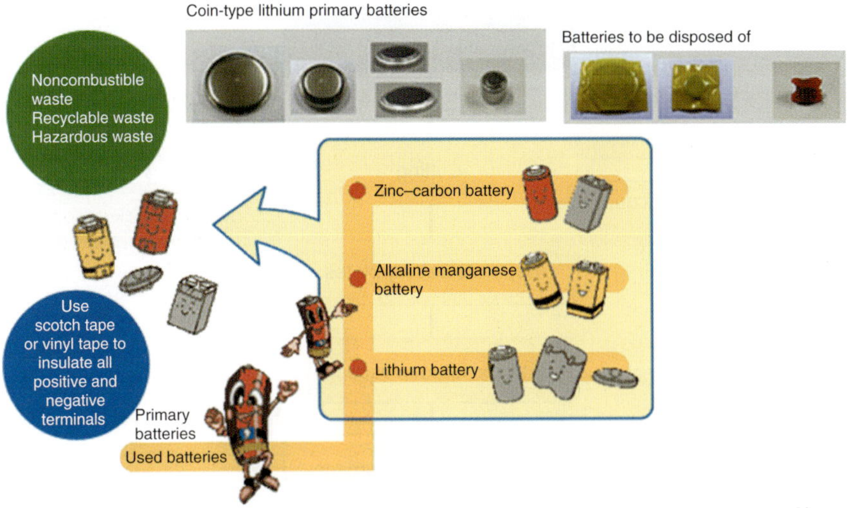

Figure 4.12 Recycling of primary battery.

4.4.2 Recycling Rechargeable Batteries

Rechargeable batteries are also called "accumulator" batteries because they can be recharged by pumping electric current through the cell reversibly. The average secondary battery can be charged about 1 000 times before disposal. Unlike single-use batteries, rechargeable batteries often contain valuable heavy metal elements that work on the reversible chemical reactions. These heavy metals can be hazardous to the people and pose threat to the environment upon irresponsible disposal. Thus, the material should be recycled and reused properly (Figures 4.13 and 4.14).

Lead-acid batteries could cause the largest environmental concerns because no suitable replacement is available (Figure 4.14). There are few types of this battery such as the regular lead-acid, sealed lead-acid, gel type, or absorbed glass mat battery. It is widely used in cars, golf carts, and motorcycles. The recycling effort has been started way back in 1912 as it was too toxic to dispose in waste stream. Car batteries containing lead should be brought only to waste management centers or back to the retailers, where they can eventually be recycled. The recycling process is actually simple by extracting the lead to its elemental form before purification. The process is cost-effective and can be reused multiple times.

Another secondary battery is Ni based. When NiCd batteries are disposed to the landfill, poisonous cadmium will dissolve and seep into the water supply, leading to huge environment concern (Figure 4.13). Fortunately, successful substitution of NiCd by less toxic NiMH lessens the harm to the environment. Upon recycling, nickel and iron could be recovered by rotary hearth and electric furnaces vacuum and used to feed stainless steel production, making it on track for successful recycling.

Several factors had contributed in making LIB recycling process complicated. Reactivity and toxicity of lithium in the battery, making it dangerous when

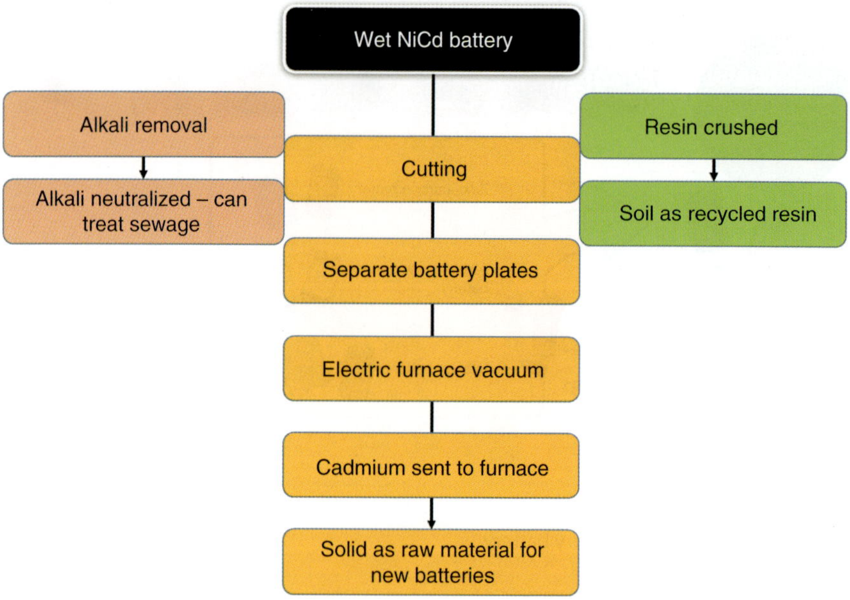

Figure 4.13 Wet NiCd battery recycling process.

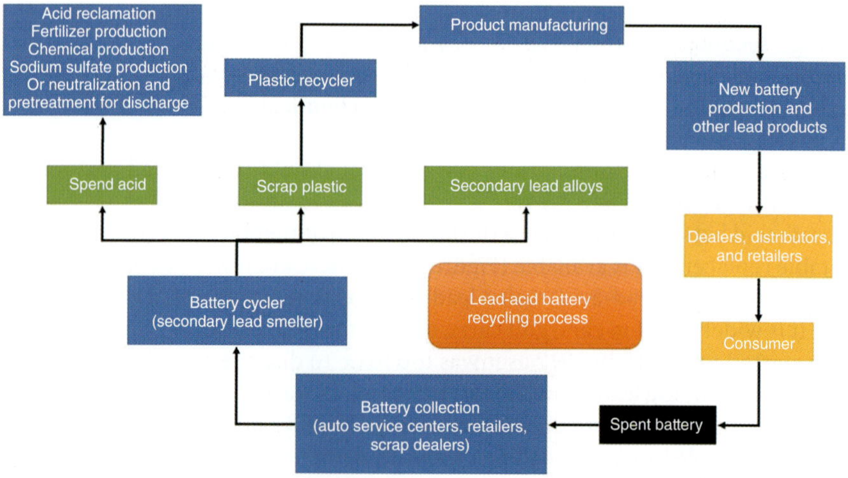

Figure 4.14 Battery recycling of lead-acid battery.

overheated, should be managed properly according to the manufacturer's instructions. On the other hand, battery assembly is also complicated. The powder form active materials are coated onto metal foil before sandwiching, rolling, and sealing processes. These make the recycling procedure harder as different materials need to be separated individually. In addition, there are many chemical compositions of the cathode active materials, $LiCoO_2$. Various doping amounts of Ni, Al, and Mn are used to replace Co by respective manufacturers.

This factor makes it hard to have a standardized recycling method or regulation, resulting in the extraction of lithium from used battery that is five times more costly than mined lithium. With the surge amount of disposal Li-based battery recently, new initiative to commercialize recycling industry is emerging.

Acknowledgments

The authors are thankful for the financial support by the Ministry of Higher Education (MOHE) of Malaysia [FP019-2014B] and University of Malaya [GPF038B-2018].

References

1 Yen Chew, S. (2009). *Advanced Materials for Electrodes and Electrolyte in Rechargeable Lithium Batteries*. University of Wollongong.
2 Morus, I.R. (1998). Galvanic cultures: electricity and life in the early nineteenth century. *Endeavour* 22 (1): 7–11.
3 Zhang, Z. (2015). *Development of Advanced Materials for Rechargeable Lithium-Ion Batteries*. University of Wollongong.
4 Gao, X. (2014). *Development of Novel Materials for Rechargeable Lithium Batteries*. University of Wollongong.
5 Linden, D. (1995). Handbook of batteries. In: *Fuel and Energy Abstracts*, vol. 4, 36, 265. Elsevier Science.
6 Pop, V., Bergveld, H.J., Danilov, D. et al. (2008). State-of-the-art of battery state-of-charge determination. In: *Battery Management Systems: Accurate State-of-Charge Indication for Battery-Powered Applications*, 11–45. Dordrecht: Springer.
7 Goetzberger, A., Knobloch, J., and Voss, B. (1998). *Crystalline Silicon Solar Cells*, vol. 1. Hoboken, NJ: Wiley.
8 David, L. and Thomas, B.R. (2001). *Handbook of Batteries*. McGraw-Hill Professional.
9 Scrosati, B. (2011). History of lithium batteries. *J. Solid State Electrochem.* 15 (7–8): 1623–1630.
10 Yoshio, M., Brodd, R.J., and Kozawa, A. (2009). *Lithium-Ion Batteries*, vol. 1. Springer.
11 Kibria, M. T., Ahammed, A., Sony, S. M., et al. (2014). A review: comparative studies on different generation solar cells technology. *Proceedings of 5th International Conference on Environmental Aspects of Bangladesh*. University of Dhaka, Bangladesh.
12 Pearce, J. and Lau, A. (2002). Net energy analysis for sustainable energy production from silicon based solar cells. In: *ASME Solar 2002: International Solar Energy Conference*, 181–186. American Society of Mechanical Engineers.
13 Krebs, F.C. et al. (2013). Freely available OPV—the fast way to progress. *Energy Technol.* 1 (7): 378–381.

14 Marshall, V. (2014). Spray-deposition steers perovskite solar cells towards commercialisation. https://www.chemistryworld.com/research/spray-deposition-steers-perovskite-solar-cells-towards-commercialisation/7600.article (accessed 29 July 2014).
15 Noerochim, L. (2012). *Advanced Materials for Flexible Lithium Rechargeable Batteries*. University of Wollongong.
16 Sabin, K.C. *Solid-State Electrolyte for Supercapacitors*, vol. 2016. Notodden: Høgskolen i Sørøst-Norge.
17 Andújar, J.M. and Segura, F. (2009). Fuel cells: history and updating. A walk along two centuries. *Renew. Sust. Energy Rev.* 13 (9): 2309–2322.
18 Al-Alawi, B.M. and Bradley, T.H. (2013). Review of hybrid, plug-in hybrid, and electric vehicle market modeling studies. *Renew. Sustain. Energy Rev.* 21: 190–203.
19 Requia, W.J. et al. (2017). Carbon dioxide emissions of plug-in hybrid electric vehicles: a life-cycle analysis in eight Canadian cities. *Renew. Sustain. Energy Rev.* 78: 1390–1396.
20 MacMath, J. (2017). Will America be next to phase out gasoline- and diesel-powered vehicles? https://www.accuweather.com/en/weather-news/will-america-be-next-to-phase-out-gasoline-and-diesel-powered-vehicles/70002278 (accessed 25 July 2017).
21 Asthana, A. and Taylor, M. (2017). Britain to ban sale of all diesel and petrol cars and vans from 2040. https://www.theguardian.com/politics/2017/jul/25/britain-to-ban-sale-of-all-diesel-and-petrol-cars-and-vans-from-2040 (accessed 25 July 2017).
22 Labarotary, A.N. Planning for the future: battery recycling and life-cycle analysis. https://access.anl.gov/projects/recycling (accessed 12 June 2019).
23 unimog-international.com (2017). Battery voltage chart filebatteries comparison 45 D C Aa Aaa Aaaa A23 9v. https://www.unimog-international.com/24-battery-voltage-chart/battery-voltage-chart-filebatteries-comparison-45-d-c-aa-aaa-aaaa-a23-9v (accessed 27 April 2017).
24 Tran, Y.-V. (2016). Understanding recycling batteries. https://designlifehacks.com/2016/10/06/understanding-recycling-batteries-bulbs (accessed 6 October 2016).

5

Insight on Polymer Electrolytes for Electrochemical Devices Applications

Maria Manuela Silva[1], *Verónica de Zea Bermudez*[2], *and Agnieszka Pawlicka*[3]

[1] *Universidade do Minho, Centro de Química e Departamento de Química, Gualtar, 4710-057 Braga, Portugal*
[2] *Universidade de Trás-os-Montes e Alto Douro, Departamento de Química e CQ-VR, Escola de Ciências da Vida e do Ambiente, Apartado 1013, 5001-801 Vila Real, Portugal*
[3] *Universidade de São Paulo, Instituto de Química de São Carlos, Av. Trabalhador Sãocarlense 400, 13566-590 São Carlos-SP, Brazil*

5.1 Introduction

Solid polymer electrolytes (SPEs) are composed of an ionic salt dissociated in a neutral polymer matrix [1, 2]. Since the first report on salt/polymer complexes by Wright [3] and the subsequent recognition by Armand et al. [4] of the potential use of these materials as electrolytes in solid state electrochemical devices [5], the interest in SPEs has not ceased to increase. In fact, electrolytes are indispensable in all electrochemical devices, and their basic function is independent of the much-diversified chemistries and applications.

SPEs offer many advantages, such as good mechanical strength, good electrolyte/active material contact, good flexibility, and processability. The latter advantage is particularly important in the choice of the polymeric matrix, because the electrochemical performance is dependent on the thermal and potential stability windows of the SPE. SPEs are useful for small, portable, and electronic devices. For example, in a battery the electrolyte represents a significant portion of its size and weight [6]. SPEs may also be used for fabrication of solid-state electrochemical devices, such as sensors, displays, and electrochromic or electroluminescent devices [3, 4]. SPEs should also exhibit adequate ionic conductivity for practical purposes; low electronic conductivity; good mechanical properties, chemical, electrochemical, and thermal stability; and ease to handle and manufacture [6]. Moreover, some scientific works have shown that comparative with traditional liquid electrolytes, SPEs present a wider operating temperature range, no leakage problems, good electrode/electrolyte contact, and bigger shelf life [1, 2].

Despite their technological potential, SPEs suffer from a series of drawbacks that have delayed their application in commercial devices. In particular, they display a marked tendency to crystallize, and the ionic guest species tend to salt out at high salt concentration. Consequently, they show ionic conductivities 100–1000 times lower than those exhibited by liquid electrolytes [2, 4, 6].

Polymer Electrolytes: Characterization Techniques and Energy Applications,
First Edition. Edited by Tan Winie, Abdul K. Arof, and Sabu Thomas.

Still, the low ionic conductivity of SPEs, which is typically between 10^{-8} and 10^{-5} S cm^{-1} at room temperature [2, 4], can be counterbalanced by a thin film cell configuration, which increases the respective energetic density [6]. Furthermore, as liquid electrolytes pose significant safety and environmental concerns, considerable efforts have been devoted to developing SPEs with high ionic conductivity.

The ionic conduction of SPEs is due to a presence of polar groups or atoms (heteroatoms) with lone electron pairs (e.g. –O–, –NH–, and –CN–, among others) in polymer chains that may complex the cations of the added salts. The distance between the coordinating groups should maximize the polymer–cation interaction, and the polymer should be able to adopt low energy conformations in order to allow inter- and intramolecular coordination [7, 8]. Therefore, the ideal SPE polymer host should have high molecular weight and be amorphous.

Owing to its extraordinary solvating ability toward salts, poly(ethylene oxide) (PEO) has been the polymer host most widely employed to synthesize SPEs. Most studies carried out in the area of SPEs are focused on complexes based on PEO as host polymer and guest lithium salts. These systems incorporating lithium ions have been the most extensively investigated to date, because of the foreseen applications in advanced primary and secondary batteries and electrochromic devices (ECDs) [9]. These materials fulfil the prerequisites of environmental compatibility, offer no hazard to the consumer, are of low density, and can be produced as thin films. However, their practical application in solid state electrochemical devices has been limited by two major drawbacks: their poor processability and marked tendency to crystallize. The latter aspect is particularly limiting as it usually results in a decrease of the ionic conductivity [10]. The room temperature ionic conductivities of some SPEs systems based on PEO, poly(propylene oxide) (PPO), and related synthetic copolymers doped with lithium salts are summarized in Table 5.1.

Table 5.1 Ionic conductivities of lithium-doped SPEs at room temperature.

Polymer	SPE	Ionic conductivity at 25 °C (S cm^{-1})	References
PEO	PEO_8LiClO_4	~10^{-5}	[101]
	PEO_5LiBF_4	~10^{-5}	[102]
PPO	PPO_8LiTFSI	~10^{-6}	[103]
aPEO	$aPEO_{30}$LiTFSI	~10^{-4}	[104]
	$aPEO_{25}LiClO_4$	~10^{-5}	[105]
	$aPEO_{25}LiCF_3SO_3$	~10^{-5}	[106]
PEO–PPO	PEO-$PPO_{16}LiAsF_6$	~10^{-4}	[107]
P(EO-MEEGE)	P(EO-MEEGE) LiTFSI	~10^{-5}	[108]
	P(EO-MEEGE) $LiClO_4$	~10^{-5}	
PEO–PPO–PMMA	PEO–PPO–$LiCF_3SO_3$	~10^{-5}	[109]

Table 5.2 Ionic conductivity of biopolymers-based SPEs at room temperature.

Polymer	Salt/dopant	Ionic conductivity at 25 °C (S cm^{-1})	References
Starch	$LiClO_4$	10^{-5} to 10^{-4}	[110–112]
Agar	CH_3COOH	$\sim 10^{-4}$	[113, 114]
Pectin	$LiClO_4$	10^{-5} to 10^{-4}	[52, 115]
	KCl	10^{-4} to 10^{-3}	[115]
Carrageenan	H_3PO_4	$\sim 10^{-4}$	[116]
Gellan gum	$LiCF_3SO_3$	$\sim 10^{-4}$	[117]
Chitosan	CH_3COOH	$\sim 10^{-5}$	[53]
	$LiCF_3SO_3$	$\sim 10^{-5}$	[118–120]
	HCl	$\sim 10^{-4}$	[121]
Alginate	$LiClO_4$	$\sim 10^{-4}$	[122]
	CH_3COOH	$\sim 10^{-5}$	
Gelatin	$LiClO_4$	$\sim 10^{-5}$	[67, 123, 124]
DNA	$LiClO_4$	$\sim 10^{-4}$	[125, 126]

To overcome PEO-based electrolytes' bad processability and tendency to crystallize and to account for the rise of environmental regulations, recent attention has been devoted to the use of natural macromolecules as SPE polymer hosts.

The interest in natural polymers is big due to the depletion of oil and natural gas resources. Furthermore, the use of biodegradable biopolymers reduces the environmental impact of the disposal and long-term degradation of synthetic polymers. Studies of the use of natural polymers also take into account the low production cost due to the great variety and low prices of raw materials [11]. Several natural macromolecules-based SPEs have been developed, and some of them applied in small electrochemical devices in the last 20 years. Besides the many advantages listed earlier, these materials are also in most cases almost predominantly amorphous and their ionic conductivity values are of about 10^{-4} to 10^{-5} S cm^{-1} at room temperature. Table 5.2 outlines the room temperature ionic conductivities of SPEs based on natural polymers.

Different salts are used for SPEs formulation. In general, salts with a highly polarizing cation and a large anion, with a well-delocalized charge, and a low lattice energy are the most suitable for use in SPEs [12, 13], but the most commonly employed are lithium salts, such as lithium perchlorate ($LiClO_4$), lithium trifluorosulfonylimide (LiTFSI), lithium tetrafluoroborate ($LiBF_4$), lithium hexafluorophosphate ($LiPF_6$), or lithium hexafluoroarsenate (V) ($LiAsF_6$). On the other hand, some other organic or inorganic molecules are added to increase, not only the ionic conductivity but also the mechanical stability. For example, plasticizers are added to increase the amorphous phase of the polymeric matrix and increase conductivity values and inorganic fillers, such as barium titanate ($BaTiO_3$) [14], alumina (Al_2O_3) [15], silica (SiO_2) [16], or titania (TiO_2) [17], to improve the mechanical stability of these materials.

After all the attention devoted to lithium salts, in the last few years, ionic liquids (ILs) have gained momentum. ILs are room temperature molten salts that possess unique properties, such as almost null volatility at room temperature [18], good thermal stability [19], and nonflammability [20], together with high ionic conductivity and a wide electrochemical stability window. Due to their unique properties, ILs have been investigated as potential electrolytes for application in electrochemical devices, including lithium ion batteries [21], fuel cells [22, 23], and solar batteries [24]. However, their liquid nature limits their use in actual devices due to problems of leakage, non-portability, and the impossibility of miniaturization. The combination of ILs and suitable polymers makes possible to obtain SPEs in the desired shape while retaining the main properties of ILs, i.e. improved safety, stability, and higher operating temperature range. Another advantage of ILs is their design ability, i.e. the interchangeability between thousands of possible cations and anions, which can be used to tune their properties [25]. Common ILs cations include 1-alkyl-3-methyl-imidazolium, *N*-alkyl-pyridinium, *N*-alkyl-*N*-methyl-piperidinium, tetraalkyl-ammonium, tetraalkyl-phosphonium, *N*-alkyl-*N*-pyrrolidinium, 1,2-dialkyl-pyrazolium, *N*-alkyl-thiazolium, and trialkyl-sulfonium [26]. ILs anions can be selected from a broad range of inorganic anions (e.g. halides (Cl^-, Br^-, I^-), polyatomic (PF_6^-, BF_4^-), and polyoxometalates) or from organic anions (e.g. acetate (OAc^-), triflate (OTf^-), isocyanate (SCN^-)). Therefore, the properties of ILs in terms of conductivity, hydrophobicity, melting point, viscosity, and solubility can be tuned by altering the cation or its substitutive groups and/or the anion [27].

Recently, it has been successfully demonstrated that the addition of ILs increases the ionic conductivity of PEO-based SPEs. Shin et al. [28–30] have reported that PEO–LiTFSI electrolytes incorporating the IL *N*-methyl-*N*-propylpyrrolidinium bis(trifluoromethanesulfonyl)imide enhanced the room temperature ionic conductivity to more than $10^{-4}\,S\,cm^{-1}$ for IL contents higher than 50 wt%, together with a wide electrochemical stability and good compatibility toward the lithium anode, even after prolonged storage time. A PEO–LiTFSI electrolyte incorporating the IL 1-butyl-3-methylimidazolium bis(trifluoromethanesulfonyl)imide ([Bmim] [TFSI]) has also been prepared by Ahn and coworkers [31]. The addition of the IL resulted in a very significant increase of the ionic conductivity of the electrolyte, the effect being more pronounced at lower temperature. Ionic conductivity increased with the increase of IL content and reached $3.2 \times 10^{-4}\,S\,cm^{-1}$ at 25 °C for an IL concentration of ~44 wt%. Yongxin et al. [32] doped PEO-LiTFSI electrolytes with *N*-methyl-*N*-propylpiperidinium bis(trifluoromethanesulfonyl)imide to enhance their electrochemical stability. The resulting SPE presented the highest conductivity of the series (~$2.06 \times 10^{-4}\,S\,cm^{-1}$ for an IL content of ~32 wt%), while the conductivity value of PEO-LiTFSI was only $3.96 \times 10^{-6}\,S\,cm^{-1}$ at ~20 °C. Other PEO-IL polymer electrolytes containing ILs 1-butyl-3-methylimidazolium hexafluorophosphate and 1-ethyl-3-methylimidazolium tosylate, lacking lithium salts, were prepared by Ribeiro's and Singh's groups, respectively [33, 34]. The same order of magnitude for the ionic conductivity, i.e. ~$10^{-5}\,S\,cm^{-1}$ was obtained for PEO-IL with >35 wt% IL content at room temperature (25–30 °C). The ionic conductivities were further increased to $\sim 10^{-3}$ S cm^{-1} at temperatures higher than 80 °C. Enhanced ionic conductivity values have also been obtained by introducing the IL tetrahexylammonium iodide into the PEO matrix [35], which yielded an optimum conductivity of 4.6×10^{-4} S cm^{-1} for an IL content of ~13 wt%, at 30 °C.

Following the successful incorporation of ILs into PEO-based SPEs, more attention has been devoted to SPEs based on natural polymers doped with ILs, e.g. gelatin [36], agar [37], DNA [38, 39], starch [40], and kappa-carrageenan [41].

SPEs are obtained by synthetic routes either by the solvent casting or sol–gel methods. The solvent casting method is one of the simplest methods of obtaining membranes. It is easy to perform and does not require sophisticated equipment or vessel. It consists in the dissolution of the polymer and salt in an appropriate solvent to form a homogeneous solution, which is then cast into a Petri dish and dried to remove the solvent. This procedure results in a free-standing membrane that is ready to be analyzed and/or applied in electrochemical devices.

The sol–gel method is a handy synthetic chemical process, which provides a facile and versatile route for the preparation of highly pure and essentially amorphous materials under mild conditions. This method involves the formation of a colloidal suspension (sol), which is subsequently converted into a viscous gel and then to a solid material. Traditionally, the formation of a sol occurs through hydrolysis and condensation of metal alkoxide precursors. The gel can be modified by a variety of dopants to obtain unique properties. The final product (xerogel or aerogel) varies according to the applied drying process [42, 43]. Among the strategies that have been proposed, the combination of the sol–gel method with the hybrid concept and the development of modified electrolytes is considered to be a very attractive approach [27]. There are also proposed different polymeric blends or mixtures.

5.2 Theory: Ionic Conductivity

The total ionic conductivity of an SPE is an important characterization parameter, which has been used as the criterion for possible applications in devices. SPEs intended to be used in diverse electrochemical applications must have adequate ionic conductivity together with negligible electronic conductivity. SPE is considered to be a promising candidate for commercial application if its ionic conductivity is as high as 10^{-5} S cm^{-1} at room temperature [44–46].

In general, the ionic conductivity of SPEs is measured as a function of the salt composition and temperature. The goal of this characterization is to identify the electrolyte with the most favorable behavior for use as a component of a practical device [10].

The total ionic conductivity (σ_{total}) of SPE is given by the sum of the product of the concentration of ionic charge carriers and their mobility (Eq. (5.1)):

$$\sigma_{\text{total}} = \sum_i n_i \mu_i z_i \tag{5.1}$$

where n_i is the number of charge carriers, μ_i is the ionic mobility, and z_i is the ionic charge.

In the last decades, attention has been devoted to the comprehension of the conductivity transport process. Although Gadjourova et al. [47] demonstrated that ionic conductivity in the crystalline domains of the host polymer can be significantly higher than in the corresponding amorphous regions, in practice, nowadays the prevailing procedure follows the view of Berthier et al. [48]. In semicrystalline SPEs ionic transport is restricted to the amorphous phases and is associated with the segmental motion of the polymer chains, at temperatures above the polymer glass transition temperature (T_g).

Armand et al. [4, 49] explored the influence of the temperature in the ionic conductivity of predominantly amorphous SPEs. These authors concluded that the ionic conductivity behavior could be described by the free volume theory and by the Vogel–Fulcher-Tammann (VFT) equation (Eq. (5.2)):

$$\sigma(T) = \sigma_0 e^{\frac{-B}{(T-T_0)}} \tag{5.2}$$

where σ_0 is constant, B is pseudo energy of activation, and T_0 is the temperature at which the configurational entropy of the polymer becomes zero and is close to the T_g [50]. Graphically, the ionic conductivity behavior is described by the VFT equation when a curved correlation between log σ and $1/T$ is observed. However, in several polymeric systems the preferred path for ionic movement is a jump from one to another site of complexation (so-called Grotthuss mechanism [51]). This model of ionic conductivity as a function of temperature is described by the Arrhenius equation (Eq. (5.3)):

$$\sigma(T) = \sigma_0 e^{\frac{-E_a}{T}} \tag{5.3}$$

where σ_0 is constant, E_a is energy of activation, and T is the temperature in Kelvin [1]. Therefore, the energy of activation of the sample can be simply obtained from the slope of log σ versus $1/T$ if this plot is linear. Two examples of linear plots are shown in Figure 5.1 for samples of pectin-based electrolytes plasticized with 37 and 54 wt% of glycerol [52]. Three other samples plotted in Figure 5.1, with 64–70 wt% of glycerol, reveal nonlinear behavior of log σ versus $1/T$, suggesting the VFT model. This means that the quantity of plasticizer influences the movement of charge carriers. This happens because the plasticizer increases the fraction of free volume by better separation of the polymer chains. The free volume theory suggests that, as temperature increases, the polymer chains acquire faster internal modes in which bond rotation produces segmental motion of the polymer chains. Consequently, the free-volume available increases too and favors interchain hopping and intrachain ion movements. Finally, the sample becomes more amorphous and more conductive, as shown in Figure 5.1 [1, 2]. In summary, it was proposed that both Arrhenius (Grotthuss) and VFT (vehicular) mechanisms govern ionic conductivity in SPEs, but which one is predominant in a given system depends on its structure, constituents, and temperature. Results obtained by NMR technique also confirm these findings [53].

The application of the VFT equation implies that the ionic conductivity in polymers is strongly coupled to the flow behavior of the polymer. Lower T_g produces higher polymer flow and greater ionic diffusivity: i.e. high ionic conductivity is

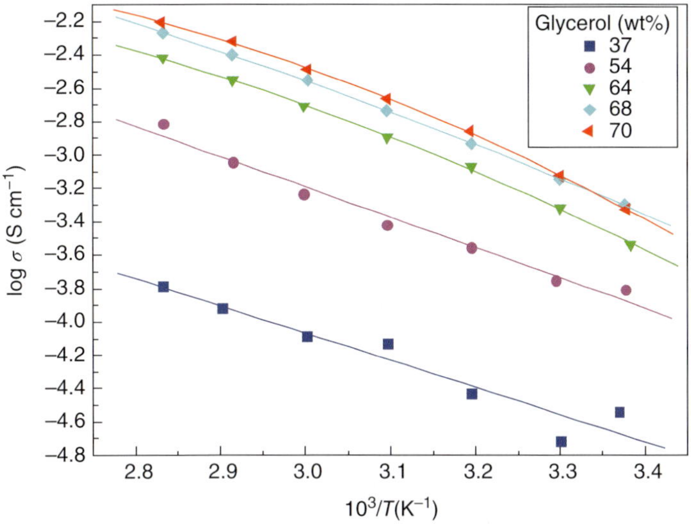

Figure 5.1 Arrhenius plot of pectin-based electrolytes for the samples with different glycerol concentrations. Source: Andrade et al. 2009 [52]. Reproduced with permission of Elsevier.

obtained in polymers that have highly flexible backbones and thus low T_gs. The crystallinity that exists in many polymers serves as a barrier to ionic motion and tends to reduce the conductivity of SPEs. Therefore, in addition to having flexible backbones and low T_gs, polymer hosts for highly conductive SPEs should be substantially amorphous [50].

The SPEs total ionic conductivity is also influenced by the concentration and mobility of charge carriers. When an ionic salt is incorporated into a polymer matrix, the ionic conductivity tends to increase due to the larger number of charged ionic species. However, a further increase of the salt content beyond a certain concentration can result in a decrease of the total ionic conductivity. This observation can be explained by the formation of associated ionic species (e.g. ion pairs with neutral charge (contact ion pairs) or ion aggregates with reduced mobility) and by the increased tendency for ions to form bridging interactions between adjacent polymer chains, a process designated as "ionic crosslinking" [10].

Associated ionic species may be expected to show lower mobility than dissociated ions, and the restriction of host matrix segment mobility also contributes to a reduction in the rate of ion transport and electrolyte conductivity [10]. Typically, "free" or weakly coordinated ions have high mobility, ionic aggregates have low to moderate mobility, and cations bonded strongly to the host polymer have very low mobility [54]. The formation of ionic pairs is also influenced by the dielectric constant (ε) of the medium. The ions aggregation is favored in mediums with low ε [1, 55].

The total ionic conductivity of an SPE can be measured by means of complex impedance spectroscopy. It is noteworthy that the ionic conductivity measurements provide information about the total charge transport, and, consequently, it is quite difficult to establish the exact contribution of the anions and cations. Nevertheless, the ionic conductivity measurements also provide information about the mechanisms of ion motion, which can be due to diffusion (concentration gradient) or migration (electric field) [1].

High ionic conductivity is also fundamental for SPEs and should have a wide electrochemical window: i.e. they exhibit a broad potential range and electrochemical stability (above 1.0 V) [56]. Finally, the thermal stability is also very important and must be adequate according to the envisaged applications [57–60]. In summary, many different polymer electrolytes have been developed worldwide and tested in practical devices as described in the following few examples.

5.3 Applications

The chemical, mechanical, and electrochemical properties of SPEs maximize the possibility of their application in a broad variety of electrochemical devices. SPEs can play three functions: ionic conductor, separator, and adhesive [1], allowing for the construction of devices with innovative architectures.

In order to achieve a successful application, SPE should satisfy another three requirements: performance, durability, and cost. The performance is defined by (i) the ionic conductivity of the material, which should be the highest possible; (ii) the electrochemical stability window, which is especially important for energy storage devices; and (iii) specificity for sensors. The durability of the device will be as long as the capability of the SPE to work under the operating conditions of the device, namely, the temperature, the electrochemical potential, and the presence of contaminants in the medium. Lastly, the cost of the device should be low to enable its commercialization. In general, the causes for higher cost are the use of materials with limited abundance (e.g. Li or Pd), a complex preparation process of the SPE (e.g. unstable intermediates), and a complex assembly of the device (e.g. assembly in inert atmosphere) [61].

5.3.1 Conventional Batteries and Transient Batteries

One of the great challenges of the twenty-first century is undoubtedly the production and storage of energy. The early research on lithium-based batteries can be traced back to 1960–1970 due to the energy crisis and the growing interest in energy sources for mobile applications [62]. The market for lithium batteries is mainly focused in portable electronic devices, in particular notebooks, computers, and mobile phones. Currently, lithium-based batteries are by far the most significant storage system available on the market [63]. Rechargeable lithium batteries are very promising candidates for next-generation power sources because of their high gravimetric and volumetric energies with respect to other cell chemistries [64]. The nonaqueous solvents mostly used in electrolyte solutions belong to organic esters and ethers class [65]. Ethylene carbonate (EC), propylene carbonate (PC), dimethyl carbonate (DMC), diethyl carbonate (DEC), and ethyl methyl carbonate (EMC) are representative examples. Therefore, a new focus is to fabricate SPEs by incorporating salts directly into the polymer matrix [66].

In a battery, the separation between the cathode and anode is obtained through the electrolyte, which allows the passage of lithium ions, without letting lithium atoms or free electrons pass. Current collectors of cathode and anode only allow the passage of electrons. When an external circuit is connected to the battery, an oxidation reaction occurs at battery anode. This reaction will stimulate the electrons, through external circuit, and the ions, through the electrolyte, to flow to the battery cathode. The lithium ions move into the gaps created in the cathode during battery charge. The electrons and ions will recombine again in the battery cathode [67]. The electrolyte should possess low flammability, high diffusion of Li^+ ions, and chemical and electrochemical stability at about 5.0 V [63]. The use of an SPE is advantageous because it can tolerate impacts and vibrations; it has exceptional shelf life; it can operate in a wide temperature range; and it is flexible to the desired design of the battery.

Very recently a new hot topic emerged in the area of batteries. It regards the development of transient batteries for transient electronic devices or disposables (e.g. disposable biomedical devices with no need for removal from the body after operation). Transient implantable medical bionics offer great promise in the field of smart controlled release and tissue regeneration. Integrated energy storage is the ideal power source to drive them. Transient battery is a new type of technology that allows the battery to disappear by an external trigger at any time. The requirements of such batteries are completely opposed to those of classical batteries, since electrochemical features, such as long-term stability, long-term operation, high cycling rate, and long-lifespan, are no longer critical. Instead, the major requirements to be fulfilled by transient batteries are [68]: (i) All the components (i.e. electrodes, electrolytes, current collectors, and packaging) must physically or chemically disappear. (ii) They must be transient at a controllable rate. (iii) They must have high battery performance. (iv) They must have appropriate battery size and mass. (v) They must have flexible design of battery that is compatible with other transient electronics.

Research on degradable batteries is still at its early infancy. Only a few reports of such systems may be found in the literature. These will be mentioned as follows, although in most of them the electrolyte employed in not strictly an SPE, but rather an organic electrolyte embedded in a polymer that acts as a separator.

Yin et al. [69] designed a fully transient primary battery entirely made of biocompatible and biodegradable materials and triggered by biofluids and groundwater. The new silk-based compact Mg battery included (i) Mg as anode; (ii) biodegradable molybdenum (Mo) as cathode; (iii) a composite polymer electrolyte membrane composed of biocompatible IL choline nitrate ([Ch][NO_3]) immobilized in silk fibroin, exhibiting an ionic conductivity of 3.4 mS cm^{-1} and a two-day degradation profile in concentrated buffered protease solution; and (iv) biodegradable polymers (e.g. poly(anhydrides)) as packaging. A single Mg–Mo cell provided a voltage of 0.45 V, and stacked cells gave a stable voltage output around 1.6 V at a constant current density (0.1 mA cm^{-1}). The degradation of the battery began once the battery was made due to the dissoluble nature of the metals. The encapsulated thin film battery offered a capacity of 0.06 mAh cm^{-2} at a current density of 10 μA cm^{-2} and an almost full degradation profile of 45 days when incubated in the protease solution. The battery lifetime upon exposure to buffer solution phosphate-buffered saline (PBS) could be tuned using silk protection layers.

Fu et al. [70] were the first to design rechargeable transient lithium ion batteries incorporating vanadium oxide (V_2O_5) as the cathode, Li foil as the anode, a poly(vinylpyrrolidone) (PVP) membrane made of 600 nm diameter electrospun fibers as separator, a 1 M $LiPF_6$ in EC/DEC (1 : 1 vol%) solution as electrolyte, sodium alginate film as battery encasement, and Al and copper (Cu) as the current collectors. The battery remained stable with the electrolyte and in an ambient environment but exhibited fast transience when triggered by water. It fully disintegrated and vanished through cascade reactions within 10 minutes. The alginate substrate was first dissolved into the water, followed by the PVP separator dissolution, the solution reacted with the Li anode releasing hydrogen gas and generating lithium hydroxide (LiOH). The as-formed LiOH rapidly reacted with V_2O_5 cathode. Eventually all the battery components, including the Cu current collector, disappeared within a few minutes without any visible residue. In addition, the battery had a good electrochemical performance, with a working voltage of ~2.8 V, energy delivered of ~0.29 mW h, and great cycling performance with a high Coulombic efficiency.

Fu et al. [71] introduced a novel lithium ion battery consisting of three components (Figure 5.2): (i) a tri-layer anode including a lithiated aluminum (LiAl) alloy anode to provide high capacity, a PEO coating on the LiAl alloy surface as a reservoir of a liquid electrolyte (1 M $LiPF_6$ in EC/DEC (1 : 1 vol%), an unlithiated Al foil conducting strip as current collector and a poly(propylene) (PP) membrane as separator; (ii) a high-capacity origami perforated V_2O_5 cathode with high mass loading; and (iii) a bilayer battery packaging design based on an electrospun poly(vinyl alcohol) (PVA) nonwoven membrane and

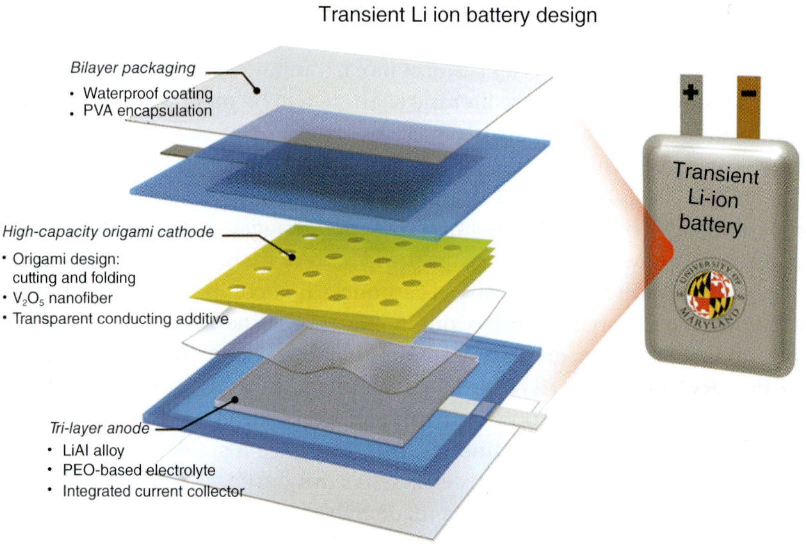

Figure 5.2 Scheme of a transient LiAl-V_2O_5 battery. Source: Fu et al. 2016 [71]. Reproduced with permission of Wiley-VCH.

coated with waterproof polycarbonate (PC). When these components were integrated together, the transient batteries exhibited a high capacity of about 3 mAh cm^{-2} with a high working voltage above 2.0 V. Potassium hydroxide (KOH) was used to decompose this transient battery system. Water, selected as the trigger material, induced the release of KOH into the system. When the as-formed alkali solution met the battery, the latter dissolved rapidly. In this design, all of the components dissolve when submerged in alkali solution.

Jia et al. [72] developed a true biodegradable SPE composed of silk fibroin–choline nitrate. Its efficiency was demonstrated when deployed in a biodegradable thin-film magnesium battery, encapsulated in silk, which offered a specific capacity of 0.06 mAh cm^{-2}. The enzymatic degradation of the entire device occurred over 45 days in the buffered protease XIV solution. A programmed battery lifetime was achieved using silk protection layers.

5.3.2 Fuel Cells

Fuel cells are among the promising technological alternatives to reduce our dependence on fossil fuels and emission of pollutants. They are electrochemical devices able to convert chemical energy directly into electrical energy [73], and they are considered clean because of using hydrocarbon fuel resources [74]. Fuel cells are generally used for portable applications and transportation because they hold several advantages over conventional technologies, such as their high electrical efficiency, silence, no vibrations, and low or null pollutant emissions [74]. Apart from alkaline, phosphoric acid, molten carbonate, solid oxide, reversible, or direct methanol fuel cells (DMFCs), there are polymer electrolyte membrane fuel cells (PEMFCs), also called proton exchange membrane fuel cells, which have been attracting considerable scientific and technological attention in the last decades. However, nowadays, not only PEMFCs but also other ones can incorporate polymer electrolytes in their composition [74]. Nonetheless, these devices seem to be better than other ones for stationary and mobile applications because of their high energy production, power density, and low emissions [75]. For example, DMFCs have higher energy density than PEMFCs, but its oxidation kinetics is slower and there is a decrease of efficiency because of fuel permeation. PEMFC's main drawback consists in the use of expensive Pt-based electrocatalysts at the cathode, which are required for the oxygen reduction reaction. Therefore, many studies have been done to find non-noble substitutes for Pt-catalyst, but this is still a challenge to be overcome [75]. The operation of a PEMFC, shown in Figure 5.3, is based on catalytic separation of hydrogen into protons and electrons at the Pt-based catalyst anode. While the protons pass through the electrolyte membrane to the cathode, the electrons travel through an external circuit, generating an electrical current. At the cathode, with usually Pt- or other noble metal-based catalyst, a combination of electrons, protons, and oxygen atoms occurs, resulting in water synthesis. Oxygen for this reaction can be obtained directly from air, and the only waste product of this process is water, thus an eco-friendly substance and process.

Research on first membranes for fuel cells goes back to 1959 [74]. First, there were phenolic ones, then partially sulfonated polystyrene sulfonic acid

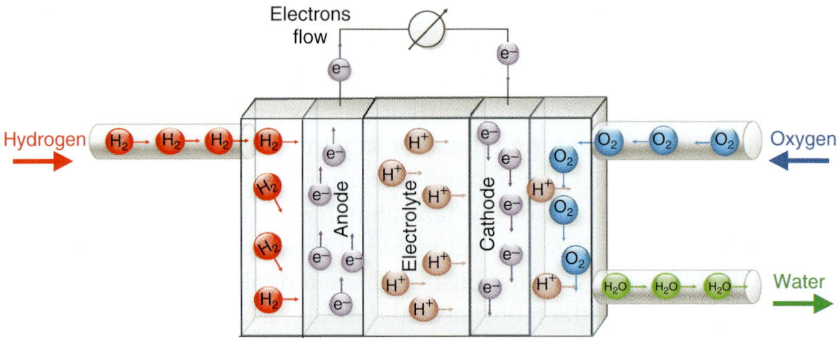

Figure 5.3 Schematic representation of the processes occurring during polymer electrolyte membrane fuel cell (PEMFC) operation.

membranes, followed by cross-linked polystyrene-divinylbenzene sulfonic acid membrane/polymer in an inert matrix, etc. In 1970, DuPont patented Nafion®, a perfluorosulfonic acid membrane, which has been successfully used until now. A comparison of thickness, conductivity, composition, and ion exchange capacity (IEC) of some different commercial membranes can be found in a contribution by Smitha et al. [74]. Besides those listed previously, there also some others, for example, composed of a blend of chitosan (CS) and sulfonated polyvinylidene fluoride (SPVDF) [76]. In the latter contribution, the authors showed that these membranes combine good thermal stability, mechanical properties, dimensional stability, and methanol barrier properties and that the sample CS-SPVDF with 90 : 10 ratio had the highest proton conductivity of $2.85 \times 10^{-2}\ S\,cm^{-1}$ at 90 °C. They also reported on CS/sulfonated polyaniline (PANI)/silica (SiO_2) hybrid membrane, with 3 wt% of PANI/SiO_2 in CS matrix, for fuel cell [77] that had proton conductivity of $8.39 \times 10^{-3}\ S\,cm^{-1}$ at 80 °C. They attributed this high conductivity in fully hydrated state due to its excellent water retention properties.

Several studies have arisen in literature demonstrating the advantages of the use of ILs in fuel cells. An innovative approach to generate SPEs from ILs is the polymerization of the IL monomers. Polymeric ionic liquids (PILs) can be designed to form different systems, such as polymer electrolytes as the membranes to separate the electrodes and the oxygen and hydrogen (or methanol) gas streams [78]. In fuel cells, the SPE acts as a medium for proton exchange between electrodes [79].

5.3.3 Supercapacitors

Supercapacitors are the ideal product for temporary energy storage, for capturing and storing the energy from regenerative braking, and for providing a booster charge in response to sudden power demands [80]. They store energy using either ion adsorption (electrochemical double-layer capacitors) or fast surface redox (pseudocapacitors) [81] reactions.

In double layer capacitors, an electric double layer is formed at the interface between the electrode surface and the SPE. In a pseudocapacitor, most of the

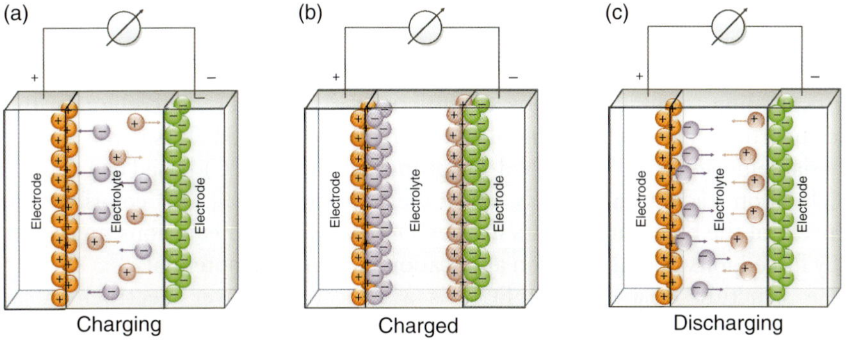

Figure 5.4 Schematic representation of the process occurring during capacitor charging (a), completely charged (b), and discharging (c) processes.

charge is transferred at the interface or by the material near the surface of the electrode [80]. A schematic picture of capacitor operation modes is shown in Figure 5.4.

Supercapacitors as other electrochemical devices need an electrolyte to operate, and the commonly used ones are liquid. However, as already mentioned earlier, these electrolytes are difficult to handle during the device assembly and are susceptible to leakage in case of device damage. Consequently, many researchers are investigating the use of polymer or gel polymer electrolytes (GPEs) for supercapacitor applications. For example, Ortega et al. [82] proposed a novel nanocomposite polymer electrolyte prepared by using poly(vinylidene fluoride) (PVDF), SiO_2, and 1-ethyl-3-methylimidazolium bis(fluorosulfonyl) imide (EMIFSI) IL. This new electrolyte exhibited Arrhenius type ionic conductivity of 10^{-4} S cm^{-1} at room temperature and energy of activation (E_a) of 0.157 eV. Supercapacitors with an area of 1 cm^2 were assembled by sandwiching GPE disc in between two symmetrical carbon nanotube electrodes. The authors observed that the device capacitance was about 72% higher when compared with a capacitor prepared with the GPE without SiO_2 at ambient temperature. Moreover, it was electrochemically stable during 2000 high current density (1.0 A g^{-1}) cycles losing only 9% of its initial capacitance and had an energy of 32.2 Wh kg^{-1} and a power density of 0.9 kW kg^{-1}. Polymer electrolytes containing PVA, ammonium acetate (CH_3COONH_4), and IL 1-butyl-3-methylimidazolium chloride (BmImCl) were synthesized by Liew et al. [83] and used for supercapacitor assembly. The ionic conductivity of transparent SPEs increased from 10^{-4} to 10^{-3} S cm^{-1} with IL content ranging from 20 to 50 wt%. Electrical double layer capacitors (EDLCs) were prepared by sandwiching SPEs between two symmetrical activated carbon electrodes with PVDF binder. The energy density of this cell was of 2.39 Wh kg^{-1} and power density of 19.79 W kg^{-1} with Coulombic efficiency above 90%.

5.3.4 Electrochromic Devices

An electrochromic material shows reversible color changes when a potential difference is applied across it, i.e. a visible and reversible variation of optical properties, namely, light transmittance and/or reflectance, is shown by a material upon its electrochemical oxidation/reduction. Color changes typically range between a transparent (bleached) state and one or two-colored states, corresponding to different redox conditions [84]. However, depending on the electrochromic material, the initial state can also be colored. This is the case of Prussian blue, in an ECD with hydroxypropyl cellulose (HPC)-based electrolyte, which upon reduction process undergoes bleaching [85]. Its coloring, i.e. recovery of the blue state, occurs in open circuit by spontaneous ions migration or by forced ions migration upon application of a positive potential.

The structure of a classical ECD is a typical multilayer electrochemical cell composed of a glass substrate, a transparent conducting oxide that is usually indium-tin oxide (ITO), an electrochromic coating that can be tungsten oxide (WO_3) or niobium oxide (Nb_2O_5), an ion conductor (SPE) that can be liquid, gel, or solid, and an ion storage coating that can be cerium oxide–titania (CeO_2–TiO_2), also referred to as counter electrode [84, 86]. A typical configuration of this kind of device is shown in Figure 5.5.

These devices can be assembled with many different materials and can have many different configurations, but most of them describe transparent/blue ECDs. Nonetheless, the comparison of the described systems is very difficult because of the variety of used materials. These devices are mostly assembled with different SPEs, where those based on natural macromolecules are also included. For example, Alves et al. [87] have proposed ECD with chitosan–thulium triflate ($Tm(CF_3SO_3)_3$) that showed 5% of transmittance change at 633 nm between colored and bleached states. There are also some few reports on ECDs with different colors as, for example, a green-yellow reflective ECD with poly(vinyl butyral) (PVB)-based electrolyte [88] or double ECD [89].

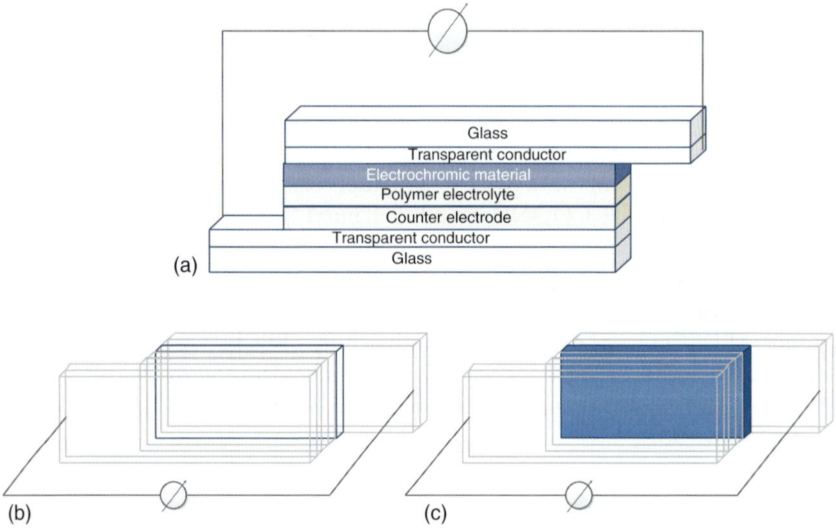

Figure 5.5 Schematic representation of an ECD (a) in its transparent (b) and blue-colored states (c).

The electrochromic effect is induced by electrochemical redox processes experienced by the electrochromic material. Following its oxidation or reduction, the material experiences an insertion of anions or cations, respectively, which flow into it from the ion storage layer through the SPE [61]. The practical applications are common in cars, such as rear-view mirrors or in Boeing 787 Dreamliner both by Gentex [90].

5.3.5 Dye-Sensitized Solar Cells

Usually, dye-sensitized solar cells (DSSCs) consist of a transparent electrode coated with a mesoporous film of nanocrystalline particles of TiO_2 sensitized with a dye, an electrolyte containing a suitable redox couple (usually I^-/I_3^- in acetonitrile), and a Pt-coated counter electrode (Figure 5.6) [91].

The efficiency of a DSSC in the process for energy conversion depends on the relative energy levels and the kinetics of electron transfer processes at the sensitized semiconductor/electrolyte interface. It also depends on the penetration of the polymer network into TiO_2 film [92]. For example, a mixture of poly(acrylonitrile), EC, PC, acetonitrile, and a specific concentration of NaI was used as GPE to assemble DSSC, but the energy conversion efficiency was low compared to liquid electrolytes [93]. Nogueira et al. [92] described some other polymer electrolytes used for DSSC. Among them, they cited very promising polymer electrolyte (PE), produced by Daiso Co.,

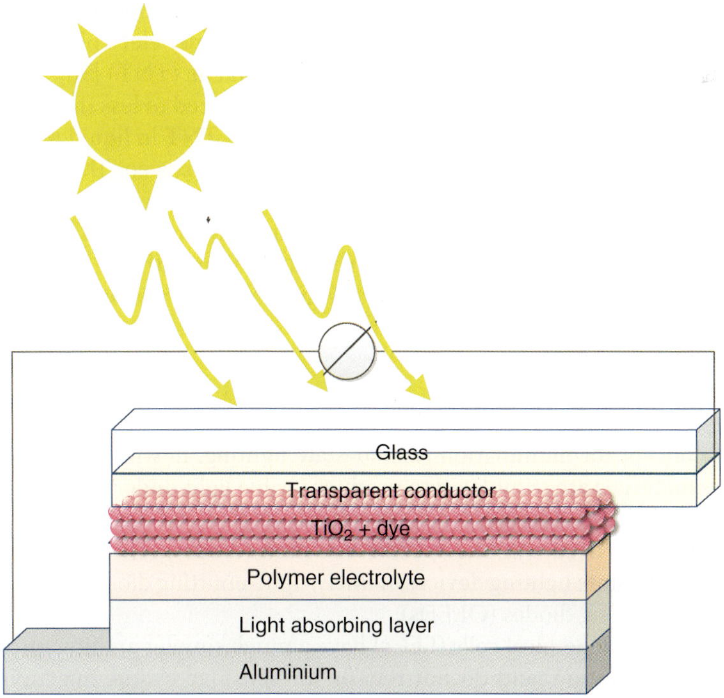

Figure 5.6 Schematic representation of a DSSC.

Ltd. (Osaka, Japan), which was based on the copolymer poly(ethylene oxide-*co*-epichlorohydrin) (P(EO–EPI)84 : 16) complexed with sodium or lithium iodide salts. DSSC of 1 cm^2 active area assembled with this PE and using Ru^{II}(2,2′-bipyridyl-4,4′-dicarboxylate)$_2$-$(NCS)_2$ as dye revealed the energy conversion efficiency of $\eta = 2.6\%$ under light intensity of 10 mW cm^{-2} (1.6% under 100 mW cm^{-2}). The use of PEO and PEO copolymers, coordinated with $LiClO_4$, in electrolytes for DSSC, with some more recent data can be also found in Freitas et al. [91] contribution.

5.3.6 Sensors

An electrochemical sensor is a device that detects and responds to an external stimulus, e.g. temperature, chemical, or ionic gradients. In the assembly of a sensor, the SPE acts as ionic conductor, ensuring a good contact between the active layer (membrane or electrode) and the collector electrode [94]. Electrochemical sensors are mainly developed for oxygen, hydrogen, and other gases, including toxic ones, detection and are made of noble metals or semiconducting materials as electrodes immersed in a liquid, gel, or solid electrolyte [95–97]. These devices have also a diffusion barrier that allows gas to permeate into the device and adsorb on a sensing electrode. Some schematic structures of sensors are presented in Ref. [97]. The gas detection occurs because of redox reactions that occur on sensing electrode. This results in a shift of potential and indicates a presence of a gas. More recently, sensors also have a reference electrode, which is used to disregard side reactions that may occur at counter electrode. There are also developed solid-state electrochemical sensors with polymer electrolytes for explosive compounds as, for example, for 2,4,6-trinitrotoluene (TNT) [98]. This device containing IL in its GPE formulation can be miniaturized to less than half square centimeter and disposable. The average detection of TNT in liquid phase was of 0.37 μg ml^{-1}, very fast, and unaffected by presence of oxygen. Therefore, this very small device is very promising for TNT detection in ground waters. Another example of recently developed sensors are capacitive-type ones for relative humidity (RH) detection [99]. This sensor used PVA and lithium chloride (LiCl) thick films as electrolyte and showed very good sensing properties over 33–94% RH range.

5.3.7 Light-Emitting Electrochemical Cells

The emerging concept for illumination is solid-state lighting, in which selected semiconductor materials are stimulated to produce visible light under the action of an electrical field (electroluminescence) in suitably engineered devices where the transport of charge occurs in one specific direction (diodes). There are two main families of solid-state lighting devices, namely, light-emitting diodes (LEDs) and organic light-emitting diodes (OLEDs).

Light-emitting electrochemical cells (LECs) have a much simpler architecture, are processed from solution, and do not rely on air-sensitive charge injection layers or metals for electron injection. LECs are solid-state electroluminescent devices that convert electric current into light and are especially attractive for use in display technology [100].

References

1 Gray, F.M. (1991). *Solid Polymer Electrolytes: Fundamentals and Technological Applications*. New York: VCH Publishers Inc.
2 MacCallum, J.R. and Vincent, C.A. (1989). *Polymer Electrolyte Reviews 1, 2*. London: Elsevier.
3 Wright, P.V. (1975). Electrical conductivity in complexes of poly(ethylene oxide). *Br. Polym. J.* 7: 319.
4 Armand, M., Duclot, M.T., and Chabagno, J.M. Polymeric solid electrolytes. In: *Second International Meeting on Solid State Electrolytes*, vol. 2309, 6. Scotland: St. Andrews.
5 Gray, F.M. (1997). *Polymer Electrolytes*. Great Britain: Royal Society of Chemistry.
6 Sequeira, C.A.C. and Santos, D.M.F. (2010). *Polymer Electrolytes: Properties and Applications*. Cambridge: Woodhead Publishing Limited.
7 Gray, F. and Armand, M. (2000). Polymer electrolytes. In: *Energy Storage Systems in Electronics* (eds. T. Osaka and M. Datta). New York: Taylor & Francis.
8 Vincent, C.A. (1990). *Electrochemical Science and Technology of Polymers* (ed. R.G. Linford). London: Elsevier Applied Science.
9 Tarascon, J.M. and Armand, M. (2001). Issues and challenges facing rechargeable lithium batteries. *Nature* 414: 359.
10 de Zea Bermudez, V. and Silva, M.M. (2010). Lithium-doped hybrid polymer electrolytes. In: *Polymer Electrolytes: Fundamentals and Applications* (eds. C. Sequeira and D. Santos). Oxford, Cambridge, Philadelphia, PA, New Delhi: Woodhead Publishing Limited.
11 Pawlicka, A. and Donoso, J.P. (2010). Polymer electrolytes based on natural polymers. In: *Polymer Electrolytes: Fundamentals and Applications* (eds. C. Sequeira and D. Santos), 95. Oxford, Cambridge, Philadelphia, PA, New Delhi: Woodhead Publishing Limited.
12 Vincent, C.A. (1987). Polymer electrolytes. *Prog. Solid State Chem.* 17: 145.
13 Watanabe, M., Rikukawa, M., Sanui, K., and Ogata, N. (1986). Effects of polymer structure and incorporated salt species on ionic conductivity of polymer complexes formed by aliphatic polyester and alkali metal thiocyanate. *Macromolecules* 19: 188.
14 Itoh, T., Ichikawa, Y., Uno, T. et al. (2003). Composite polymer electrolytes based on poly(ethylene oxide), hyperbranched polymer, $BaTiO_3$ and $LiN(CF_3SO_2)_2$. *Solid State Ionics* 156: 393.
15 Muhammad, F.H., Subban, R.H.Y., Majid, S.R. et al. (2009). Characterisation of Al_2O_3 doped hexanoyl chitosan-$LiCF_3SO_3$-EC polymer electrolytes. *Mater. Res. Innovations* 13: 249.

16 Pehlivan, I.B., Granqvist, C.G., Marsal, R. et al. (2012). [PEI-SiO_2]:[LiTFSI] nanocomposite polymer electrolytes: Ion conduction and optical properties. *Sol. Energy Mater. Sol. Cells* 98: 465.

17 Buraidah, M.H., Teo, L.P., Yusuf, S.N.F. et al. (2011). TiO_2/chitosan-$NH_4I(+I_2)$-BMII-based dye-sensitized solar cells with anthocyanin dyes extracted from black rice and red cabbage. *Int. J. Photoenergy* 2011: 2707.

18 Earle, M.J., Esperança, J.M., Gilea, M.A. et al. (2006). The distillation and volatility of ionic liquids. *Nature* 439: 831.

19 Baranyai, K.J., Deacon, G.B., MacFarlane, D.R. et al. (2004). Thermal degradation of ionic liquids at elevated temperatures. *Aust. J. Chem.* 57: 145.

20 Smiglak, M., Reichert, W.M., Holbrey, J.D. et al. (2006). Combustible ionic liquids by design: is laboratory safety another ionic liquid myth? *Chem. Commun.* 5828: 2554.

21 Park, M., Zhang, X., Chung, M. et al. (2010). A review of conduction phenomena in Li-ion batteries. *J. Power Sources* 195: 7904.

22 De Souza, R.F., Padilha, J.C., Gonçalves, R.S., and Dupont, J. (2003). Room temperature dialkylimidazolium ionic liquid-based fuel cells. *Electrochem. Commun.* 5: 728.

23 Sun, J., MacFarlane, D., and Forsyth, M. (2003). Novel alkaline polymer electrolytes based on tetramethyl ammonium hydroxide. *Electrochim. Acta* 48: 1971.

24 Wang, P., Zakeeruddin, S.M., Exnar, I., and Grätzel, M. (2002). High efficiency dye-sensitized nanocrystalline solar cells based on ionic liquid polymer gel electrolyte. *Chem. Commun.* 5832: 2972.

25 Holbrey, J. and Seddon, K. (1999). Ionic liquids. *Clean Technol. Environ. Policy* 1: 223.

26 Plechkova, N.V. and Seddon, K.R. (2008). Applications of ionic liquids in the chemical industry. *Chem. Soc. Rev.* 37: 123.

27 Ye, Y.-S., Rick, J., and Hwang, B.-J. (2013). Ionic liquid polymer electrolytes. *J. Mater. Chem. A* 1: 2719.

28 Shin, J.-H., Henderson, W.A., Scaccia, S. et al. (2006). Solid-state Li/$LiFePO_4$ polymer electrolyte batteries incorporating an ionic liquid cycled at 40 °C. *J. Power Sources* 156: 560.

29 Shin, J.-H., Henderson, W.A., and Passerini, S. (2005). PEO-based polymer electrolytes with ionic liquids and their use in lithium metal-polymer electrolyte batteries. *J. Electrochem. Soc.* 152: A978.

30 Shin, J.-H., Henderson, W.A., and Passerini, S. (2003). Ionic liquids to the rescue? Overcoming the ionic conductivity limitations of polymer electrolytes. *Electrochem. Commun.* 5: 1016.

31 Choi, J.-W., Cheruvally, G., Kim, Y.-H. et al. (2007). Poly(ethylene oxide)-based polymer electrolyte incorporating room-temperature ionic liquid for lithium batteries. *Solid State Ionics* 178: 1235.

32 Yongxin, A., Xinqun, C., Pengjian, Z. et al. (2012). Improved properties of polymer electrolyte by ionic liquid $PP_{1.3}$TFSI for secondary lithium ion battery. *J. Solid State Electrochem.* 16: 383.

33 Costa, L.T., Lavall, R.L., Borges, R.S. et al. (2007). Polymer electrolytes based on poly(ethylene glycol) dimethyl ether and the ionic liquid 1-butyl-3-methylimidazolium hexafluorophosphate: preparation, physico-chemical characterization, and theoretical study. *Electrochim. Acta* 53: 1568.

34 Chaurasia, S.K., Singh, R.K., and Chandra, S. (2011). Structural and transport studies on polymeric membranes of PEO containing ionic liquid, EMIM-TY: evidence of complexation. *Solid State Ionics* 183: 32.

35 Bandara, T.W.J., Ekanayake, P., Dissanayake, M.L. et al. (2010). A polymer electrolyte containing ionic liquid for possible applications in photoelectrochemical solar cells. *J. Solid State Electrochem.* 14: 1221.

36 Leones, R., Sentanin, F., Rodrigues, L.C. et al. (2012). Novel polymer electrolytes based on gelatin and ionic liquids. *Opt. Mater.* 36: 187.

37 Leones, R., Sentanin, F., Rodrigues, L.C. et al. (2012). Investigation of polymer electrolytes based on agar and ionic liquids. *Express Polym. Lett.* 6: 1007.

38 Leones, R., Rodrigues, L.C., Pawlicka, A. et al. (2012). Characterization of flexible DNA films. *Electrochem. Commun.* 22: 189.

39 Leones, R., Esperança, J.M.S.S., Pawlicka, A. et al. (2015). Polymer electrolyte based on DNA and N,N,N-trimethyl-N-(2-hydroxyethyl)ammonium bis(trifluoromethylsulfonyl)imide. *J. Electroanal. Chem.* 748: 70.

40 Ramesh, S., Liew, C.-W., and Arof, A.K. (2011). Ion conducting corn starch biopolymer electrolytes doped with ionic liquid 1-butyl-3-methylimidazolium hexafluorophosphate. *J. Non-Cryst. Solids* 357: 3654.

41 Nunes, S.C., Pereira, R.F.P., Sousa, N. et al. (2017). Eco-friendly red seaweed-derived electrolytes for electrochemical devices. *Adv. Sustain. Syst.* 1: 1700070.

42 Wright, J.D. and Sommerdijk, N.A.J.M. (2000). *Sol-Gel Materials: Chemistry and Applications*. London, New York: Taylor & Francis Ltd.

43 Brinker, C.J. and Sherer, G.W. (1990). *Sol-Gel Science: The Physics and Chemistry of Sol-Gel Processing*. San Diego, CA: Academic Press Inc.

44 Catlow, C., Chadwick, A., Greaves, G. et al. (1983). An EXAFS study of the structure of rubidium polyethyleneoxide salt complexes. *Solid State Ionics* 9: 1107.

45 Bruce, P.G., Krok, F., Evans, J., and Vincent, C.A. (1988). Preliminary results on a new polymer electrolyte, poly(ethylene oxide)-$Hg(ClO_4)_2$. *Br. Polym. J.* 20: 193.

46 Armstrong, R. and Clarke, M. (1984). Lithium ion conducting polymeric electrolytes based on poly(ethylene adipate). *Electrochim. Acta* 29: 1443.

47 Gadjourova, Z., Andreev, Y.G., Tunstall, D.P., and Bruce, P.G. (2001). Ionic conductivity in crystalline polymer electrolytes. *Nature* 412: 520.

48 Berthier, C., Gorecki, W., Minier, M. et al. (1983). Microscopic investigation of ionic-conductivity in alkali-metal salts poly(ethylene oxide) adducts. *Solid State Ionics* 11: 91.

49 Armand, M.B., Chabagno, J.M., and Duclot, M.J. (1979). *Fast Ion Transport in Solids: Electrodes and Electrolytes* (eds. P. Vashishta, J.N. Mundy and G.K. Shenoy), 131. Amsterdam: North-Holland.

50 Abraham, K.M. (1993). Highly conductive polymer electrolytes. In: *Applications of Electroactive Polymers* (ed. B. Scrosati), 75. Dordrecht: Springer Netherlands.

51 Kreuer, K.D. (1997). Fast proton conductivity: a phenomenon between the solid and the liquid state? *Solid State Ionics* 94: 55.

52 Andrade, J.R., Raphael, E., and Pawlicka, A. (2009). Plasticized pectin-based gel electrolytes. *Electrochim. Acta* 54: 6479.

53 Pawlicka, A., Mattos, R.I., Tambelli, C.E. et al. (2013). Magnetic resonance study of chitosan bio-membranes with proton conductivity properties. *J. Membr. Sci.* 429: 190.

54 Gonçalves, M., Fernandes, I., Hümmer, J., and de Zea Bermudez, V. (2011). FT-IR and FT-Raman spectroscopy study of di-urethanesil hybrids doped with $Mg(CF_3SO_3)_2$. *Vib. Spectrosc.* 57: 187.

55 Chen-Yang, Y., Hwang, J., and Chang, F. (1997). Polyphosphazene electrolytes. 1. Preparation and conductivities of new polymer electrolytes based on poly[bis (amino) phosphazene] and lithium perchlorate. *Macromolecules* 30: 3825.

56 Lu, W., Fadeev, A.G., Qi, B. et al. (2002). Use of ionic liquids for π-conjugated polymer electrochemical devices. *Science* 297: 983.

57 Costa, L., Gad, A.M., Camino, G. et al. (1992). Thermal and thermooxidative degradation of poly(ethylene oxide)-metal salt complexes. *Macromolecules* 25: 5512.

58 Cameron, G.G., Ingram, M.D., Qureshi, M.Y. et al. (1989). The thermal degradation of poly(ethylene oxide) and its complex with NaCNS. *Eur. Polym. J.* 25: 779.

59 Shodai, T., Owens, B.B., Ohtsuka, H., and Yamaki, J.i. (1994). Thermal stability of the polymer electrolyte (PEO) $_8LiCF_3SO_3$. *J. Electrochem. Soc.* 141: 2978.

60 Costa, L., Camino, G., Luda, M. et al. (1995). The thermal degradation of poly(propylene oxide) and its complexes with LiBr and LiI. *Polym. Degrad. Stab.* 48: 325.

61 Fu, X. (2010). Polymer electrolytes for electrochromic devices. In: *Polymer Electrolytes: Fundamentals and Applications* (eds. C. Sequeira and D. Santos), 471. Oxford, Cambridge, Philadelphia, PA, New Delhi: Woodhead Publishing Limited.

62 Buchmann, I. (1997). *Batteries in a Portable World: A Handbook on Rechargeable Batteries for Non-Engineers.* Ec & M Books.

63 Bini, M., Capsoni, D., Ferrari, S. et al. (2015). 1 - Rechargeable lithium batteries: key scientific and technological challenges A2 - Franco, Alejandro A. In: *Rechargeable Lithium Batteries* (ed. A. Franco), 1. Woodhead Publishing.

64 Montanino, M., Passerini, S., and Appetecchi, G.B. (2015). 4 - Electrolytes for rechargeable lithium batteries A2 - Franco, Alejandro A. In: *Rechargeable Lithium Batteries*, 73. Woodhead Publishing.

65 Janz, G. and Tomkins, R. (1972). *Nonaqueous Electrolytes Handbook*. New York: Academic Press, Vol. 1, Google Scholar 5982 (1976).

66 Arora, P. and Zhang, Z. (2004). Battery separators. *Chem. Rev.* 104: 4419.

67 Silva, M.M., Barbosa, P.C., Rodrigues, L.C. et al. (2010). Gelatin in electrochromic devices. *Opt. Mater.* 32: 719.

68 Fu, K.K., Wang, Z., Dai, J. et al. (2016). Transient electronics: materials and devices. *Chem. Mater.* 28: 3527.

69 Yin, L., Huang, X., Xu, H. et al. (2014). Materials, designs, and operational characteristics for fully biodegradable primary batteries. *Adv. Mater.* 26: 3879.

70 Fu, K., Liu, Z., Yao, Y. et al. (2015). Transient rechargeable batteries triggered by cascade reactions. *Nano Lett.* 15: 4664.

71 Fu, K.K., Wang, Z., Yan, C. et al. (2016). All-component transient lithium-ion batteries. *Adv. Energy Mater.* 6 (10): 1502496.

72 Jia, X., Wang, C., Ranganathan, V. et al. (2017). A biodegradable thin-film magnesium primary battery using silk fibroin–ionic liquid polymer electrolyte. *ACS Energy Lett.* 2: 831.

73 Wang, Y., Chen, K.S., Mishler, J. et al. (2011). A review of polymer electrolyte membrane fuel cells: technology, applications, and needs on fundamental research. *Appl. Energy* 88: 981.

74 Smitha, B., Sridhar, S., and Khan, A. (2005). Solid polymer electrolyte membranes for fuel cell applications—a review. *J. Membr. Sci.* 259: 10.

75 Kiani, M., Zhang, J., Luo, Y. et al. (2018). Recent developments in electrocatalysts and future prospects for oxygen reduction reaction in polymer electrolyte membrane fuel cells. *J. Energy Chem.* 27: 1124–1139.

76 Vijayalekshmi, V. and Khastgir, D. (2018). Chitosan/partially sulfonated poly(vinylidene fluoride) blends as polymer electrolyte membranes for direct methanol fuel cell applications. *Cellulose* 25: 661.

77 Vijayakumar, V. and Khastgir, D. (2018). Hybrid composite membranes of chitosan/sulfonated polyaniline/silica as polymer electrolyte membrane for fuel cells. *Carbohydr. Polym.* 179: 152.

78 Itoh, T. (2010). Hyperbranched polymer electrolytes for high temperature fuel cells. In: *Polymer Electrolytes: Fundamentals and Applications* (eds. C. Sequeira and D. Santos), 524. Oxford, Cambridge, Philadelphia, PA, New Delhi: Woodhead Publishing Limited.

79 Latham, R., Linford, R., and Schlindwein, W. (2003). Pharmaceutical and medical applications of polymer electrolytes. *Ionics* 9: 41.

80 Samui, A.B. and Sivagnanam, P. (2010). Solid polymer electrolytes for supercapacitors. In: *Polymer Electrolytes: Fundamentals and Applications* (eds. C. Sequeira and D. Santos), 431. Oxford, Cambridge, Philadelphia, PA, New Delhi: Woodhead Publishing Limited.

81 Simon, P. and Gogotsi, Y. (2008). Materials for electrochemical capacitors. *Nat. Mater.* 7: 845.

82 Ortega, P.F., Trigueiro, J.P.C., Silva, G.G., and Lavall, R.L. (2016). Improving supercapacitor capacitance by using a novel gel nanocomposite polymer electrolyte based on nanostructured SiO_2, PVDF and imidazolium ionic liquid. *Electrochim. Acta* 188: 809.

83 Liew, C.-W., Ramesh, S., and Arof, A.K. (2014). Good prospect of ionic liquid based-poly(vinyl alcohol) polymer electrolytes for supercapacitors with

excellent electrical, electrochemical and thermal properties. *Int. J. Hydrogen Energy* 39: 2953.
84 Monk, P.M.S., Mortimer, R.J., and Rosseinsky, D.R. (2007). *Electrochromism and Electrochromic Devices*. New York: Cambridge University Press.
85 Assis, L.M.N., Sabadini, R.C., Santos, L.P. et al. (2015). Electrochromic device with Prussian blue and HPC-based electrolyte. *Electrochim. Acta* 182: 878.
86 Fernandes, M., Leones, R., Costa, A.M.S. et al. (2015). Electrochromic devices incorporating biohybrid electrolytes doped with a lithium salt, an ionic liquid or a mixture of both. *Electrochim. Acta* 161: 226.
87 Alves, R., Sentanin, F., Sabadini, R. et al. (2017). Innovative electrolytes based on chitosan and thulium for solid state applications: synthesis, structural, and thermal characterization. *J. Electroanal. Chem.* 788: 156.
88 Assis, L.M.N., Ponez, L., Januszko, A. et al. (2013). A green-yellow reflective electrochromic device. *Electrochim. Acta* 111: 299.
89 Reynolds, J.R., Dyer, A.L., and Unur, E. (2010), Dual active film electrochromic displays device, Patent Application US2010265562-A1.
90 Aerospace overview. http://www.gentex.com/aerospace/aerospace-overview (accessed 11 June 2013).
91 Freitas, J.N., Benedetti, J.E., Freitas, F.S. et al. (2010). Polymer electrolytes for dye-sensitized solar cells. In: *Polymer Electrolytes: Fundamentals and Applications* (eds. C. Sequeira and D. Santos), 381. Oxford, Cambridge, Philadelphia, PA, New Delhi: Woodhead Publishing Limited.
92 Nogueira, A.F., Longo, C., and De Paoli, M.A. (2004). Polymers in dye sensitized solar cells: overview and perspectives. *Coord. Chem. Rev.* 248: 1455.
93 Cao, F., Oskam, G., and Searson, P.C. (1995). A solid state, dye sensitized photoelectrochemical cell. *J. Phys. Chem.* 99: 17071.
94 Munichandraiah, N., Scanlon, L., Marsh, R. et al. (1995). Influence of zeolite on electrochemical and physicochemical properties of polyethylene oxide solid electrolyte. *J. Appl. Electrochem.* 25: 857.
95 Weppner, W. (1987). Solid-state electrochemical gas sensors. *Sens. Actuators* 12: 107.
96 Rheaume, J.M. (2010). Solid state electrochemical sensors for nitrogen oxide (NOx) detection in lean exhaust gases. In: *Engineering – Mechanical Engineering*, 164. Berkeley: University of California.
97 Kim, S., Jung, B., Park, C.O., and Rapp, R.A. (2017). New solid-state electrochemical method of measuring dissolved hydrogen in Al melt. *Sens. Actuators, B* 239: 374.
98 Yu, H.A., Lee, J., Lewis, S.W., and Silvester, D.S. (2017). Detection of 2,4,6-trinitrotoluene using a miniaturized, disposable electrochemical sensor with an ionic liquid gel-polymer electrolyte film. *Anal. Chem.* 89: 4729.
99 Imawan, C., Bai, T., and Budiawanti, S. (2016). A capacitive-type humidity sensor using polymer electrolytes of PVA-LiCl thick films. *2016 International Seminar on Sensors, Instrumentation, Measurement and Metrology (ISSIMM)*, Vol. 5906 (10–11 August 2016), p. 48.

100 Meier, S.B., Tordera, D., Pertegas, A. et al. (2014). Light-emitting electrochemical cells: recent progress and future prospects. *Mater. Today* 17: 217.

101 Gorecki, W., Andreani, R., Berthier, C. et al. (1986). NMR, DSC, and conductivity study of a poly(ethylene oxide) complex electrolyte: PEO ($LiClO_4$) x. *Solid State Ionics* 18: 295.

102 Chiodelli, G., Ferloni, P., Magistris, A., and Sanesi, M. (1988). Ionic conduction and thermal properties of poly(ethylene oxide)-lithium tetrafluoroborate films. *Solid State Ionics* 28: 1009.

103 Brown, S., Greenbaum, S., McLin, M. et al. (1994). Complex impedance, DSC and lithium-7 NMR studies of poly(propylene oxide) complexed with LiN $(SO_2CF_3)_2$ and with $LiAsF_6$. *Solid State Ionics* 67: 257.

104 Ismail, I. and Owen, J. (1996). Microcell conductivity measurements of polymer electrolytes. *Solid State Ionics* 85: 163.

105 Nagae, S.-I., Nekoomanesh, H.M., Booth, C., and Owen, J.R. (1992). The effect of salt concentration on the properties of poly[oxymethylene-oligo (oxyethylene)]/$LiClO_4$ polymer electrolytes. *Solid State Ionics* 53: 1118.

106 Nicholas, C.V., Wilson, D.J., Booth, C., and Giles, J.R. (1988). Improved synthesis of oxymethylene-linked poly(oxyethylene). *Br. Polym. J.* 20: 289.

107 Andrieu, X., Fauvarque, J., Goux, A. et al. (1995). Solid polymer electrolytes based on statistical poly(ethylene oxide-propylene oxide) copolymers. *Electrochim. Acta* 40: 2295.

108 Nishimoto, A., Watanabe, M., Ikeda, Y., and Kohjiya, S. (1998). High ionic conductivity of new polymer electrolytes based on high molecular weight polyether comb polymers. *Electrochim. Acta* 43: 1177.

109 Wieczorek, W. and Stevens, J. (1997). Impedance spectroscopy and phase structure of polyether– poly(methyl methacrylate)– $LiCF_3SO_3$ blend-based electrolytes. *J. Phys. Chem. B* 101: 1529.

110 Lopes, L.V.S., Dragunski, D.C., Pawlicka, A., and Donoso, J.P. (2003). Nuclear magnetic resonance and conductivity study of starch based polymer electrolytes. *Electrochim. Acta* 48: 2021.

111 Marcondes, R.F.M.S., D'Agostini, P.S., Ferreira, J. et al. (2010). Amylopectin-rich starch plasticized with glycerol for polymer electrolyte application. *Solid State Ionics* 181: 586.

112 Pawlicka, A., Sabadini, A.C., Raphael, E., and Dragunski, D.C. (2009). Ionic conductivity and thermogravimetry measurements of starch-based polymeric electrolytes (vol. 485, pg 804, 2008). *Mol. Cryst. Liq. Cryst.* 506: 179.

113 Raphael, E., Avellaneda, C.O., Manzolli, B., and Pawlicka, A. (2010). Agar-based films for application as polymer electrolytes. *Electrochim. Acta* 55: 1455.

114 Lima, E., Mattos, R., Sentanin, F. et al. (2011). Functional novel polymer electrolytes containing europium picrate. *Mater. Res. Innovations* 15: 3.

115 Leones, R., Botelho, M.B.S., Sentanin, F. et al. (2014). Pectin-based polymer electrolytes with Ir(III) complexes. *Mol. Cryst. Liq. Cryst.* 604: 117.

116 Arof, A.K., Shuhaimi, N.E.A., Alias, N.A. et al. (2010). Application of chitosan/iota-carrageenan polymer electrolytes in electrical double layer capacitor (EDLC). *J. Solid State Electrochem.* 14: 2145.

117 Noor, I.S.M., Majid, S.R., Arof, A.K. et al. (2012). Characteristics of gellan gum - $LiCF_3SO_3$ polymer electrolyte. *Solid State Ionics* 225: 649.
118 Winie, T. and Arof, A.K. (2006). Transport properties of hexanoyl chitosan-based gel electrolyte. *Ionics* 12: 149.
119 Winie, T., Ramesh, S., and Arof, A.K. (2009). Studies on the structure and transport properties of hexanoyl chitosan-based polymer electrolytes. *Physica B* 404: 4308.
120 Arof, A.K., Osman, Z., Morni, N.M. et al. (2001). Chitosan-based electrolyte for secondary lithium cells. *J. Mater. Sci.* 36: 791.
121 Pawlicka, A., Danczuk, M., Wieczorek, W., and Zygadlo-Monikowska, E. (2008). Influence of plasticizer type on the properties of polymer electrolytes based on chitosan. *J. Phys. Chem. A* 112: 8888.
122 Iwaki, Y., Escalona, M.H., Briones, J., and Pawlicka, A. (2012). Sodium alginate-based ionic conducting membranes. *Mol. Cryst. Liq. Cryst.* 554: 221.
123 Avellaneda, C.O., Vieira, D.F., Al-Kahlout, A. et al. (2008). All solid-state electrochromic devices with gelatin-based electrolyte. *Sol. Energy Mater. Sol. Cells* 92: 228.
124 Vieira, D.F., Avellaneda, C.O., and Pawlicka, A. (2007). Conductivity study of a gelatin-based polymer electrolyte. *Electrochim. Acta* 53: 1404.
125 Ohno, H. and Takizawa, N. (2000). Ion conductive and flexible DNA films. *Chem. Lett.* 29: 642.
126 Pawlicka, A., Firmino, A., Sentanin, F. et al. (2015). Bio-inspired materials for electrochemical devices. *SPIE 9652*, 96520U1.

6

Polymer Electrolyte Application in Electrochemical Devices

Siti Nor Farhana Yusuf and Abdul K. Arof

Centre for Ionics University of Malaya, Department of Physics, Faculty of Science, University of Malaya, 50603 Kuala Lumpur, Malaysia

6.1 Introduction

An electrolyte conducts ions, but not electrons. In the 1830s Michael Faraday reported electrical conduction in Ag_2S and PbF_2 at high temperatures. Between 1910 and 1930, Tubandt and his group proved that electrical charge in these materials are carried by ions [1]. The electrolytes used in many energy devices are in liquid form even to this day. Liquid electrolytes have high room temperature conductivity, but may vaporize at elevated temperatures. Devices using liquid electrolytes are bulkier and may require extra space. This can be a problem if space is limited. Wright and coworkers [2] discovered ionic conduction in polymer–salt complexes during the 1970s and sometime later, Michel Armand declared the technological importance of this discovery as a "solid electrolyte" during the ionic community meeting in St Andrews, Scotland, in 1978. This announcement opened new ideas among the members of the ionic community [3].

6.2 Properties of Polymer Electrolytes (PEs)

Before we list the properties of polymer electrolytes (PEs), it is important to know that the polymer should have electron-donating or -withdrawing atoms for complexation with cation of the salts and a low glass transition temperature for segmental motion to assist ion transport. PEs whether in solid or gel form separate the anode and cathode of a battery. They are the medium channeling Li^+ ions during charge and discharge. A PE is one of the main components that influence the magnitude of the current density, number of cycles, and capacity of electrochemical devices. PEs are an important factor for battery safety and should have the following criteria:

1. High ionic conductivity, σ, in a wide temperature range.
2. Good thermal stability for good device operation at any desired temperature.

Polymer Electrolytes: Characterization Techniques and Energy Applications,
First Edition. Edited by Tan Winie, Abdul K. Arof, and Sabu Thomas.

3. Wide electrochemical window as it should not react with the electrodes.
4. Good mechanical properties for enhanced safety.
5. Easy to manufacture and low cost.
6. Material components have high flashing point for safety.
7. PEs must be nontoxic and environment friendly.

6.3 Review of Polymer Electrolytes

Electrolytes exist in the liquid, solid, and gel states. Electrolyte in the liquid state can be made by dissolving a lithium salt such as $LiClO_4$ [4], $LiCF_3SO_3$ [5], lithium bis(oxalato) borate or simply LiBOB [6], and lithium hexafluorophosphate ($LiPF_6$) [7] in an organic solvent, e.g. an organic alkyl carbonate. The conductivity of 0.82 M $LiClO_4$ in equal ratio of ethylene carbonate (EC) and propylene carbonate (PC) is ~6.2 mS-cm^{-1} at ambient. The conductivity of 0.88 M $LiClO_4$ in equal ratio of EC and diethyl carbonate (DEC) is about 5.1 mS-cm^{-1} [8]. Unfortunately, the perchlorate anion is a strong oxidant. Thus, $LiClO_4$ is unsuitable for use in electrolytes [9].

Entrapping the liquid electrolyte in a polymer to form PEs is a solution to problems that emerged from the use of ionic conductors in the liquid state. The polymer serves as a host matrix. PEs are candidate materials for application in electrochemical devices. These devices include electrochromic windows (ECWs), fuel cells, electrical double layer capacitors (EDLCs), rechargeable batteries, and dye-sensitized solar cells (DSSCs) [10]. An important property of PEs is its ionic conductivity. PEs should exhibit negligible electronic conductivity. The development of PEs has gone through the stages of (i) dry solid polymer electrolytes (SPEs) and (ii) gel polymer electrolytes (GPEs). The dry SPEs comprise a salt and a polymer dissolved in a common solvent, which is evaporated. Gel polymer electrolytes can contain ionic liquids (ILGPEs) and nanoparticles and referred to as composite gel polymer electrolytes (CGPEs) [11].

6.3.1 Dry Solid Polymer Electrolytes (SPEs)

Dry SPEs are thin films without liquid solvents. Dry SPEs exhibit poor conductivity (~10^{-6} S-cm^{-1}) at room temperature (RT). Dry SPEs have little practical applications [11]. They may also contain additives (plasticizers and fillers). For example, Croce et al. [12] reported conductivities of ~10^{-5} S-cm^{-1} at 30 °C and ~10^{-4} S-cm^{-1} at 50 °C in a PEO–$LiClO_4$ (PEO, poly(ethylene oxide)) complex containing nano-TiO_2 fillers. In another example, Das and Ghosh [13] reported a conductivity of ~2.0 μS-cm^{-1} for PEO–$LiClO_4$ (EO/Li = 18) at 30 °C. The conductivity increased to ~80.0 μS-cm^{-1} when added with 40 wt% PC at the same temperature. Table 6.1 shows other examples of SPEs.

Wang et al. [20] also studied PEO–$LiClO_4$–TiO_2 SPEs and the best conductivity value of ~1.0×10^{-4} S-cm^{-1} obtained from 40 to 60 °C. Since SPEs can only achieve such high conductivities at elevated temperatures, research "changed direction" toward GPEs that have the better of solid and liquid electrolytes combined.

Table 6.1 Characteristics of some SPEs.

Sample	σ (S-cm^{-1})	Characteristics				References
69.4 wt% PAN–23.1 wt% PMMA–7.4 wt% $LiClO_4$	5.62×10^{-6} (at 30 °C)	TS = 219 °C				[14]
PEO-$LiClO_4$	1.31×10^{-6}	$T_m = 54$ °C; $T_g = -33$ °C; $X_c = 25\%$				[15]
PEO–$LiClO_4$–FGnP	2.53×10^{-5}	$T_m = 51$ °C; $T_g = -36.5$ °C; $X_c = 19\%$				
		T_g (°C)	T_m (°C)	ΔH_m (J g^{-1})	X_c (%)	[16]
95 wt%PEO/PVP–5 wt% $LiNO_3$	1.20×10^{-4}	−68.21	66.23	102.32	48.08	
90 wt%PEO/PVP–10 wt% $LiNO_3$	3.70×10^{-4}	−70.01	64.12	92.89	43.46	
85 wt%PEO/PVP–15 wt% $LiNO_3$	1.13×10^{-3}	−73.09	58.35	74.75	34.04	
80 wt%PEO/PVP–20 wt% $LiNO_3$	2.60×10^{-4}	−71.23	63.24	84.82	40.00	
75 wt%PEO/PVP–25 wt% $LiNO_3$	1.22×10^{-4}	−69.15	66.41	99.12	44.80	
PEO	5.23×10^{-9}	T_g: −58.1 °C; T_m: 72.6 °C; ΔH: 151.3 J g^{-1}; X_c: 100%				[17]
PEO_{20}LiDFOB	2.46×10^{-6}	T_g: −42.8 °C; T_m: 57.6 °C; ΔH: 69.1 J g^{-1}; X_c: 45.7%				
PEO_{20}–LiDFOB–40% EMImTFSI	1.85×10^{-4} (all at 30 °C)	T_g: −51.4 °C; T_m: 41.8 °C; ΔH: 19.9 J g^{-1}; X_c: 13.2%				
		Film strength				[18]
90 wt% PVdF–HFP–10 wt% LiTf	6×10^{-7}	Very good				
85 wt% PVdF–HFP–15 wt% LiTf	2.7×10^{-6}	Very good				
80 wt% PVdF–HFP–20 wt% LiTf	8.1×10^{-6}	Very good				
75 wt% PVdF–HFP–25 wt% LiTf	5.76×10^{-5}	Very good				
70 wt% PVdF–HFP–30 wt% LiTf	3.21×10^{-5}	Good				
65 wt% PVdF–HFP–35 wt% LiTf	2.33×10^{-5}	Good				
		T_{onset} (°C)				[19]
Chitosan[C_2mim][C(CN)$_3$]	1.35×10^{-4}	138				
Chitosan[C_4mim][SCN]	3.55×10^{-4}	137				
Chitosan[C_2mim][N(CN)$_2$]	7.15×10^{-4}	232				
Chitosan[C_2mim][SCN]	1.61×10^{-3}	144				
Chitosan matrix	~1.6×10^{-7} (all at 25 °C)	135				

FGnP, polyethylene glycol-grafted graphene; PVP, poly(vinyl pyrrolidone); TS, thermal stability; T_g, glass transition temperature; T_m, melting point; X_c, degree of crystallinity; EMImTFSI, 1-ethyl-3-methylimidazolium bis(trifluoromethylsulfonyl)imide.

As can be seen from Table 6.1, a fixed ratio of poly(acrylonitrile) (PAN) and poly(methyl methacrylate) (PMMA) polymer blend electrolytes complexed with $LiClO_4$ that has been prepared via the solution cast method [14] exhibited conductivity values of the order 10^{-6} S-cm^{-1} at 30 °C. The decrease in degree of crystallinity observed in [15] showed that conductivity increased with increased amorphousness of the SPEs. Blending polyethylene glycol (PEG) grafted onto graphene (FGnP) with PEO–$LiClO_4$ complexes reduced PEO crystallinity and enhanced ionic conduction. The glass transition temperature, T_g was minimum at 0.5 wt% FGnP concentration with value −36.2 °C. At this composition, FGnP have produced maximum disruption to the molecular packing and crystallinity of PEO. The PEO chain was most mobile and free volume increased at this composition. In addition, the PEG molecules grafted to graphene exhibited effective plasticizing action and dissociated contact ion pairs to free mobile ions. However, at more than 0.5 wt% FGnP, T_g increased due to increased intensity of hydrogen bonding between the FGnP and PEO chains and the SPE became more rigid. The increase in conductivity is due to the energy pathways formed along the interface between the carbon-based filler and the polymer. The strong affinity between lithium ions and the electron cloud from the filler enabled the Li^+ ions to hop along the FGnP sites and enhanced lithium ion transport within the electrolyte.

Grafting or attaching PEG molecules onto graphene nanoplatelets also has an appreciable effect on the elongation and toughness of the SPE. These properties improved markedly in samples containing FGnP. At 0.5 wt% FGnP, the tensile strength increased by 104%, toughness increased by 292%, and elongation at break increased by 101% compared to neat PEO. This work [15] showed how FGnP affects the properties of an SPE.

Jinisha et al. [16] studied the blended PEO/PVP (poly(vinyl pyrrolidone)) host incorporated with $LiNO_3$. The sample 85 wt% PEO/PVP–15 wt% $LiNO_3$ showed thermal durability up to 410 °C. This work showed that SPEs should be thermally stable because heat generated from internal shorts in a cell can damage the SPE.

Introduction of 1-ethyl-3-methylimidazolium bis(trifluoromethylsulfonyl) imide or for short EMImTFSI [17] was able to increase the amorphousness of the SPE membranes containing PEO and lithium difluoro(oxalato)borate (LiDFOB). The SPE exhibited a conductivity of 0.19 mS-cm^{-1} at 30 °C and indicated the potential use of the electrolyte in secondary lithium ion batteries (LIBs).

As reported in [18], the incorporation of ionic liquids (ILs) in chitosan led to the increment of conductivity. After a certain optimum salt content, the conductivity will decrease. Conductivity increased with temperature for all salt concentrations. The increase in free volume due to polymer expansion enhanced segmental motion and movement of the free ions. These factors led to conductivity increase with temperature. The ILs included 1-ethyl-3-methylimidazolium thiocyanate ([C_2mim][SCN]), 1-ethyl-3-methylimidazolium dicyanamide ([C_2mim][$N(CN)_2$]), 1-ethyl-3-methylimidazolium tricyanomethanide ([C_2mim][$C(CN)_3$]), and 1-butyl-3-methyl imidazolium thiocyanate ([C_2mim][SCN]).

Chitosan[C_2mim][SCN] exhibited the highest conductivity of 1.61 mS-cm^{-1} at 25 °C and ~13 mS-cm^{-1} at 100 °C [19]. Even after two years, conductivity was still of the order of 10^{-4} S-cm^{-1} and might be due to the loss of mobile ions and decreasing mobility. However, it is still sufficient for use in electrochemical devices. The results indicate that the chitosan [$C_n$mim][$X(CN)_m$] SPEs ($n = 2$, $m = 1, 2, 3$, and X = S, C, N) are reasonably stable in time and have potential for practical applications.

6.3.2 Gel Polymer Electrolytes (GPEs)

Polymer host materials for GPEs have included PEO [21–23], PAN [24, 25], PMMA [26, 27], PVdF [28, 29], PVdF-HFP [30–33], and phthaloylchitosan [34–36]. These are depicted in Figure 6.1.

Thin film PEO-based electrolyte was the first system studied and used in solid-state batteries. Conduction in PEO-based electrolytes, even in gel form,

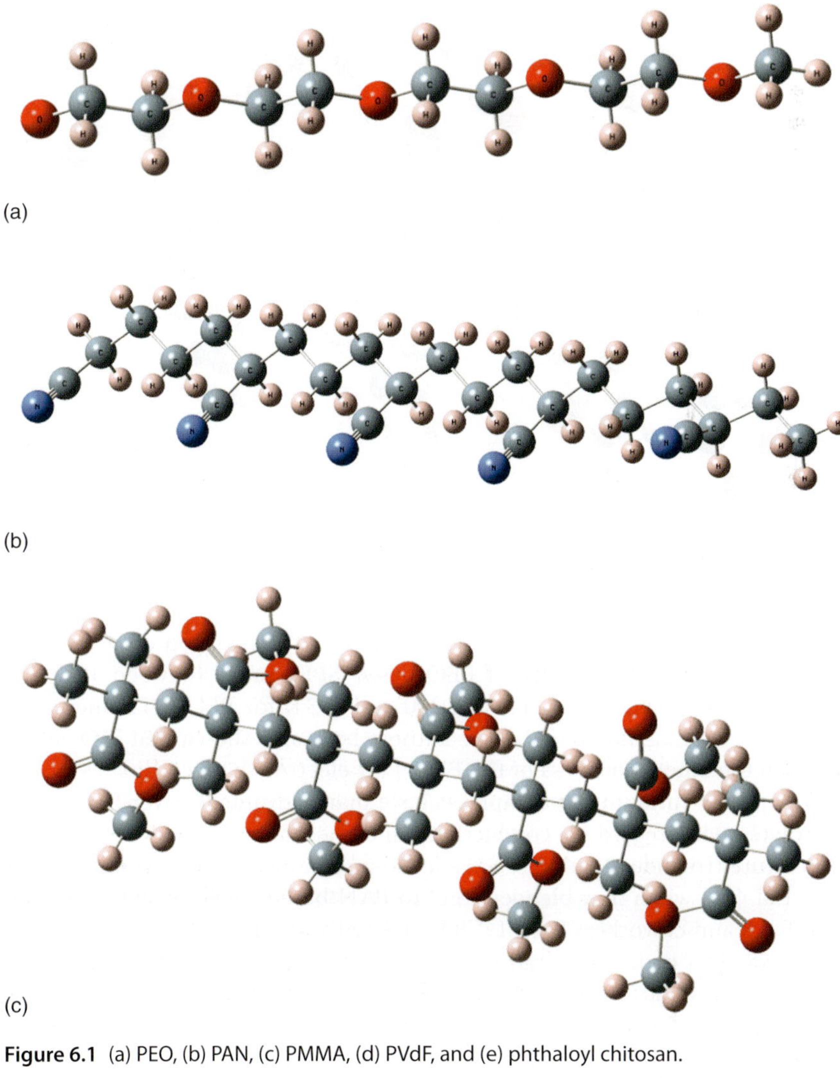

Figure 6.1 (a) PEO, (b) PAN, (c) PMMA, (d) PVdF, and (e) phthaloyl chitosan.

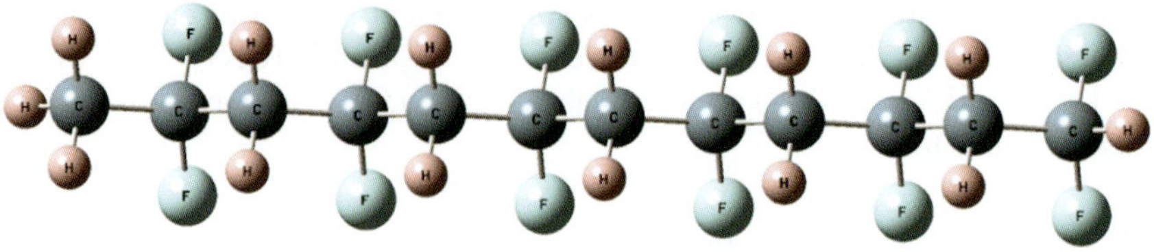

(d)

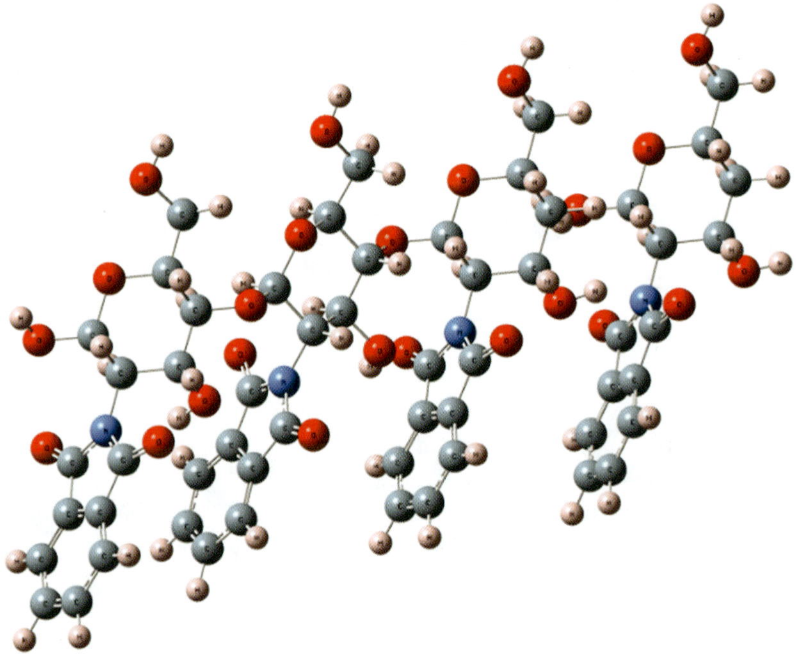

(e)

Figure 6.1 (*Continued*)

occurs through the complexes between cation and the oxygen atom of PEO. However, the low conductivity of PEO-based SPEs varies between 10^{-6} and 10^{-4} S-cm^{-1} at temperatures between 300 and 373 K for PEO_9LiTF due to the PEO semicrystalline nature [37]. The authors have also shown that TiO_2 fillers increased the amorphousness of the PEO_9LiTf electrolyte. According to Huang et al. [38], PAN–lithium salt complexes have many advantages over PEO-based electrolytes with regard to conductivity and mechanical properties at room temperature. To understand this, consider some conductivity results reported by Choi et al. [39] when they blended PEO to PAN-based gel electrolytes. In their work, Choi and coworkers fixed $LiClO_4$ at 12 wt% and EC and γ-butyrolactone (GBL) mixed in equal mole ratio, at 38 wt%. The percentage of PEO to PAN was 0 : 50, 10 : 40, 20 : 30, and 40 : 10. The electrolyte with the higher percentage of PAN exhibited a higher conductivity, implying that PAN-based electrolytes have an advantage over PEO-based electrolytes. The blended host exhibited a higher conductivity and displayed better dimensional and wider electrochemical stability.

Patel et al. [40] revealed that the PAN-based composite electrolyte with 92.5% of 1 M $LiClO_4$–succinonitrile solution exhibited a conductivity of 7.0 mS cm^{-1} at 25 °C. Even at sub-ambient temperature of −20 °C and conductivity of (100 − x)%-[$LiClO_4$–SN]:x%-PAN (x = 5–20 wt%), composites still exhibited a reasonably high conductivity from ~30.0 μS cm^{-1} to 0.45 mS-cm^{-1}. This shows the possibility of application at low temperatures. However, nitrile containing PEs are said to have low stability. The electrolyte is not that compatible with lithium or lithiated graphite anode [41]. These factors restrict their use in various devices.

Yuan et al. [42] synthesized PAN–PEO copolymer using the macro-monomer method. Dimethyl formamide (DMF) dissolved the copolymer and $LiClO_4$ salt. At 30 °C and 10 wt% PEO conductivity was 0.76 μS-cm^{-1}. This increased to ~3 mS-cm^{-1} at 20 wt% PEO for the same temperature. Conductivity also increased with temperature due to free volume increase in the polymer. The SPEs were dimensionally and mechanically stable throughout the temperature range studied. The SPEs also exhibited good electrochemical stability. At voltages below ~4.8 V versus Li/Li^+, current response was negligible. This indicated the possible use of the electrolyte in a 5 V cell.

PVdF is another type of polymer used as a host for ionic conduction. It contains the strong electron-withdrawing C–F group that can induce a net dipole moment and enables the polymer to exhibit high electrochemical stability. Its dielectric constant, $\varepsilon_{(PVdF)}$, at 1 kHZ is ~10.4 [43]. From the results reported in [44], it is deducible that at 1 kHZ, $\varepsilon_{(PVdF)} \approx 12$ and at 1 MHz, $\varepsilon_{(PVdF)} \approx 10$. However, according to Choe et al. [45], $\varepsilon_{(PVdF)} \approx 8.4$. Choe and coworkers also observed high ionic conductivity of 1.74 mS-cm^{-1} at 30 °C in the PVdF-PC-$LiN(SO_2CF_3)_2$ electrolyte system with 3 mol% salt. As a second example on the use of PVdF as polymer host, the GPE 24 wt% PVdF–68 wt% EC/PC–2 wt%-$LiCF_3SO_3$–6 wt% $LiClO_4$ exhibited a conductivity of 2.8 mS-cm^{-1} at 28 °C [46]. The high conductivity is not only because of the high ε of PVdF but is also attributable to the weak ion–polymer coupling that led to enhancement of ionic mobility.

A copolymer, poly(vinylidenefluoride-*co*-hexafluoropropylene) or (PVdF-*co*-HFP) or PVdF-HFP for short, is another useful polymer host for GPEs [32, 47, 48]. PVdF-*co*-HFP has higher dielectric constant [49], $\varepsilon \approx 10$ at 1 kHZ compared to PVdF at the same frequency [43] and is higher than PVdF when compared to that reported in [45]. PVdF and PVdF-HFP have high ε due to the existence of fluorine, which is highly electronegative and also the random alignment of C–F dipoles in the crystalline phases. P(VdF–TrFE–CTFE) terpolymers exhibit a ε of ~50 at 1 kHZ when the terpolymer has composition ~79 mol% VdF, ~7 mol% TrFE, and ~14 mol% CTFE [50]. Here, CTFE is short for chlorotrifluoroethylene and TrFE is trifluoroethylene. Costa et al. [51] have blended polyvinylidene fluoride–trifluoroethylene or P(VDF-TrFE) together with PEO for electrolyte application in LIBs. The room temperature conductivity was 25 mS-cm^{-1} for the blend incorporating 40 wt% PEO soaked in 1 M $LiClO_4{\cdot}3H_2O$-PC solution. In another work, Costa et al. [52] soaked PVdF-TrFE, PVdF-HFP, and PVdF-TrFE/PEO blend films in 1 M $LiPF_6$ solution of equal weights of EC and dimethyl carbonate (DMC). The ionic conductivity at 297 K for all membranes was in the $mS\,cm^{-1}$ range with PVdF-HFP membrane exhibiting the highest at $3.5\,mS\,cm^{-1}$.

6.3.2.1 Ionic Liquid Gel Polymer Electrolytes (ILGPEs)

Liquids containing only ions are termed as ionic liquids (ILs). Broadly speaking, the term embodies all molten salts. Nowadays, the term "ionic liquid" represents salts with melting point below 100 °C and salts that exist as liquid at ambient are described as room temperature ionic liquids (RTILs). ILs have excellent conductivity and a wide voltage window. They have high thermal stability, high heat capacity, low volatility, and negligible vapor pressure [53–55]. ILs can improve conductivity when added to PEs [56]. Conductivity enhancement comes from the extra ions provided by the IL [57]. The plasticizing characteristic of ILs also softened the polymer chain, increased polymer segmental motion, and led to increase in amorphousness of the PE [57, 58]. The low viscosity of ILs also enhanced ion mobility [59, 60]. These properties make ILs potential candidates for electrolytes in advanced LIBs and electrochemical devices in general [56–60].

Kuo et al. [61] had prepared an IL from phenolic epoxy resin and blended it with PVdF-*co*-HFP + $LiPF_6$ + EC + DEC to produce a good GPE. The material is reported to be nonflammable since its limiting oxygen index (LOI) value is constant at 29. LOI is the percentage of oxygen in the O_2–N_2 mix to allow ignition which determined the flammability of the materials. Generally, a material is not easily flammable or fire retarded if its LOI is more than 27. The GPE exhibited conductivity of 2.0 and $6.6\,mS\text{-}cm^{-1}$ at 303 and 353 K, respectively, although the organic electrolyte uptake was less than 50%.

Sirisopanaporn et al. [62] prepared dimensionally stable and flexible PVdF-HFP-based GPEs from solutions of *N*-*n*-butyl-*N*-ethylpyrrolidinium *N*,*N*-bis (trifluoromethane) sulfonamide (Py_{24}TFSI) and LiTFSI or lithium *N*,*N*-bis(trifluoro-methane)sulfonamide salt. The ambient conductivity for GPE incorporated with 56 wt% IL–salt solution and 14 wt% EC–PC mixture is $0.8\,mS\text{-}cm^{-1}$. Conductivity dropped to $0.4\,mS\text{-}cm^{-1}$ at IL concentration above 70 wt%. The conductivity exhibited is sufficient for LIBs. The higher conductivity for the EC/PC containing gels is attributable to the large ε and the viscosity of EC/PC mixture that is relatively low. The large ε helped to shield Li^+ cation and $TFSI^-$ anion interaction, which enhanced ionic separation of LiTFSI. Consequently, the number of free ions increased, which contributed to the conductivity. The low viscosity helped to enhance ionic mobility. These IL-based gels are thermally stable up to 383 K where they exhibit conductivity of $\sim 10\,mS\text{-}cm^{-1}$.

Stepniak et al. [63] (see Table 6.2) prepared an ILGPE in situ by photoinduced hardening of *N*-methyl-*N*-propylpiperidinium bis(trifluoromethanesulfonyl) imide or MPPipTFSI IL, lithium bis(trifluoro-methane-sulfonyl)imide or LITFSI and ethoxylated bisphenol A diacrylate mixture. The photocured ILGPE containing 80 wt% of LiTFSI in MPPipTFSI solution exhibited conductivity of $64\,\mu S\text{-}cm^{-1}$ that increased to $4.8\,mS\text{-}cm^{-1}$ at 368 K. The ILGPE exhibited an electrochemical stability window (ESW) of ~4.8 V versus Li/Li^+. The ILGPE membrane is mechanically strong and dimensionally stable.

Table 6.2 Characteristics of GPEs with IL(s).

Sample	σ (mS-cm^{-1})	Characteristics			References
Bisphenol A diacrylate(lbisAEA)4–0.4 M LiTFSIMPPipTFSI)	0.064 (at 25 °C) 4.8 (at 95 °C)	ESW = 4.8 V versus Li^{+}/Li			[63]
		$D_{diff}P_{13}^{+}$ ($m^2 s^{-1}$, 33 °C)	$D_{diff}FSI^{-}$ ($m^2 s^{-1}$, 33 °C)		[64] [64]
$P_{13}FSI$	9.9	3.92×10^{-11}	4.63×10^{-11}		
16 wt% PEGMA–4 wt% PEGDMA–80 wt%$P_{13}FSI$	1.3	1.79×10^{-11}	2.12×10^{-11}		
		T_g (°C)	T_o (°C)	T_m (°C)	[65]
$PyR_{24}TFSI/LiTFSI$	1.6	−81	−107	—	
20 wt% PVdF–HFP + 80 wt% (LiTFSI/$PyR_{24}TFSI$)	0.5	−79	−111	96	
30 wt% PVdF–HFP + 70 wt% (LiTFSI/$PyR_{24}TFSI$)	0.5	−78	−112	97	
$PyR_{14}TFSI/LiTFSI$	1.6	−79	−109	5	
20 wt% PVdF–HFP + 80 wt% (LiTFSI/$PyR_{14}TFSI$)	0.8	−76	−115	97	
30 wt% PVdF–HFP + 70 wt% (LiTFSI/$PyR_{14}TFSI$)	0.6 (all at 27 °C)	−76	−115	99	
60% LiNfO/EMImNfO + 40% PVdF-HFP	2.9 (at 100 °C)	$t_{Li}^{+} = 0.37$; TS = 332 °C			[66]
80% LiNfO/EMImNfO + 20% PVdF-HFP	18 (at 100 °C)	$t_{Li}^{+} = 0.50$; TS = 345 °C; ESW = 5.4 V			
P(VdF-HFP)–LiTFSI–$PYR_{14}TFSI$ (mass ratio of P(VdF-HFP)–LiTFSI is 1 : 1)	0.23 (at 20 °C) 2.1 (at 80 °C)	ESW = 5.0 V versus Li^{+}/Li TS = 150 °C			[67]
80 wt% PEO + 20 wt% LiFSI	0.004	$t_{Li}^{+} = 0.11$; $T_g = -47$ °C			[68]
77.5 wt% PEO + 20 wt% LiFSI + 2.5 wt% EMIMFSI	0.011	$t_{Li}^{+} = 0.16$; $T_g = -53$ °C			
PEO + 20 wt% LiFSI + 7.5 wt% EMIMFSI	0.289	$t_{Li}^{+} = 0.28$; $T_g = -55$ °C			
PEO + 20 wt% LiFSI + 12.5 wt% EMIMFSI	0.021	$t_{Li}^{+} = 0.21$; $T_g = -57$ °C			
P(MMA-AN-EA) + LiTFSI + EMITFSI + SiO_2 + Al_2O_3	3.20	Mechanical strength: 160 MPa			[69]

P(MMA-AN-EA), poly(methyl methacrylate-acrylonitrile-ethyl acrylate).

Chaudoy et al. [64] developed a PEGMA/PEGDMA/P_{13}FSI GPE via thermal free radical polymerization. PEGMA is short for poly(ethylene glycol) methyl ether methacrylate, PEGDMA stands for poly(ethylene glycol) dimethacrylate, and P_{13}FSI is 1-propyl-1-methylpyrrolidinium bis(fluorosulfonyl)imide RTIL. Pure P_{13}FSI has a conductivity of 9.9 mS-cm^{-1}. The diffusion coefficient of the pyrrolidinium cation (P_{13}^+) is 3.92×10^{-7} cm^2 s^{-1} and for the bis(fluorosulfonyl)imide anion (FSI^-) is 4.63×10^{-7} cm^2 s^{-1}. For PEGMA/PEGDMA/P_{13}FSI with weight ratio 16 : 4 : 80 conductivity decreased to 1.6 mS-cm^{-1}. The RTIL cation and anion diffusion coefficients were 1.79×10^{-7} and 2.12×10^{-7} cm^2-s^{-1}, respectively. The GPE was able to store a large amount of liquid.

Pitawala and coworkers [65] investigated the thermal, structural, and transport properties of a PVdF-HFP-based GPE containing IL and LiTFSI. The pure PVdF-HFP became more amorphous when added with IL/LiTFSI mixtures. The two ILs used have in common the bis(trifluoro-methane-sulfonyl)imide ($TFSI^-$) anion. The cations were *N*-butyl-*N*-methyl (PyR_{14}^+) and *N*-butyl-*N*-ethyl (PyR_{24}^+) pyrrolidinium cations. The ionic conductivity of the GPEs indicated their potential for use in electrochemical devices.

6.3.2.2 Gel Polymer Electrolytes with Nanomaterials

The addition of nanoparticles to GPEs enhances ionic conductivity. Ortega and coworkers [29] have prepared a PVdF-based composite GPE containing 1-ethyl-3-methylimidazolium bis(fluorosulfonyl) imide or EMIFSI IL and also SiO_2 nanoparticles. EMIFSI is nonvolatile and chemically, thermally, and electrochemically stable.

Li et al. [70] studied PVdF/SiO_2 and PVdF/SiO_2-PAALi composite GPEs after activation in 1 M $LiPF_6$ with equal volumes of EC, DMC, and ethyl methyl carbonate (EMC). Electrolyte uptake was higher in PVdF-SiO_2-PAALi compared to PVdF-SiO_2. However, in terms of thermal stability, PVdF-SiO_2 is better than PVdF-SiO_2-PAALi (PAA is polyacrylic acid), but both doped membranes have better thermal stability compared to pure PVdF. The interaction between dopants and PVdF help strengthen the GPEs thermally. For pure PVdF gel, ~20% weight loss occurred at ~433 °C; for PVdF-SiO_2-PAALi, this occurred at ~472 °C; and for PVdF-SiO_2, the amount of weight loss occurred at ~475 °C. The composites are therefore suitable for application in electrochemical devices. The dopants also improved the mechanical strength of PVdF. SiO_2 nanoparticles increased the tensile stress of pure PVdF by ~52% and SiO_2-PAALi ion complexes increased the stress by 69%. When subjected to a force, the dopants will help the nanofibrous PVdF membrane by absorbing part of the load. The CGPEs also possess good ionic conductivity with a broad electrochemical window.

Vishwakarma and Jain [71] presented measurements of ionic conductivities in a GPE based on PVdF. GPEs with high thermal conductivity can minimize temperature increase of an energy storage cell above ambient temperature. Hence, the incorporation of fillers with high thermal conductivity should help in disseminating heat from an energy storage device. Al_2O_3 and boron nitride (BN) have high thermal conductivity. Their thermal conductivity is between 18 and 34 W m^{-1} K^{-1}, low electrical conductivity ~10^{-14} S-cm^{-1} and they are of lightweight. Compared to other ceramics Al_2O_3 and BN are cheaper. Listed in Table 6.3 are some characteristics of selected GPEs added with nanoparticles.

Table 6.3 Characteristics of GPEs added with nanoparticles.

Sample	σ (mS-cm^{-1}) at 30 °C	Characteristics				References
(25 wt% PVdF–75 wt% EMIFSI)–x wt% SiO_2 ($x = 0$ and 5 wt% relative to the mass of PVdF)	~0.32 (0 wt% SiO_2) ~0.50 (5 wt% SiO_2)	5 wt% silica showed good dispersion of nanofiller and transparency T_m decreased from 168 to 134 °C (0 wt% SiO_2) and 135 °C (5 wt% SiO_2) Decomposition temperature 181 °C (0 wt% SiO_2); 180 °C (5 wt% SiO_2)				[29]
		Electrolyte uptake (%)	Stress (MPa)	Strain (%)	ECW (V)	[70]
PVdF-SiO_2	~3.02	368	43.9 ± 0.1	25.8 ± 0.1	4.93	
PVdF-SiO_2-PAALi	~3.55	377	48.8 ± 0.1	25.6 ± 0.1	5.05	
X = 1.0 M $LiPF_6$ mixed in EMITF:EC:PC volume ratio of 2 : 1 : 1		GPE thermal conductivity, k_{eff} (W m^{-1} K^{-1})				[71]
X + PVdF	4.1	0.09				
X + PVdF + BN	~2.1	0.83				
X + PVdF + Al_2O_3	~2.2	0.98				
X + PVdF + Al_2O_3 + BN	1.4	1.2				
BN is boron bitride. BN and Al_2O_3 are of 800 nm size, but BN mixed with Al_2O_3 has particle size 2.5 μm						
PVdF-HFP/PMMA/TiO_2 in 1 M $LiPF_6$ EC: DMC (v/v = 1)		R_b (Ω)	Electrolyte uptake (%)			[72]
0 wt% TiO_2	1.35	3.671	160			
2 wt% TiO_2	1.62	3.075	211			
5 wt% TiO_2	2.49	1.996	267			
7 wt% TiO_2	1.93	2.574	240			

6.4 Application of PEs in Electrochemical Devices

Electrochemical devices dealt in this chapter are limited to the branches shown in Figure 6.2.

6.4.1 Dye-Sensitized Solar Cells (DSSCs)

DSSCs consist of a photoanode on a conducting substrate, an electrolyte (a GPE may be favorable) and a counter electrode on another conducting substrate. The photoanode comprises a wide bandgap semiconductor soaked in a sensitizing dye. An example of a semiconducting oxide is TiO_2. The GPE that also contains a redox mediator is sandwiched in between the counter electrode and photoanode (Figure 6.3).

DSSC may make a suitable choice for affordable low power generation in housing and isolated areas. The working principle of a DSSC is as follows. When the

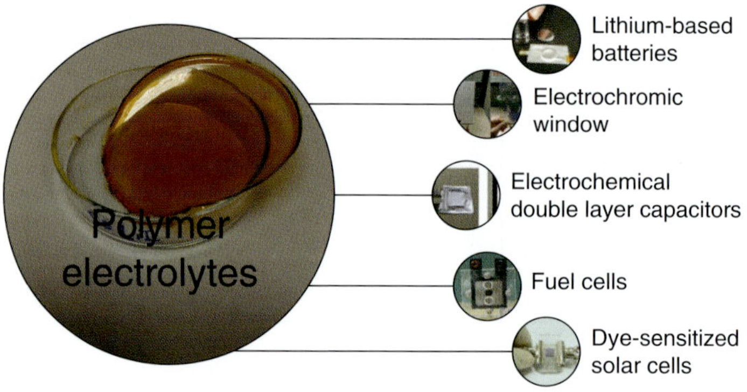

Figure 6.2 Electrochemical devices discussed in this chapter.

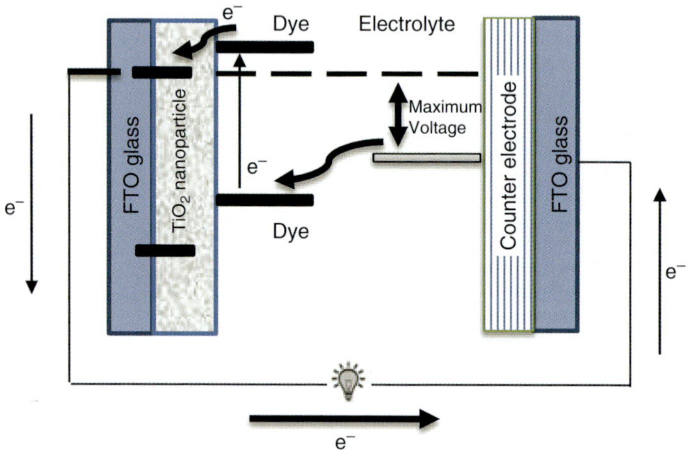

Figure 6.3 Structure of a DSSC.

dye molecules adsorbed on TiO_2 component of the photoanode get excited on absorption of light, electrons are also released leaving a hole in each dye molecule. On transfer of electrons to the wide bandgap semiconductor, the electrons percolate through the photoanode and exit to the external load. At the counter electrode, the redox electrolyte transports the electrons back to the dye for reunion with the holes and the circuit is completed. The electrons that reached the counter electrode reduced the triiodide ions into iodide ions, which diffused to the photoanode. The iodide ions released the electrons to the holes in the dye molecules and get oxidized back into triiodide ions that diffused to the counter electrode. The device efficiency depends on the electrolyte and dye used.

To explain the use of GPEs in DSSCs, we need to understand the role of the wide bandgap semiconductor oxides and we limit the discussion to TiO_2 semiconducting oxide. Actually there are two layers of TiO_2 deposited one over the other on the transparent oxide substrate.

The first layer is a compact nanoparticle TiO_2 layer and the particles are of a smaller size, 14 nm. The second layer, on top of the first layer, is composed of TiO_2 nanoparticles of a bigger size (21 nm). The second layer is a mesoporous layer. It is made mesoporous so that when it is soaked in the dye solution, there will be more surface area for the dye molecules to adsorb to the TiO_2 since the dye molecules can enter the pores of the second layer. Likewise, liquid electrolyte (LE) can also penetrate the pores and reach the compact layer. Hence, there will be a TiO_2/dye/LE interface. For GPEs, it is doubtful if the gel can penetrate deep into the pores as in the LE. However, DSSCs utilizing GPEs work reasonably well as shown in Table 6.4.

Table 6.4 Some characteristics of DSSCs with PEs.

Electrolyte	Dye	J_{SC} (mA cm^{-2})	V_{OC} (V)	η (%)	References
PVdF-HFP + PVA + TiO_2	N719	4.39	0.80	3.03	[73]
Chitosan + NaI	—	1.05	0.35	0.13	[74]
Chitosan + NaI + EMImSCN	—	2.62	0.53	0.73	[74]
Chitosan + BMII + NH_4I	ABR	0.90	0.36	0.15	[75]
Chitosan + BMII + NH_4I	ARC[a)]	1.59	0.45	0.29	[75]
PhCh + PEO + BMII + NH_4I	ARC[b)]	3.50	0.34	0.46	[75]
PhCh + EC + PC + TPAI + LiI + I_2	N719	7.25	0.77	3.71	[36]
PhCh + EC + DMF + TPAI + I_2	N3	12.72	0.60	5.00	[35]
PhCh + EC + DMF + TPAI + LiI + I_2	N3	17.29	0.59	6.36	[34]
PMMA + EC + PC + TPAI + I_2	N719	7.42	0.71	3.21	[76]
PMMA + EC + PC + TPAI + KI + I_2	N719	10.70	0.66	3.99	[76]
PMMA + EC + PC + KI + I_2	N719	7.55	0.70	3.39	[76]

ABR, anthocyanin from black rice; ARC, anthocyanin from red cabbage; MP, mangosteen peel; PhCh, phthaloylchitosan (wherever necessary).

a) Added with HCl.

b) Added with tartaric acid.

Tiautit et al. [73] used TiO_2 and SiO_2 as fillers in 10 wt% PVdF-HFP/polyvinyl alcohol (PVA) (ratio 8 : 2) and added 2 ml of liquid electrolyte comprising DMPII + N-butyl-1H-benzimidazole (NBB) dissolved in 1-methyl-2-pyrrolidone (NMP). The DSSC with TiO_2 filler decreased in efficiency with increased TiO_2 content. The SiO_2-fillered electrolyte content showed a lower efficiency at the same TiO_2 concentration (0.25 wt%). The efficiency increased to 3.03% at 0.5 wt% SiO_2 concentration.

In reference [74], the effect of adding ethyl methyl imidazolium thiocyanate (EMImSCN) IL in large amounts increased the efficiency of the DSSC employing a chitosan–NaI electrolyte from 0.13% to 0.73%. The source of the natural pigment used in a DSSC employing the 80 wt% BMII + 20 wt% (0.55 chitosan–0.45 NH_4I) electrolyte is also a factor controlling the efficiency. For the same pigment source [75], but using a modified chitosan host, i.e. phthaloylchitosan (PhCh) blended with PEO as in 80 wt% BMII + 20 wt% (0.55 chitosan–0.45 NH_4I) and 80 wt% BMII + 20 wt% (0.55 PhCh:PEO [30 : 70]–0.45 NH_4I), there is no significant change. Finally consider the electrolyte comprising PhCh:EC/PC:TPAI:LiI:I_2 (TPAI, tetrapropyl ammonium iodide) electrolyte [36]. The efficiency of the DSSC depends on whether there are multiple salts and the size of the cation of the salts used. With incorporation of a second salt with a smaller size of cation, LiI helps to improve the efficiency. Photogeneration of electrons will be improved as small cations, namely, Li^+ or K^+, possess high charge density [76]. With increased LiI content, the efficiency increased from 3.50% (only TPAI salt) to 3.71% for 6.33 wt% LiI [36]. The similar effect can also be observed in [34] as the efficiency of the DSSC improved from 5.00% (TPAI) [35] to 6.36% (TPAI + LiI). Further increase in LiI content led to the decline in efficiency indicating that the there is an optimized content of salt with a large cation.

6.4.2 Lithium Ion Batteries

An electrochemical storage device that has received much attention is the LIB. LIBs can reversibly convert chemical energy into electrical via redox reactions [66]. LIBs have high energy density between 250 and 610 $W\text{-}h\,l^{-1}$. Electronic devices, for example, laptops and mobile phones, use LIBs. Hybrid electric vehicles (HEVs) observed on the road use LIBs. Careful handling of LIBs is important. Leakage of toxic organic carbonates in LIBs may lead to explosion and fire. To prevent leakage in LIBs, it is necessary to substitute the liquid electrolyte with GPEs. These GPEs should, however, comply with the requirement of good dimensional stability and mechanical and electrical properties.

For LIB application, the cationic transference number $t_{Li}{}^+$ of the electrolyte must be high [66]. Minimization and if possible total immobilization of anion mobility is important for better performance. The free anions can take part in reactions at electrodes, which are detrimental to battery health. Using a salt with a bulky anion can help reduce the anionic transference number. In the case of PVdF-*co*-HFP-based electrolyte containing LiNfO or lithium nanofluoro-1-butanesulfonate salt and 1-ethyl-3-methyl imidazolium nanofluoro-1-butanesulfonate IL, the matrix limits the NfO^- movement, making Li^+ ion transport easier [66]. The PVdF-*co*-HFP (80%):

LiNfO-EMImNfO (20%) provides more mobile Li^+ with $t_{Li}{}^+$ value of 0.50. The Li/80% LiNfO-EMImNfO + 20% PVdF-HFP/$LiCoO_2$ half-cell discharged at C/10 rate displayed a first discharge cycle capacity of 164 mA-h g^{-1}. As an indication of its good performance, after 45 cycles, capacity is maintained at 161 mA-h g^{-1}.

The performance of PEO-based SPE containing LiDFOB and EMImTFSI was studied at 298 K [17]. The initial specific capacity reported was 155 mA-h g^{-1}. A plateau appeared near 3.5 V during charging and near 3.38 V during discharge. After 50 cycles, the initial discharge capacity reduced by 13%. The performance indicates that the 60 wt% PEO–LiDFOB–40 wt% EMImTFSI electrolyte is compatible with the electrodes.

The Li/poly(bisAEA4 + LiTFSI + MPPipTFSI)/$LiFePO_4$ cell, which used ILGPE showed good capacity and cyclability indicating good electrode/electrolyte contact [63]. This implied good compatibility between ILGPE and the electrode material. Even after 50 cycles and at C/20 rate, the discharge retained 98% initial capacity. This is equivalent to about 95% of $LiFePO_4$ theoretical capacity. The results indicated that the poly(bisAEA4 + LiTFSI + MPPipTFSI) ILGPE prepared in situ by photoinduced polymerization is a potential electrolyte for bendable LIBs.

The Li/P(VdF-HFP)–LiTFSI–PP_{14}TFSI/$LiFePO_4$ [67] were initially charged and discharged at different temperatures. At 293 K, the capacity of the cell when charged to 4.2 V is 130 mA-h g^{-1}. However, at this temperature, the electrolyte conductivity was low. This led to a strong concentration polarization and resulted in a low discharge capacity of 98.65 mA-h g^{-1} at 1 C rate. The charge/discharge capacities increased with temperature. The Li/P(VdF-HFP)–LiTFSI–PP_{14}TFSI/$LiFePO_4$ electrochemical cells exhibit good cyclability at 60 °C and 1 C. The cell exhibited a capacity of 131 mA-h g^{-1} upon discharge. Charge retention was 90% of the initial capacity after 100 cycles. The ILGPE showed promise for use in LIBs especially for high temperature applications. Table 6.5 lists some LIB systems and their specific discharge capacities.

While the GPE is getting the attention of many researchers, there are still arguments concerning their limitations. Mechanical strength of GPEs still does not reach some degree of satisfaction [78]. Thus, various approaches have been implemented such as adding a filler into the GPE. Li et al. have obtained good cyclability and C-rate performance of $LiCoO_2$/Li cells using a composite PE composed of PVdF-HFP + PMMA + TiO_2 dipped in $LiPF_6$ in equal volumes of EC and DMC [70]. Even after 50 cycles, the specific capacity remained more than 90% of the initial capacity when discharged at 0.2 C. The composite GPE has the potential to meet the working and thermal safety requirements of high performance LIBs. The high diffusion coefficient of lithium ion for the Li/$LiCoO_2$ cell with 5 wt% TiO_2 nanoparticles in PVdF-HFP/PMMA PE also demonstrated the benefit of adding fillers to the PE.

During first charge of an LIB with graphite electrode, the electrolyte goes through reduction and begins to decompose at the negatively polarized electrode. A passive or solid electrolyte interphase (SEI) layer forms [79]. The SEI consists of inorganic and organic electrolyte breakdown products and is a protective layer that prevents further electrolyte decomposition. Efficient

Table 6.5 Some characteristics of PE batteries.

Battery	Specific discharge capacity (mA-h g^{-1})			References
Li/poly(bisAEA4–0.4 M LiTFSI–MPPipTFSI)/$LiFePO_4$	162 at C/20 134 at C/2			[63]
Li/80% LiNfO–EMImNfO + 20% PVdF–HFP/LiCoO2	164 (initial discharge at C/10 rate)			[66]
Li/P(VdF-HFP)–LiTFSI–PYR_{14} TFSI (1 : 1 : 1)/$LiFePO_4$	145.77 (at 60 °C) 158.75 (at 80 °C)			[67]
	Discharge capacity (mA-h g^{-1})			[70]
Li/PVdF/$LiCoO_2$	149.1			
Li/PVdF + SiO_2/$LiCoO_2$	152			
Li/PVdF + (SiO_2-PAALi)/$LiCoO_2$	156.5			
Li/PVdF-HFP/PMMA/TiO_2 immersed in 1M $LiPF_6$ with EC:DMC (v/v = 1)/$LiCoO_2$	Discharge capacity (mA-h g^{-1})		D(Li^+) (cm^2 s^{-1})	[72, 77]
	1st cycle	50th cycle		
0 wt% TiO_2	~144	~114	1.32×10^{-13}	
2 wt% TiO_2	~164	~142	2.88×10^{-13}	
5 wt% TiO_2	~188	~173	3.98×10^{-13}	
7 wt% TiO_2	~178	~154	3.17×10^{-13}	

NBB, N-butyl-1H-benzimidazole; BMII, 1-butyl-3-methyl imidazolium iodide; TPAI, tetrapropyl ammonium iodide; NMP, 1-methyl-2-pyrolidone.

passive layers assure good cycling reversibility. Fong et al. [80] reported that Li/electrolyte/petroleum coke and Li/electrolyte/graphite cells exhibited reactions that are irreversible only during the first discharge. The electrolyte was 1 M $LiAsF_6$ in a mixture that contained equal ratios of PC and EC. The SEI formation on the carbon surface ensured irreversibility of Li^+ entry and exit of the negative electrode. When the SEI layer has stopped electrolyte breakdown, the cells showed reversible cycling with no loss in capacity.

At present rechargeable LIBs in the market frequently used $LiPF_6$ as Li salt in the electrolyte [81]. However, its thermal and chemical stabilities are limited in lithium polymer systems possibly due to the high electronegativity of fluorine in the structure of $LiPF_6$. According to Schmidt et al. [82], P–F bond can be stabilized by substituting some of the fluorine with perfluorinated alkyl groups.

6.4.3 Electrical Double Layer Capacitors (EDLCs)

EDLCs are another type of energy storage device that have shown promise for applications such as in HEVs. An EDLC comprises an electron-insulating electrolyte sandwiched between two electrodes. Depending on the applied potential, a corresponding amount of ions "build up" on the electrode surface to form the electrical double layer (EDL). The EDL comprised an electrical space charge on the side of the electrode and an ion space charge on the electrolyte side of the electrode/electrolyte interface. Charge is stored electrostatically in EDLCs. There will be no charge transfer across the interface during operation and during charge. The energy density is dependent on the positive and the negative capacitances, i.e. C_+ and C_- of the electrodes, respectively, and the maximum operating voltage of the device, i.e.

$$E = \left(\frac{C_- C_+}{C_- + C_+}\right)(V_{\text{max}})^2 \tag{6.1}$$

For maximizing EDLC energy, the two electrodes should possess the same or almost the same capacitance. In this case,

$$E = \frac{1}{2}C(V_{\text{max}}^{\text{EDLC}})^2 \tag{6.2}$$

The maximum voltage of the EDLC usually determines the stability window. Figure 6.4 shows a schematic of the EDL structure formed near an electrode. Due to electrostatic attraction, adsorbance of the solvated cations to the negatively charged electrode surface occurs. Likewise, due to the charge on the anions and the negatively charged electrode, they are repelled. Near the electrode surface, there are no free charges. This layer is the inner compact layer or also known as the Stern layer. The anions and cations are mobile beyond the Stern layer. The ions are under the combined influence of diffusion and electrostatic forces. The layer containing the mobile ions is the diffuse layer.

An EDLC comprises porous carbonaceous electrodes. Here, we focus on carbonaceous electrodes. The porosity of the electrodes ensures a large surface area for capacitance and storage energy enhancement. The pore size distribution also affects the EDLC performance. Micropores with less than 2 nm are known to greatly contribute to the EDL formation. However, for ion accessibility and fast dynamic charge, mesopores of 50 nm diameter are necessary. Pores of size smaller than the ions usually do not contribute to double layer capacitance. Table 6.6 lists EDLCs utilizing GPEs and activated carbon electrodes. Kumar et al. [77], compared EDLC performance fabricated with activated carbon and multiwalled carbon nanotubes (MWCNT) electrodes. Two types of gel

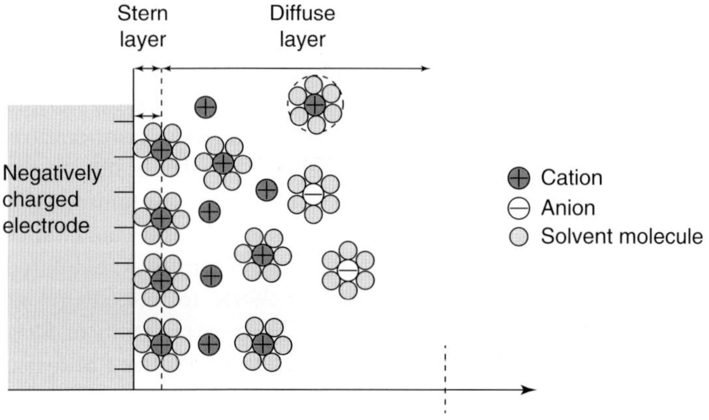

Figure 6.4 Stern and diffuse layers in EDLC.

electrolyte used were (i) LiTf + EMITf + PVdF-HFP designated as ILGPE-1 and (ii) LiTf + EMITf + PVdF-HFP + EC + PC designated as ILGPE-2. EC and PC are in equal volumes. The room temperature conductivity for ILGPE-1 and ILGPE-2 was 4.5 and 8.0 mS cm^{-1}, respectively. EMITf is 1-ethyl-3-methylimidazolium trifluoromethanesulfonate. Table 6.6 lists some EDLCs that used GPEs.

The MWCNT electrodes in EDLC exhibited a capacitance of 32 F g^{-1}, while the one based on activated carbon offered a higher capacitance value of 157 F g^{-1} [77]. However, the EDLC using MWCNT electrodes exhibit a higher rate capability than EDLCs using AC electrodes. The number of cycles for the MWCNT-based EDLC exceeded 50 000, whereas for the AC-based device, the number of cycles was only 5000.

The Nyquist plots of the AC/P_{13}FSI/AC and AC/PEGMA-PEGDMA-P_{13}FSI/AC supercapacitors [64] show capacitive and porous interface characteristics. A vertical line almost parallel to the imaginary impedance axis followed a depressed semicircle on viewing from high to low frequencies. At intermediate or medium frequencies, a set of points corresponding to the real and imaginary impedances formed a line with a 45° slope. This represents a Warburg line attributable to the semi-infinite diffusion impedance of ions through the electrode pores. The Warburg line also segregates the resistive domain from the purely capacitive domain of the supercapacitor. The resistive domain is in the region of high and medium frequencies, and the capacitive domain is in the low frequency region. The impedance at high and medium frequencies is higher for the AC/PEGMA-PEGDMA-P_{13}FSI/AC EDLC than it is for the AC/P_{13}FSI/AC supercapacitor. The plot of imaginary capacitance against frequency indicated a relaxation peak that separates the resistive and capacitive domains of the supercapacitor. The inverse of the relaxation frequency gives the relaxation time. Capacitance decreases with specific current. For AC/P_{13}FSI/AC EDLC, this is from ~34 F g^{-1} at 0.15 A g^{-1} to ~21 F g^{-1} at 2 A g^{-1}. For the same specific currents, the EDLC with cross-linked GPE, i.e. PEGMA-PEGDMA-P_{13}FSI, exhibited a decrease in capacitance from ~29 to ~11 F g^{-1}. Capacitance was quite stable at ~21 F g^{-1} for about 2400 cycles at 1 A g^{-1} current drain for AC/PEGMA-PEGDMA-P_{13}FSI/AC EDLC.

Ortega et al. [29] fabricated EDLCs employing nanostructured SiO_2-fillered GPE (PVdF + EMIFSI) that exhibited higher capacitance compared with that using filler-free GPE. At 10 mV s^{-1} scan rate and 25 °C, the filler-free EDLC displayed a capacitance of 41.6 F g^{-1} and the EDLC with SiO_2 filler GPE showed a capacitance of 71.7 Fg^{-1}. Capacitance also increased with temperature. Capacitance of both devices increased from 25 to 100 °C probably due to the increased mobility of the conducting ions. At 298 K, the EDLC containing GPE without the SiO_2 filler obtained a capacitance of 41.6 F g^{-1}. The capacitance increased to 76.0 F g^{-1} at 373 K. The EDLC with GPE containing SiO_2 exhibited a larger increase in capacitance from 71.7 F g^{-1} at 298 K to 124.6 F g^{-1} at 373 K, compared at the same scan rate, 10 mV s^{-1}. The incorporation of SiO_2 facilitates the dissociation of salt, which would lead to an increase in σ and also

Table 6.6 EDLCs utilizing GPEs.

Device	Characteristics			References
AC/P_{13}FSI/AC	Capacitance decreases from 35 to 20 F g^{-1} between 2 and 100 mV s^{-1}			[64]
AC/PEGMA–PEGDMA–P_{13}FSI/AC	Capacitance decreases from 31 to 13 F g^{-1} between 2 and 100 mV s^{-1}			
GPE:PVdF + EMIFSI				[29]
X = GPE + 0 wt% SiO_2				
Y = GPE + 5 wt% SiO_2				
MWCNT/X/MWCNT	41.6 F g^{-1} at 25 °C, 10 mV s^{-1}			
MWCNT/Y/MWCNT	71.7 F g^{-1} at 25 °C, 10 mV s^{-1}			
	R_b (Ω cm^2)	R_{ct} (Ω cm^2)	$R_{10\,mHz}$ (Ω cm^2)	[77]
X| ILGPE-1 |X	4.9	8.5	172	
X| ILGPE-2 |X	3.8	2.4	133	
Y| ILGPE-1 |Y	6.8	4.5	46	
Y| ILGPE-2 |Y	5.3	3.0	24	
	$C_{10\,mHz}$ (mF cm^{-2})		$C_{10\,mHz}$ (F g^{-1})	
X| ILGPE-1 |X	19		19	
X| ILGPE-2 |X	32		32	
Y| ILGPE-1 |Y	138		92	
Y| ILGPE-2 |Y	236		157	
X = MWCNT: Y = AC				
Ti|PAM + Li_2SO_4|Ti	Capacitance: 15.0 μF cm^{-2} at 1 V s^{-1}			[83]
Ti/CNT|PAM + Li_2SO_4|CNT/Ti	ESR = 2.4 Ω cm^2			
Carbon nanotube (CNT) deposited on Ti foil	Capacitance: 7.08 mF cm^{-2} at 100 mV s^{-1} 5.24 mF cm^{-2} at 5 V s^{-1}			
AC|10EMImTFSI-PEO-FPC|AC, where the mass ratio of ionic liquid to PEO is 10. σ = 6.7 mS cm^{-1}	At 1 A-g^{-1} Capacitance: 70.84 F-g^{-1} ECW: 3.5 V Energy density: 30.13 Wh-kg^{-1} Power density: 874.8 W-kg^{-1} Can drive a 3.0 V light-emitting diode (LED)			[84]

ESR, equivalent series resistance.

amorphousness. These results verify that the GPEs are suitable for application in EDLCs. The EDLCs may liberate heat while in use. Thus, the supercapacitors must tolerate changes in the environmental temperature and still maintain their performance. The low volatility of the EMIFSI and good mechanical integrity of the GPE are value-added features for safety consideration. The capacitance for the EDLCs decreased steadily with increasing scan rate. However, the EDLC with SiO_2-added GPE still exhibited a higher capacitance.

Song et al. [72] synthesized PVdF-HFP/PMMA-based CGPE that comprised up to 7 wt% TiO_2 nanoparticles. The TiO_2 nanoparticles (i) improved thermal stability of the GPE, (ii) reduced shrinkage of the GPE, and (iii) enhanced ionic conductivity and electrochemical stability.

Virya and Lian [83] prepared polyacrylamide (PAM) transparent solid PEs composed of (i) 9.1 wt% and (ii) 16.7 wt% of Li_2SO_4. SPE containing 9.1 wt% Li_2SO_4 showed better stability and utilized in solid symmetric electrochemical cells. The 2.0 V window for the electrolyte is twice the voltage window displayed by many acid electrolytes. The voltage window exhibited by the Li_2SO_4-PAM system was also higher than that exhibited by LiCl-PVA and LiCl-PAM electrolytes. The EDLC exhibited good rate capability (~5 V s^{-1}) and excellent recyclability of more than 10 000 cycles.

Zhong et al. [84] have improved the mechanical properties and ionic conductivity of electrolyte in EDLC by dissolving benzophenone (Bp) into IL. In this case, IL used is EMImTFSI. Following this was the dissolution of PEO into the IL-Bp mixture. After obtaining a homogeneous semisolid gel by annealing the PEO-IL-Bp mixture, the GPEs were UV irradiated for cross-linking. The designation for the GPE used by the authors was *n*IL-PEO, with *n* as variable denoting the mass ratio of IL to PEO. A nonwoven cellulose separator (FPC) holds the *n*IL-PEO to form an *n*IL-PEO-FPC electrolyte film. The use of FPC improved the mechanical strength of the GPE. A FPC film with $n = 10$ exhibited a conductivity of 6.7×10^{-3} S-cm^{-1}. The film also showed an excellent mechanical property for the flexible EDLC. The cell showed the same performance at various bending angles, implying good flexibility.

6.4.4 Polymer Electrolyte Fuel Cells

In this section, proton conductors will be the PEs discussed as they are used in polymer electrolyte fuel cells (PEFCs) and direct methanol fuel cells (DMFCs).

A fuel cell converts hydrogen and oxygen into electricity, water, and heat. The motivation to develop fuel cells is the uniqueness of the technology to provide minimal emissions and therefore is a clean or green technology. As no combustion is involved, a fuel cell converts fuel to electricity more efficiently compared to other technologies since there is no combustion involved.

A PEFC converts the chemical energy released during the reaction between H and O into electrical energy. At the anode, the reaction will be

$$H_2 \rightarrow 2H^+ + 2e^- \tag{6.3}$$

This is an oxidation reaction. At the anode, the Pt catalyst causes the hydrogen molecules to split into protons and electrons. The PE provides the passage to the cathode. The electrons will have to reach the cathode through an external circuit. At the cathode, the following reaction occurs:

$$\frac{1}{2}O_2 + 2H^+ + 2e^- \rightarrow H_2O \tag{6.4}$$

The overall reaction is

$$H_2 + \frac{1}{2}O_2 \rightarrow H_2O \tag{6.5}$$

It is obvious there is no combustion involved and the only reaction product is water. Therefore, fuel cell technology is a green technology. Ideally, the theoretical voltage is about 1.3 V since water is the reaction product. The practical voltage obtained is less than the theoretical voltage; hence, the fuel cell must be stacked to fulfill the requirement of the application.

Similarly, a DMFC depends upon methanol oxidation on a catalyst (platinum or platinum ruthenium alloy) layer to form CO_2. Water ingestion occur at the anode. However, water production takes place at the cathode when protons (H^+) react with oxygen as illustrated above. For DMFC, the electrode reactions are as follows:

At the anode:

$$CH_3OH + H_2O \rightarrow 6H^+ + 6e^- + CO_2 \tag{6.6}$$

At the cathode:

$$\frac{3}{2}O_2 + 6H^+ + 6e^- \rightarrow 3H_2O \tag{6.7}$$

The overall reaction:

$$CH_3OH + \frac{3}{2}O_2 \rightarrow 2H_2O + CO_2 \tag{6.8}$$

It is without doubt that the electrolyte film is the central component in DMFCs and PEFCs. For high coulomb efficiency, the electrolyte must have (i) high proton conductivity, (ii) sufficient mechanical integrity, and chemical and electrochemical stability [85]. Polymers used for PEFCs are as follows:

(a) Perfluorinated ionomers, e.g. perfluorosulfonic acid (PFSA), perfluorocycloalkene (PFCA), and perfluorosulfonylimide (PFSI). PFSA membranes consist of C–F backbone with sulfonic acid containing perfluoro groups as side chains. PFSA is the product of tetrafluoroethylene and a perfluorovinyl ether terminating in a sulfonyl fluoride group.
(b) Partially fluorinated polymers, e.g. sulfonated trifluorostyrene-grafted poly(tetrafluoroethylene) (PTFE-*g*-TFS) and polyvinylidene difluoride-grafted polystyrene sulfonic acid (PVdF-*g*-PSSA).
(c) Non-fluorinated membranes with aromatic backbone. They are a class of ionomer membranes with ionogenic groups attached to a fluorine-free polymeric backbone. The ionogenic groups are responsible for the H^+ conduction. Non-fluorinated polymers are generally composed of units of polyaromatic or polyheterocyclic units, such as polysulfones, poly(ether

sulfone)s, poly(ether ketone)s, poly(phenylquinoxaline), polybenzimidazole (PBI), and polyimides. They are thermally, mechanically, and chemically stable at temperatures of at least 80 °C [86]. On doping, polybenzimidazole (PBI)-based electrolytes exhibit high proton conductivity. The following example shows the PBI doped with H_3PO_4 and the conducting ions involved:

$$H_3PO_4 + PBI \rightarrow H_2PO_4^- \rightarrow +PBI\text{-}H^+ \tag{6.9}$$

Protons travels along the polymer backbone through hydrogen bonds assisted by PO_4^- anions [87].

(d) Hydrocarbon membranes. These offer some definite advantages. They are cheap, available in the market, and exhibit good water uptake. As an example, PVA membranes are good methanol barriers and therefore suitable for DMFCs.

(e) Acid–base blends. The concept of acid–base blends involves blending of basic and acidic polymers through ionic cross-linking. Ionic cross-links occur between these polymers when protons are transferred from the acidic to the basic polymer. The non-covalent linking of the polymers also resulted in thermal stability improvement. Figure 6.5 shows the structure of typical polymers used in polymer electrolyte membrane fuel cells (PEMFCs) and DMFCs.

Acid–base blends include sulfonated poly(ether ether ketone) (SPEEK)/polyethyleneimine (PEI), sulfonated polyphenylsulfone (SPPSU)/PEI, SPEEK/PBI/poly (4-vinylpyridine) (P4VP), SPPSU/PEI, and PVA/H_3PO_4. Among the natural polymers that are often used as electrolyte hosts are chitosan. The benefits of chitosan that has been frequently reported are non-immunogenic, nontoxic, and biocompatible [88]. Chitosan also exhibited its polyelectrolyte behavior as it possess protonated amino group in its chemical structure [89]. Table 6.7 lists some examples of DMFCs using chitosan-based PEs.

Membranes composed of PVA and different concentrations of chitosan were prepared using solution casting method [90]. The blended chitosan was cross-linked to improve the thermal and chemical stabilities of the membrane and simultaneously reduce the swelling ratio. For cross-linking, a solution containing glutaraldehyde, hydrochloric acid, and acetone was used. The solution was stirred at 313 K for different periods ranging from one to four hours. Compared to methanol permeability of Nafion®115 (18×10^{-7} cm^{-2} s^{-1}), samples that were heated for one hour have a slightly lower value of 16×10^{-7} cm^{-2} s^{-1}. However, in terms of selectivity, all of PVA–chitosan samples possess higher value than Nafion 115.

Another work of crosslinked chitosan was done by using sulfuric acid. The prepared membrane was applied in DMFCs. The sample was proved to be excellent as the chitosan membrane water uptake was higher compared to the commercial, Nafion 117 at temperature 293–333 K. Thermal stability of the membrane was reported up to 503 K with methanol permeability value of 1.4×10^{-6} cm^{-2} s^{-1}. The proton resistance and fluxes across the membrane is 4.9 mS cm^{-1} and 2.73 mol cm^{3} s^{-1}, respectively.

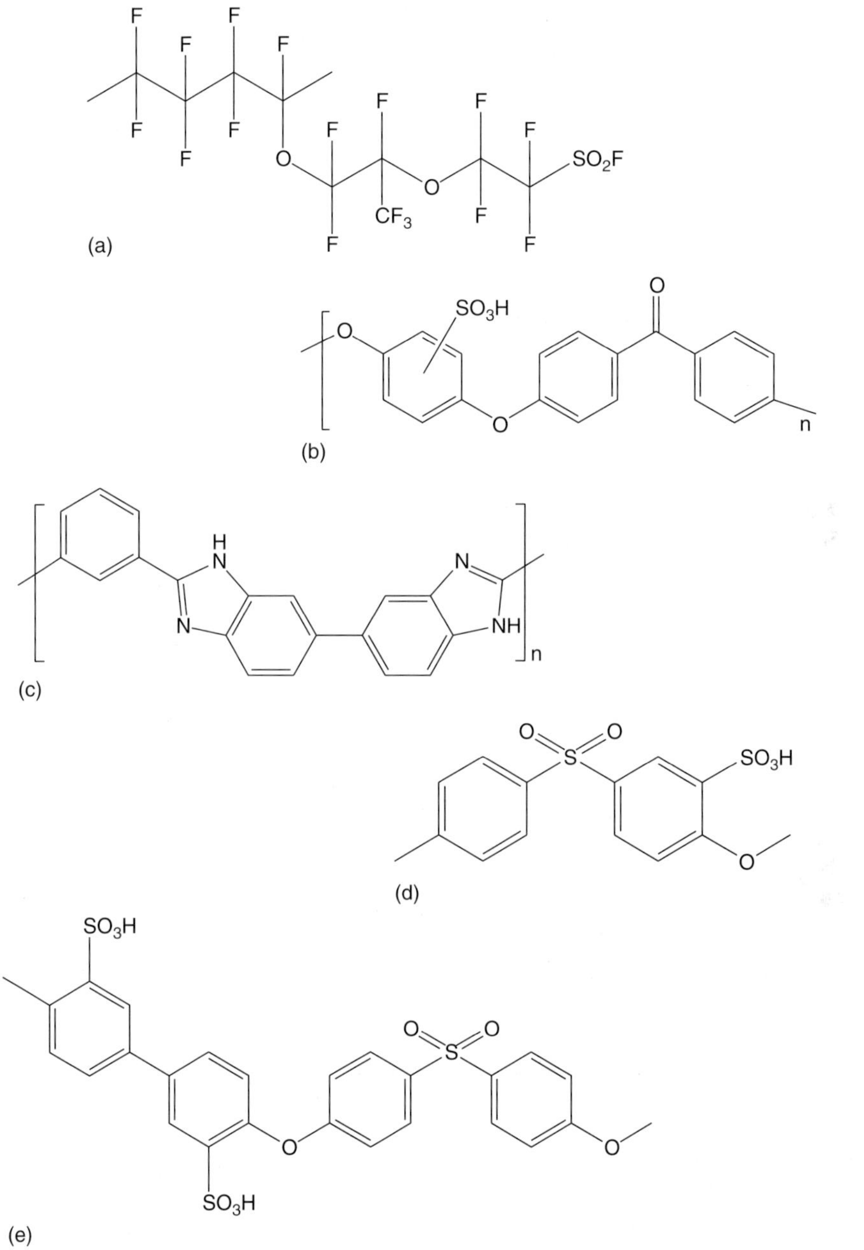

Figure 6.5 Structure of some common polymers used as host for ionic conduction in fuel cells. (a) Perfluorosulfonic acid ionomers. (b) Sulfonated poly(ether ether ketone) (SPEEK). (c) Poly[(2,2′-(*m*-phenylene)-5,5′-(bibenzimidazole)]. Most studied PBI membrane used as an electrolyte for high temperature PEMFCs. (d) Sulfonated polyethersulfone (PES). (e) Sulfonated polyphenylsulfone (SPPSU).

Table 6.7 Various polymer electrolytes for DMFCs.

Electrolyte	Tensile strength ($N\ mm^{-2}$)	Elongation (%)	Liquid uptake (%)	σ ($mS\ cm^{-1}$)	Methanol permeability ($\times 10^{-7}\ cm^2\ s^{-1}$)	Selectivity ($10^3\ S\ cm^{-3}\ s$)	References
PCS91	43 ± 2	115 ± 8	157.3 ± 3	—	24.98	—	[90]
PCS91-G1h	44 ± 2	71 ± 12	99.8 ± 12	20	16.24	12.52	
PCS91-G2h	45 ± 1	30 ± 9	89.6 ± 7	13	9.421	13.46	
PCS91-G3h	46 ± 2	26 ± 8	75.3 ± 8	10	7.421	13.23	
PCS91-G4h	45 ± 1	21 ± 5	58.9 ± 8	—	—	—	
Nafion 115	—	—	—	14	18	7.8	
			(alkaline uptake)				
CS flakes					14.0		[91]
Nafion 117					39.0		
CS/M(0)-25			129		8.7		[92]
CS/S(30)-25			—		7.8		
CS/M(15)-25			84		5.3		
CS/M(15)/S(30)-60			90		4.9 (lowest)		
			(methanol–water uptake)				
CS			90		10	26	[93]
CS-NaY(10)			100		8.5	31	
CS-H_2NY(10)			90		6.5	28	
CS-HO_3SY(10)			87		5.5	43	

Nafion 117			—		39	17.5	
			(water uptake)		(12 M methanol)	(12 M methanol)	
CS			20.8				
CS-NaY(10)			23.5				
CS-H2NY(10)			19.7				
CS-HO3SY(10)			19.0				
			(methanol uptake)				
Nafion 117					39		[94]
CS					10		
CS/3A(30)			78 (21)		—		
CS/4A(30)			69 (24)	18.3	8	18	
CS/M(30)			52 (28.5)		15.5	20	
CS/Z(30)			54 (31)	17.7	9		
CS/13X(30)			H_2O (CH_3OH) uptakes				
Nafion 117	—	—	15.2	84	24	35	[95]
CS	20.1	7.4	11.6	3.5	—	—	
CS/PMMA	38.7	4.3	18.6	15	2.7	56	
CS/PWA	36.5	4.7	17.0	12.2	3.3	37	
CS/SiWA	32.3	5.2	17.4	10	3.8	26	

Chitosan (CS)-controlled membrane (CS-T), CS/mordenite (CS/M(*X*)-T) hybrid membranes, sorbitol-plasticized CS (CS/S(*Y*)-T) membranes, and sorbitol-plasticized CS/mordenite (CS/M(*X*)/S(*Y*)-T) hybrid membranes (thickness 50–80 μm) were prepared by Yuan et al. [92] for DMFC. Various weight percentage *X* of modernite to CS was varied from 5 to 20 in steps of 5 wt%. For the weight percentage of sorbitol to CS, *Y* was varied as 20, 30, or 40 wt%. The membrane preparation temperatures were carried out at $T = 298$ or 333 K. This work showed that mordenite has made the chitosan chain favorably more rigid and created interfacial voids. Sorbitol, on the other hand, helped to make the chitosan flexible and suppressed interfacial voids. Increasing the temperature of formation further enhanced the plasticization action. The combined effect of mordenite and sorbitol and also the formation temperature of 333 K gave the best methanol permeability value of 4.9×10^{-7} cm^2 s^{-1} in 12 M water–methanol solution at 298 K. This is lower than the methanol permeability of Nafion 117 membrane (2.3×10^{-6} cm^2 s^{-1}) under similar conditions. The methanol resistance together with the small free volume cavity (FVC) size contributed to the improvement.

Wu et al. [93] filled the CS membrane with surface-modified Y zeolite for DMFC application. The modification was to improve interfacial morphology at the CS matrix and inorganic filler interface. CS, CS-NaY(X), CS-H_2NY(X), and CS-HO_3SHY(X) were prepared (X = zeolite mass content). As reported by the authors, the proton conductivity of Nafion 117 and pure chitosan are 6.91×10^{-2} and 2.61×10^{-2} S cm^{-1}, respectively. The composite samples have conductivity 2.30×10^{-2}, 2.09×10^{-2}, and 2.58×10^{-2} S cm^{-1}, for CS-NaY(10), CS-H_2NY(10), and CS-HO_3SHY(10), respectively. The selectivity of the composite membrane is comparable to Nafion 117 at 2 M methanol concentration. The transit state formed between chitosan and zeolite reduced the interfacial voids.

The solution casting method was performed to prepare pure hybrid chitosan–zeolite membrane for DMFCs [94]. CS was cross-linked in H_2SO_4. Samples were designated CSN(*Y*) where *N* represents the types of zeolites 3A, 4A, 5A (all these have Si:Al = 1 : 1), 13X (Si:Al = 1.3 : 1), and M (mordenite) with Si:Al = 6.5 : 1 and Z is HZSM-5 (Si:Al = 25.0 : 1) zeolite (hydrophobic and hydrophilic). *Y* represents zeolite to CS weight ratio. The nature of zeolites affect membrane performance as the hydrophilic site will improved the FVC. Nevertheless, FVC can be reduced by the hydrophobic site of zeolites. The hydrogen bonding formed between chitosan and zeolites with low Si/Al ratios are weak, which led to increase in FVC size. The FVC size is smaller in zeolites with high Si/Al ratios (mordenite and HZSM-5). Methanol transport occurred via solution diffusion mechanism. High Si/Al ratio zeolites (hydrophobic) exhibit small-sized FVC, low swelling, water uptake, and methanol permeability, but higher methanol uptake. In CS membranes with high Si/Al ratios, the strong hydrogen bonding considerably "tightened" the CS chains that decreased the FVC. The opposite occurred in CS with hydrophilic zeolites. The cross-linked chitosan chains contain many amino and hydroxyl groups. They also contain sulfate SO_4^{2-} and protonated amino groups, NH_3^+ in the chitosan structure. Hence, transportation of the H^+ ions could occur via the hopping mechanism. Apart from hopping in a coordinated manner across a collection of H_2O molecules, the protons could transport via the SO_4^{2-} and $-NH_3^+$ sites. Composites with hydrophilic zeolites also allowed vehicle mechanism of proton transport in the form of H_3O^+ ions. Thus, protons travel mainly within chitosan bulk phase rather than zeolite. However, the proton conductivity will be affected if plenty of water involved. The zeolite particles in CS increased proton pathways, thus leading to conductivity decrease compared to pure CS membrane. However, water in composites filled with hydrophilic zeolite increased with filler content, which is favorable for proton conductivity. In membranes filled with hydrophobic zeolites the opposite occurred. The authors believed that the proton conductivity of chitosan and CS/zeolite composites is sufficient for application in DMFCs as proton exchange membranes.

Cui et al. [95] prepared proton conductors from chitosan and phosphomolybdic (PMA), phosphotungstic (PWA), and silicon-tungstic (SiWA) acids. The activation energy for the chitosan–PMA, chitosan–PWA and chitosan–SiWA membranes were ~0.12, ~0.13, and ~0.11 eV, implying that proton conduction occurred by way of the Grotthus and vehicular mechanisms based on the criterion for hopping or jump mechanism that requires energy between 0.145 and 0.415 eV. The low methanol permeability suggests the possible use of these membranes for DMFCs.

6.4.5 Electrochromic Windows

Electrochromism is a situation where a material exhibits a change in color when subjected to a voltage. The electrochromic technology began after S.K. Deb [96] described WO_3 coloration mechanism during the 1970s. However, the history of coloration could have started when Johann Jacob Diesbach first synthesized hexacyanoferrate or Prussian blue (PB) around 1706 [97]. PB changed color from transparent to blue on oxidation of iron. A schematic diagram of an ECW is depicted in Figure 6.6.

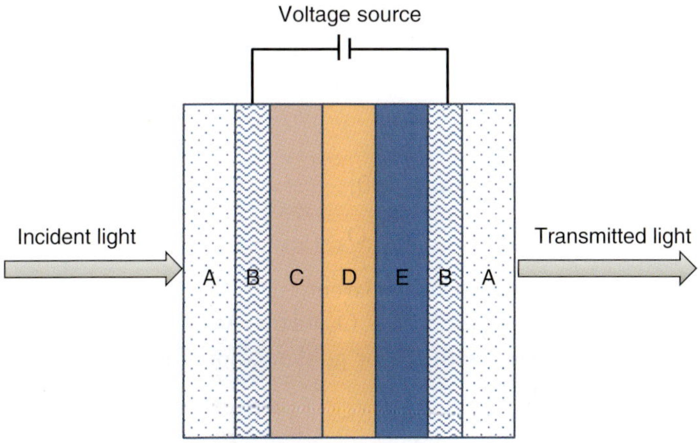

Figure 6.6 Electrochromic device structure (**A**, glass/plastic substrate; **B**, transparent conducting oxide; **C**, ion storage layer; **D**, electrolyte; and **E**, electrochromic layer).

The optical properties of electrochromic materials such as absorbance, transmittance, emittance, and reflectance change when charge intercalates or de-intercalates from them. A schematic representation of an electrochromic device (ECD) as shown in Figure 6.6 consists of substrates that are either glass or polyester (**B**). A TCO or transparent conducting oxide coats these substrates, which can be either indium-doped tin oxide (ITO) or fluorine-doped tin oxide (FTO). Deposited on one of the TCOs on one side of the ECD is an electrochromic film. WO_3 is an example. The WO_3 layer is a mixed conductor that is able to conduct both electrons and ions. On the other side, TCO is deposited an ion storage layer with electrochromic properties complementary to the former electrochromic layer. Sandwiched between these layers is the PE, which is the subject of this chapter. The preferred ions are protons and Li^+ ions since the ions should be small for better mobility.

On application of a small DC voltage of about one volt between the TCOs, ion transport occurs between the ion storage layer (**C**) and electrochromic layers (**E**). Electrons from the TCOs alter the optical properties. Voltage reversal brings back the original properties [98].

ECDs can operate in the transmission or reflectance mode. Smart window application uses the transmission mode. Smart windows can reduce entrance of light and heat to a comforting level. This can help to minimize the use of air-conditioning, save energy, and cut cost of electricity bills. In the reflectance mode, a reflecting surface (that of Ag, Al, and Au) replaces one of the transparent conducting electrodes (TCEs). The surface controls the intensity of the reflected light. An application of this type of ECD is in car mirrors for rear viewing.

6.4.6 Electrochromic Materials

A large number of electrochromic materials are available. In a review by Mortimer [99], examples include PB systems, transition metal oxides, conducting polymers, the viologens, metallopolymers, and metal phthalocyanines. In this chapter, we focus on transition metal oxides.

6.4.6.1 Transition Metal Oxides

(i) Tungsten trioxide (WO_3)

The electrochromic reaction occurs as [100]

$$WO_3 + yLi^+ + ye^- \leftrightarrow Li_yW^{6+}_{(1-y)}W^{5+}_yO_3 \tag{6.10}$$

At low "*y*," WO_3, which was initially transparent, turns blue on intercalation of Li^+ ions when the product $Li_yW^{6+}_{(1-y)}W^{5+}_yO_3$ forms. At high "*y*," Li^+ insertion forms a metallic "bronze," with red or golden color. The formation of metallic "bronze" is irreversible.

(ii) Molybdenum oxide (MoO_3)

The electrochromic reaction occurs as [101]

$$MoO_3 + y(M^+ + e^-) \leftrightarrow M_yMo^{6+}_{(1-y)}Mo^{5+}_yO_3 \tag{6.11}$$

MoO_3 also turned from transparent to blue on M^+ intercalation. Charge transfer between neighboring W^{5+} (Mo^5) and W^{6+} (Mo^{6+}) sites is responsible for the blue coloration.

(iii) Vanadium oxide (V_2O_5)
The electrochromic reaction occurs as [102]

$$V_2O_5 + y(M^+ + e^-) \leftrightarrow M_yV_2O_5 \tag{6.12}$$

V_2O_5 turned pale blue from yellow on M^+ intercalation and the formation of $M_yV_2O_5$. In these examples, reduction (cathodic ion insertion) yields the more intensely absorbing redox state.

(iv) Iridium hydroxide $Ir(OH)_3$
The proposed mechanisms of coloration, with both proton extraction and anion insertion routes are

$$Ir(OH)_3 \leftrightarrow IrO_2 \cdot H_2O + H^+ + e^- \tag{6.13}$$

$$Ir(OH)_3 + OH^- \leftrightarrow IrO_2 \cdot H_2O + H_2O + e^- \tag{6.14}$$

The color change that occurred is from transparent to blue-black.

A schematic representation of a smart or ECW is as shown in Figure 6.7. There are five layers illustrated in Figure 6.7:

1. A transparent plastic or glass substrate coated with FTO or ITO – the TCO substrate.
2. A transparent electroactive cathode layer coated on the TCO (in 1) that becomes colored when intercalated with Li^+ ions.
3. A Li^+ conducting PE.
4. An electroactive anodic layer or counter electrode coated on another TCO (see 5) into which Li^+ ions intercalate and bleach the cathode.
5. Another TCO substrate.

On coating the TCO (in 1) with an electrochromic film (e.g. WO_3), the cathode is formed [105]. As in LIBs, a Li^+ ion electrolyte connects the transparent electrodes. On applying another oxide layer (e.g. CeO_2), the anode is formed and acts as a storage for ions. On applying a voltage of about 1–2 V across the

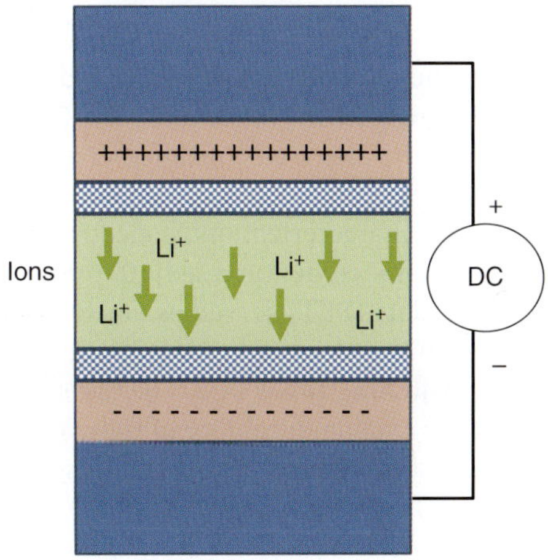

Figure 6.7 Structure of an ECW [103, 104].

electrodes, Li^+ ions in the anode will travel to the cathode through the electrolyte and intercalate in the WO_3 film. The initially transparent WO_3 will be colored: i.e. the transparency of the film is altered [103]. On reversing the voltage polarity, the Li^+ ions de-intercalate the WO_3 film deposited on the TCO substrate that served as the cathode. The Li^+ ions migrate through the electrolyte and intercalate into the ion storage film. The cathode bleaches and the device regains its original properties until all Li^+ ions have deintercalated the cathode. Continuous bleached or colored states are achievable depending on the duration of incident light.

In reference [106], cyclic voltammetry performed on the ECD was from −2.5 to 2.0 V at 10 mV s^{-1}. On decreasing the voltage to 0 V, the ECD was colorless (in the bleached state) indicating that the electrochromic layer that forms the cathode was oxidized to WO_3, Eq. (6.15).

$$\underset{\text{dark blue}}{Li_xWO_3} \leftrightarrow \underset{\text{colourless}}{WO_3} + xLi^+ + xe^- \tag{6.15}$$

On the other hand, the anodic coloring reduced to $Li_2Fe^{2+}[Fe^{2+}(CN)_6]$ (Everitt's salt), Eq. (6.16).

$$\underset{\text{Prussian Blue}}{LiFe^{2+}[Fe^{3+}(CN)_6]} + Li^+ + e^- \leftrightarrow \underset{\text{Everitt's salt}}{Li_2Fe^{2+}[Fe^{2+}(CN)_6]} \tag{6.16}$$

On application of a more negative voltage, $Fe^{2+}(CN)_6$ in $Li_2Fe^{2+}[Fe^{2+}(CN)_6]$ undergo oxidization to $LiFe^{2+}[Fe^{3+}(CN)_6]$ that turned the color to PB. The reduction and oxidation peaks were observed at approximately −1.2 and −0.4 V, respectively.

The process occurs on reduction of the cathode layer containing WO_3 to Li_xWO_3 and the ECD turned from no color to dark blue when reduction current increased. A small peak due to reduction was observed in the range from −2.0 to −2.2 V in the cyclic voltammogram trace and attributed to the reversible process as shown below:

$$\underset{\text{PB}}{LiFe^{2+}[Fe^{3+}(CN)_6]} \leftrightarrow \underset{\text{Prussian Green (PG)}}{Fe^{3+}[Fe^{3+}(CN)_6]} + Li^+ + 2e^- \tag{6.17}$$

The application of 1.8 and −2.0 V led to the reversal of the optical transmittance of the device from bleached with 57.9% to darkened state (5.5%) at 695 nm, respectively. The transmittance change (ΔT) at 695 nm remained 44.5% after 2250 cycles.

For the characterization of electrochromic without ITO, the reader can refer to the reference denoted as [107]. Screen-printed silver grids on polyethylene terephthalate (PET) served as electrically conductive electrodes. The ECDs performed 100 cycles before degradation. Between redox states ΔT was 50% with faster switching times of 4.5 seconds. The authors have provided some insights about the cycling durability of these devices.

The best of hydroxypropyl cellulose (HPC)-based PEs [108] exhibited conductivity of 35 μS-cm^{-1} at 298 K and 0.11 mS-cm^{-1} at 323 K. The T_g was −37 °C.

The electrolyte was thermally stable up to 130 °C. PB, PANI (polyaniline), and PEDOT:PSS (poly(3,4-ethylenedioxythiophene):poly(styrene sulfonate)) were chosen as electrochromic materials to characterize the ECDs because during oxidation $Li_2Fe^{2+}[Fe^{2+}(CN)_6]$ turned to blue, from the bleached or transparent state. In the case of PANI, transparent yellow changed to green and blue but blue from PEDOT:PSS changed to transparent. Therefore, PEDOT:PSS, PB, and PANI were blue at the same time. The results obtained showed that electrolytes based on HPC are potential candidates for application in ECDs with PEDOT:PSS as primary and PANI and PB as ion storage materials. Table 6.8 lists some ECDs and their characteristics.

Lin et al. [109] reported an ITO/MoO_3/ethylcellulose-$LiClO_4$-PC/NiO/ITO device with two electrochromic layers deposited on an ITO conducting glass substrate. As in the device configuration, these are the NiO anodically and an MoO_3 cathodic colored layers. A GPE comprising ethyl cellulose, $LiClO_4$, and PC separates the electrochromic layers. The bleaching and coloration of both oxides occur at the same time. This increased coloration efficiency and optical contrast.

With the suitable voltage, the dual electrochromic layer ECD is able to effectively filter visible light up to 2500 nm in the near-infrared (NIR). Thus, the device is able to prevent radiated heat from passing through it. As more ions intercalate into the MoO_3, the device becomes darker in color. The ECD showed good electrochromic and optical properties between bleached and colored states in the investigated wavelength.

6.5 Challenges and Improvements

6.5.1 In Electrolytes

Electrolytes are usually in liquid form. Liquid electrolytes (LEs) provide good contact area with electrodes, absorb volume changes in electrode size during charge and discharge of a device, and have high conductivity at room temperature, but are prone to leakages if the device accidentally damaged. LEs may contain volatile materials [65], and the volatile materials can harm the environment if released during damage. These drawbacks can cause safety concerns. To ensure safety, electrolytes need to be solid either and ceramic or polymers are potential electrolyte candidates. However, ceramic electrolytes do not exhibit good electrode/electrolyte contacts and volume changes during Li^+ ion intercalation and de-intercalation can lead to poor device performance and failure. SPEs also suffer from low room temperature ionic conductivity. To benefit from the mechanical strength of solid electrolytes and the high conductivity of LEs, PEs need to be prepared as gels. The polymer chosen must have good mechanical, thermal, and electrochemical stabilities. Researchers studied PAN [111], PMMA [112], PEO [113], PVdF [114], and PVdF-HFP [115] to name a few in the effort to obtain the appropriate polymer host. A wide electrochemical stability range is required. Other than ILs, nanofillers can improve the battery performance. The issue of

Table 6.8 Some characteristics of ECDs using GPEs.

Device	Characteristics				References
ITO/WO_3/PMMA–$LiClO_4$–PC-SN/Prussian Blue/ITO	$\sigma_{electrolyte}$ = 1.13 mS-cm^{-1}; response time = 10 s; ΔT = 52.4% at 695 nm; ΔV = −2.0 to 1.8 V				[106]
PET/Ag grid/PEDOT:PSS/PEO–$LiClO_4$-based electrolyte/PEDOT:PSS/Ag grid/PET	Response time 4.5 s; ΔT = 50%; ΔV = −1.5 to 1.5 V				[107]
ITO/PB/HPC + CH_2Cl_2 + PEG + Bu_4NBF_4 + DBSA/PEDOT:PSS/ITO (31 wt% DBSA)	ΔT = 35% at 650 nm				[108]
ITO/PB/HPC + CH_2Cl_2 + PEG + Bu_4NBF_4 + DBSA/PEDOT:PSS/ITO (31 wt% DBSA)	ΔT = 33%				
ITO/MoO_3/ethyl cellulose–$LiClO_4$–PC/NiO/ITO	ΔT ranged from 55.1% to 50.3% ΔOD ranged from 0.891 to 0.693 ΔV = −1.6 to −1.3 at 550 nm				[109]
	Switching time (s)		ΔT% (at 0 and 1000 cycles)		[110]
	C to B	B to C			
ITO/PEDOT/P_{14}TFSI/PPy/ITO	0.7	1.9	47	43	
ITO/PEDOT/$(PEO)_8$LiTFSI + 25%EC/PC/PPy/ITO	1.3	2.0	45	37	
ITO/PEDOT/$(PEO)_8$LiTFSI + 25%P_{14}TFSI/PPy/ITO	1.1	4.0	44	43	
ITO/PEDOT/$(PEO)_8LiClO_4$ + 25%EC/PC/PPy/ITO	1.0	1.4	44	36	
ITO/PEDOT/$(PEO)_8LiClO_4$ + 25%P_{14}TFSI/PPy/ITO	3.8	6.5	45	43	

ΔT, transmission change; ΔV, applied voltage.

Li^+ cation transference number is also another challenge. LIBs require the Li^+ transference number to be high. To enhance this parameter, additives were added into the electrolyte to immobilize the large anion of the salts [116].

Researchers are also improving electrolytes for DSSCs. One of the problems is the liquid electrolyte that contains I^-/I_3^- redox couple, which is corrosive, volatile, and prone to leaks. It can also react with the dye, all of which limit long-term stability.

To avoid problems connected with corrosion and reactivity of the liquid electrolyte, researchers at Northwestern University used fluorine-doped cesium tin iodide ($CsSnI_{2.95}F_{0.05}$), an inorganic semiconductor to replace the liquid electrolyte. The fabricated cells can achieve photon to electricity efficiencies of about 10%. Grätzel's group have also searched for compatible dyes and redox couples. They found that using DSSCs with zinc porphyrin dye and electrolyte with Co^{2+}/Co^{3+} redox mediators efficiencies exceeding that of DSSCs utilizing the iodide/ruthenium systems are achievable. However, implementing a laboratory finding in a manufacturing process may take years [117].

6.5.2 In Devices

6.5.2.1 DSSCs

The major challenges to overcome before DSSCs may enter the consumer market are issues pertaining to their low efficiency and low stability. These challenges do not depend on a single factor only. Factors responsible include dark current, poor dye performance in the NIR region, poor electrode/electrolyte contacts, and electrolyte degradation due to UV absorption. This chapter is only concerned with the electrolytes.

The return of electrons for dye regeneration from the counter electrode to the oxidized dye molecules occurs by way of the electrolyte through a process of redox diffusion. The highest photon to electricity efficiency ever achieved by a DSSC used an electrolyte solution comprising iodine ions and iodine dissolved in acetonitrile. However, acetonitrile evaporates easily. This led to efficiency drop at high temperatures and instability with prolonged use. Although the electrolyte can be sealed and contained in a cell, for safety reasons, it is better to solidify the electrolyte and prevent accidental damages to the cell. Solid or quasi-solid electrolytes have better mechanical strength than liquids. A DSSC that utilized a quasi-solid electrolyte comprising a symmetrically alkyl-substituted imidazolium-based ILs exhibited light to electricity stable efficiency between 7% and 8% for 1000 hours under outdoor conditions and 1 sun [118].

With the intention to prepare an all solid-state cell, research has focused on the use of CuI [119–121], CuSCN [122], and certain conducting polymers, for instance, PANI [123] and poly(3,4-ethylenedioxythiophene) (PEDOT) [124]. PEDOT is a good hole-conducting polymer that can substitute the iodide/triiodide redox LE. The hole conductors, namely, Spiro-OMeTAD, which is short for 2,2′,7,7′-tetrakis[*N*,*N*-di(4-methoxyphenyl)amino]-9,9′-spirobifluorene and poly(3-hexylthiophene) (P3HT), were studied by Yang et al. were compared in solid-state DSSCs [125].

6.5.2.2 Fuel Cell

The trade name of the perfluorinated polymer membranes extensively used for fuel cells is Nafion, a product of DuPont. Nafion is a sulfonated tetrafluoroethylene-based fluoropolymer-copolymer and Figure 6.8 depicts its structure. H^+ on SO_3H jumps from one acid site to another. For Nafion 120, $m = 1$; $x = 5–13.5$; $n = 2$; $y = 1$. Its equivalent weight is 1200 and thickness is 260 μm [127]. Nafion 117, Nafion 115, and Nafion 112 all have the same equivalent weight of 1100 and their thicknesses are 175, 125, and 80 μm, respectively.

Asahi Glass has produced perfluorinated polymer named Flemion®. Asahi Chemical produces Aciplex-S®. Dow Chemical produces the membrane that goes by the trade name Dow®. According to Smitha et al. [128], the DuPont product is more superior attributable to the excellent conductivity, chemical stability, and mechanical strength.

However, there are limitations. Nafion dehydrates above 80 °C, leading to conductivity loss [129]. This is a worrying issue because for a better performance [130] and CO tolerance of the platinum catalyst in the cathode [131], higher temperatures are required. Incorporation of TiO_2 [132, 133], silica [134, 135], and zirconium phosphate [136] into Nafion water channels through in situ chemical reactions can enhance the working temperature up to 373 K.

Another disadvantage of Nafion membranes is that the membranes allow methanol and water crossover [137, 138]. These react chemically with oxygen at the cathode and forms H_2O and CO_2 without producing electrons. Nafion requires water to exhibit good proton conductivity, but too much water uptake can reduce ionic conduction. The reaction at the cathode is as follows:

$$\frac{3}{2}O_2 + 6H^+ + 6e^- \rightarrow 3H_2O \tag{6.18}$$

The reaction is slow, but is still faster than methanol reduction. This leads to a meaningful drop in voltage and performance. The methanol reaction with oxygen is

$$CH_3OH + \frac{3}{2}O_2 \rightarrow CO_2 + 2H_2O \tag{6.19}$$

The methanol permeating via the membrane thus aggravated the cathode over potential. No net electrons are produced.

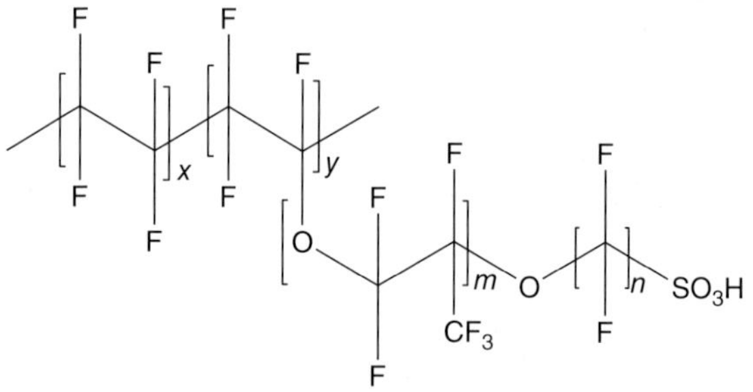

Figure 6.8 Basic chemical structure of Nafion [126].

Although Nafion possesses good H^+ conductivity, displays good thermal and chemical stability, and is used in PEFCs and DMFCs, there is a need to modify it to reduce the methanol crossover [139] that can result in open-circuit potential loss by 0.15–0.2 V and poisoning of the Pt electrocatalyst [140]. Methanol migration from anode to cathode delayed the industrial materialization of DMFCs and a lot of research carried out to modify Nafion membranes. These activities include Nafion polymerization with poly(1-methylpyrrole) [141] and the establishment of composite membranes of Nafion with silica [134, 135], zirconium phosphate [136], cesium ions [142], and poly(furfuryl alcohol) [143].

With reduction in methanol crossover, the DMFCs have improved in performance. To arrive at a meaningful reduction in crossover, the oxide content, e.g. in Nafion silica composite is limited ~20 wt% silica [134]. However, this will limit the proton conduction and the mechanical properties due to changes in membrane microstructure. Several approaches to reduce methanol crossover that have been attempted include (i) inserting a Pd thin film between Nafion membranes [144], depositing a Pd and/or Pd–Cu alloy thin film on the Nafion surface [145–147], and putting a layer of Pd nanoparticles via ion exchange succeeded by chemical reduction [148]. Unfortunately, this will reduce the conductivity of the cell. The scattered Pd particles modified the membrane's microstructure causing the cell's performance and stability to decrease.

Yang and Manthiram [149] have used SPEEK thin film multilayers to curb methanol crossover. Si et al. [150] prepared 3-layered membranes comprising two conductive layers and also a methanol barrier film in order to curb the methanol crossover. It can also be done by casting a non-conductive polymer, for example, PVA, onto the Nafion membranes [151]. Unexpectedly a significant reduction in proton conductivity accompanied the suppression of methanol crossover.

6.5.2.3 Batteries

The central issue about batteries is electrolyte leakage. According to EU Regulation (EG) Act 1272/2008, DEC and EMC are flammable liquids. EC, DEC, and EMC are harmful and can cause skin and eye irritations [152]. Besides toxicity, there are also worries about battery safety particularly fire outbreaks and explosions. Fires associated with the Boeing 787 and Tesla Model S caused by LIBs have signaled the importance of battery safety [153]. Tidblad and coauthors [154] suggested that upcoming research should use solid or gel-type electrolytes to eradicate the danger of electrolyte leakage and improve resilience to abuse.

One of the important properties of GPEs for lithium batteries is the ESW. If lithium metal oxide is the active material for cathode, the ESW of the GPE must be at least 4.5 V (versus Li metal) [155]. The authors reported a gel electrolyte with ESW of over 4.8 V.

Although there is renewed interest in the use of lithium metal anode in batteries, manufacturers are reluctant to move away from the established LIB design [156]. The use of Li metal anode may be required to realize the advanced 5 V LIBs with high power and energy density. If dendrite formation is the reason for not using lithium metal anode, the addition of ILs in the GPEs or the use of CGPEs may provide some solutions. The addition of ILs to GPEs not only can ensure high conductivity but also can suppress dendrite growth that will ensure long-term stability of the anode. The uniformly dispersed ceramic fillers in the GPE may provide stronger and more complicated interfaces that can restrict the dendrite penetration.

6.5.2.4 EDLCs

Many EDLCs function at potentials above the electrolyte limit of thermodynamic stability. ESW of aqueous acidic (1 M H_2SO_4) or basic (6 M KOH) electrolytes is about 1 V, limited by the maximum thermodynamic stability of water, i.e. 1.23 V. Neutral aqueous electrolytes have a higher stability between 1.6 and 2.2 V [157]. Organic electrolytes can give cell voltage between 2.7 and 2.8 V [158]. Hence, industrial EDLCs use organic electrolytes since the high voltage that they help to supply gives rise to high electrical energy storage. The use of IL electrolytes can result in a higher cell voltage between 3 and 5 V. These are electrolytes in the liquid state. However, an alginate-based gel electrolyte with 1-ethyl-3-methylimidazolium tetrafluoroborate ($EMImBF_4$) IL used in EDLC with activated carbon (AC) electrodes exhibited thermal (~553 K) and electrochemical (~3.5 V) stability [159]. The charge/discharge characteristics also showed that the internal resistance of the EDLC with configuration AC/alginate-$EMImBF_4$/AC is smaller than AC/$EMImBF_4$/AC EDLC.

6.5.2.5 Electrochromic Windows (ECWs)

Apart from high transmittance change between colored and bleached states and good electrical stability, durability is another important property for ECW applications in smart windows, rear-view mirrors for cars, information displays, electrochromic sunglasses, and variable emittance infrared (IR) skins [160].

Durability is a concern when GPEs are used because the gels will in due course solidify. The change in forms affect the transmittance and reflectance of light and both decrease as the gel solidifies. Nguyen et al. [161] studied the effects of GPE on the movement of ions and also performance of the devices between 860 and 2500 nm of ECDs using GPEs based on P(VdF-TrFE) and the conducting polymer PANI. As an electrolyte changed in form from gel to semisolid and finally to solid, ionic conductivity decreased. Better performance is exhibited if the polymer host has a high dielectric constant. Thus, device performance depends on electrolyte form. Light modulation is an important issue to address and is related to conductivity of electrolyte. The amount of ion dissociation affects the magnitude of light modulation for devices with gel electrolytes. Ionic transport through the electrolyte and the interface influences light modulation. In solid-state devices, the number density of mobile ions, dissociation in the electrolyte, and mobile ion interaction with molecules of the electrochromic materials and electron leakage between the electrodes limit the device performance. Solving these issues successfully can produce devices that are more stable and have higher light modulation.

6.6 Future Aspects

6.6.1 Electrochromic Windows

Electrochromic or smart window technology is still in its infancy stage. Once matured it would bring great environmental benefits [162]. A light sensor can easily operate and control these windows. They do not use much electricity to switch from colored to bleach and this contributes to overall electricity saving. Colored to bleached and bleached to colored states can now take place in seconds. In the colored state, they block almost all incident light. This reduces cost in buildings where air-conditioning is a necessity.

It is expensive to install ECWs, as self-contained units. Application of thin electrochromic films on existing windows can save cost. However, scientists from Princeton University have developed a smart window operated using solar power [163]. The smart window comprises electrochromic polymers to regulate the color change.

Combining ECWs with other devices into a more reliable system is another futuristic happening. Light reflected by the ECWs can run DSSCs to produce electricity to charge LIBs. The energy stored is useful at night. How efficient the coupled device system will be is out of the scope of this chapter. We may be seeing a lot of electrochromic technology in the years to come that may not use liquid electrolytes.

6.6.2 Batteries

The startup company Seeo has come up with a solid-state LIB with improved battery lifetime, power storage, and safety [164]. Seeo's DryLyte™ batteries use a nonflammable nanostructured PE. The electrolyte in Seeo's DryLyte battery is totally in the solid-state and therefore safe. A DryLyte solid-state battery can provide higher energy density, which should allow vehicles to cover a longer distance. They are reliable with extended lifetime.

New battery chemistries (e.g. sodium and magnesium technologies) have emerged as possible high-energy alternatives to LIBs. Sodium is ranked sixth and magnesium eighth, and lithium 33 in abundance inside the Earth's crust. The production of sodium in 2012 was 280 million tons. For magnesium, production was 750 000 tons and for lithium 37 000 tons. The greater abundance of sodium and magnesium in the earth compared with lithium has probably also led to the development of different battery chemistries. The number of published items on magnesium batteries has increased since 1998 and in 2016, the total number of published items totaled more than 250. For sodium batteries, it is about 1300.

6.6.3 DSSCs

What will be the future of solar cell technology? According to [117], despite the rapid growth in solar energy, the percentage of photovoltaic power generation power is very small. An estimate by the US Energy Information Administration (EIA) revealed that in 2015, only 0.6% of electricity comes from solar energy. As a comparison, in Malaysia, the average monthly solar energy per square meter of area received is between 400 and 600 MJ m^{-2} [165]. Yet solar contribution to the total electricity generation in West Malaysia is only 0.007% [166]. The government of Malaysia has also emphasized dependency on renewable energy especially solar energy since the ninth Malaysia Plan.

Although the price of silicon photovoltaic has dropped, the photovoltaic technologies that are emerging promise to be even cheaper. Silicon photovoltaic is not as cheap as the non-silicon technologies [117], because pure silicon is required for the silicon cells currently in production. The manufacturing process is also tedious and expensive compared to DSSCs. Moreover, silicon is not a strong sunlight absorber and, therefore, the silicon layer in silicon cells needs to be reasonably thick. This will make the layer brittle. To avoid this, a rigid piece of heavy glass support is necessary. All these contribute to cost, which explains why the emerging technologies are cheaper. Moreover, the emerging photovoltaic fabricated on flexible supports only use cheap materials and inexpensive roll-to-roll printing techniques.

DSSCs have achieved a maximum certified efficiency of 11.9%. A major setback in DSSC is the usage of electrolyte in liquid state that limits their stability and endurance. This problem is partly solvable using polymer or gel electrolyte. It can be expected that future DSSCs use PEs as one of their components.

Several companies such as Dongjin Semichem in South Korea and G24 Power in Newport, Wales, are commercializing the DSSC technology. G24 Power offers a number of products for sale under the trade name GCell. However, there is no knowledge about the nature of the electrolyte used.

6.6.4 Fuel Cells

The PEMFC is an efficient energy conversion means for transportation application [167]. The use of PEMFC in vehicles has been demonstrated during the early 1990s. However, according to the authors [167], unsolved issues and the expensive cost for residential applications have delayed its application. The establishment of a "water-free" PE membrane could contribute to cost reduction. Water-free electrolytes do not need hydration and the cell can even function at temperatures above 373 K. This would also improve cell efficiency. Acid–base polymer networks created by attaching a strong acid to a highly basic polymer make good proton conductors that do not require hydration. Examples of basic polymers are poly(benzimidazole), poly(aminosilicates), poly(acrylamide), poly(ethyleneimine), PVP, poly(vinylalchohol), and even PEO. When combined with sulfuric, phosphoric, and halide acids as proton sources, they can be potential membranes to replace Nafion provided that their performance can match that of Nafion.

In Malaysia, PEMFC application is limited. However, there is a possibility of increased usage of PEMFCs in electrical cars [127]. The vehicle industry has the widest application in store for fuel cells. According to *Malaysia Automotive Info*, the number of passenger and commercial vehicles produced and assembled in Malaysia in 2016 is more than half a million. The price of fuel cell can be cheap if only 5% of the total vehicles on the Malaysian road uses this technology.

A roadmap for the Malaysian fuel cell application (see [127]) sets a 50% reduction in technology cost by 2025. It also predicts a 100% fuel cell use in motorbikes and 2% for passenger cars by 2035.

Providing electricity to residential areas is the ultimate aim for fuel cell application. Integrating the power generating system with fuel cells curbs power losses delivered through transmission lines. However, the system is expensive and residential application in Malaysia may be confined to high-income earners. The Malaysian Institute of Architects or *Institut Arkitek Malaysia* is working on the criteria on Green Building Indexing and have identified fuel cell as a suitable alternative for batteries currently used in power backup systems for buildings.

References

1 Sequeira, C.A.C. and Hooper, A. (eds.) (1985). *Solid State Batteries*. Martinus Nijhoff Publishers.

2 Wright, P.V. (1975). Electrical conductivity in ionic complexes of poly(ethylene oxide). *Br. Polym. J.* 7: 319–327.

3 Armand, M.B., Chabagno, J.M., and Duclot, M. (1979). Poly(ethylene oxide) is a medium capable of dissolving salt to form a conductive polymer. In: *Fast ion Transport in Solids, Electrodes and Electrolytes* (eds. P. Vashitshta, J.N. Mundy and G.K. Shenoy). Amsterdam: North Holland Publishers.

4 Choudhary, S. and Sengwa, R. (2012). Ionic conductivity of lithium perchlorate salt in polymeric electrolyte solutions and MMT nano-sheets dispersed colloids. *Indian J. Eng. Mater. Sci.* 19: 245–252.

5 Klongkan, S. and Pumchusak, J. (2015). Effects of the addition of $LiCF_3SO_3$ salt on the conductivity, thermal and mechanical properties of PEO-$LiCF_3SO_3$ solid polymer electrolyte. *Int. J. Chem. Eng. Appl.* 6: 165–168.

6 Xu, K., Zhang, S., Jow, T.R. et al. (2002). LiBOB as salt for lithium-ion batteries: a possible solution for high temperature operation. *Electrochem. Solid-State Lett.* 5: A26–A29.

7 Ding, M., Xu, K., Zhang, S. et al. (2001). Change of conductivity with salt content, solvent composition, and temperature for electrolytes of LiPF6 in ethylene carbonate-ethyl methyl carbonate. *J. Electrochem. Soc.* 148: A1196–A1204.

8 Klassen, B., Aroca, R., Nazri, M., and Nazri, G. (1998). Raman spectra and transport properties of lithium perchlorate in ethylene carbonate based binary solvent systems for lithium batteries. *J. Phys. Chem. B* 102: 4795–4801.

9 Verma, P., Maire, P., and Novák, P. (2010). A review of the features and analyses of the solid electrolyte interphase in Li-ion batteries. *Electrochim. Acta* 55: 6332–6341.

10 Agrawal, R.C. and Pandey, G.P. (2008). Solid polymer electrolytes: materials designing and all-solid-state battery applications: an overview. *J. Phys. D: Appl. Phys.* 41: 223001.

11 Stephan, A.M. and Nahm, K. (2006). Review on composite polymer electrolytes for lithium batteries. *Polymer* 47: 5952–5964.
12 Croce, F., Appetecchi, G.B., Persi, L., and Scrosati, B. Nanocomposite polymer electrolytes for lithium batteries. *Nature* 394: 456–458.
13 Das, S. and Ghosh, A. (2015). Ionic conductivity and dielectric permittivity of PEO-$LiClO_4$ solid polymer electrolyte plasticized with propylene carbonate. *AIP Adv.* 5: 027125.
14 Flora, X.H., Ulaganathan, M., and Rajendran, S. (2012). Influence of lithium salt concentration on PAN-PMMA blend polymer electrolytes. *Int. J. Electrochem. Sci.* 7: 7451–7462.
15 Gomari, S., Esfandeh, M., and Ghasemi, I. (2017). All-solid-state flexible nanocomposite polymer electrolytes based on poly(ethylene oxide): lithium perchlorate using functionalized graphene. *Solid State Ionics* 303: 37–46.
16 Jinisha, B., Anilkumar, K., Manoj, M. et al. (2017). Development of a novel type of solid polymer electrolyte for solid state lithium battery applications based on lithium enriched poly(ethylene oxide) (PEO)/poly(vinyl pyrrolidone) (PVP) blend polymer. *Electrochim. Acta* 235: 210–222.
17 Polu, A.R. and Rhee, H.-W. (2017). Ionic liquid doped PEO-based solid polymer electrolytes for lithium-ion polymer batteries. *Int. J. Hydrogen Energy* 42: 7212–7219.
18 Abreha, M., Subrahmanyam, A., and Kumar, J.S. (2016). Ionic conductivity and transport properties of poly(vinylidene fluoride-*co*-hexafluoropropylene)-based solid polymer electrolytes. *Chem. Phys. Lett.* 658: 240–247.
19 Leones, R., Sabadini, R.C., Esperança, J.M. et al. (2017). Effect of storage time on the ionic conductivity of chitosan-solid polymer electrolytes incorporating cyano-based ionic liquids. *Electrochim. Acta* 232: 22–29.
20 Wang, G.X., Yang, L., Wang, J.Z. et al. (2005). Enhancement of ionic conductivity of PEO based polymer electrolyte by the addition of nanosize ceramic powders. *J. Nanosci. Nanotechnol.* 5: 1135–1140.
21 Ahn, J.H., Wang, G.X., Liu, H.K., and Dou, S.X. (2003). Tin-based composite materials as anode materials for Li-ion batteries. *J. Power Sources* 402: 119–121.
22 Zhang, S., Lee, J.Y., and Hong, L. (2004). Visualization of particle distribution in composite polymer electrolyte systems. *J. Power Sources* 126: 125–133.
23 Pandey, G., Kumar, Y., and Hashmi, S. (2011). Ionic liquid incorporated PEO based polymer electrolyte for electrical double layer capacitors: a comparative study with lithium and magnesium systems. *Solid State Ionics* 190: 93–98.
24 Nicotera, I., Coppola, L., Oliviero, C. et al. (2004). Some physicochemical properties of PAN-based electrolytes: solution and gel microstructures. *Solid State Ionics* 167: 213–220.
25 Bandara, T., Jayasundara, W., Dissanayake, M. et al. (2013). Effect of cation size on the performance of dye sensitized nanocrystalline TiO_2 solar cells based on quasi-solid state PAN electrolytes containing quaternary ammonium iodides. *Electrochim. Acta* 109: 609–616.

26 Sil, A., Sharma, R., and Ray, S. (2015). Mechanical and thermal characteristics of PMMA-based nanocomposite gel polymer electrolytes with CNFs dispersion. *Surf. Coat. Technol.* 271: 201–206.

27 Osman, Z., Ghazali, M.M., Othman, L., and Isa, K.M. (2012). AC ionic conductivity and DC polarization method of lithium ion transport in PMMA–$LiBF_4$ gel polymer electrolytes. *Results Phys.* 2: 1–4.

28 Li, W., Xing, Y., Xing, X. et al. (2013). PVDF-based composite microporous gel polymer electrolytes containing a novelsingle ionic conductor SiO_2 (Li^+). *Electrochim. Acta* 112: 183–190.

29 Ortega, P.F., Trigueiro, J.P.C., Silva, G.G., and Lavall, R.L. (2016). Improving supercapacitor capacitance by using a novel gel nanocomposite polymer electrolyte based on nanostructured SiO_2, PVDF and imidazolium ionic liquid. *Electrochim. Acta* 188: 809–817.

30 Liu, T., Chang, Z., Yin, Y. et al. (2018). The PVDF-HFP gel polymer electrolyte for Li-O_2 battery. *Solid State Ionics* 318: 88–94.

31 Cznotka, E., Jeschke, S., Grünebaum, M., and Wiemhöfer, H.-D. (2016). Highly-fluorous pyrazolide-based lithium salt in PVDF-HFP as solid polymer electrolyte. *Solid State Ionics* 292: 45–51.

32 Chou, H.-T., Hsu, H.-C., Lien, C.-H., and Chen, S.-T. (2015). The effect of various concentrations of PVDF-HFP polymer gel electrolyte for dye-sensitized solar cell. *Microelectron. Reliab.* 55: 2174–2177.

33 Isa, K.B.M., Osman, Z., Arof, A.K. et al. (2014). Lithium ion conduction and ion–polymer interaction in PVdF-HFP based gel polymer electrolytes. *Solid State Ionics* 268: 288–293.

34 Yusuf, S., Azzahari, A., Selvanathan, V. et al. (2017). Improvement of *N*-phthaloylchitosan based gel polymer electrolyte in dye-sensitized solar cells using a binary salt system. *Carbohydr. Polym.* 157: 938–944.

35 Yusuf, S., Azzahari, A., Yahya, R. et al. (2016). From crab shell to solar cell: a gel polymer electrolyte based on *N*-phthaloylchitosan and its application in dye-sensitized solar cells. *RSC Adv.* 6: 27714–27724.

36 Yusuf, S.N.F., Aziz, M.F., Hassan, H.C. et al. (2014). Phthaloylchitosan-based gel polymer electrolytes for efficient dye-sensitized solar cells. *J. Chem.* 2014: 1–8.

37 Vignarooban, K., Dissanayake, M., Albinsson, I., and Mellander, B.-E. (2014). Effect of TiO_2 nano-filler and EC plasticizer on electrical and thermal properties of poly(ethylene oxide) (PEO) based solid polymer electrolytes. *Solid State Ionics* 266: 25–28.

38 Huang, B., Wang, Z., Li, G. et al. (1996). Lithium ion conduction in polymer electrolytes based on PAN. *Solid State Ionics* 85: 79–84.

39 Choi, B., Kim, Y., and Shin, H. (2000). Ionic conduction in PEO–PAN blend polymer electrolytes. *Electrochim. Acta* 45: 1371–1374.

40 Patel, M., Chandrappa, K.G., and Bhattacharyya, A.J. (2008). Increasing ionic conductivity and mechanical strength of a plastic electrolyte by inclusion of a polymer. *Electrochim. Acta* 54: 209–215.

41 Hu, P., Chai, J., Duan, Y. et al. (2016). Progress in nitrile-based polymer electrolytes for high performance lithium batteries. *J. Mater. Chem. A* 4: 10070–10083.

42 Yuan, F., Chen, H.-Z., Yang, H.-Y. et al. (2005). PAN–PEO solid polymer electrolytes with high ionic conductivity. *Mater. Chem. Phys.* 89: 390–394.
43 Yan, X., Liu, E., Sun, L. et al. (2016). Improved cycling performance of prussian blue cathode for sodium ion batteries by controlling operation voltage range. *Electrochim. Acta* 225: 235–242.
44 Sun, L., Li, B., Zhao, Y. et al. (2010). Structure-induced high dielectric constant and low loss of CNF/PVDF composites with heterogeneous CNF distribution. *Nanotechnology* 21: 305702.
45 Choe, H.S., Giaccai, J., Alamgir, M., and Abraham, K.M. (1995). Preparation and characterization of poly(vinyl sulfone)- and poly(vinylidene fluoride)-based electrolytes. *Electrochim. Acta* 40: 2289–2293.
46 Rajendran, S., Sivakumar, P., and Babu, R.S. (2006). Investigation on poly(vinylidene fluoride) based gel polymer electrolytes. *Bull. Mater. Sci.* 29: 673–678.
47 Tafur, J.P. and Romero, A.J.F. (2014). Electrical and spectroscopic characterization of PVdF-HFP and TFSI—ionic liquids-based gel polymer electrolyte membranes. Influence of $ZnTf_2$ salt. *J. Membr. Sci.* 469: 499–506.
48 Ferrari, S., Quartarone, E., Mustarelli, P. et al. (2010). Lithium ion conducting PVdF-HFP composite gel electrolytes based on *N*-methoxyethyl-*N*-methylpyrrolidinium bis (trifluoromethanesulfonyl)-imide ionic liquid. *J. Power Sources* 195: 559–566.
49 Xie, L., Huang, X., Yang, K. et al. (2014). "Grafting to" route to PVDF-HFP-GMA/$BaTiO_3$ nanocomposites with high dielectric constant and high thermal conductivity for energy storage and thermal management applications. *J. Mater. Chem. A* 2: 5244–5251.
50 Zhu, L. and Wang, Q. (2012). Novel ferroelectric polymers for high energy density and low loss dielectrics. *Macromolecules* 45: 2937–2954.
51 Costa, C., Nunes-Pereira, J., Rodrigues, L. et al. (2013). Novel poly(vinylidene fluoride-trifluoroethylene)/poly(ethylene oxide) blends for battery separators in lithium-ion applications. *Electrochim. Acta* 88: 473–476.
52 Costa, C., Ribelles, J.G., Lanceros-Méndez, S. et al. (2014). Poly (vinylidene fluoride)-based, co-polymer separator electrolyte membranes for lithium-ion battery systems. *J. Power Sources* 245: 779–786.
53 Appetecchi, G.B., Montanino, M., Balducci, A. et al. (2009). Lithium insertion in graphite from ternary ionic liquid-lithium salt electrolytes: I. Electrochemical characterization of the electrolytes. *J. Power Sources* 192: 599–605.
54 Gerbaldi, C., Nair, J.R., Ahmad, S. et al. (2010). UV-cured polymer electrolytes encompassing hydrophobic room temperature ionic liquid for lithium batteries. *J. Power Sources* 195: 1706–1713.
55 Armand, M., Endres, F., MacFarlane, D.R. et al. (2009). Ionic-liquid materials for the electrochemical challenges of the future. *Nat. Mater.* 8: 621–629.
56 Galiński, M., Lewandowski, A., and Stępniak, I. (2006). Ionic liquids as electrolytes. *Electrochim. Acta* 51: 5567–5580.
57 Moreno, J.S., Jeremias, S., Moretti, A. et al. (2015). Ionic liquid mixtures with tunable physicochemical properties. *Electrochim. Acta* 151: 599–608.

58 Patel, M., Gnanavel, M., and Bhattacharyya, A.J. (2011). Utilizing an ionic liquid for synthesizing a soft matter polymer "gel" electrolyte for high rate capability lithium-ion batteries. *J. Mater. Chem.* 21: 17419–17424.

59 Zhang, H., Zhang, P., Li, Z. et al. (2007). A novel sandwiched membrane as polymer electrolyte for lithium ion battery. *Electrochem. Commun.* 9: 1700–1703.

60 Appetecchi, G., Kim, G.-T., Montanino, M. et al. (2010). Ternary polymer electrolytes containing pyrrolidinium-based polymeric ionic liquids for lithium batteries. *J. Power Sources* 195: 3668–3675.

61 Kuo, P.-L., Tsao, C.-H., Hsu, C.-H. et al. (2016). A new strategy for preparing oligomeric ionic liquid gel polymer electrolytes for high-performance and nonflammable lithium ion batteries. *J. Membr. Sci.* 499: 462–469.

62 Sirisopanaporn, C., Fernicola, A., and Scrosati, B. (2009). New, ionic liquid-based membranes for lithium battery application. *J. Power Sources* 186: 490–495.

63 Stepniak, I., Andrzejewska, E., Dembna, A., and Galinski, M. (2014). Characterization and application of N-methyl-N-propylpiperidinium bis (trifluoromethanesulfonyl) imide ionic liquid–based gel polymer electrolyte prepared in situ by photopolymerization method in lithium ion batteries. *Electrochim. Acta* 121: 27–33.

64 Chaudoy, V., Van, F.T., Deschamps, M., and Ghamouss, F. (2017). Ionic liquids in a poly ethylene oxide cross-linked gel polymer as an electrolyte for electrical double layer capacitor. *J. Power Sources* 342: 872–878.

65 Pitawala, J., Navarra, M.A., Scrosati, B. et al. (2014). Structure and properties of Li-ion conducting polymer gel electrolytes based on ionic liquids of the pyrrolidinium cation and the bis(trifluoromethanesulfonyl) imide anion. *J. Power Sources* 245: 830–835.

66 Karuppasamy, K., Reddy, P.A., Srinivas, G. et al. (2017). An efficient way to achieve high ionic conductivity and electrochemical stability of safer nonaflate anion-based ionic liquid gel polymer electrolytes (ILGPEs) for rechargeable lithium ion batteries. *J. Solid State Electrochem.* 21: 1145–1155.

67 Li, L., Wang, J., Yang, P. et al. (2013). Preparation and characterization of gel polymer electrolytes containing N-butyl-N-methylpyrrolidinium bis(trifluoromethanesulfonyl) imide ionic liquid for lithium ion batteries. *Electrochim. Acta* 88: 147–156.

68 Balo, L., Gupta, H., Singh, S.K. et al. (2018). Performance of EMIMFSI ionic liquid based gel polymer electrolyte in rechargeable lithium metal batteries. *J. Ind. Eng. Chem.* 65: 137–145.

69 Li, M., Liao, Y., Liu, Q. et al. (2018). Application of the imidazolium ionic liquid based nano-particle decorated gel polymer electrolyte for high safety lithium ion battery. *Electrochim. Acta* 284: 188–201.

70 Li, W., Xing, Y., Wu, Y. et al. (2015). Study the effect of ion-complex on the properties of composite gel polymer electrolyte based on Electrospun PVdF nanofibrous membrane. *Electrochim. Acta* 151: 289–296.

71 Vishwakarma, V. and Jain, A. (2017). Enhancement of thermal transport in gel polymer electrolytes with embedded BN/Al_2O_3 nano-and micro-particles. *J. Power Sources* 362: 219–227.

72 Song, D., Xu, C., Chen, Y. et al. (2015). Enhanced thermal and electrochemical properties of PVDF-HFP/PMMA polymer electrolyte by TiO_2 nanoparticles. *Solid State Ionics* 282: 31–36.

73 Tiautit, N., Puratane, C., Panpinit, S., and Saengsuwan, S. (2014). Effect of SiO_2 and TiO_2 nanoparticles on the performance of dye-sensitized solar cells using PVDF-HFP/PVA gel electrolytes. *Energy Procedia* 56: 378–385.

74 Singh, P.K., Bhattacharya, B., Nagarale, R.K. et al. (2010). Synthesis, characterization and application of biopolymer-ionic liquid composite membranes. *Synth. Met.* 160: 139–142.

75 Buraidah, M., Teo, L., Yusuf, S. et al. (2011). TiO_2/chitosan-NH4I-BMII-BasedDye-sensitized solar cells with anthocyanin dyes extracted from black rice and red cabbage. *Int. J. Photoenergy* 2011: 1–11.

76 Dissanayake, M.A.K.L., Jayathissa, R., Seneviratne, V.A. et al. (2014). Polymethylmethacrylate (PMMA) based quasi-solid electrolyte with binary iodide salt for efficiency enhancement in TiO_2 based dye sensitized solar cells. *Solid State Ionics* 265: 85–91.

77 Kumar, Y., Pandey, G., and Hashmi, S. (2012). Gel polymer electrolyte based electrical double layer capacitors: comparative study with multiwalled carbon nanotubes and activated carbon electrodes. *J. Phys. Chem. C* 116: 26118–26127.

78 Yan, X., Xiuli, W., Xinhui, X. et al. (2017). A newly designed composite gel polymer electrolyte based on poly(vinylidene fluoride-hexafluoropropylene) (PVDF-HFP) for enhanced solid-state lithium-sulfur batteries. *Chem. Eur. J.* 23: 15203–15209.

79 Xu, K. (2004). Nonaqueous liquid electrolytes for lithium-based rechargeable batteries. *Chem. Rev.* 104: 4303–4418.

80 Fong, R., von Sacken, U., and Dahn, J.R. (1990). Studies of lithium intercalation into carbons using nonaqueous electrochemical cells. *J. Electrochem. Soc.* 137: 2009–2013.

81 Younesi, R., Veith, G.M., Johansson, P. et al. (2015). Lithium salts for advanced lithium batteries: Li–metal, Li–O_2, and Li–S. *Energy Environ. Sci.* 8: 1905–1922.

82 Schmidt, M., Heider, U., Kuehner, A. et al. (2001). Lithium fluoroalkylphosphates: a new class of conducting salts for electrolytes for high energy lithium-ion batteries. *J. Power Sources* 97: 557–560.

83 Virya, A. and Lian, K. (2017). Li_2SO_4-polyacrylamide polymer electrolytes for 2.0 V solid symmetric supercapacitors. *Electrochem. Commun.* 81: 52–55.

84 Zhong, X., Tang, J., Cao, L. et al. (2017). Cross-linking of polymer and ionic liquid as high-performance gel electrolyte for flexible solid-state supercapacitors. *Electrochim. Acta* 244: 112–118.

85 Mohamed, W., Atan, R., and Yiap, T. (2009). Current and possible future applications of hydrogen fuel cells in Malaysia. *Proceedings of the International Conference on Advances in Mechanical Engineering (ICAME).*

86 Pica, M. (2014). Non-fluorinated membranes: fuel cell applications. In: *Encyclopedia Membranes* (eds. E. Drioli and L. Giorno), 1–2. Berlin, Heidelberg: Springer Verlag.

87 Quartarone, E., Angioni, S., and MMustarelli, P. (2017). Polymer and composite membranes for proton-conducting, high-temperature fuel cells: a critical review. *Materials* 10: 687–703.

88 Dodane, V. and Vilivalam, V.D. (1998). Pharmaceutical applications of chitosan. *Pharm. Sci. Technol. Today* 1: 246–253.

89 Klotzbach, T., Watt, M., Ansari, Y., and Minteer, S.D. (2006). Effects of hydrophobic modification of chitosan and Nafion on transport properties, ion-exchange capacities, and enzyme immobilization. *J. Membr. Sci.* 282: 276–283.

90 Ming Yang, J. and Chih Chiu, H. (2012). Preparation and characterization of polyvinyl alcohol/chitosan blended membrane for alkaline direct methanol fuel cells. *J. Membr. Sci.* 419–420: 65–71.

91 Osifo, P.O. and Masala, A. (2010). Characterization of direct methanol fuel cell (DMFC) applications with H_2SO_4 modified chitosan membrane. *J. Power Sources* 195: 4915–4922.

92 Yuan, W., Wu, H., Zheng, B. et al. (2007). Sorbitol-plasticized chitosan/zeolite hybrid membrane for direct methanol fuel cell. *J. Power Sources* 172: 604–612.

93 Wu, H., Zheng, B., Zheng, X. et al. (2007). Surface-modified Y zeolite-filled chitosan membrane for direct methanol fuel cell. *J. Power Sources* 173: 842–852.

94 Wang, J., Zheng, X., Wu, H. et al. (2008). Effect of zeolites on chitosan/zeolite hybrid membranes for direct methanol fuel cell. *J. Power Sources* 178: 9–19.

95 Cui, Z., Xing, W., Liu, C. et al. (2009). Chitosan/heteropolyacid composite membranes for direct methanol fuel cell. *J. Power Sources* 188: 24–29.

96 Deb, S.K. (1973). Optical and photoelectric properties and colour centres in thin films of tungsten oxide. *Philos. Mag. J. Theor. Exp. Appl. Phys.* 27: 801–822.

97 Kraft, A. (2008). On the discovery and history of Prussian Blue. *Bull. Hist. Chem.* 33: 61–67.

98 Granqvist, C.G. (2005). Electrochromic devices. (Electroceramics cherbourg). *J. Eur. Ceram. Soc.* 25: 2907–2912.

99 Mortimer, R.J. (1997). Electrochromic materials. *Chem. Soc. Rev.* 26: 147–156.

100 Jiao, Z., Wang, J., Ke, L. et al. (2012). Electrochromic properties of nanostructured tungsten trioxide (hydrate) films and their applications in a complementary electrochromic device. *Electrochim. Acta* 63: 153–160.

101 Sivakumar, R., Shanthakumari, K., Thayumanavan, A. et al. (2009). Molybdenum oxide (MoO_3) thin film based electrochromic cell characterisation in 0.1 M $LiClO_4$.PC electrolyte. *Surf. Eng.* 25: 548–554.

102 Oliveira, R., Nono, N., Alves, O. et al. (2015). Electrochromic behavior of vanadium oxide nanostructures synthesized by melt sonoquenching. *Rev. Virtual Quim.* 7: 1876–1892.

103 Granqvist, C.G., Azens, A., Smulko, J., and Kish, L. (2007). Oxide-based electrochromics for energy efficient buildings: materials, technologies, testing, and perspectives. *Journal of Physics: Conference Series*, IOP Publishing.

104 Tavares, P.F., Gaspar, A.R., Martins, A.G., and Frontini, F. (2014). Evaluation of electrochromic windows impact in the energy performance of buildings in Mediterranean climates. *Energy Policy* 67: 68–81.

105 Ulrich, S., Szyszko, C., Jung, S., and Vergöhl, M. (2017). Electrochromic properties of mixed oxides based on titanium and niobium for smart window applications. *Surf. Coat. Technol.* 314: 41–44.

106 Wang, J.-Y., Wang, M.-C., and Jan, D.-J. (2017). Synthesis of poly(methyl methacrylate)-succinonitrile composite polymer electrolyte and its application for flexible electrochromic devices. *Sol. Energy Mater. Sol. Cells* 160: 476–483.

107 Califórnia, A., Silva, A.S., Gonçalves, J. et al. (2016). Silver grid electrodes for faster switching ITO free electrochromic devices. *Sol. Energy Mater. Sol. Cells* 153: 61–67.

108 Ledwon, P., Andrade, J.R., Lapkowski, M., and Pawlicka, A. (2015). Hydroxypropyl cellulose-based gel electrolyte for electrochromic devices. *Electrochim. Acta* 159: 227–233.

109 Lin, S.-Y., Chen, Y.-C., Wang, C.-M. et al. (2012). Study of MoO_3–NiO complementary electrochromic devices using a gel polymer electrolyte. *Solid State Ionics* 212: 81–87.

110 Desai, S.S. (2012). Fabrication of dual layer conducting polymer electrochrimic devices using novel electrolytes. School of Chemistry, University of Wollongong.

111 Bandara, T.M.W.J., Aziz, M.F., Fernando, H.D.N.S. et al. (2015). Efficiency enhancement in dye-sensitized solar cells with a novel PAN-based gel polymer electrolyte with ternary iodides. *J. Solid State Electrochem.* 19: 2353–2359.

112 Pal, P. and Ghosh, A. (2018). Investigation of ionic conductivity and relaxation in plasticized PMMA-$LiClO_4$ solid polymer electrolytes. *Solid State Ionics* 319: 117–124.

113 Li, W., Pang, Y., Liu, J. et al. (2017). A PEO-based gel polymer electrolyte for lithium ion batteries. *RSC Adv.* 7: 23494–23501.

114 Wang, F., Li, L., Yang, X. et al. (2018). Influence of additives in a PVDF-based solid polymer electrolyte on conductivity and Li-ion battery performance. *Sustainable Energy Fuels* 2: 492–498.

115 Faxing, W., Xiaowei, W., Zheng, C. et al. (2015). A quasi-solid-state sodium-ion capacitor with high energy density. *Adv. Mater.* 27: 6962–6968.

116 Wang, S.-H., Lin, Y.-Y., Teng, C.-Y. et al. (2016). Immobilization of anions on polymer matrices for gel electrolytes with high conductivity and stability in lithium ion batteries. *ACS Appl. Mater. Interfaces* 8: 14776–14787.

117 Jacoby, M. (2016). The future of low-cost solar cells. *Chem. Eng. News* 94: 30–35.

118 Lennert, A., Sternberg, M., Meyer, K. et al. (2017). Iodine-pseudohalogen ionic liquid-based electrolytes for quasi-solid-state dye-sensitized solar cells. *ACS Appl. Mater. Interfaces* 9: 33437–33445.

119 Christians, J.A., Fung, R.C., and Kamat, P.V. (2013). An inorganic hole conductor for organo-lead halide perovskite solar cells. Improved hole conductivity with copper iodide. *J. Am. Chem. Soc.* 136: 758–764.

120 Dematage, N., Premalal, E., and Konno, A. (2014). Employment of CuI on Sb_2S_3 extremely thin absorber solar cell: N719 molecules as a dual role of a recombination blocking agent and an efficient hole shuttle. *Int. J. Electrochem. Sci.* 9: 1729–1737.

121 Hemamali, G. and Kumara, G. (2013). Solid state solar cells based on TiO_2 sensitized with natural pigment extracted from the anthurium. *Int. J. Sci. Eng. Res.*: 440.

122 Kumara, G., Konno, A., Senadeera, G. et al. (2001). Dye-sensitized solar cell with the hole collector p-CuSCN deposited from a solution in n-propyl sulphide. *Sol. Energy Mater. Sol. Cells* 69: 195–199.

123 Tan, S., Zhai, J., Wan, M. et al. (2004). Influence of small molecules in conducting polyaniline on the photovoltaic properties of solid-state dye-sensitized solar cells. *J. Phys. Chem. B* 108: 18693–18697.

124 Xia, J., Masaki, N., Lira-Cantu, M. et al. (2008). Influence of doped anions on poly(3, 4-ethylenedioxythiophene) as hole conductors for iodine-free solid-state dye-sensitized solar cells. *J. Am. Chem. Soc.* 130: 1258–1263.

125 Yang, L., Cappel, U.B., Unger, E.L. et al. (2012). Comparing spiro-OMeTAD and P3HT hole conductors in efficient solid state dye-sensitized solar cells. *Phys. Chem. Chem. Phys.* 14: 779–789.

126 Kwon, K. and Kim, D. (2016). Polymer electrolyte membrane and methanol fuel cell. In: *Nanostructured Polymer Membranes, Volume 2: Applications* (eds. P.M. Visakh and N. Olga), 209–250. Wiley.

127 Li, Q., He, R., Jensen, J.O., and Bjerrum, N.J. (2003). Approaches and recent development of polymer electrolyte membranes for fuel cells operating above 100 C. *Chem. Mater.* 15: 4896–4915.

128 Smitha, B., Sridhar, S., and Khan, A. (2005). Solid polymer electrolyte membranes for fuel cell applications—a review. *J. Membr. Sci.* 259: 10–26.

129 Sahu, A.K., Ketpang, K., Shanmugam, S. et al. (2016). Sulfonated graphene–Nafion composite membranes for polymer electrolyte fuel cells operating under reduced relative humidity. *J. Phys. Chem. C* 120: 15855–15866.

130 Esfeh, H.K. and Hamid, M.K.A. (2014). Temperature effect on proton exchange membrane fuel cell performance part II: parametric study. *Energy Procedia* 61: 2617–2620.

131 Yang, C., Costamagna, P., Srinivasan, S. et al. (2001). Approaches and technical challenges to high temperature operation of proton exchange membrane fuel cells. *J. Power Sources* 103: 1–9.

132 Wehkamp, N., Breitwieser, M., Büchler, A. et al. (2016). Directly deposited Nafion/TiO_2 composite membranes for high power medium temperature fuel cells. *RSC Adv.* 6: 24261–24266.

133 Amjadi, M., Rowshanzamir, S., Peighambardoust, S. et al. (2010). Investigation of physical properties and cell performance of Nafion/TiO_2 nanocomposite membranes for high temperature PEM fuel cells. *Int. J. Hydrogen Energy* 35: 9252–9260.

134 Miyake, N., Wainright, J., and Savinell, R. (2001). Evaluation of a sol-gel derived Nafion/silica hybrid membrane for polymer electrolyte membrane

fuel cell applications: II. Methanol uptake and methanol permeability. *J. Electrochem. Soc.* 148: A905–A909.

135 Staiti, P., Arico, A., Baglio, V. et al. (2001). Hybrid Nafion–silica membranes doped with heteropolyacids for application in direct methanol fuel cells. *Solid State Ionics* 145: 101–107.

136 Yang, C., Srinivasan, S., Arico, A. et al. (2001). Composite Nafion/zirconium phosphate membranes for direct methanol fuel cell operation at high temperature. *Electrochem. Solid-State Lett.* 4: A31–A34.

137 Wang, Y., Li, L., Hu, L. et al. (2003). A feasibility analysis for alkaline membrane direct methanol fuel cell: thermodynamic disadvantages versus kinetic advantages. *Electrochem. Commun.* 5: 662–666.

138 Wang, X., Hu, J.-M., and Hsing, I.-M. (2004). Electrochemical investigation of formic acid electro-oxidation and its crossover through a Nafion® membrane. *J. Electroanal. Chem.* 562: 73–80.

139 Jiang, S.P., Liu, Z., and Tian, Z.Q. (2006). Layer-by-layer self-assembly of composite polyelectrolyte–Nafion membranes for direct methanol fuel cells. *Adv. Mater.* 18: 1068–1072.

140 Arico, A., Srinivasan, S., and Antonucci, V. (2001). DMFCs: from fundamental aspects to technology development. *Fuel cells* 1: 133–161.

141 Jia, N., Lefebvre, M.C., Halfyard, J. et al. (2000). Modification of Nafion proton exchange membranes to reduce methanol crossover in PEM fuel cells. *Electrochem. Solid-State Lett.* 3: 529–531.

142 Tricoly, V. (1998). Proton and methanol transport in poly(perfluorosulfonate) membranes containing Cs^+ and H^+ cations. *J. Electrochem. Soc.* 145: 3798.

143 Liu, J., Wang, H., Cheng, S., and Chan, K.-Y. (2005). Nafion–polyfurfuryl alcohol nanocomposite membranes for direct methanol fuel cells. *J. Membr. Sci.* 246: 95–101.

144 Pu, C., Huang, W., Ley, K.L., and Smotkin, E.S. (1995). A methanol impermeable proton conducting composite electrolyte system. *J. Electrochem. Soc.* 142: L119–L120.

145 Yoon, S., Hwang, G., Cho, W. et al. (2002). Modification of polymer electrolyte membranes for DMFCs using Pd films formed by sputtering. *J. Power Sources* 106: 215–223.

146 Prabhuram, J., Zhao, T., Liang, Z. et al. (2005). Pd and Pd-Cu alloy deposited Nafion membranes for reduction of methanol crossover in direct methanol fuel cells. *J. Electrochem. Soc.* 152: A1390–A1397.

147 Hejze, T., Gollas, B., Sauerbrey, R. et al. (2005). Preparation of Pd-coated polymer electrolyte membranes and their application in direct methanol fuel cells. *J. Power Sources* 140: 21–27.

148 Kim, Y.J., Choi, W.C., Woo, S.I., and Hong, W.H. (2004). Evaluation of a palladinized Nafion™ for direct methanol fuel cell application. *Electrochim. Acta* 49: 3227–3234.

149 Yang, B. and Manthiram, A. (2004). Multilayered membranes with suppressed fuel crossover for direct methanol fuel cells. *Electrochem. Commun.* 6: 231–236.

150 Si, Y., Lin, J.-C., Kunz, H.R., and Fenton, J.M. (2004). Trilayer membranes with a methanol-barrier layer for DMFCs. *J. Electrochem. Soc.* 151: A463–A469.

151 Shao, Z.-G., Wang, X., and Hsing, I.-M. (2002). Composite Nafion/polyvinyl alcohol membranes for the direct methanol fuel cell. *J. Membr. Sci.* 210: 147–153.

152 Nedjalkov, A., Meyer, J., Köhring, M. et al. (2016). Toxic gas emissions from damaged lithium ion batteries—analysis and safety enhancement solution. *Batteries* 2: 5.

153 Deng, D. (2015). Li-ion batteries: basics, progress, and challenges. *Energy Sci. Eng.* 3: 385–418.

154 Tidblad, A.A., Berg, H., Edström, K. et al. *Batteries – Present and Future Challenges*. Swedish Hybrid Vehicle Centre.

155 Oh, B., Sun, Y.K., and Kim, D.W. (2001). Electrochemical properties of gel polymer electrolytes prepared by poly(vinyl acetate) copolymers. *Bull. Korean Chem. Soc.* 22: 1136–1140.

156 Varzi, A., Raccichini, R., Passerini, S., and Scrosati, B. (2016). Challenges and prospects of the role of solid electrolytes in the revitalization of lithium metal batteries. *J. Mater. Chem. A* 4: 17251–17259.

157 Abbas, Q., Ratajczak, P., Babuchowska, P. et al. (2015). Strategies to improve the performance of carbon/carbon capacitors in salt aqueous electrolytes. *J. Electrochem. Soc.* 162: A5148–A5157.

158 Ruan, D. and Zuo, F. (2015). High voltage performance of spiro-type quaternary ammonium salt based electrolytes in commercial large supercapacitors. *Electrochemistry* 83: 997–999.

159 Soeda, K., Yamagata, M., and Ishikawa, M. (2015). Outstanding features of alginate-based gel electrolyte with ionic liquid for electric double layer capacitors. *J. Power Sources* 280: 565–572.

160 Wang, M.-C., Chen, Y.-C., Hsieh, M.-H. et al. (2016). The improvement of all-solid-state electrochromic devices fabricated with the reactive sputter and cathodic arc technology. *AIP Adv.* 6: 115009.

161 Nguyen, C.A., Xiong, S., Ma, J. et al. (2009). Toward electrochromic device using solid electrolyte with polar polymer host. *J. Phys. Chem. B* 113: 8006–8010.

162 Schiller, B. (2011). Smart glass becomes more or less transparent depending on the weather. https://www.fastcompany.com/1784221/smart-glass-becomes-more-or-less-transparent-depending-weather (accesed 30 July 2018).

163 Adarlo, S. (2017). Self-powered system makes smart windows smarter. https://www.princeton.edu/news/2017/06/30/self-powered-system-makes-smart-windows-smarter (accesed 30 July 2018).

164 Sigler, D. (2014). Two "new" battery contenders. http://sustainableskies.org/two-new-battery-contenders (accesed 30 July 2018).

165 Mekhilef, S., Safari, A., Mustaffa, W. et al. (2012). Solar energy in Malaysia: current state and prospects. *Renewable Sustainable Energy Rev.* 16: 386–396.

166 Sung, C.T.B. (2012). Electricity from solar energy in Malaysia: clean, renewable, and abundant energy source, so what's the problem? http://www.christopherteh.com/blog/2012/05/solar-malaysia (accesed 30 July 2018).

167 Sopian, K. and Daud, W.R.W. (2006). Challenges and future developments in proton exchange membrane fuel cells. *Renewable Energy* 31: 719–727.

7

Polymer Electrolytes for Lithium Ion Batteries and Challenges: Part I

Shishuo Liang, Wenqi Yan, Minxia Li, Yusong Zhu, Lijun Fu, and Yuping Wu

Nanjing Tech University, State Key Laboratory of Materials-Oriented Chemical Engineering and School of Energy Science and Engineering, No. 30, Puzhu South Road, Nanjing 211816, China

The global energy crisis and an increase in environmental pollution in the recent years have drawn the attention of the scientific community toward the development of efficient electrochemical devices. Polymer electrolytes (PEs) are an important component for the well-used lithium ion batteries (LIBs) because they are inherently safer than some classical liquid electrolytes. In this chapter, PEs for LIBs are selected and discussed. In particular, the challenging issues and possible solutions of PEs are emphasized. The ongoing trends in PEs for LIBs are also pointed out.

7.1 Introduction

The world's energy consumption will reach >20 TW in the next 10 years, most of which comes from the burning of fossil fuels: coal, natural gas, and petroleum (crude oil). Petroleum supplies may be seriously depleted in the near future. Development of alternative energy sources to fossil fuels is an international urgent issue to solve the worldwide energy problems, which extremely spurs the scientific and technological endeavor for renewable and green energy [1–3]. Electrochemical power sources include the following categories: primary (single use) and secondary (rechargeable) batteries. In this chapter, we mainly concentrate on the most mainstream rechargeable batteries – LIBs.

LIBs are important for energy storage in a wide variety of applications including consumer electronics, transportation, and large-scale energy production [4]. The first LIB was produced in the early 1990s by the Sony company (Japan) with the negative electrode of a carbon material capable of incorporating lithium ions [5]. Nowadays, it is impossible to imagine modern society without LIBs because they have been widely used in highly sophisticated portable devices such as cellular phones with amazing applications (including laptops, video cameras, and more) [6].

Generally, electrolytes for LIBs could be classified as three types: liquid electrolytes, solid electrolytes, and gel electrolytes. Properties and key parameters

Polymer Electrolytes: Characterization Techniques and Energy Applications,
First Edition. Edited by Tan Winie, Abdul K. Arof, and Sabu Thomas.

Table 7.1 Summary of the primary electrolyte properties desired for various lithium batteries.

Electrolyte properties	Liquid	Solid	Gel
Fundamental properties			
Ionic conductivity	High $>10^{-3}$ S cm^{-1}	Low $<10^{-4}$ S cm^{-1}	Medium $>10^{-4}$ S cm^{-1}
Contact/interfacial properties	Good	Poor	Medium
Electrochemical stability	Poor	Good	Poor
Thermal stability	Poor	Good	Poor
Dimension stability	Poor	Good	Medium
Safety	Poor	Good	Medium
Additional properties required by a specific battery			
Resistance to forming or dissolving by-products (e.g. polysulfides for Li–S)	Low	High	Medium
Ability to dissolve oxygen (e.g. Li–O_2)	Good	Poor	Medium

for each kind of electrolytes are shown in Table 7.1, from which we could know that a qualified electrolyte should possess high ionic conductivity, low contact resistance, wide electrochemical stability, good thermal ability, high safety, etc.

Liquid electrolytes have achieved great success in practical applications, especially in high-performance energy storage devices (ESDs), with good capability rate. However, with a growing demand on safety, design flexibility, mechanical performances (such as flexibility), and multi-functionality for advanced ESDs, solid electrolytes are becoming increasingly attractive, owing to their excellent performance in terms of safety, mechanical properties, and design flexibility for batteries. PEs coupling a polymer with a lithium salt is one of the common solid electrolytes. However, solid electrolytes face several challenging issues, including low ionic conductivity and poor interfacial properties (high interfacial resistance, poor/unstable contact with electrodes) [7]. These issues prevent them from wide practical applications.

7.2 Classification of Polymer Electrolytes

As mentioned earlier, PEs can be classified into mainly two kinds: solid polymer electrolytes (SPEs) and gel polymer electrolytes (GPEs).

7.2.1 Solid Polymer Electrolytes (SPEs)

SPEs refer to the solid polymeric materials with ability to transport lithium ions. Generally, SPEs are organic polymer matrices dissolved with lithium salts, which is a solvent-free system. The transfer mechanism is that the transfer ions first coordinate with the groups on the polymer chain. Under the action of electric field, the coordination and desorption between ions and coordination groups occur, so as to realize the migration of ions [8]. SPEs are believed to be the most promising candidate for high-performance electrolytes in next-generation energy devices, primarily because of their excellent mechanical and safety properties [9]. As electrolytes, SPEs are free of the safety issues for traditional organic liquid electrolytes and enable design flexibility for LIBs. For example, SPEs can support the design of flexible all-solid-state batteries and microbatteries for microelectronics [10, 11]. According to their structures, they can be divided into homogeneous and heterogeneous SPEs. Homogeneous SPEs are pure polymeric solid solutions of ions, whereas heterogeneous SPEs can be any solid polymeric materials comprising different phases/structures with different abilities to transport ions. Owing to the complexity of polymeric materials, SPEs are usually heterogeneous.

Among the various SPEs, electrolytes based on polyethylene oxide (PEO) are the most attractive, owing to the fact that PEO shows excellent solubility for lithium salts [12]. In the following, we mainly introduce PEO-based SPEs. However, the ionic conductivity of PEO-based SPEs is usually in the range of 10^{-8} to $10^{-4}\,S\,cm^{-1}$ at room temperature [13]. Therefore, improving the ionic conductivity is the most challenging and critical issue for PEO-based SPEs. PEO is a typical semicrystalline polymer and can form various complexes with lithium salts, owing to the coupling effects between Li^+ and the oxygen atoms on the PEO chains [14–17]. On the one hand, these strong coupling effects enable the dissolution of lithium salts, but on the other hand, they act as an anchor, impeding the transportation of ions. The coupling effects between PEO chains and Li^+ ions have been well observed, for example, the ionic conductivity decreasing with increasing molecular weight of the PEO, the addition of a plasticizer remarkably increasing the ionic conductivity, and so on [18–20].

SPEs are preferred over liquid electrolyte and GPEs due to many advantages, such as high durability, long shelf life, high energy density, light weight, great flexibility for cell design, low reactivity toward the electrodes, free from the problems of solvent leakage and harmful gases for above ambient temperature operations, and reduced packaging cost. It also shows wider electrochemical and thermal stability range as well as low volatility [21]. SPEs consist of dissolution of salt in a polymer matrix with electron donor group and have various advantages like design flexibility, miniaturization feasibility, improved safety, ease of device fabrication, etc.

Furthermore, an ionically conducting polymer membrane for electrochemical device applications should have a proper balance between its ionic conductivity and mechanical strength (i.e. high ionic conductivity and excellent mechanical strength) under the conditions desirable for a particular application. SPEs having optimized conductivity and stability properties under ambient and sub-ambient conditions may be expected to have the potential for applications in solid-state ionic devices such as the high energy density LIBs and supercapacitors [22]. When lithium salt is added in the polymer system, it gets dissociated by interaction with the polar group of polymer and lithium ions move along the long polymer chains by hopping process in amorphous phase of the polymers [23].

7.2.2 Gel Polymer Electrolytes (GPEs)

GPEs, also known as plasticized PEs, were first introduced by Feuillade and Perche in 1975 [24]. GPEs are plasticized or gelled polymer matrices, wherein the addition of plasticizer results in a polymer matrix swollen in a liquid electrolyte [25]. Plasticizers increase the conductivity of a PE as they promote the movements of molecular chain segments. The introduction of a plasticizer can effectively improve the dielectric constant of the whole system. The increase of the dielectric constant of the system reduces the electrostatic interaction between positive and negative ions, which makes the dissociation energy of lithium ions lower and facilitates the movement of ions [26]. At the same time, the interaction between the plasticizer and the polymer segment can increase the amorphous region in the polymer, weaken the complexation of the segment, and enhance the mobility of the polymer segment, thereby reducing the glass transition temperature (T_g) of the PE and facilitating the motion of lithium ions between the segments [27]. Typical plasticizers are mainly solvents for liquid electrolytes such as ethylene carbonate (EC), propylene carbonate (PC), diethyl carbonate (DEC), dimethyl carbonate (DMC), and ethyl methyl carbonate (EMC) [28]. These plasticizers have the common feature of having a lower melting point and a higher dielectric constant.

GPEs can be prepared easily by mixing an appropriate polymer matrix like PEO, an alkali metal salt such as lithium salt, and a solvent. Gels can be classified into two categories from the sample preparation methods [29, 30]:

(1) *Physical gels*: liquid electrolyte is confined in a polymer matrix without many bond formations between polymer and solvent, e.g. $LiClO_4$/EC/PC in polymethyl methacrylate (PMMA).
(2) *Chemical gels*: a cross-linker leads to the formation of chemical bond between the functional group of polymer and cross-linking agent.

Several polar polymers, such as PEO [31–33], poly(vinylidene fluoride) (PVDF) [34, 35], poly(vinylidene fluoride)–hexafluoropropylene (PVDF-HFP) [36, 37], PMMA [38, 39], and polyacrylonitrile (PAN) [40], have been employed as the polymer host (solid component) for GPEs, owing to their good affinity with liquid components. At the same time, in addition to traditional organic liquid electrolytes, new liquid electrolytes such as ionic liquid electrolytes [40, 41] have also been of great interest for recent fabrications of GPEs.

7.3 Performance and Improvements

In lithium polymer batteries, SPEs act as both electrolyte and separator. Compared with both inorganic solid electrolytes and liquid ones, SPEs have better flexibility and higher safety. However, the low lithium ionic conductivity of

SPE inhibits its further application in lithium batteries. Strategies have been developed to enhance conductivity with good understanding of the conducting mechanisms.

Currently, most results indicate that the amorphous phase with activated chain segments (above the glass transition temperature T_g) can aid ion transportation. Therefore, increasing the percentage of the amorphous phase of a PEO electrolyte is believed to be one of the most effective ways of improving ionic conductivity. The addition of nanoparticles, such as Al_2O_3 [42–45], SiO_2 [46–48], and TiO_2 [49, 50], was also found to be very effective for many reasons. Firstly, nanoparticles are enhancement additives for mechanical properties, so they can compensate for the deterioration of mechanical properties that result from lower crystallinity. Secondly, the nanoparticles can form special pathways in the interphase for ion transportation, which further increases the ion conductivity. Blending with other polymers is another easy approach to increase ionic conductivity. Polymers with low crystallinity are promising candidates, including PMMA, polypropylene oxide (PPO), polysiloxane (PSi), poly[bis(methoxy-ethoxy-ethoxy)phosphazene] (MEEP), etc. Copolymers with conductive blocks (usually PEO block) are also promising candidates for high-performance SPEs [51–53]. The most interesting attribute of the copolymer electrolytes is the self-assembled microstructures, which provides a good balance between ion conductivity and mechanical performance. Specifically, the ion-conducting PEO block provides a pathway for ion transportation, and other blocks, such as polyethylene (PE) or polystyrene (PS) blocks, form three-dimensional (3D) connected frameworks for good mechanical properties. More importantly, the two blocks can form unique patterns with special pathways for ion conduction. For example, by adjusting the fraction of the PEO block in the PEO-PS copolymer and by using the flow alignment technique, the ionic conductivity of PEO-PS electrolytes was found to be dependent on specific self-assembled nanostructures such as hexagonally perforated lamellae, hexagonally packed cylinders, and parallel lamellae [52]. Although copolymers enable the control of nanoscale pathways for ion conduction, the mechanism for the conduction inside the conductive phase is basically the same with those in conventional SPEs, which determines that the ionic conductivity of copolymer-based SPEs is usually at the same level as it is for common SPEs.

Furthermore, SPEs face another problem – there is a huge interface impedance between electrodes and the dry-state SPEs. To solve this trouble, strategies have been developed as follows. In situ growth of electrode layer on the interface between electrode and SPEs effectively reduces the interface impedance [54, 55]. Meanwhile, mixing polymer materials into the electrode material (especially positive electrode material) is an easier solution to address the problem [56–58]. In situ polymerization is an effective means to solve the large interface resistance. In this method, a lithium salt is doped into a polymer monomer, and a PE is prepared by in situ polymerization of a polymer monomer by adding an initiator. In this way, the electrode is tightly bonded to the PE, the interface resistance is small, and the interface chemistry is stable.

To some extent, GPEs can be seen as the development of SPEs. As an electrolyte with a new state between a liquid and a solid, and as a combination of liquid and solid components, GPEs combine the merits of the individual components, that is, high ionic conductivity and good interfacial properties from the liquid phase and good mechanical properties from the solid component [59]. Furthermore, GPEs possess low volatility, low reactivity, good operation safety, and good chemical, mechanical, electrochemical, and structural stabilities [25, 60]. The attractive properties of GPEs enhance safety and its applicability in electrochemical devices.

The addition of the plasticizer increases the ionic conductivity, but the increase in ionic conductivity is usually at the expense of the decrease in mechanical strength [61, 62]. Thus, an ionically conducting polymer membrane for electrochemical device applications should have a proper balance between its ionic conductivity and mechanical strength (i.e. high ionic conductivity and excellent mechanical strength) under the conditions desirable for application.

As introduced earlier, the ionic conductivity for SPEs is the key issue that limits its wide use for LIBs. As an electrolyte with a new state between a liquid and a solid, and as a combination of liquid and solid components, GPEs combine the merits of the individual components, that is, high ionic conductivity and good interfacial properties from the liquid phase and good mechanical properties from the solid component. Although there are some inevitable trade-offs in the properties of the components, the combination of the properties from the different components is believed to be a simple and effective way of achieving all-round properties. Therefore, GPEs have been of great interest as high-performance electrolytes for the fabrication of advanced ESDs with enhanced safety and flexibility [59, 63]. As GPEs are, in essence, not a new type of electrolyte, studies on GPEs are focusing on how to optimize their compositions and structures for better electrolyte properties, such as ionic conductivity, mechanical strength, formability, and contact with electrodes. In the following, we will briefly summarize the studies on improving these properties.

(1) *Ionic conductivity and mechanical properties*: For GPEs, achieving high ionic conductivity as well as good mechanical properties is challenging, as the addition of the liquid component will deteriorate the mechanical strength of the polymer host through swelling/wetting caused by the liquid [64, 65]. To achieve a good balance, the key is to design structures with the ability to absorb a large amount of the liquid component while maintaining structural integrity. For example, Kil et al. employed cross-linking and nanoparticles to fabricate a high-performance composite GPE with excellent flexibility [66]. For example, in a polymer Al_2O_3 GPE, the cross-linking of the polymer matrix (ethoxylated trimethylolpropane triacrylate, ETPTA) can maintain the structural integrity and give rise to a good affinity to liquid electrolyte, whereas the introduction of nano-Al_2O_3 particles can increase the mechanical properties as well as provide interfaces/pathways for fast ion transportation. At the same time, the electrolyte has a malleable shape, implying good contact with the electrode surface. This study is representative, as cross-linking and addition of nanoparticles represent the most effective ways to improve the ionic conductivity and mechanical properties [65, 67, 68].

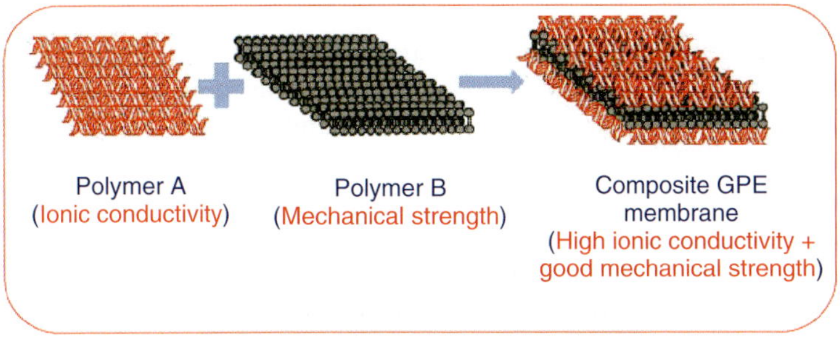

(a)

(b)

Figure 7.1 (a) Schematic illustration of the preparing process for composites as GPE membranes. Source: Courtesy from DKJ New Energy Tech. Co. Ltd. (b) SEM micrograph of a commercial GPE membrane. Source: Courtesy from DKJ New Energy Tech. Co. Ltd.

Another promising method is to form composites. For example, one component provides good gelling performance and another one provides mechanical strength [69, 70]. As a result, both the ionic conductivity and mechanical strength are good enough (Figure 7.1a). The scanning electron microscopy (SEM) morphology of one commercial product from DKJ New Energy Tech. Co. Ltd. is shown in Figure 7.1b.

(2) *Formability/contact with electrodes*: A high-performance GPE has to take the formability/contact with the electrodes into account. Although the incorporation of a liquid component will facilitate the wetting of GPEs on the electrode surface, achieving full contact between GPEs and electrode requires special design of the electrode/electrolyte structures, processing, and device assembly. The most common strategy is to pre-wet the electrode surface with the GPE solution several times. This will ensure full penetration of the GPE solution and finally good contact between the electrolyte and the electrodes [6].

(3) *Flammability*: Unlike SPEs, GPEs have the component of plasticizers, and most of which are flammable organic solvents. Once the liquid leaks, it is likely to cause the battery to burn. In order to reduce the flammability of the PE, nonflammable plasticizers have been developed, such as ionic

liquids. Ionic liquids, also known as room temperature molten salts, consist primarily of larger organic cations and smaller inorganic anions. In addition to nonflammability, ionic liquids have several advantages: high ionic conductivity, negligible volatility, high chemical and thermal stability, and wide electrochemical window [71–73]. More importantly, some ionic liquids are electrochemically stable to lithium metal [74]. Now, several ionic liquids comprising cations based on pyridinium, piperidinium, imidazolium, quaternary ammonium, and anions based on $[PF_6]^-$, $[BF_4]^-$, $[CF_3SO_3]^-$, $[C_4F_9SO_3]^-$, $[CF_3CO_2]^-$, $[N(CF_3SO_2)_2]^-$, $[N(CN)_2]^-$, and $[CF_3CONCF_3SO_2]^-$ have been investigated for LIBs.

(4) *Li^+ transference number*: For PEO-based PEs, the cation can form a coordinate bond with the oxygen on the PEO chain, while the interaction between the anion and the polymer chain is much weaker [75], so the anion is more likely to migrate on the polymer segment, which brings one defect to PEs: low cation (Li^+) transference number. The Li^+ transference number of a common Li-PEO type PE is generally 0.2–0.3 [76, 77]. The migration of anions toward the negative electrode causes serious concentration polarization, which results in a substantial decay in ionic conductivity or a time-dependent increase in cell impedance [75, 78]. The measures to increase Li^+ transference number are (i) to anchor the anion to the polymer backbone via covalent bonds, which is currently the most common method, and (ii) to add an anion receptor that preferentially interacts with anions.

7.4 Application and Performance of Polymer Lithium Ion Batteries

Due to the flexibility and low density of PEs, polymer LIBs can be thinned and arbitrarily shaped and have higher energy density. These advantages will enable polymer LIBs to be used in portable electronic products and wearable electronic devices such as cell phones, laptops, watches, etc.

When it is used for LIBs by using the coated $LiCoO_2$ as the positive electrode, its safety is superior to those from the ceramic-coated separators. Under normal safety testing such as heavy pressing, short circuit, and nailing after full charge to 4.4 V, there is no fire or explosion. At the charge cut-off voltage of 4.4 V, it can still be very safe even when it is short-circuited at 60 °C. In addition, the as-assembled lithium ion batteries can pass through the needle crush test without fire or explosion. The highest temperature during the needle test is below 150 °C. In contrast, those from the ceramic-coated porous polyalkylene separators could not pass through the needle test, which was recently removed from the international standard, since the highest temperature can be above 900 °C. The rate capability of the prepared LIBs can be comparable with those from liquid electrolytes (Table 7.2). These primary results show that (quasi) solid LIBs for electric vehicles are not far away.

Compared with liquid electrolytes, solid electrolytes are nonvolatile and generally nonflammable. Based on this advantage, the SPE can be used for lithium

Table 7.2 Comparison of rate capability of lithium ion batteries (model: 416271PU-3020 mAh, cut-off charge voltage is 4.4 V) from (a) GPE membrane of DKJ New Energy Tech Co. Ltd. and (b) a ceramic-coated separator.

LIBs from different membranes	Capacity at 0.2 C (mAh)	Capacity at 0.5 C (mAh)	Capacity at 1 C (mAh)	Capacity at 2 C (mAh)	Capacity ratio for 0.5 C/0.2 C (%)	Capacity ratio for 1 C/0.2 C (%)	Capacity ratio for 2 C/0.2 C (%)
a (No. 1)	3134.4	3100.2	3025.5	2678.3	98.91	96.53	85.45
a (No. 2)	3122.6	3095.2	3056.1	2717.8	99.12	97.87	87.04
a (No. 3)	3138.7	3105.2	3034.7	2680.0	98.93	96.69	85.39
b (No. 1)	3133.4	3108.0	3025.9	2680.6	99.19	96.57	85.55
b (No. 2)	3127.1	3094.6	3029.3	2707.0	98.96	96.87	86.57
b (No. 3)	3136.7	3109.2	3042.4	2712.2	99.12	96.99	86.47

Source: Courtesy from DKJ New Energy Tech. Co. Ltd.

batteries operating in high temperature environments without worrying about batteries burning. High temperature polymer LIBs can be used under extreme temperature conditions.

Some PE materials have a wide electrochemical window, which makes high-voltage electrode materials promising, thereby increasing battery energy density. The emergence of polymer LIBs will greatly increase the mileage of electric vehicles.

Compared with the porous gel electrolyte and the porous membrane immersed in the liquid electrolyte, the SPE is dense and has high strength and hardness, which can effectively prevent the penetration of lithium dendrites, thereby improving the safety of the battery and making the use of metal Li as a negative electrode possible, which will greatly increase the battery energy density (because the specific capacity of lithium metal is much higher than that of the currently used graphite negative electrode).

7.5 Future Trends

For SPEs, studies toward achieving good/stable interfacial properties are in critical need. Firstly, interfacial properties are more significant than ionic conductivity, owing to the fact that the measurement of ionic conductivity itself is notably affected by interfacial properties of the electrolyte and because good/stable interfacial properties between electrolyte and electrodes are a basic prerequisite for achieving high-performance devices. Tremendous efforts are being made to improve interfaces in composites (e.g. solid-state grafting), which should be instructive for solving interfacial issues in batteries. In addition, developing adhesive electrolytes and new fabrication techniques for electrodes/electrolytes are other effective ways of achieving better contact/interfacial properties. With regard to ionic conductivity, constructing continuous pathways with low energy barriers to ion conduction is important to obtain high ionic conductivity. Therefore, new macromolecular structures or special additives for SPEs need to be designed, such as nanoparticles, to create the desired pathways for the fast conduction of ions.

GPEs will continue to be more and more attractive because of its high performance. In addition to the conventional study on achieving high ionic conductivity and good mechanical properties, different GPEs have their particular issues to be solved. For GPEs composed of organic liquid electrolytes, safety enhancement is a critical issue. For GPEs with ionic liquid electrolytes as the liquid phase, improving the ionic conductivity and forming stable solid electrolyte interphase (SEI) will be important tasks. The solution to these issues is to develop composite GPEs with multifunctional components and structures. For example, the nanofillers of composite GPEs can be designed not only to improve the ionic conductivity and mechanical properties but also to overcome safety problems, whereas the polymer matrix can be designed with improved elastic and adhesive properties. Finally, to achieve good contact between the electrolyte and the electrodes, composite GPEs should be fabricated such that they are compatible with the fabrication of the electrodes.

At present, raw materials of the PEs include PEO, PVDF, PAN, PMMA, etc. These chemical products are derived from the petrochemical industry, which not only is limited by the cost of crude oil but also inevitably produces pollution during the production process. For the sustainable development of the earth, more and more natural polymers have been developed for the preparation of PEs such as cellulose, lignin, starch, and so on [79, 80] to replace the artificial polymers from fossil oils. These raw materials are abundant in source, low in price, and environmentally friendly and have broader application prospects.

Optimization of the preparation method is also one of the current research trends. So far, the preparation method of the PE is mainly that the lithium salt and the polymer are dissolved in an organic solvent, and then the solvent is evaporated by heating (i.e. solution casting method). Evaporation of organic solvents in the solution casting method requires a large amount of thermal energy, and the process is slow. Both the efficiency and benefits are low. It is an irresistible trend to adopt a more efficient, environmentally friendly, and energy-saving preparation method. For example, electrospinning technology can rapidly evaporate a solution under the action of an electrostatic field and is currently being studied by a large number of researchers.

Acknowledgments

Financial support from the National Materials Genome Project (2016YFB0700600) and the Distinguished Young Scientists Program of the National Natural Science Foundation of China (NSFC51425301, 21374021, 51673096, and U1601214) is gratefully appreciated.

References

1 Li, X., Zhang, H., Mai, Z. et al. (2011). *Energy Environ. Sci.* 4: 1147–1160.
2 Wang, W., Luo, Q., Li, B. et al. (2013). *Adv. Funct. Mater.* 23: 970–986.
3 Yin, B., Li, Z., Dai, W. et al. (2015). *J. Power Sources* 285: 109–118.
4 Baskakova, Y.V., Ol'ga, V.Y., and Efimov, O.N. (2012). *Russ. Chem. Rev.* 81: 367.
5 Yoo, H.D., Markevich, E., Salitra, G. et al. (2014). *Mater. Today* 17: 110–121.
6 Wang, Y. and Zhong, W.H. (2015). *ChemElectroChem* 2: 22–36.
7 Verma, P., Maire, P., and Novák, P. (2010). *Electrochim. Acta* 55: 6332–6341.
8 Long, L., Wang, S., Xiao, M., and Meng, Y. (2016). *J. Mater. Chem. A* 4: 10038–10069.
9 Hallinan, D.T. Jr. and Balsara, N.P. (2013). *Annu. Rev. Mater. Res.* 43: 503–525.
10 Wright, P.V. (2002). *MRS Bull.* 27: 597–602.
11 Croce, F., Appetecchi, G., Persi, L., and Scrosati, B. (1998). *Nature* 394: 456.
12 Christie, A.M., Lilley, S.J., Staunton, E. et al. (2005). *Nature* 433: 50.
13 Ibrahim, S. and Johan, M.R. (2011). *Int. J. Electrochem. Sci.* 6: 5565–5587.
14 Agapov, A.L. and Sokolov, A.P. (2011). *Macromolecules* 44: 4410–4414.
15 Borodin, O. and Smith, G.D. (2006). *Macromolecules* 39: 1620–1629.
16 Chintapalli, S. and Frech, R. (1996). *Macromolecules* 29: 3499–3506.
17 Diddens, D., Heuer, A., and Borodin, O. (2010). *Macromolecules* 43: 2028–2036.
18 Teran, A.A., Tang, M.H., Mullin, S.A., and Balsara, N.P. (2011). *Solid State Ionics* 203: 18–21.
19 Chandrasekaran, R. and Selladurai, S. (2001). *Polym. Int.* 50: 89–94.
20 Huq, R., Koksbang, R., Tonder, P., and Farrington, G.C. (1992). *Electrochim. Acta* 37: 1681–1684.
21 Shin, J.-P., Chang, B.-J., Kim, J.-H. et al. (2005). *J. Membr. Sci.* 251: 247–254.
22 Nasef, M. (2009). Fuel cell membranes by radiation-induced graft copolymerization: current status, challenges, and future directions. In: *Polymer Membranes for Fuel Cells* (eds. S.M.J. Zaidi and T. Matsuura). New York, NY: Springer Science.
23 Bhattacharya, A. and Misra, B. (2004). *Prog. Polym. Sci.* 29: 767–814.
24 Feuillade, G. and Perche, P. (1975). *J. Appl. Electrochem.* 5: 63–69.
25 Saikia, D., Chen-Yang, Y., Chen, Y. et al. (2008). *Desalination* 234: 24–32.
26 Zhang, H., Zhang, X., Shiue, E., and Fedkiw, P.S. (2008). *J. Power Sources* 177: 561–565.
27 Peng, X., Zhou, L., Jing, B. et al. (2016). *J. Solid State Electrochem.* 20: 255–262.
28 Bouchet, R., Maria, S., Meziane, R. et al. (2013). *Nat. Mater.* 12: 452.
29 Sequeira, C. and Santos, D. (2010). *Polymer Electrolytes: Fundamentals and Applications.* Elsevier.

30 Hellio, D. and Djabourov, M. (2006). Physically and chemically crosslinked gelatin gels. In: *Macromolecular Symposia*, vol. 241, 23–27. Wiley Online Library.
31 Wang, Y., Li, B., Ji, J. et al. (2013). *Adv. Energy Mater.* 3: 1557–1562.
32 Kang, Y., Cheong, K., Noh, K.-A. et al. (2003). *J. Power Sources* 119: 432–437.
33 Ren, Y., Zhang, Z., Fang, S. et al. (2002). *Sol. Energy Mater. Sol. Cells* 71: 253–259.
34 Li, H., Chen, Y.-M., Ma, X.-T. et al. (2011). *J. Membr. Sci.* 379: 397–402.
35 Shen, Y., Reddy, M.J., and Chu, P.P. (2004). *Solid State Ionics* 175: 747–750.
36 Park, M.-S., Hyun, S.-H., Nam, S.-C., and Cho, S.B. (2008). *Electrochim. Acta* 53: 5523–5527.
37 Jeong, H.-S., Kim, J.H., and Lee, S.-Y. (2010). *J. Mater. Chem.* 20: 9180–9186.
38 Sharma, J.P. and Sekhon, S. (2007). *Solid State Ionics* 178: 439–445.
39 Subramania, A., Sundaram, N.K., Kumar, G.V., and Vasudevan, T. (2006). *Ionics* 12: 175–178.
40 Ye, Y.-S., Rick, J., and Hwang, B.-J. (2013). *J. Mater. Chem. A* 1: 2719–2743.
41 Visentin, A.F., Alimena, S., and Panzer, M.J. (2014). *ChemElectroChem* 1: 718–721.
42 Johan, M.R., Shy, O.H., Ibrahim, S. et al. (2011). *Solid State Ionics* 196: 41–47.
43 Dissanayake, M., Jayathilaka, P., Bokalawala, R. et al. (2003). *J. Power Sources* 119: 409–414.
44 Fullerton-Shirey, S.K. and Maranas, J.K. (2010). *J. Phys. Chem. C* 114: 9196–9206.
45 Dissanayake, M., Jayathilaka, P., and Bokalawela, R. (2005). *Electrochim. Acta* 50: 5602–5605.
46 Suarez, S., Abbrent, S., Greenbaum, S. et al. (2004). *Solid State Ionics* 166: 407–415.
47 Tang, C., Hackenberg, K., Fu, Q. et al. (2012). *Nano Lett.* 12: 1152–1156.
48 Liu, Y., Lee, J., and Hong, L. (2002). *J. Power Sources* 109: 507–514.
49 Kumar, B. and Scanlon, L.G. (1999). *Solid State Ionics* 124: 239–254.
50 Adebahr, J., Best, A., Byrne, N. et al. (2003). *PCCP* 5: 720–725.
51 Gomez, E.D., Panday, A., Feng, E.H. et al. (2009). *Nano Lett.* 9: 1212–1216.
52 Young, W.-S. and Epps, T.H. III (2012). *Macromolecules* 45: 4689–4697.
53 Choi, I., Ahn, H., and Park, M.J. (2011). *Macromolecules* 44: 7327–7334.
54 Yamamoto, K., Iriyama, Y., Asaka, T. et al. (2012). *Electrochem. Commun.* 20: 113–116.
55 Amiki, Y., Sagane, F., Yamamoto, K. et al. (2013). *J. Power Sources* 241: 583–588.
56 Woo, J.H., Trevey, J.E., Cavanagh, A.S. et al. (2012). *J. Electrochem. Soc.* 159: A1120–A1124.
57 Sakuda, A., Nakamoto, N., Kitaura, H. et al. (2012). *J. Mater. Chem.* 22: 15247–15254.
58 Ohtomo, T., Hayashi, A., Tatsumisago, M. et al. (2013). *J. Power Sources* 233: 231–235.
59 Stephan, A.M. (2006). *Eur. Polym. J.* 42: 21–42.
60 Ramesh, S., Liew, C.-W., and Ramesh, K. (2011). *J. Non-Cryst. Solids* 357: 2132–2138.

61 Song, J., Wang, Y., and Wan, C.C. (1999). *J. Power Sources* 77: 183–197.
62 Zhang, P., Yang, L., Li, L. et al. (2011). *J. Membr. Sci.* 379: 80–85.
63 Chen, Z., Zhang, L., West, R., and Amine, K. (2008). *Electrochim. Acta* 53: 3262–3266.
64 Miranda, D.F., Versek, C., Tuominen, M.T. et al. (2013). *Macromolecules* 46: 9313–9323.
65 Gu, Y., Zhang, S., Martinetti, L. et al. (2013). *J. Am. Chem. Soc.* 135: 9652–9655.
66 Kil, E.H., Choi, K.H., Ha, H.J. et al. (2013). *Adv. Mater.* 25: 1395–1400.
67 Li, Y. and Fedkiw, P.S. (2007). *J. Electrochem. Soc.* 154: A1140–A1145.
68 Appetecchi, G., Alessandrini, F., Passerini, S. et al. (2005). *Electrochim. Acta* 50: 4396–4404.
69 Zhu, Y., Wang, F., Liu, L. et al. (2013). *Energy Environ. Sci.* 6: 618–624.
70 Zhu, Y., Wang, F., Liu, L. et al. (2013). *Sci. Rep.* (3): 3187.
71 Wetjen, M., Kim, G.-T., Joost, M. et al. (2013). *Electrochim. Acta* 87: 779–787.
72 Kim, G.-T., Appetecchi, G.B., Carewska, M. et al. (2010). *J. Power Sources* 195: 6130–6137.
73 Rupp, B., Schmuck, M., Balducci, A. et al. (2008). *Eur. Polym. J.* 44: 2986–2990.
74 Choi, J.-W., Cheruvally, G., Kim, Y.-H. et al. (2007). *Solid State Ionics* 178: 1235–1241.
75 Guhathakurta, S. and Min, K. (2010). *Polymer* 51: 211–221.
76 Mazor, H., Golodnitsky, D., Peled, E. et al. (2008). *J. Power Sources* 178: 736–743.
77 Li, J., Daniel, C., and Wood, D.L. III, (2011). Cathode processing for lithium-ion batteries. In: *Handbook of Battery Materials* (eds. C. Daniel and J.O. Besenhard), 939–960.
78 Sinha, K. and Maranas, J.K. (2011). *Macromolecules* 44: 5381–5391.
79 Xiao, S., Wang, F., Yang, Y. et al. (2014). *RSC Adv.* 4: 76–81.
80 Li, M., Wang, X., Yang, Y. et al. (2015). *J. Membr. Sci.* 476: 112–118.

8

Polymer Electrolytes for Lithium Ion Batteries and Challenges: Part II

Siti Nor Farhana Yusuf and Abdul K. Arof

Centre for Ionics University of Malaya, Department of Physics, Faculty of Science, University of Malaya, 50603 Kuala Lumpur, Malaysia

8.1 Introduction

The Italian physicist, Alessandro Volta, created the first battery based on pairs of copper and zinc discs. However, useful batteries known as the Daniel cells came 36 years after the voltaic pile. The Daniel cells comprised a copper jar with copper sulfate solution and produced reliable currents. The first rechargeable battery was invented in 1859 by a French physicist, Gaston Planté. This meant that the battery could be reused after restoring its drained power by a process called "charging." His work formed the basis for the lead (Pb)-based battery industry.

Since then there were many developments in battery technology. New battery chemistries were established. In 1887, the Zn-C dry cell was created. The nickel–cadmium battery technology emerged in 1899 and four years later witnessed the appearance of the nickel–iron battery. The nickel–hydrogen and nickel–metal hydride (Ni–MH) chemistries emerged in the 1970s, which also saw the development in lithium and lithium ion batteries (LIBs). These have undergone many improvements until today.

The lithium technology exhibits the highest capacity. One gram of Li contains 3800 mAh of charge. One gram of Cd contains 238.4 mAh of charge and one gram of Pb contains 129.3 mAh of charge. Cd and Pb are also toxic elements. The size of batteries also varies depending on usage from work and leisure to communications and travel. To surf the Internet on our netbooks and laptops, batteries are required. It is undeniable how the Internet has influenced our lives. Batteries are therefore an essential part of our daily life.

LIBs must not be abused. At present, liquid electrolytes (LEs) are used in LIBs. These contain volatile and flammable components. The electrolytes may not be very stable. When there is heat whether through internal shorts or high temperature, the chemical components in the electrolyte can react and create gases and more heat. Uncontrolled reactions can also lead to thermal runaways and end in a fire.

Batteries can also explode. They can fail when overcharged or charged too fast. Overcharging results when a lot of lithium enters the anode. It is a manufacturer's

Polymer Electrolytes: Characterization Techniques and Energy Applications,
First Edition. Edited by Tan Winie, Abdul K. Arof, and Sabu Thomas.

fault when the protective circuit preventing overcharging is defective. Charging too fast is sending too much current into the battery. This may result if the battery and charger are incompatible. Overcharging or charging too fast can lead to a phenomenon known as plating. On charging a battery, lithium ions enter the anode and take their positions at the respective sites in the anode material. If the battery is charged slowly, lithium will enter safely at their proper sites. If charging is too fast or too much, lithium can be deposited at other regions and not at their sites. Lithium can also be deposited onto themselves. So on each recharge, lithium accumulates on itself to form needle-like structures called dendrites. These dendrites can make the electrodes touch each other resulting in internal shorts, thermal runaways, and explosion.

As mentioned earlier, the electrolyte in LIBs is liquid. LEs are prone to leakage. As batteries discharge, the chemistry changes and gases generated. Thus pressure in the battery builds up. The excess pressure can eventually puncture the battery and the electrolyte may leak out by evaporation or some other processes.

It is for these reasons that scientists have developed solid polymer electrolytes (SPEs). SPEs have good mechanical strength, which is an advantage over LEs. Unfortunately, the conductivity σ is rather low. Attempts to increase the conductivity include the addition of plasticizers and fillers. Blending two polymer hosts has also been tried. In order to take advantage of the good mechanical strength of polymer electrolytes (PEs) in the solid form and the high conductivity of electrolytes in the liquid form, gel or quasi-solid PEs were prepared. To further increase conductivity, the gels have been added with "ionic liquids." Battery technology is continuously progressing. There are still a lot of challenges to be faced, e.g. to have batteries that have higher open-circuit voltages, electrolyte that can withstand higher voltage, faster battery charging, and longer lifetime.

8.2 Structure and Operation of Lithium Ion Batteries

A battery has two electrodes, viz. the anode and cathode. These electrodes are linked by an electrolyte. In a cell that uses LE, a polypropylene separator separates the anode and cathode to prevent internal shorts. In cells fabricated with solid or quasi-solid electrolytes, a separator material is not required. The solid or quasi-solid electrolyte can also act as the separator. Electrolytes are ion conductors, but electronic insulators. In LIBs, the cathode is a lithium compound. The anode is a carbonaceous material and the electrolyte is a lithium ion (Li^+) conductor. The cell is in the discharged state when assembled. When charged, the cathode is depleted of Li^+ ions and the anode is full of intercalated lithium. During discharge, the Li^+ ions will exit the anode and enter and travel within the electrolyte to the cathode. The corresponding electrons also travel to the cathode, but through an external circuit and power a load or electrical device. During discharge, the Li^+ ions commute from anode to cathode and vice versa during charge.

The theoretical cell voltage, ΔE, can be estimated from the equation:

$$\Delta E = -\frac{\Delta G}{nF} \tag{8.1}$$

Here ΔG is Gibb's free energy of the reaction; $F = eN_A$ is the product of electron charge and Avogadro's number, which gives being Faraday's constant; and n is the valency of the mobile cation. The Li^+ ion batteries can be benchmarked using the following criteria:

(a) Specific capacity, $C_{specific}$

This parameter is associated with the ratio of the total electrons liberated during the electrochemical reactions to the host atomic mass. It is a measure of how much charge is stored reversibly per unit mass of the battery. The theoretical $C_{specific}$ in units of $mAh\,g^{-1}$ is given by

$$C_{specific} = \frac{xF}{nM} \tag{8.2}$$

Here x is amount of electrons released in the reaction. n is the number of active material atoms that will be lithiated with 1 Li. M is the molar mass of graphite (carbon). As an example, the reversible intercalation reaction between lithium and graphite to form LiC_6 is

$$Li^+ + e^- + 6C \leftrightarrow LiC_6$$

Graphite has a molar mass of $12\,g\,mol^{-1}$, $x = 1$, and 6 C atoms are lithiated with 1 Li. Plugging in the values gives graphite a theoretical capacity of $372\,mAh\,g^{-1}$. Based on the formation on $Li_{4.4}Sn$, the theoretical reversible capacity or specific charge of tin metal is $994\,mAh\,g^{-1}$, about thrice the specific charge of graphite.

Capacity can also be expressed in volume basis. The unit for volumetric capacity is $mAh\,cm^{-3}$ [1, 2].

(b) Specific energy

Specific energy is the amount of energy the battery can release and store for every unit battery mass, $Wh\,kg^{-1}$. Specific energy can also be expressed as $Wh\,l^{-1}$, where l is the volume of the battery. It is given by

$$\text{Specific energy} = \text{specific capacity} \times \text{operating battery voltage} \tag{8.3}$$

(c) Cyclability

Cyclability is the measure of how reversible is the Li^+ ion intercalation into the electrodes and deintercalation from the electrodes. Total reversibility would mean that if 10 Li^+ ions enter an electrode, the same amount leaves the electrode. It accounts for how many charge–discharge cycles will take place before the battery fails to power a load. Cycle life depends on battery chemistry, state of charge (SOC), operating temperature, and depth of discharge (DOD). Li dendrites can form on graphite anodes at low temperature charge.

(d) Discharging and charging rates

The rate of fully loading or dispensing charge also known as C-rate indicates how quickly the battery can be discharged and recharged. At 1 C, the battery undergoes maximum discharge or dispenses all its charge in one hour. If an 8 A-h battery is charged at 8 A, then it is charging at 1 C and would finish charging in one hour. If it is charged at 2 A, then it would be charging at one-quarter C and would charge in four hours. If the battery is drained at 8 A, it will last for an hour, that is, a drain of 1 C.

8.2.1 Anode Materials

Carbon [3], silicon [4], $LiTi_4O_5$ [5], and Li_2SnO_3 [6] have been used as anodes, among others. Lithium has a specific charge of 3862 mAh g^{-1}. There are renewed interests in using lithium metal as anode material, but cycling behavior needs improvement and dendritic growth suppressed. This can result in short circuits and thermal runaways. Due to their relatively low cost and availability, carbonaceous materials are used as anodes, although the theoretical C_{specific} is only 372 mAh g^{-1}. This is less than one-tenth that of lithium. Silicon has high C_{specific} of 4200 mAh g^{-1} [7], corresponding to a composition of $Li_{22}Si_5$.

From literature [8] for SnO anodes, the two-step reaction is

$$SnO + 2Li^+ + 2e^- \rightarrow Sn + Li_2O \tag{8.4}$$

$$Sn + xLi^+ + xe^- \leftrightarrow Li_xSn \quad (0 \leq x \leq 4.4) \tag{8.5}$$

Equation (8.4) is not reversible. The electrochemical reaction between SnO and Li^+ ion leads to the formation of metallic tin and Li_2O. The metallic tin formed reacts reversibly with Li^+ ion to form Li_xSn alloy as shown in Eq. (8.5). SnO has a molecular weight of 134.71 g mol^{-1}. From the equations, if the total number of electrons is released, x is taken as 6.4, and then the theoretical C_{specific} will be 1273 mAh g^{-1}. However, the reversible theoretical C_{specific} (see the equation 8.5) is only 875 mAh g^{-1}.

SnO_2 will also form an alloy with lithium when used as anode in a Li/SnO_2 half-cell [9]. The reactions are

$$SnO_2 + 4Li^+ + 4e^- \rightarrow Sn + 2Li_2O \tag{8.6}$$

$$Sn + xLi^+ + xe^- \leftrightarrow Li_xSn \quad (0 < x < 4.4) \tag{8.7}$$

From Eq. (8.6), tin dioxide reduces to tin and Li_2O during the first discharge. This reaction is not reversible. The oxide Li_2O alleviates the volume change, but does not prolong cycling. On the other hand, Eq. (8.7) shows reversible alloying/dealloying reactions where the metallic tin reacts with lithium ion to form $Li_{4.4}Sn$ alloy with maximum uptake of 4.4 Li per Sn atom. This leads to a reversible theoretical C_{specific} of 783 mAh g^{-1}.

The electrochemical reactions involving Li_2SnO_3 [10] are

$$Li_2SnO_3 + 4Li^+ + 4e^- \rightarrow 3Li_2O + Sn \tag{8.8}$$

$$Sn + xLi^+ + xe^- \leftrightarrow Li_xSn \quad (x \leq 4.4) \tag{8.9}$$

Lithium tin oxide, Li_2SnO_3, has a molecular weight of 180.71 g mol^{-1}. The initial irreversible capacity with $x = 8.4$ is 1246 mAh g^{-1} and the reversible capacity is 652 mAh g^{-1}. During initial discharge, two kinds of reaction have taken place, i.e. the destruction of the crystal structure, which caused irreversible formation of Sn atom embedded in Li_2O matrices followed by the formation of reversible Li–Sn alloy. Even for non-lithium tin oxide compounds [11], Y–Sn–O (Y = Ba,

Sr, Ca, and Co), the irreversible and reversible reactions still occur as shown in the following:

$$YSnO_3 + 4Li^+ + 4e^- \rightarrow YO + 2Li_2O + Sn \quad (Y = Ba, Sr, Ca, \text{and } Co) \tag{8.10}$$

$$Sn + xLi^+ + xe^- \leftrightarrow Li_xSn \quad (0 < x < 4.4) \tag{8.11}$$

8.2.2 Cathode Materials

Advanced cathode materials for lithium ion cells include $LiCoO_2$ [12], $LiMn_2O_4$ [13], $LiNiO_2$ [14], $LiNi_{0.5}Mn_{1.5}O_4$, and $LiFePO_4$ [15]. Layered oxides having Co and Ni are frequently used as the cathode materials. They are highly stable in the high-voltage range. However, the natural abundance of cobalt is limited and it is toxic. Manganese is cheaper. The material has good thermal stability and rate capabilities. Unfortunately, its cycling behavior is rather limited. Therefore, to minimize the weaknesses, Mn, Co, and Ni combination is regularly used. Vanadium oxide is also added and has shown to exhibit a large capacity [16]. However, vanadium oxide tends to become amorphous due to Li insertion and extraction. This factor limits the cyclability of the cell. Although olivines [17] have moderate capacity and low fade, their conductivity is low. They are also nontoxic. To make up for the poor electrochemical behavior, researchers attempted partial substitution, nanosizing, and applied conductive coatings to the olivines. This however incurred costs to the battery. $LiMnPO_4$ is one of the olivine candidates for cathode in lithium ion cells. It is safe even in the charged state. Its energy density is at par with the commercial materials. In addition, the raw material costs are low. The redox process is

$$LiM^{2+}PO_4 \leftrightarrows Li^+ + 2e^- + M^{3+}PO_4 \tag{8.12}$$

The cathode reaction for $LiCoO_2$ [18] is

$$LiCoO_2 \leftrightarrow Li^+ + e^- + CoO_2 \tag{8.13}$$

Here, the number of electrons transferred per $LiCoO_2$ molecule, x, is 1. The number of active material that will be lithiated with 1 Li is $n = 1$. The theoretical $C_{specific}$ is 274 mAh g^{-1}. Hence, it is critical to know the complete reaction that takes place. The practical $C_{specific}$ is less than the theoretical capacity because not all of the Li can be removed from the host material lattice.

The cathode reaction for $LiMn_2O_4$ [19] is

$$LiMn_2O_4 \leftrightarrow Li^+ + e^- + Mn_2O_4 \tag{8.14}$$

Here, the amount of electrons transferred per $LiMn_2O_4$ molecule, x, is 1. The number of active material that will be lithiated with 1 Li is $n = 1$. The molecular weight of $LiMn_2O_4$ is 180.816 g mol^{-1}. Thus, the theoretical $C_{specific}$ is 148 mAh g^{-1}. The practical $C_{specific}$ is 120 mAh g^{-1}.

As another example, consider the reaction:

$$LiFePO_4 \rightarrow Li^+ + e^- + FePO_4 \tag{8.15}$$

The molecular weight of $LiFePO_4$ is 158 g mol^{-1}. Here x and n are equal to 1. The theoretical capacity is 170 mAh g^{-1}.

Consider the reaction:

$$LiNi_{0.5}Mn_{1.5}O_4 \rightarrow Li^+ + e^- + Ni_{0.5}Mn_{1.5}O_4 \quad (8.16)$$

In this case, the molecular weight of $LiNi_{0.5}Mn_{1.5}O_4$ is 176 g mol^{-1}. Again x and n are unity. The theoretical capacity is 152 mAh g^{-1}.

8.2.3 Electrolytes

All batteries require an electrolyte. The electrolyte must have high Li^+ transference number and good Li^+ ion conductivity. Apart from these, the electrolyte must be stable mechanically, thermally, and electrochemically. Thus, the electrolyte must have a rigid structure, be able to withstand elevated temperatures, and have a high decomposition voltage. The high Li^+ ion transference number is a requirement in LIBs. There has been increasing interest for high energy density Li^+ batteries employing PEs. PEs can exist in dry solid and gel forms. Dry SPEs exhibit low conductivity (generally $\leq$0.1 μS cm^{-1} at ambient temperature), but in gel form the conductivity can reach up to 10^{-3} S cm^{-1} [20]. PEs comprise the polymer host, salt, and solvent. The host materials can also consist of blends of two polymers. Poly(ethylene oxide) (PEO) has been used many times as polymer host. SPEs based on PEO have garnered much attention as they are compatible with Li metal electrode. Reports have shown that PEO–salt complexes can have ionic conductivities suitable for device use at elevated temperatures approaching 373 K, but below 333 K, the conductivity is too low for RT devices. This is because of the crystalline–amorphous phase transition [21]. In PEs, ions are the main conducting species. Often the addition of plasticizers and fillers results in conductivity enhancement. Plasticizers should have low viscosity and high dielectric constant. It is for these factors that plasticizers are used in pairs. Using one plasticizer with high dielectric constant and one with lower viscosity will produce better conductivity. Fillers can provide additional pathways for ion movement via their surface.

8.2.4 Li^+ Ion Transport in Polymer Electrolytes

For utilization of a PE in an LIB, a crucial condition to consider is the conductivity. Several ion transport equations have been suggested to understand ionic conduction of PEs. These are the Arrhenius [22], Williams–Landel–Ferry (WLF) [23], and Vogel–Tammann–Fulcher (VTF) [24] equations. The VTF equation is used to analyze highly amorphous electrolytes and the Arrhenius equation for crystalline ionic conductivity [25]. Since high conductivity PEs are favored and high conductivity implies that the electrolyte is highly amorphous, we concentrate only on the VTF equation. It was established to explain the viscosity of super cooled liquids [26]:

$$\eta = C \exp\left[-\frac{B}{T - T_o}\right] \quad (8.17)$$

The pre-exponent factor C is proportional to $T^{1/2}$, B is associated with the activation energy, and T_o is the ideal temperature where all “free volume,” segmental motion, or excess configurational entropy of the PE disappears. Combining the previous equation with the Stokes–Einstein equation,

$$\sigma = \frac{ne^2}{kT}D \tag{8.18}$$

and the Nernst–Einstein equation,

$$D = \frac{kT}{6\pi r_i \eta} \tag{8.19}$$

we will obtain the following equation:

$$\sigma = \sigma_o \exp\left[-\frac{B}{k(T - T_o)}\right] \tag{8.20}$$

In Eq. (8.18), n is the free ion number density, e is the charge on each ionic carrier, k is Boltzmann's constant, and D is the ionic diffusion coefficient. r_i from Eq. (8.19) is a diffusion radius. The pre-exponential factor $\sigma_o \propto T^{-1/2}$.

Equation (8.20) has been corrected as

$$\sigma = AT^{-1/2} \exp\left[-\frac{E_a}{k(T - T_o)}\right] \tag{8.21}$$

and applied to the ionic transport behavior in PEs [27–29]. E_a is the energy required for ion transport in the electrolytes.

8.3 Polymer Electrolyte for Lithium Ion Batteries

Ion conducting electrolytes influence significantly the performance of an LIB. An excellent electrolyte should have high conductivity and cationic transference number of one. Thus, the anion of the doping salt should be immobile. This is a challenge because a transference number of one would result in low conductivity. Anions move faster than lithium ions and therefore have a greater contribution to the conductivity compared with lithium ions. The anion does not react with the electrode. This results in anion accumulation at the boundary between the electrode and electrolyte that leads to concentration polarization [30, 31]. The lithium ion transference number is less than 0.5. Concentration polarization will result in conductivity decay limiting the cell life, charging rate, and energy density [32, 33]. Hence, there has to be a compensation between conductivity and the lithium ion transference number. The degree of amorphousness of the PE influences conductivity and glass transition temperature. The degree of amorphousness depends on the amount of doping salt, plasticizer, and filler added. Tables 8.1–8.4 list some properties of lithium ion conducting electrolytes that maybe suitable for use in LIBs.

Plasticizer addition is an efficient strategy to enhance the conductivity of a PE. The choice of plasticizer is on its dielectric constant and viscosity. As already mentioned, plasticizers should exhibit high dielectric constant to facilitate salt dissociation and low viscosity to assist the polymer host dilution and ease ionic movement, in other words, the enhancement of ionic mobility, μ. The separation

Table 8.1 Examples of solid polymer electrolytes with plasticizers.

Solid polymer electrolyte	σ (mS cm^{-1})	T_g (°C)	Thermal stability (°C)	Electro-chemical stability (V)	Mechanical stability (MPa)	References
(PVA)/PMMA–$LiBF_4$–EC	1.29	—	336	—	—	[8]
(PVA)/PMMA–$LiBF_4$–PC	0.47	—	274	—	—	[8]
(PVA)/PMMA–$LiBF_4$–DEC	0.049	—	349	—	—	[8]
(PVA)/PMMA–$LiBF_4$–GBL	0.10	—	307	—	—	[8]
PAN–NH_4NO_3–PC	7.48	42.6	—	—	—	[6]
P(GMA-*co*-MMA)–EC–$LiClO_4$	0.30	—	—	3.8	—	[7]
PAN–EC–DMP–$LiBF_4$	10.7	—	—	3.6	—	[12]
PAN–SN–LiTFSI	11.7	67	—	3.6	—	[13]
PEO–$LiCF_3SO_3$–DOP	0.76	−63.86	—	—	1.2	[10]
PEO–$LiCF_3SO_3$–PEG	0.017	−55.82	—	—	4.2	[10]
PMMA–$Li_2B_4O_7$–PC	0.51	—	373	—	0.12	[34]
PEO–$LiCF_3SO_3$–EC	0.081	−70.5	—	—	—	[11]
$(PEO)_9LiCF_3SO_3$–EC	0.002	−48	—	—	—	[18]
PVdF–PVC–EC/PC–$LiClO_4$	4.68	—	250	—	—	[19]
PVC–PEMA–PC–$LiBF_4$	6.73	—	254	—	—	[3]
PVC–PEMA–PC–$LiClO_4$	3.45	—	231	—	—	[3]
PVC–PEMA–PC–$LiCF_3SO_3$	2.18	—	248	—	—	[3]
PEO–DBP–$CdSO_4$	1.45×10^{-5}	−130.3	—	—	—	[4]
PEO–DBP–$CdCl_2$	2.06×10^{-5}	−128.0	—	—	—	[4]
PVA–H_3PO_4–PEG	0.081	56	—	—	—	[5]

of the salt into ions increases the number density, n, of the charge carriers. The increase in number density and mobility enhances the ionic conductivity, σ, since $\sigma = ne\mu$ where e is the ionic charge. The plasticizer decreased the glass transition temperature and increased amorphousness of the PE. Many theories have been suggested to explain the plasticizer action [67]. According to the "lubricity theory," the plasticizers serve as a lubricant to accelerate the motion of charge carriers. The "gel theory" takes into account the disruption of polymer–polymer interactions. With increased disorder, the free volume also increased. The free volume is the empty space in a solid or liquid unoccupied by the atoms or molecules. Following this description, free volume will decrease when cooled. At the glass transition temperature, free volume will "disappear" and the polymer chains are in fixed positions. Hence, the more the free volume, the lower will be the glass transition temperature. This is the "free volume theory."

The (PVA)/PMMA–$LiBF_4$ blended host system has been plasticized with ethylene carbonate (EC), propylene carbonate (PC), γ-butyrolactone (GBL), and dimethyl carbonate (DMC) [68]. The conductivities of these SPEs are 1.29, 0.47, 0.10, and 0.049 mS cm^{-1}, respectively. The dielectric constant of these plasticizers in the same respective order is 89.78 [69], 64.92 [69], 39.00 [70],

Table 8.2 Examples of SPEs with fillers.

Solid polymer electrolyte	σ (mS cm^{-1})	T_g (°C)	Thermal stability (°C)	Electro-chemical stability (V)	Mechanical stability (MPa)	References
PEO–LiTf–Al_2O_3	0.021	−50	—	—	—	[14]
PEO–$MgCl_2$–B_2O_3	0.0072	−18	—	—	—	[15]
PEO–LiTf–EC–Al_2O_3	0.51	−72	—	—	—	[16]
PEG–$LiClO_4$–δ-Al_2O_3	0.0045	−37	—	—	—	[17]
PEO–KI–ZnS	0.31	−52	—	—	—	[20]
PEO–LiTf–TiO_2	0.049	−46	—	—	—	[21]
PEO–LiTf–TiO_2–EC	0.16	−50	—	—	—	[21]
PEO–LiTDI–TiO_2	0.021	—	250	4.5	0.62	[35]
PEG–$LiClO_4$–hydrotalcite	0.011	−60	~327	—	—	[36]
PEO–$LiClO_4$–$ZnAl_2O_4$	0.0022	—	—	~5.5	—	[37]
PEO–$LiClO_4$–^{f}SBA-15	0.0033	−36	—	4.8	—	[38]
Cross-linked poly(EO/PO)–$LiNTf_2N$–$BaTiO_3$	0.045	—	433	4.4	1.73	[39]
Cross-linked poly(EO/PO)–$LiNTf_2N$–γ-$LiAlO_2$	0.057	—	450	4.4	2.34	[39]
PEO–$LiClO_4$–SiO_2	0.023	−40	—	—	—	[40]
PEO–$CuTf_2$–Al_2O_3	0.0009	−40	—	—	—	[41]
PEO–$LiClO_4$–$Li_{1.3}Al_{0.3}Ti_{1.7}(PO_4)_3$	0.014	−54	—	—	—	[42]
PAN–$LiClO_4$–α-Al_2O_3	0.57	−67	—	—	—	[43]
PEO–$LiClO_4$–SiO_2	0.095	−35	—	—	—	[32]
PEO–$LiNTf_2$–SiO_2	0.14	−43	—	—	—	[32]
PEO–NH_4HSO_4–SiO_2	0.25	−65, −4	—	—	—	[44]

Tf: triflate (CF_3SO_3).

and 3.10 [69]. The respective viscosity is 1.93 mPa-s at 313 K [71], 2.50 mPa-s at 298 K [72], 1.72 Pa-s [70], and 0.585 mPa-s at 298 K [73]. The structures of these plasticizers are shown in Figure 8.1.

The decrease in conductivity, σ, i.e. $\sigma(\text{EC}) > \sigma(\text{PC}) > \sigma(\text{GBL}) > \sigma(\text{DMC})$, follows the trend of decrease in dielectric constant ε_r, i.e. $\varepsilon_r(\text{EC}) > \varepsilon_r(\text{PC}) > \varepsilon_r(\text{GBL}) > \varepsilon_r(\text{DMC})$. Hence, it can be expected that the number density of Li^+ and BF_4^- ions, n, follows the order $n(\text{EC}) > n(\text{PC}) > n(\text{GBL}) > n(\text{DMC})$. However, the viscosity values do not follow the trend, and it may be inferred that the value of viscosity is not critical in influencing the conductivity value.

The SPE 85 wt% PEO–15 wt% $LiCF_3SO_3$ exhibited the highest conductivity of 1.00×10^{-6} S cm^{-1} [74]. In addition of 20 wt% dioctyl phthalate (DOP), the conductivity increased to 0.76 mS cm^{-1}. After adding 20 wt% PEG (M.W. 6000, BASF Aktiengesellschaft), the value increased to 8.73×10^{-6} S cm^{-1}. The dielectric constant of PEG 6000 as reported by Abdali and Abdulazeez is between 1.9 and 1.95

Table 8.3 Properties of GPEs with various ILs.

Gel polymer electrolyte	σ (S cm^{-1})	T_g (°C)	Thermal stability (°C)	Electro-chemical stability (V)	Mechanical stability (MPa)	References
PEO–MGTf$_2$–EMITf	5.6×10^{-4}	−71	~340–350	—	—	[45]
(PVdF-HFP)–ZnTf$_2$–(EMIM TFSI)	3.8×10^{-3}	−35	323	3.5	—	[46]
(PVdF-HFP)–LiTFSI–(PYR$_{14}$ TFSI)	2.8×10^{-4}	—	400	4.8	—	[47]
(PVdF-HFP/PMMA)–([BMIM]BF$_4$)	1.4×10^{-3}	−87	~180–310	4.5	—	[48]
(PVdF-HFP)–LiTFSI–(B$_4$MePyTFSI)	2.0×10^{-4}	—	400	5.5	—	[49]
(PVdF-HFP)–NaTFSI–imidazolium	2.2×10^{-4}	−78.5	380	—	—	[50]
PVA/PAA–PYR$_{14}$TFSI–LiTFSI	2.2×10^{-3}	—	—	5.0	—	[51]
PUA/PMMA DEEYTFSI–LiTFSI	2.8×10^{-4}	—	—	4.7	—	[52]
PEO-LiTFSI–EMIMTFSI	2.1×10^{-4}	−53.3	363	4.6	—	[53]
PMMA-LiTFSI–PP$_{13}$TFSI	3.4×10^{-4}	—	400	6.0	—	[54]
PMMA–PVC–LiTFSI–BmImTFSI	1.6×10^{-4}	112	255	—	—	[55]
(PVdF-HFP)–ZnTf$_2$–(EPy TFSI)	1.2×10^{-1}	—	—	~2.5	—	[56]
(PVdF-HFP)–ZnTf$_2$–(EMIM TFSI)	1.7×10^{-1}	—	—	~3.0	—	[56]
(PVdF-HFP)–ZnTf$_2$–(P$_{6,6,6,14}$TFSI)	9.5×10^{-3}	—	—	~4.0	—	[56]
(PVdF-HFP)–LiTFSA–(EMIM–TFSA)	1.7×10^{-3}	—	—	5.1	—	[57]
(PVdF-HFP)–LiTFSA–(EMPyrr–TFSA)	6.6×10^{-5}	—	—	6.1	—	[57]
(PVdF-HFP)–LiTFSA–(MPPyrr–TFSA)	1.0×10^{-3}	—	—	5.8	—	[57]
(PVdF-HFP)–LiTFSA–(DMMA–TFSA)	3.7×10^{-4}	—	—	5.3	—	[57]

[75]. The viscosity of PEG 6000 is 0.894 mPa-s at 298 K [76]. The structure of DOP is depicted in Figure 8.2, its viscosity is 56 mPa-s [77], and dielectric constant is 5.1 at 297 K [78]. This demonstrates that dielectric constant has a more pronounced effect on the conductivity compared with viscosity.

The conductivities of PEO–$CdSO_4$ and PEO–$CdSO_4$–DBP (with 48.22 wt% DBP) were reported to be in the order of 10^{-7} and 1.45×10^{-6} S m^{-1},

Table 8.4 Properties of GPEs with various fillers.

Gel polymer electrolyte	σ (S cm^{-1})	T_g (°C)	Thermal stability (°C)	Electro-chemical stability (V)	Mechanical stability (MPa)	References
PMMA–(PC–DEC)–$LiClO_4$–(CNFs)	3.9×10^{-5}	—	220	—	5.28	[58]
PMMA–(PC–DEC)–$LiClO_4$–(MMT)	1.3×10^{-3}	−75	—	4.9	—	[59]
(PVAc/PVdF-*co*-HFP)–(EC–DEC)–$LiCF_3SO_3$)–$LiAlO_2$	5.0×10^{-3}	—	362	—	—	[42]
PEO–LiTf–EC–TPGS-S	7.8×10^{-4}	—	—	4.5	—	[60]
PVdF-HFP/TPU/PMMA–EC–$LiClO_4$–SiO_2	8.5×10^{-3}	—	400	5.9	10.8	[22]
PVdF-*co*-HFP–(EC–DEC–$LiClO_4$)–$LiAlO_2$	8.1×10^{-3}	—	—	5.2	—	[25]
P(VdF-HFP)/P(EO-EC)–$LiCF_3SO_3$–TiO_2	5.1×10^{-5}	−56.3	—	—	—	[26]
P(VdF-*co*-HFP)–(EC–DEC)–$LiPF_6$–montmorillonite/smectite clay	9.0×10^{-4}	—	~422–428	—	5.9	[61]
(PVAc/PVdF-*co*-HFP)–(EC–DEC)–$LiPF_6$–ZnO	3.7×10^{-4}	—	325	—	—	[27]
PVdF–(PC–DEC)–$LiClO_4$–MMT	2.3×10^{-3}	—	—	4.6	—	[28]
P(VdF-HFP)–(EC–DMC)–$LiPF_6$–Al_2O_3	1.8×10^{-3}	—	—	—	—	[29]
P(VdF-HFP)–LiTFSI–BMITFSI–Al_2O_3	3.6×10^{-3}	—	—	5.5	—	[62]
P(VdF-HFP)–LiTFSI–BMITFSI–SiO_2	4.4×10^{-3}	—	—	5.9	—	[62]
P(VdF-HFP)–LiTFSI–BMITFSI–$BaTiO_3$	5.2×10^{-3}	—	—	6.0	—	[62]
(PAN/TEGDA–BA)–Al_2O_3	2.4×10^{-3}	—	110	4.5	1.03	[63]
P(VdF-HFP)–(EC–DEC–DMC)–$LiPF_6$–SiO_2	1.1×10^{-3}	—	—	4.8	—	[64]
P(VdF-HFP)–LiTFSI–$PYRA_{12O1}$TFSI–SBA–15 (silica)	2.5×10^{-4}	~180	—	>4.7	—	[65]
Poly(EO/PO)–$LiN(CF_3SO_2)_2$–$BaTiO_3$	4.5×10^{-5}	—	433	4.4	1.73	[66]
Poly(EO/PO)–$LiN(CF_3SO_2)_2$–γ-$LiAlO_2$	5.7×10^{-4}	—	450	4.4	2.34	[66]

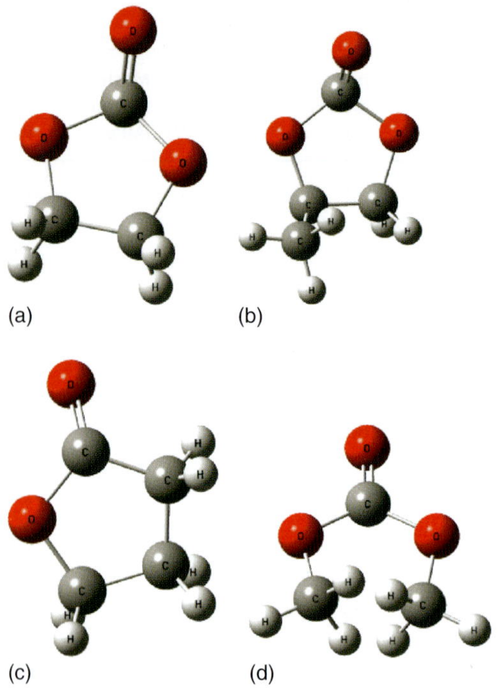

Figure 8.1 Structural formulas for (a) ethylene carbonate, (b) propylene carbonate, (c) γ-butyrolactone (GBL), and (d) dimethyl carbonate.

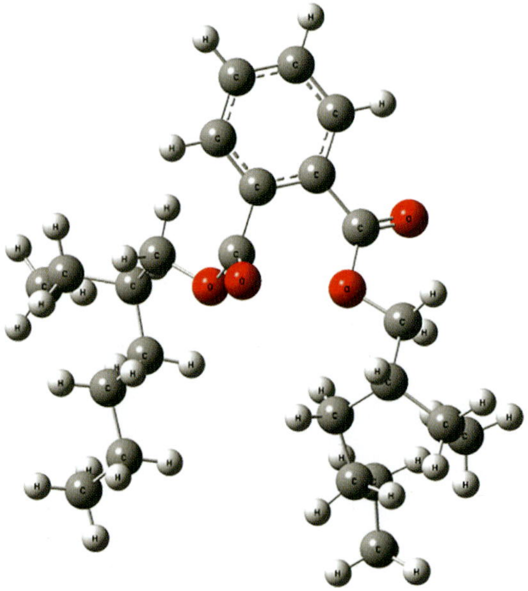

Figure 8.2 Structural formula for DOP.

respectively [79]. The structure of DBP is shown in Figure 8.3. The viscosity of DBP at 298 K is 17 mPa-s [77, 80]. Its dielectric constant at 293 K is 6.58 [80]. The high dielectric constant of DBP helps to improve the conductivity.

Table 8.2 lists some properties of SPEs added with fillers such as Al_2O_3, B_2O_3, ZnS, etc.

Figure 8.3 Structural formula for DBP.

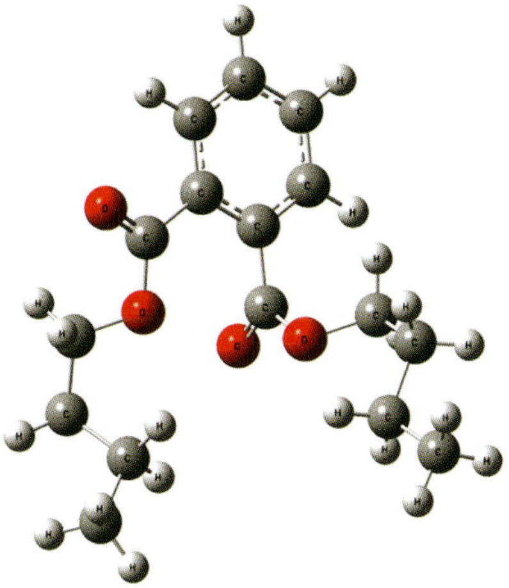

The conductivity of $(PEO)_9LiTf$, $(PEO)_9LiTf + EC$, and $(PEO)_9$ $LiTf + EC + Al_2O_3$ is 3.5×10^{-7}, 1.6×10^{-6}, and $1.5 \times 10^{-4}\,S\,cm^{-1}$, respectively [81]. The conductivity is enhanced by the addition of EC and is further enhanced by the addition of Al_2O_3 to the order of $10^{-4}\,S\,cm^{-1}$. The glass transition decreased from 229 K in $(PEO)_9LiTf$ to 225 K in $(PEO)_9LiTf + EC$ and to 217 K in $(PEO)_9$ $LiTf + EC + Al_2O_3$. This indicates that EC has increased the amorphousness of $(PEO)_9LiTf$ and increased the concentration of free ions per unit volume. Al_2O_3 can increase the amorphousness of $(PEO)_9LiTf + 50$ wt% EC. Not only did T_g decreased, the melting temperature, T_m, also decreased indicating the softening of the polymer backbone and increase in segmental motion. The increase in amorphousness segmental motion led to the enhancement in Li^+ and Tf^- conductivity.

The conductivity for $(PEG)_{46}LiClO_4$ is $0.73\,\mu S\,cm^{-1}$ and increased to $\sim 4.5\,\mu S\,cm^{-1}$ with 10 mol% δ-Al_2O_3 added at 300 K [82]. With more filler added, the conductivity decreased. At low filler concentration, conductivity enhancement was due to the increase in the sample's amorphousness and to the increase in mobility of Li^+ charge carriers. The latter was inferred from the narrowing of the ^7Li signals. The decrease in conductivity after the addition of more than 10 mol% was attributed to blockage of conduction pathways as more of the insulating fillers were added. From temperature-programmed desorption (TPD) studies, pH of the filler surface can be determined. The filler surface has significant impact on the conductivity [83]. TPD analysis revealed that the surface groups on δ-Al_2O_3 fillers are acidic. It will interact with both ClO_4^- ions and PEG segments, resulting in more $LiClO_4$ dissociation and increase in amorphousness of the host PEG [82]. The increase in $LiClO_4$ implied the increase in free ion concentration. Hence, increase in ionic mobility, number density of ions, and amorphousness led to enhancement of the conductivity, in the case of $(PEG)_{46}LiClO_4$ added with δ-Al_2O_3.

Dey et al. [84] studied the electrical and thermal properties of PEO–KI complexes containing ZnS. The 80 wt% PEO and 20 wt% KI system were prepared via the solution cast technique. Sample with 5 wt% of ZnS exhibited the highest room temperature conductivity of 0.31 mS cm^{-1}, an order higher compared with the PEO–KI sample. In the presence of ZnS fillers, conductivity increased as the KI salt actively decomposed into ions. Further addition of ZnS, beyond 5 wt%, decreased the conductivity due to the ion aggregation that lessens the number of conductivity contributing ions. The increase in conductivity by addition of fillers can be attributed to increase in amorphousness and increase in concentration of mobile ions.

The PEO–$LiCF_3SO_3$ electrolyte was investigated by Vignarooban et al. [85]. The conductivity of PEO_9LiTf filler-free PE was enhanced from ~2.0 to 49 μS cm^{-1} by introducing 10 wt% TiO_2 nano-filler. TiO_2 was the filler of choice instead of Al_2O_3, SiO_2, and ZrO_2 since it has high Lewis acid character and dielectric constant of 85 at 1 MHz compared with PEO. This would lead to active dissociation of LiTf. The conductivity was further improved to 0.16 μS cm^{-1} at 303 K on addition of EC. The results are similar to that reported by Pitawala et al. [81].

Borgohain et al. [36] prepared the system PEG–$LiClO_4$ and added hydrotalcite, an anionic clay, as the filler. $(PEG)_{46}LiClO_4$ exhibited a conductivity of 0.73 μS cm^{-1}. At 3.6 wt% hydrotalcite, the conductivity was enhanced to 11.14 μS cm^{-1}. The authors attributed the conductivity enhancement to the high conducting pathways at the boundaries between the PE and the hydrotalcite nanoparticles (average 50 nm).

The PEO-$LiClO_4$ system has been added with $ZnAl_2O_4$ (diameter: 10–15 nm) as filler by Wang et al. [86]. $ZnAl_2O_4$ increased amorphousness of PEO via Lewis acid–base interactions with the polymer. These resulted in remarkable improvement in the conductivity of $PEO_{16}LiClO_4$–$ZnAl_2O_4$. The room temperature maximum conductivity 2.23×10^{-4} S m^{-1} was achieved by the electrolyte with 8 wt% $ZnAl_2O_4$. $ZnAl_2O_4$ also enhanced the cationic transference number, t_{Li}. The t_{Li} of $PEO_{16}LiClO_4$ was only 0.191. This is possibly due to the Li^+ ion coordination with the C–O–C in PEO and the O atoms in ClO_4^-, thus restricting the transport ability of the Li^+ ions. At 8 wt% $ZnAl_2O_4$ loading, t_{Li} increased to 0.498. One of the potential factors for this is the interactions of Lewis acid–base between $ZnAl_2O_4$ (Lewis acid) and O atoms in PEO and ClO_4^- (Lewis bases). This weakened the interactions between lithium ions and the O atoms. Besides, the high specific surface area of $ZnAl_2O_4$ might be contributed to increment of conductivity as it provides more sites for polymer–filler interaction. Thus, conducting pathways for the Li^+ ions transport increases. However, the t_{Li} values remain essentially constant (~0.5) after more $ZnAl_2O_4$ was added to the $PEO_{16}LiClO_4$–x wt% $ZnAl_2O_4$ system possibly because of the aggregation of $ZnAl_2O_4$ particles. The nanocomposite polymer electrolyte (NCPE) demonstrated good electrochemical stability (>5.5 V versus Li^+/Li). The large stability window and good transport properties suggest that $PEO_{16}LiClO_4$-$ZnAl_2O_4$ NCPE is an excellent electrolyte material option for all-solid-state lithium batteries.

Wen et al. [66] studied the thermal, electrical, and mechanical properties of cross-linked poly(ethylene oxide-*co*-propylene oxide) composite PEs. The salt was $LiN(CF_3SO_2)_2$. The ratio [Li]/[O] was 1/16. The fillers used were $BaTiO_3$

with two different diameters (0.5 and 0.1 μm) and γ-$LiAlO_2$ (0.1 μm in diameter). $BaTiO_3$ and γ-$LiAlO_2$ addition to the chemically cross-linked poly(EO/PO) and $LiN(CF_3SO_2)_2N$ resulted in a fully amorphous complex. $BaTiO_3$ decreased the decomposition temperature from 713 K for $BaTiO_3$-free polymer matrix to as low as 703 K. The nano γ-$LiAlO_2$ filler improved the PE thermally, which increased with filler content. The electrolyte-free conductivity was 3.45×10^{-5} and 2.98×10^{-4} S cm^{-1} at 303 and 353 K. $BaTiO_3$ of 0.5 μm diameter decreased the conductivity, but the conductivity increased to 4.53×10^{-5} and 3.98×10^{-4} S cm^{-1} after 10 wt% $BaTiO_3$ of 100 nm diameter was introduced at 303 and 353 K, respectively. At 353 K, 10 wt% nano γ-$LiAlO_2$ fillers increased the conductivity to 5.76×10^{-4} S cm^{-1}. Only PE with 10 wt% nano $LiAlO_2$ showed an increase in tensile strength from 2.25 for filler-free electrolyte to 2.34 MPa at 303 K. The fillers were able to increase the breakdown voltage. However, the ion transference number of the pristine and filler containing electrolytes was low between 0.10 and 0.14.

The role of functional nano-inorganic fillers in PEO-based PEs has been studied by Ji et al. [87]. The conductivity of PEO–$LiClO_4$ (EO:Li = 16) increased from $\sim 2 \times 10^{-7}$ to 2.3×10^{-5} S cm^{-1} at 298 K on addition of 10 wt% nanoSiO_2 (~12 nm). The increase in conductivity can be related to mobile charge carrier density, n, and to the ionic mobility, μ, based on the equation, conductivity $\sigma = ne\mu$. The dielectric constant of SiO_2 is between 3.5 and 4.2, which is not so high compared with PEO (between 2.8 and 3.3). Thus, the dissociation of $LiClO_4$ is not so dominant and the mobility increase would be more effective compared with the mobile ion concentration for Li^+ conductivity. Conductivity was observed to increase with temperature due to the softening of the polymer backbone since the polymer segmental motion increased with amorphousness of the polymer and not because of the increment of carrier concentration. The authors have shown that above 313 K, the conductivity of the pristine electrolyte and that of the PE added with 10 wt% SiO_2 are almost the same. The decrease in T_g with increasing SiO_2 content indicated that the electrolyte has become more amorphous. The decrease in conductivity beyond 10 wt% SiO_2 can be explained as a result of the PE becoming more crystalline.

Dissanayake et al. [88] have reported the effects of Al_2O_3 filler on $PEO_9Cu(CF_3SO_3)_2$ SPE. The system PEO_9CuTf_2 was incorporated with 10 wt% of Al_2O_3 for various specific surface area of grains: 0.17, 45, 100, and 150 m^2 g^{-1}. Although conductivity increased with specific surface area of grains, conductivity only about doubled from ~0.47 to 0.91 μS cm^{-1} at 301 K. Conductivity increased by at least two orders of magnitude from 301 to 353 K for every grain specific surface area. Alumina grains can consist of surface acid groups. As reported by Jayathilaka et al. [89], the negative ions have a greater attraction toward surface acid groups of Al_2O_3. Thus, Tf^- ions and the polar groups can react via hydrogen bonding. The surface groups thus provide the ions additional conduction pathways facilitating ionic migration compared with the electrolyte without filler. Gel polymer electrolyte (GPE) with grains of 150 m^2 g^{-1}

specific surface area displayed the highest conductivity and maybe because of the increase in polar groups with increase in grain surface area.

A high conducting transparent/opaque and strong PAN + $LiClO_4$ + α-Al_2O_3 PE was prepared by Chen-Yang et al. using dimethylformamide (DMF) as the solvent [90]. The best conductivity obtained at 303 K was 5.7×10^{-4} S cm^{-1} for the sample with 0.6 mole ratio of the salt and 7.5 wt% filler. The addition of α-Al_2O_3 into the PAN–$LiClO_4$ electrolyte suppressed Li^+–CN interactions. Li^+ ion decouples from the polymer chain and increased Li^+ ion transference number, t_{Li}, and additional ion–ceramic pathways. The coalition of increased t_{Li} and the fast ion transport through the ion–ceramic pathways led to a high conductivity for the as-prepared PAN/$LiClO_4$/α-Al_2O_3 constant phase elements (CPEs).

8.4 Performance Characteristics of Lithium Ion Batteries

Numerous polymers have been applied as the electrolyte base in Li ion batteries including PEO, PMMA, PVA, PAN, and PVdF. Li et al. reported a high performance battery with 150 mAh g^{-1} discharge capacity using PEO based GPE containing $LiPF_6$ and 81% capacity retained after 500 cycles of the charge–discharge process under 0.5 C [91]. Another Li battery using PEO-based GPE consists of $LiClO_4$ and $Li_{6.4}La_3Zr_{1.4}Ta_{0.6}O_{12}$ fillers achieved 140 mAh g^{-1} $C_{specific}$ and maintained 83% after 500 cycles [92]. Lithium bis(triflouromethane) sulfonamide salt was introduced to PVA electrolyte to be applied in Li battery showed a conductivity value of 4.5×10^{-4} S cm^{-1}. The device obtained a $C_{specific}$ of 136 mAh g^{-1} at 333 K [93]. Li ion battery employed PVA-PAN blended electrolyte (with $LiPF_6$ in carbonate solvent) exhibited a discharge capacity of ~150 mAh g^{-1} and maintained at 94–96% after 200 cycles [94].

Another popular choice of PE base is PVdF and its derivatives (Table 8.5). Due to the presence of C–F group and high dielectric constant, PE with PVdF as the base showed high electrochemical stability. These factors also help in salt dissociation process to enhance the amount of charge carrier [106, 107]. PVdF-based GPE consisting of ethylenediaminetetraacetic acid (EDTA), polyvinylpyrrolidone (PVP), and lithium bis(trifluoromethanesulfonyl)imide (LiTFSI) salt has been fabricated in LIB to achieve 147.9 mAh g^{-1} $C_{specific}$ at 0.1 C current rate.

However, PVdF is less effective as it is high in crystallinity that gave low conductivity values [108]. Therefore, its derivative, PVdF-HFP, started to replace PVdF as the PE base [62, 96, 97]. The conductivity values were reported around 10^{-3} S cm^{-1} as shown in Table 8.5. Battery with PVdF-HFP–$LiPF_6$–EC–DMC–$BaTiO_3$ electrolyte achieved a $C_{specific}$ value of 165.8 mAh g^{-1} at 0.1 C.

Another approach to obtain high $C_{specific}$ battery is blended PE. Combinations of PVdF and PVA electrolyte have achieved $C_{specific}$ of 123 mAh g^{-1} with retention of 97.1% after 100 cycles [109]. From Table 8.5, we can observe that PVdF has also been blended with ethyl cellulose [98], PEO [99], high density polyethylene (HDPE) [100], PAN [101], PVC [102], and poly(vinyl chloride-*co*-vinyl acetate) (P(VC-VAc)) [43]. Combination of PVdF-HFP and PMMA seems effective as they achieved $C_{specific}$ values over ~160 mAh g^{-1} [103–105].

Table 8.5 Performance of certain polymer electrolytes in lithium ion batteries.

Gel polymer electrolyte	Conductivity ($S\,cm^{-1}$)	$C_{specific}$	Current rate (C)	Voltage range (V)	Cyclability	References
PVdF/EDTA/PVP/LiTFSI	7.19×10^{-4}	$147.9\,mAh\,g^{-1}$	0.1	2.6–4.2	50 cycles	[95]
PVdF-HFP/$LiClO_4$/EC/DEC	1.06×10^{-3}	$135.1\,mAh\,g^{-1}$	0.1	4.5	—	[96]
PVdF-HFP/$LiPF_6$/EC/DMC/$BaTiO_3$	5.20×10^{-3}	$165.8\,mAh\,g^{-1}$	0.1	5.5	20 cycles	[62]
PVdF-HFP/$PYRA_{12O1}$TFSI/LiTFSI/silica	0.25×10^{-3}	$139.0\,mAh\,g^{-1}$	~0.1	4.0	174 cycles	[65]
PVdF-HFP/EMIMDCA/$LiClO_4$	6.00×10^{-4}	$300.0\,mAh\,cm^{-2}$	0.25	2.1	20 cycles	[97]
PVdF/ethyl cellulose/EC/$LiPF_6$/EC/DEC	1.33×10^{-3}	$123.0\,mAh\,g^{-1}$	0.2	5.25	94.6% retain after 100 cycles	[98]
PVdF/PEO/$LiClO_4$/EC/PC	3.03×10^{-3}	$146.8\,mAh\,g^{-1}$	0.1	5.0	90% retain after 70 cycles	[99]
PVdF/HDPE/PVP/$LiPF_6$/EC/DMC/EMC	2.36×10^{-3}	$145.0\,mAh\,g^{-1}$	0.1	4.3	50 cycles	[100]
PVdF/PAN/ESFM/$LiClO_4$/PC	7.80×10^{-3}	$120.4\,mAh\,g^{-1}$	0.1	2.8–4.25	93% retain after 150 cycles	[101]
PVdF/PVC/$LiClO_4$/PC/EC	2.25×10^{-3}	$145.1\,mAh\,g^{-1}$	0.1	5.1	90% retain after 50 cycles	[102]
PVdF/P(VC-VAc)/$LiPF_6$/EC/DMC/EMC	3.57×10^{-3}	$131.0\,mAh\,g^{-1}$	0.2	5.4	96% retain after 200 cycles	[43]
PVdF–HFP/PMMA/TPU/SiO_2	8.50×10^{-3}	$168.5\,mAh\,g^{-1}$	0.1	5.9	99.5% retain after 50 cycles	[22]
PVdF-HFP/PMMA/$ZnAl_2O_4$/$LiPF_6$/EC/DEC	4.14×10^{-3}	$132.0\,mAh\,g^{-1}$	0.1	4.5	30 cycles	[103]
PVdF-HFP/titania/PMMA/$LiPF_6$/EC/DEC	3.40×10^{-3}	$160.4\,mAh\,g^{-1}$	0.5	2.75–4.4	100 cycles	[104]
PVdF-HFP/PMMA/TiO_2/EC/DMC/$LiPF_6$	2.49×10^{-3}	$188.1\,mAh\,g^{-1}$	0.2	2.5–4.3	92% retain after 50 cycles	[105]

8.5 Challenges and Improvement

For a high discharge capacity in Li batteries, it is best that the anode is Li metal since its gravimetric and volumetric capacities are highest compared with other elements in the periodic table. The use of Li metal as the anode in Li batteries has faced safety problems caused by dendrite formation, which can, in the end, short the battery. This is due to the interaction between the LE in LIBs and the Li metal. These shortcomings have resulted in the abandoning of Li metal batteries [110]. The problems associated with LEs have been addressed using electrolytes in solid form that can demonstrate high thermodynamic stability against Li. Alternatively, the electrolytes should form a stable passivation coating on the Li metal electrode. This will enable a smooth and reversible deposition when the battery is cycled [111]. Thus, these electrolytes are able to make the battery safer.

One of the difficulties faced in PE is finding a polymer with mechanical strength and ionic conductivity required in a solid electrolyte. A PE has to be mechanically, thermally, and electrochemically stable. In this section, some strategies to overcome the stability and related issues are presented. The strategies presented here are, however, not meant to be exhaustive, but to illustrate the challenges and improvements that have been undertaken by researchers to solve the energy problem the world faces.

We start with some work that compares the thermal stability between solid and LEs. The thermal stability of the $(PEO)_8LiCF_3SO_3$ based PE using various electrode pairs was studied by Shodai et al. [112]. The electrolyte was stable with lithium metal when compared with LEs, indicating the superiority of PEs to its liquid counterpart. The electrolyte did not react with lithium metal up to 493 K and the $(PEO)_8LiCF_3SO_3$ decomposed at 503 K, thus the high thermal stability of the PE.

PEO–SiO_2–LiTf electrolytes have been studied by Chaurasia and Chandra [113]. The highest conducting free standing thin film at 303 K with good mechanical and thermal stability (~473 K) contained 20 wt% LiTf. The conductivity was ~1.24 $\mu S\ cm^{-1}$ and the Li^+ ion transference number was 0.39. Although thermal stability is high, the low value of conductivity and transference number of Li^+ ions may not make the electrolyte suitable for Li batteries.

Boaretto et al. [114] synthesized a set of networked hybrid PEs based on poly(ethylene glycol) monomethyl ether (Me-PEG_6-OH and Me-PEG_9-OH subscripts indicating ethylene oxide units), the ion source LiTf, and also filler TiO_2. The electrolytes with Li:EO ratio of 0.05 have high conductivity (~$10^{-2}\ S\ m^{-1}$) at ambient temperature. TiO_2 nanoparticles have a plasticizing effect that can improve the conductivity and transference number. Conductivity increased in the semi-crystallization region is due to the inhibition of the PEO crystallization [115]. Conductivity increased in the amorphous phase [83, 116] and the transference number is due to the TiO_2 Lewis acid nature, which puts them in competition with the Li^+ cation interaction with polyether chains and also anions. This encourages Li^+ ion dissociation and motion [117]. The composite PEs also have good interfacial properties with Li metal electrodes. It therefore seems possible to use the PE in lithium metal polymer batteries and to take advantage of the high lithium gravimetric and volumetric capacities of Li metal.

The low standard potential value of lithium (−3.040 V) [118] also results in a high electrochemical energy equivalent. However, when determining the Li^+ ion transference number Boaretto et al. realized that there was a decrease in bulk resistance when a DC current flows through the Li/PE/Li cell. This suggests that the composite electrolyte has a low resistance and low mechanical strength against lithium dendrite formation and is therefore unsuitable for the fabrication of Li^+ cells. As a verification, the Li/$LiFePO_4$ cell fabricated showed cycling instability and worsened with the length and depth of charge.

The electrolyte system based on PEO–LiBNFSI [41] has been prepared for application in LIBs. A conductivity of 0.22 $\mu S\ cm^{-1}$ at 333 K with the transference number of 0.31 was attained for the Li:EO ratio of 0.07. The electrolyte with a tensile strength of 2 MPa is mechanically robust. The electrochemical stability window of ~3.3 V indicates the possible application in rechargeable Li ion batteries.

Wu et al. [119] prepared a polymer matrix that consisted of hyperbranched polyether PHEMO and PVdF-HFP. The matrix was intended for application in Li ion batteries. To provide the conducting ions, LiBOB was added to the polymer matrix. LiBOB has high thermal stability of ~575 K. The PE exhibited good thermal stability and decomposed above 693 K. The PE with 10 wt% LiBOB obtained an ionic conductivity of $1.1 \times 10^{-5}\ S\ cm^{-1}$ at 303 K. Conductivity increased with temperature and at 80 °C the value was $2.3 \times 10^{-4}\ S\ cm^{-1}$. These results indicate that the electrolyte may be usable at high temperatures.

Li et al. [91] prepared a GPE by curing a solution containing PEO, 2-hydroxy-2-methyl-1-phenyl-1-propanone, and trimethylolpropane ethoxylate triacrylate using 365 nm UV irradiation. At room temperature, the GPE obtained a conductivity value of $3.3 \times 10^{-3}\ S\ cm^{-1}$. The Li^+ ion transference number of 0.76 was determined using the Vincent–Evans method. The improved thermal stability of the GPE ensures battery safety. No changes were observed on heating the GPE at 423 K for 30 minutes in air. A commercialized Celgard separator showed severe thermal shrinkage under the same conditions. When the treatment temperature was 473 K and the test materials were subjected for another 30 minutes, the GPE could retain its original dimension, but the Celgard separator almost burnt. The GPE decomposed at 4.9 V. The Li/GPE/$LiFePO_4$ cell exhibited an initial capacity of 150 $mAh\ g^{-1}$ on discharge at 16 $mA\ g^{-1}$ and after 500 cycles at 0.5 C rate, 81% of capacity remained. The GPE displayed potential application in LIBs.

8.6 Future Trends

Ever since Professor J. B. Goodenough invented the first LIB using LEs, there has not been any major advancement made in battery science. LEs favor dendrite formation during operation and recharge. Dendrites affect the ability of the anode to deliver power and increase the internal resistance. High internal resistance leads to more energy loss and the battery heats up while being used. Such mishaps have led to intensive research on solid electrolytes.

Solid and quasi-solid electrolytes have wide electrochemical windows that will not promote chemical reactions inside the cell, thus making the battery safe. The use of solid electrolytes allows alkali metals Li, Na, and K as anodes. Sodium (Na) metal is easier to obtain and the process is reasonably cheap. Since there was no dendritic buildup, the battery can be discharged and recharged at higher C-rates than batteries using LEs.

Batteries are very widespread and found in small and large devices. Hence, the benchmark to measure battery performance is also broad. The criteria include the following:

(a) Charging time, i.e. the time required for the battery to fully charge from when the charge has depleted.
(b) Energy density, which is a measure of how much energy that can be put inside a battery of a given size and weight.
(c) Size and weight of a battery require the battery to be small to save space when used in electric vehicles and weigh less so that the hand phone can be lighter.
(d) Cycle life, i.e. how many times a battery can be recharged from almost empty to fully charged before it drops below 80% of its ability to hold the charge.

Some of the breakthroughs that are being developed include the following:

(a) The dual carbon or dual graphite battery generates no heat during charge and discharge. Power Japan Plus in partnership with Kyushu University reported that the dual carbon cell delivers comparable energy density with that of an LIB [120]. The battery charges 20 times faster than the common LIBs. The battery has a long working lifespan of 3000 cycles and improved safety. The electrolyte consists of $LiPF_6$ salt and a mixture of EC and DMC as solvent. The battery functions at more than four volts and is recyclable. The reactions that occur are as follows:

   Positive electrode:

$$PF_6^- + nC \rightleftarrows C_n(PF_6) + e^- \tag{8.22}$$

   Negative electrode:

$$Li^+ + nC + e^- \rightleftarrows LiC_n \tag{8.23}$$

   In the dual carbon battery, on charge Li ions are inserted into the anode and the counterions (anions) are inserted into the cathode simultaneously. The ions return to the electrolyte on discharge. The electrolyte thus serves as charge carrier and active material.

(b) The lithium-air battery can deliver higher energy density than LIBs. This attracted wide awareness in 2009 as a feasible electric vehicle energy source. The Li-air cell has a theoretical energy density of 1520 $Wh\,l^{-1}$ and specific energy of 1300 $Wh\,kg^{-1}$ [121]. If successful, the battery could match the energy density of gasoline. Nevertheless, many difficulties must be resolved before the concept can become a reality. In the lithium-air system, the anode is lithium metal. Electrochemical potential forces the lithium metal to oxidize:

$$Li \leftrightarrow Li^+ + e^- \tag{8.24}$$

   The air cathode contains metal catalysts inside a mesoporous carbon substrate. The catalysts can be silver, ruthenium, platinum, cobalt, manganese, or a mixture of cobalt and manganese. Manganese- and cobalt-catalyzed cathodes have a $C_{specific}$ of 3137 and 2414 mAh g^{-1}, respectively [122]. Lithium oxides are produced in a cell containing aprotic electrolyte at cathode:

$$Li^+ + e^- + O_2 \rightarrow LiO_2 \tag{8.25}$$

$$Li^+ + e^- + LiO_2 \rightarrow Li_2O_2 \tag{8.26}$$

Since lithium oxide does not dissolve in aprotic electrolytes, this situation results in cathode blocking [123]. A new cathode material must therefore consider the formation of LiO_2 and Li_2O_2 without clogging the porous cathode, and suitable catalysts must be used to enhance electrochemical reactions.

(c) The lithium–silicon battery is expandable at the molecular level to allow for faster charging. Researchers are trying to replace graphite with silicon (Si). Si offers a higher theoretical $C_{specific}$ of ~4200 mAh g^{-1} (based on the formation of $Li_{22}Si_5$) compared with graphite, 372 mAh g^{-1}. However, lithium intercalation into the anode caused Si to undergo a large volume expansion [124]. This is detrimental to the battery, and the challenges are ongoing!
(d) Pure lithium battery is cheaper and has energy density three times higher than that of LIBs. Engineers from Stanford University, California, invented the first working lithium anode in a lithium battery [125]. The engineers made a film from carbon nanospheres to shield and prevent the anode from rupture when the lithium anode expands largely and unevenly causing dendrites. The protective film is chemically stable against reactions with the electrolyte and strong to withstand lithium volume expansion during charge.
(e) The Irvine battery, which consists of a gold nanowire coated in manganese oxide and a protective layer of electrolyte gel, can cycle through 200 000 recharges without degradation [126]. The use of the gel differentiates UC Irvine's work from others.

Apart from these, there are non-battery technologies such as capacitors, which are able to charge and discharge faster than LIBs, but hold less energy. Mazda [127] and Lamborghini [128] make use of capacitors for their cars, and Toyota is pushing hard on the green hydrogen fuel cells for electric vehicle [129].

References

1 Wang, B., Li, X., Qiu, T. et al. (2013). High volumetric capacity silicon-based lithium battery anodes by nanoscale system engineering. *Nano Lett.* 13: 5578–5584.
2 Wang, X., Sun, L., Agung Susantyoko, R. et al. (2014). Ultrahigh volumetric capacity lithium ion battery anodes with CNT–Si film. *Nano Energy* 8: 71–77.
3 Ito, Y., Kawakubo, M., Ueno, M. et al. (2012). Carbon anodes for solid polymer electrolyte lithium-ion batteries. *J. Power Sources* 214: 84–90.
4 Chan, C.K., Peng, H., Liu, G. et al. (2007). High-performance lithium battery anodes using silicon nanowires. *Nat. Nanotechnol.* 3: 31–35.

5 Alias, N.A., Kufian, M.Z., Teo, L.P. et al. (2009). Synthesis and characterization of $Li_4Ti_5O_{12}$. *J. Alloys Compd.* 486: 645–648.
6 Teo, L.P., Buraidah, M.H., Alias, N.A. et al. (2011). Characterisation of Li_2SnO_3 by solution evaporation method using nitric acid as chelating agent. *Mater. Res. Innovations* 15: s127–s131.
7 Fuchsbichler, B., Stangl, C., Kren, H. et al. (2011). High capacity graphite–silicon composite anode material for lithium-ion batteries. *J. Power Sources* 196: 2889–2892.
8 Uchiyama, H., Hosono, E., Honma, I. et al. (2008). A nanoscale meshed electrode of single-crystalline SnO for lithium-ion rechargeable batteries. *Electrochem. Commun.* 10: 52–55.
9 Sivashanmugam, A., Kumar, T.P., Renganathan, N.G. et al. (2005). Electrochemical behavior of Sn/SnO_2 mixtures for use as anode in lithium rechargeable batteries. *J. Power Sources* 144: 197–203.
10 Zhang, D.W., Zhang, S.Q., Jin, Y. et al. (2006). Li_2SnO_3 derived secondary Li–Sn alloy electrode for lithium-ion batteries. *J. Alloys Compd.* 415: 229–233.
11 Zhao, S., Bai, Y., and Zhang, W.-F. (2010). Electrochemical performance of flowerlike $CaSnO_3$ as high capacity anode material for lithium-ion batteries. *Electrochim. Acta* 55: 3891–3896.
12 Lee, S.-Y., Kim, S.K., and Ahn, S. (2008). Nano-encapsulation of $LiCoO_2$ cathodes by a novel polymer electrolyte and its influence on thermal safeties of Li-ion batteries. *Electrochem. Commun.* 10: 113–117.
13 Ding, Y.L., Xie, J., Cao, G.S. et al. (2011). Single-crystalline $LiMn_2O_4$ nanotubes synthesized via template-engaged reaction as cathodes for high-power lithium ion batteries. *Adv. Funct. Mater.* 21: 348–355.
14 Deng, D. (2015). Li-ion batteries: basics, progress, and challenges. *Energy Sci. Eng.* 3: 385–418.
15 Armstrong, G., Armstrong, A.R., Bruce, P.G. et al. (2006). TiO_2(B) nanowires as an improved anode material for lithium-ion batteries containing $LiFePO_4$ or $LiNi_{0.5}Mn_{1.5}O_4$ cathodes and a polymer electrolyte. *Adv. Mater.* 18: 2597–2600.
16 Whittingham, M.S. (2004). Lithium batteries and cathode materials. *Chem. Rev.* 104: 4271–4301.
17 Memm, M., Koentje, M., Axmann, P., and Wohlfahrt-Mehrens, M. (2015). New high-voltage step at 4.8 V in cobalt free manganese based lithium phospho olivines for lithium-ion batteries. *J. Power Sources* 276: 382–387.
18 Sharma, N., Shaju, K.M., Subba Rao, G.V., and Chowdari, B.V.R. (2002). Sol–gel derived nano-crystalline $CaSnO_3$ as high capacity anode material for Li-ion batteries. *Electrochem. Commun.* 4: 947–952.
19 Rao, M.C. (2013). $LiMn_2O_4$ cathodes for solid state lithium-ion batteries energy storage and conversion. *J. Optoelectron. Biomed. Mater.* 5: 9–16.
20 Agrawal, R.C. and Pandey, G.P. (2008). Solid polymer electrolytes: materials designing and all-solid-state battery applications: an overview. *J. Phys. D: Appl. Phys.* 41: 223001.
21 Garche, J., Dyer, C.K., Moseley, P.T. et al. (2009). *Encyclopedia of Electrochemical Power Sources*, 1e, 4538. Netherland: Elsevier.

22 He, Z., Cao, Q., Jing, B. et al. (2017). Gel electrolytes based on poly(vinylidenefluoride-co-hexafluoropropylene)/thermoplastic polyurethane/poly(methyl methacrylate) with in situ SiO_2 for polymer lithium batteries. *RSC Adv.* 7: 3240–3248.
23 Shangguan, Y., Chen, F., Jia, E. et al. (2017). New insight into time-temperature correlation for polymer relaxations ranging from secondary relaxation to terminal flow: application of a universal and developed WLF equation. *Polymers* 9: 567.
24 Aravindan, V. and Vickraman, P. (2008). Characterization of SiO_2 and Al_2O_3 incorporated PVdF-HFP based composite polymer electrolytes with $LiPF_3(CF_3CF_2)_3$. *J. Appl. Polym. Sci.* 108: 1314–1322.
25 Bandara, L.R.A.K., Dissanayake, M.A.K.L., Furlani, M., and Mellander, B.-E. (2000). Broad band dielectric behaviour of plasticized PEO-based solid polymer electrolytes. *Ionics* 6: 239–247.
26 Wang, W. and Alexandridis, P. (2016). Composite polymer electrolytes: nanoparticles affect structure and properties. *Polymers* 8: 387.
27 Masoud, E.M., Hassan, M.E., Wahdaan, S.E. et al. (2016). Gel P (VdF/HFP) / PVAc / lithium hexafluoro phosphate composite electrolyte containing nano ZnO filler for lithium ion batteries application: effect of nano filler concentration on structure, thermal stability and transport properties. *Polym. Test.* 56: 277–286.
28 Deka, M. and Kumar, A. (2011). Electrical and electrochemical studies of poly(vinylidene fluoride)–clay nanocomposite gel polymer electrolytes for Li-ion batteries. *J. Power Sources* 196: 1358–1364.
29 Shi, H. and Zhu, Q. (2014). Preparation and characterization of Al_2O_3/PVdF-HFP based polymer electrolyte. *Adv. Mater. Res.* 950: 33–37.
30 Long, L., Wang, S., Xiao, M., and Meng, Y. (2016). Polymer electrolytes for lithium polymer batteries. *J. Mater. Chem. A* 4: 10038–10069.
31 Oh, H., Xu, K., Yoo, H.D. et al. (2016). Poly(arylene ether)-based single-ion conductors for lithium-ion batteries. *Chem. Mater.* 28: 188–196.
32 Bouchet, R., Maria, S., Meziane, R. et al. (2013). Single-ion BAB triblock copolymers as highly efficient electrolytes for lithium-metal batteries. *Nat. Mater.* 12: 452.
33 Li, Y., Wong, K.W., Dou, Q., and Ng, K.M. (2016). A single-ion conducting and shear-thinning polymer electrolyte based on ionic liquid-decorated PMMA nanoparticles for lithium-metal batteries. *J. Mater. Chem. A* 4: 18543–18550.
34 Ramesh, S. and Bing, K.N. (2012). Conductivity, mechanical and thermal studies on poly(methyl methacrylate)-based polymer electrolytes complexed with lithium tetraborate and propylene carbonate. *J. Mater. Eng. Perform.* 21: 89–94.
35 Polu, A.R. and Rhee, H.-W. (2016). Effect of TiO_2 nanoparticles on structural, thermal, mechanical and ionic conductivity studies of PEO_{12}–LiTDI solid polymer electrolyte. *J. Ind. Eng. Chem.* 37: 347–353.
36 Borgohain, M.M., Joykumar, T., and Bhat, S.V. (2010). Studies on a nanocomposite solid polymer electrolyte with hydrotalcite as a filler. *Solid State Ionics* 181: 964–970.

37 He, W., Cui, Z., Liu, X. et al. (2017). Carbonate-linked poly(ethylene oxide) polymer electrolytes towards high performance solid state lithium batteries. *Electrochim. Acta* 225: 151–159.

38 Jiang, Y.-X., Xu, J.-M., Zhuang, Q.-C. et al. (2008). A novel PEO-based composite solid-state polymer electrolyte with methyl group-functionalized SBA-15 filler for rechargeable lithium batteries. *J. Solid State Electrochem.* 12: 353–361.

39 Porcarelli, L., Manojkumar, K., Sardon, H. et al. (2017). Single ion conducting polymer electrolytes based on versatile polyurethanes. *Electrochim. Acta* 241: 526–534.

40 Polu, A.R. and Rhee, H.-W. (2017). Ionic liquid doped PEO-based solid polymer electrolytes for lithium-ion polymer batteries. *Int. J. Hydrogen Energy* 42: 7212–7219.

41 Karuppasamy, K., Kim, D., Kang, Y.H. et al. (2017). Improved electrochemical, mechanical and transport properties of novel lithium bisnonafluoro-1-butanesulfonimidate (LiBNFSI) based solid polymer electrolytes for rechargeable lithium ion batteries. *J. Ind. Eng. Chem.* 52: 224–234.

42 Masoud, E.M. (2016). Nano lithium aluminate filler incorporating gel lithium triflate polymer composite: preparation, characterization and application as an electrolyte in lithium ion batteries. *Polym. Test.* 56: 65–73.

43 Zhao, M., Zuo, X., Ma, X. et al. (2017). Self-supported PVdF/P(VC-VAc) blended polymer electrolytes for $LiNi_{0.5}Mn_{1.5}O_4$/Li batteries. *J. Membr. Sci.* 532: 30–37.

44 Agrawal, R.C., Hashmi, S.A., and Pandey, G.P. (2007). Electrochemical cell performance studies on all-solid-state battery using nano-composite polymer electrolyte membrane. *Ionics* 13: 295–298.

45 Kwon, S.-J., Kim, D.-G., Shim, J. et al. (2014). Preparation of organic/inorganic hybrid semi-interpenetrating network polymer electrolytes based on poly(ethylene oxide-co-ethylene carbonate) for all-solid-state lithium batteries at elevated temperatures. *Polymer* 55: 2799–2808.

46 Sun, B., Mindemark, J., Edström, K., and Brandell, D. (2014). Polycarbonate-based solid polymer electrolytes for Li-ion batteries. *Solid State Ionics* 262: 738–742.

47 Mindemark, J., Törmä, E., Sun, B., and Brandell, D. (2015). Copolymers of trimethylene carbonate and ε-caprolactone as electrolytes for lithium-ion batteries. *Polymer* 63: 91–98.

48 Tominaga, Y. and Yamazaki, K. (2014). Fast Li-ion conduction in poly(ethylene carbonate)-based electrolytes and composites filled with TiO_2 nanoparticles. *Chem. Commun.* 50: 4448–4450.

49 Okumura, T. and Nishimura, S. (2014). Lithium ion conductive properties of aliphatic polycarbonate. *Solid State Ionics* 267: 68–73.

50 Jianjun, Z., Jianghui, Z., Liping, Y. et al. (2015). Safety-reinforced poly(propylene carbonate)-based all-solid-state polymer electrolyte for ambient-temperature solid polymer lithium batteries. *Adv. Energy Mater.* 5: 1501082.

51 Thayumanasundaram, S., Rangasamy, V.S., Seo, J.W., and Locquet, J.-P. (2017). Electrochemical performance of polymer electrolytes based on poly(vinyl alcohol)/poly(acrylic acid) blend and pyrrolidinium ionic liquid for lithium rechargeable batteries. *Electrochim. Acta* 240: 371–378.

52 Gray, F.M. (1991). *Solid Polymer Electrolytes: Fundamentals and Technological Applications*. New York: VCH Publishers.

53 Kim, S., Han, T., Jeong, J. et al. (2017). A flame-retardant composite polymer electrolyte for lithium-ion polymer batteries. *Electrochim. Acta* 241: 553–559.

54 Libo, L., Jiesi, L., Shuo, Y. et al. (2014). Gel polymer electrolytes containing ionic liquids prepared by radical polymerization. *Colloids Surf., A* 459: 136–141.

55 Ramesh, S., Liew, C.-W., and Ramesh, K. (2011). Evaluation and investigation on the effect of ionic liquid onto PMMA-PVC gel polymer blend electrolytes. *J. Non-Cryst. Solids* 357: 2132–2138.

56 Tafur, J.P. and Fernández Romero, A.J. (2014). Electrical and spectroscopic characterization of PVdF-HFP and TFSI—ionic liquids-based gel polymer electrolyte membranes. Influence of $ZnTf_2$ salt. *J. Membr. Sci.* 469: 499–506.

57 Hofmann, A., Schulz, M., and Hanemann, T. (2013). Gel electrolytes based on ionic liquids for advanced lithium polymer batteries. *Electrochim. Acta* 89: 823–831.

58 Sil, A., Sharma, R., and Ray, S. (2015). Mechanical and thermal characteristics of PMMA-based nanocomposite gel polymer electrolytes with CNFs dispersion. *Surf. Coat. Technol.* 271: 201–206.

59 Deka, M. and Kumar, A. (2010). Enhanced electrical and electrochemical properties of PMMA–clay nanocomposite gel polymer electrolytes. *Electrochim. Acta* 55: 1836–1842.

60 Mejía, A., García, N., Guzmán, J., and Tiemblo, P. (2014). Thermoplastic and solid-like electrolytes with liquid-like ionic conductivity based on poly(ethylene oxide) nanocomposites. *Solid State Ionics* 261: 74–80.

61 Raja, M., Kumar, P., Sanjeev, G. et al. (2014). Montmorillonite-based ceramic membranes as novel lithium-ion battery separators. *Ionics* 20: 943–948.

62 Raghavan, P., Zhao, X., Manuel, J. et al. (2010). Electrochemical performance of electrospun poly(vinylidene fluoride-co-hexafluoropropylene)-based nanocomposite polymer electrolytes incorporating ceramic fillers and room temperature ionic liquid. *Electrochim. Acta* 55: 1347–1354.

63 Wang, Q., Song, W.-L., Fan, L.-Z., and Song, Y. (2015). Facile fabrication of polyacrylonitrile/alumina composite membranes based on triethylene glycol diacetate-2-propenoic acid butyl ester gel polymer electrolytes for high-voltage lithium-ion batteries. *J. Membr. Sci.* 486: 21–28.

64 Zhang, P., Yang, L.C., Li, L.L. et al. (2011). Enhanced electrochemical and mechanical properties of P(VDF-HFP)-based composite polymer electrolytes with SiO_2 nanowires. *J. Membr. Sci.* 379: 80–85.

65 Ferrari, S., Quartarone, E., Mustarelli, P. et al. (2010). Lithium ion conducting PVdF-HFP composite gel electrolytes based on *N*-methoxyethyl-*N*-methylpyrrolidinium bis(trifluoromethanesulfonyl)-imide ionic liquid. *J. Power Sources* 195: 559–566.

66 Wen, Z., Itoh, T., Uno, T. et al. (2003). Thermal, electrical, and mechanical properties of composite polymer electrolytes based on cross-linked poly(ethylene oxide-co-propylene oxide) and ceramic filler. *Solid State Ionics* 160: 141–148.
67 Pandey, K., Asthana, N., Dwivedi, M.M., and Chaturvedi, S.K. (2013). Effect of plasticizers on structural and dielectric behaviour of [PEO + $(NH_4)_2C_4H_8(COO)_2$] polymer electrolyte. *J. Polym.* 2013: 12.
68 Rajendran, S., Sivakumar, M., and Subadevi, R. (2004). Investigations on the effect of various plasticizers in PVA–PMMA solid polymer blend electrolytes. *Mater. Lett.* 58: 641–649.
69 Ahmad, S. (2009). Polymer electrolytes: characteristics and peculiarities. *Ionics* 15: 309–321.
70 Arulepp, M., Permann, L., Leis, J. et al. (2004). Influence of the solvent properties on the characteristics of a double layer capacitor. *J. Power Sources* 133: 320–328.
71 Petrella, G. and Sacco, A. (1978). Viscosity and conductance studies in ethylene carbonate at 40 °C. *J. Chem. Soc., Faraday Trans. 1* 74: 2070–2076.
72 Barthel, J., Neueder, R., and Roch, H. (2000). Density, relative permittivity, and viscosity of propylene carbonate + dimethoxyethane mixtures from 25 °C to 125 °C. *J. Chem. Eng. Data* 45: 1007–1011.
73 Rodrıguez, A., Canosa, J., Domınguez, A., and Tojo, J. (2003). Viscosities of dimethyl carbonate or diethyl carbonate with alkanes at four temperatures. New UNIFAC-VISCO parameters. *J. Chem. Eng. Data* 48: 146–151.
74 Klongkan, S. and Pumchusak, J. (2015). Effects of nano alumina and plasticizers on morphology, ionic conductivity, thermal and mechanical properties of PEO-$LiCF_3SO_3$ solid polymer electrolyte. *Electrochim. Acta* 161: 171–176.
75 Abdali, O.K. and Abdulazeez, O. (2016). Optical and dielectrical properties of (PEG-CMC) films prepared by drop casting method. *World Sci. News* 46: 189–203.
76 Regupathi, I., Govindarajan, R., Amaresh, S.P., and Murugesan, T. (2009). Densities and viscosities of polyethylene glycol 6000 + triammonium citrate + water systems. *J. Chem. Eng. Data* 54: 3291–3295.
77 Abu-Dalo, M.A., Al-Rawashdeh, N.A.F., Al-Mheidat, I.R., and Nassory, N.S. (2015). Construction of uranyl selective electrode based on complex of uranyl ion with new ligand carboxybenzotriazole in PVC matrix membrane. *IOP Conf. Ser.: Mater. Sci. Eng.* 92: 012023.
78 Sahin, D., Sari, B., and Unal, H.I. (2002). An investigation of some parameters on electrorheological properties of polypyrrole suspensions. *TUrk. J. Chem.* 26: 113–124.
79 Mitra, S. and Kulkarni, A.R. (2002). Electrical conductivity studies on the plasticised PEO–DBP–CdX (X=Cl; SO_4) polymer electrolytes. *Solid State Ionics* 154–155: 37–43.
80 Pabby, A.K., Rizvi, S.S.H., and Sastre, A.M. (2015). *Handbook of Membrane Separations Chemical, Pharmaceutical, Food, and Biotechnological Applications*, 2e. Boca Raton, Florida: CRC Press, Taylor and Francis.

81 Pitawala, H.M.J.C., Dissanayake, M.A.K.L., and Seneviratne, V.A. (2007). Combined effect of Al_2O_3 nano-fillers and EC plasticizer on ionic conductivity enhancement in the solid polymer electrolyte $(PEO)_9LiTf$. *Solid State Ionics* 178: 885–888.
82 Singh, T.J. and Bhat, S.V. (2004). Increased lithium-ion conductivity in $(PEG)_{46}LiClO_4$ solid polymer electrolyte with δ-Al_2O_3 nanoparticles. *J. Power Sources* 129: 280–287.
83 Croce, F., Persi, L., Scrosati, B. et al. (2001). Role of the ceramic fillers in enhancing the transport properties of composite polymer electrolytes. *Electrochim. Acta* 46: 2457–2461.
84 Dey, A., Karan, S., and De, S.K. (2009). Effect of nanofillers on thermal and transport properties of potassium iodide–polyethylene oxide solid polymer electrolyte. *Solid State Commun.* 149: 1282–1287.
85 Vignarooban, K., Dissanayake, M.A.K.L., Albinsson, I., and Mellander, B.E. (2014). Effect of TiO_2 nano-filler and EC plasticizer on electrical and thermal properties of poly(ethylene oxide) (PEO) based solid polymer electrolytes. *Solid State Ionics* 266: 25–28.
86 Wang, L., Yang, W., Wang, J., and Evans, D.G. (2009). New nanocomposite polymer electrolyte comprising nanosized $ZnAl_2O_4$ with a mesopore network and PEO-$LiClO_4$. *Solid State Ionics* 180: 392–397.
87 Ji, K.-S., Moon, H.-S., Kim, J.-W., and Park, J.-W. (2003). Role of functional nano-sized inorganic fillers in poly(ethylene) oxide-based polymer electrolytes. *J. Power Sources* 117: 124–130.
88 Dissanayake, M.A.K.L., Jayathilaka, P.A.R.D., and Bokalawela, R.S.P. (2005). Ionic conductivity of PEO_9: $Cu(CF_3SO_3)_2$: Al_2O_3 nano-composite solid polymer electrolyte. *Electrochim. Acta* 50: 5602–5605.
89 Jayathilaka, P.A.R.D., Dissanayake, M.A.K.L., Albinsson, I., and Mellander, B.E. (2002). Effect of nano-porous Al_2O_3 on thermal, dielectric and transport properties of the $(PEO)_9LiTFSI$ polymer electrolyte system. *Electrochim. Acta* 47: 3257–3268.
90 Chen-Yang, Y.W., Chen, H.C., Lin, F.J., and Chen, C.C. (2002). Polyacrylonitrile electrolytes: 1. A novel high-conductivity composite polymer electrolyte based on PAN, $LiClO_4$ and α-Al_2O_3. *Solid State Ionics* 150: 327–335.
91 Li, W., Pang, Y., Liu, J. et al. (2017). A PEO-based gel polymer electrolyte for lithium ion batteries. *RSC Adv.* 7: 23494–23501.
92 Cheng, S.H.-S., He, K.-Q., Liu, Y. et al. (2017). Electrochemical performance of all-solid-state lithium batteries using inorganic lithium garnets particulate reinforced PEO/$LiClO_4$ electrolyte. *Electrochim. Acta* 253: 430–438.
93 Ek, G., Jeschull, F., Bowden, T., and Brandell, D. (2017). Li-ion batteries using electrolytes based on mixtures of poly(vinyl alcohol) and lithium bis(triflouromethane) sulfonamide salt. *Electrochim. Acta* 246: 208–212.
94 He, C., Liu, J., Li, J. et al. (2018). Blending based polyacrylonitrile/poly(vinyl alcohol) membrane for rechargeable lithium ion batteries. *J. Membr. Sci.* 560: 30–37.

95 Wang, F., Li, L., Yang, X. et al. (2018). Influence of additives in a PVDF-based solid polymer electrolyte on conductivity and Li-ion battery performance. *Sustain. Energy Fuels* 2: 492–498.
96 Liu, W., Zhang, X.K., Wu, F., and Xiang, Y. (2017). A study on PVDF-HFP gel polymer electrolyte for lithium-ion batteries. *IOP Conf. Ser.: Mater. Sci. Eng.* 213: 012036.
97 Li, Q. and Ardebili, H. (2016). Flexible thin-film battery based on solid-like ionic liquid-polymer electrolyte. *J. Power Sources* 303: 17–21.
98 Zuo, X., Ma, X., Wu, J. et al. (2018). Self-supporting ethyl cellulose/poly(vinylidene fluoride) blended gel polymer electrolyte for 5 V high-voltage lithium-ion batteries. *Electrochim. Acta* 271: 582–590.
99 Deng, F., Wang, X., He, D. et al. (2015). Microporous polymer electrolyte based on PVDF/PEO star polymer blends for lithium ion batteries. *J. Membr. Sci.* 491: 82–89.
100 Liu, H., Liu, J., Zhou, Q., and Wang, J. (2013). Novel polymer electrolyte based on PVDF/HDPE blending for lithium-ion battery. *Mater. Lett.* 99: 164–167.
101 Gopalan, A.I., Santhosh, P., Manesh, K.M. et al. (2008). Development of electrospun PVdF–PAN membrane-based polymer electrolytes for lithium batteries. *J. Membr. Sci.* 325: 683–690.
102 Zhong, Z., Cao, Q., Jing, B. et al. (2012). Electrospun PVdF–PVC nanofibrous polymer electrolytes for polymer lithium-ion batteries. *Mater. Sci. Eng., B* 177: 86–91.
103 Padmaraj, O., Venkateswarlu, M., and Satyanarayana, N. (2016). Effect of PMMA blend and $ZnAl_2O_4$ fillers on ionic conductivity and electrochemical performance of electrospun nanocomposite polymer blend fibrous electrolyte membranes for lithium batteries. *RSC Adv.* 6: 6486–6495.
104 Cao, J., Wang, L., Fang, M. et al. (2014). Structure and electrochemical properties of composite polymer electrolyte based on poly vinylidene fluoride–hexafluoropropylene/titania–poly(methyl methacrylate) for lithium-ion batteries. *J. Power Sources* 246: 499–504.
105 Song, D., Xu, C., Chen, Y. et al. (2015). Enhanced thermal and electrochemical properties of PVDF-HFP/PMMA polymer electrolyte by TiO_2 nanoparticles. *Solid State Ionics* 282: 31–36.
106 Tamilselvi, P. and Hema, M. (2014). Conductivity studies of $LiCF_3SO_3$ doped PVA: PVdF blend polymer electrolyte. *Physica B* 437: 53–57.
107 Chen, J., Wang, X., Yu, X. et al. (2018). High dielectric constant and low dielectric loss poly(vinylidene fluoride) nanocomposites via a small loading of two-dimensional $Bi_2Te_3@Al_2O_3$ hexagonal nanoplates. *J. Mater. Chem. C* 6: 271–279.
108 Liu, T., Chang, Z., Yin, Y. et al. (2018). The PVDF-HFP gel polymer electrolyte for Li-O_2 battery. *Solid State Ionics* 318: 88–94.
109 Ma, Y., Li, L.B., Gao, G.X. et al. (2016). Effect of montmorillonite on the ionic conductivity and electrochemical properties of a composite solid polymer electrolyte based on polyvinylidenedifluoride/polyvinyl alcohol matrix for lithium ion batteries. *Electrochim. Acta* 187: 535–542.

110 Varzi, A., Raccichini, R., Passerini, S., and Scrosati, B. (2016). Challenges and prospects of the role of solid electrolytes in the revitalization of lithium metal batteries. *J. Mater. Chem. A* 4: 17251–17259.

111 Busche, M.R., Drossel, T., Leichtweiss, T. et al. (2016). Dynamic formation of a solid-liquid electrolyte interphase and its consequences for hybrid-battery concepts. *Nat. Chem.* 8: 426–434.

112 Shodai, T., Owens, B.B., Ohtsuka, H., and Yamaki, J.I. (1994). Thermal stability of the polymer electrolyte $(PEO)_8LiCF_3SO_3$. *J. Electrochem. Soc.* 141: 2978–2981.

113 Chaurasia, S.K. and Chandra, A. (2017). Organic-inorganic hybrid electrolytes by in-situ dispersion of silica nanospheres in polymer matrix. *Solid State Ionics* 307: 35–43.

114 Boaretto, N., Horn, T., Popall, M., and Sextl, G. (2017). Optimization of the transport and mechanical properties of polysiloxane/polyether hybrid polymer electrolytes. *Electrochim. Acta* 241: 477–486.

115 Croce, F., Appetecchi, G.B., Persi, L., and Scrosati, B. (1998). Nanocomposite polymer electrolytes for lithium batteries. *Nature* 394: 456.

116 Chung, S.H., Wang, Y., Persi, L. et al. (2001). Enhancement of ion transport in polymer electrolytes by addition of nanoscale inorganic oxides. *J. Power Sources* 97–98: 644–648.

117 Liu, Y., Lee, J.Y., and Hong, L. (2003). Morphology, crystallinity, and electrochemical properties of in situ formed poly(ethylene oxide)/TiO_2 nanocomposite polymer electrolytes. *J. Appl. Polym. Sci.* 89: 2815–2822.

118 Wagman, D.D., W.H. Evans, V.B. Parker, R.H. Schumm ,I. Halow, S.M. Bailey, K.L. Churney, R.L. Nuttall (1982) The NBS tables of chemical thermodynamic properties. Selected values for inorganic and C1 and C2 organic substances in si units, In Encyclopedia of Electrochemistry. J. Phys. Chem. Ref. Data 18 1807, American Chemical Society: New York.

119 Wu, F., Feng, T., Wu, C. et al. (2010). Thermally stable hyperbranched polyether-based polymer electrolyte for lithium-ion batteries. *J. Phys. D: Appl. Phys.* 43: 035501.

120 Ishihara, T., Koga, M., Matsumoto, H., and Yoshio, M. (2007). Electrochemical intercalation of hexafluorophosphate anion into various carbons for cathode of dual-carbon rechargeable battery. *Electrochem. Solid-State Lett.* 10: A74–A76.

121 Girishkumar, G., McCloskey, B., Luntz, A.C. et al. (2010). Lithium-air battery: promise and challenges. *J. Phys. Chem. Lett.* 1: 2193–2203.

122 Abraham, K.M. and Jiang, Z. (1996). A polymer electrolyte-based rechargeable lithium/oxygen battery. *J. Electrochem. Soc.* 143: 1–5.

123 Read, J. (2002). Characterization of the lithium/oxygen organic electrolyte battery. *J. Electrochem. Soc.* 149: A1190–A1195.

124 Dash, R. and Pannala, S. (2016). Theoretical limits of energy density in silicon-carbon composite anode based lithium ion batteries. *Sci. Rep.* 6: 27449.

125 Liu, N., Lu, Z., Zhao, J. et al. (2014). A pomegranate-inspired nanoscale design for large-volume-change lithium battery anodes. *Nat. Nanotechnol.* 9: 187.

126 Thai, M.L., Chandran, G.T., Dutta, R.K. et al. (2016). 100k cycles and beyond: extraordinary cycle stability for MnO_2 nanowires imparted by a gel electrolyte. *ACS Energy Lett.* 1: 57–63.
127 Chau, K.T., Wong, Y.S., and Chan, C.C. (1999). An overview of energy sources for electric vehicles. *Energy Convers. Manage.* 40: 1021–1039.
128 Randall, D. (2017). MIT researchers collaborate with Lamborghini to develop an electric car of the future.
129 (2018). Toyota Industries to utilise Toshiba green hydrogen supply unit. *Fuel Cells Bull.* 2018: 10.

9

Polymer Electrolytes for Supercapacitor and Challenges

Safir Ahmad Hashmi, Nitish Yadav, and Manoj Kumar Singh

University of Delhi, North Campus, Department of Physics and Astrophysics, Delhi 110007, India

9.1 Introduction

From the last few decades, a tremendous amount of research work has been devoted globally to the applications of polymer-based electrolytes in solid-state/quasi-solid-state-electrochemical devices, primarily in sensors, electrochromic displays, and energy storage devices including rechargeable batteries and supercapacitors. The energy storage devices basically store and release the electrical energy as per the requirement. The role of technology for energy storage has become extremely important in recent years, particularly against global warming. One of the important examples is the lead-acid battery, which has a history of 150 years or even more and is still progressing technically and expanding in global market due to its reliability and superior output performance over a wide temperature range [1]. Among the new developments, the secondary lithium ion batteries (LIBs) have taken a large market due to high specific energy and moderately high specific power and long cycle life.

Supercapacitors have recently become the devices of global interest due to various reasons, particularly as an alternative to rechargeable batteries in applications like computer memory backups, medical appliances, etc., or as complementary power sources to batteries in hybrid electrical vehicles (HEVs) like high energy applications [2–5]. Additional attraction toward supercapacitors is due to their environmental friendliness and outstanding safety. These devices are mostly reported with liquid electrolytes in similar fashion as in batteries, and hence they are encountered with the problems similar to liquid electrolyte batteries in that they are associated with the well-known disadvantages of corrosion, self-discharge, low energy density, and bulky design [2–6]. All solid-state supercapacitors using polymer-based electrolytes have attracted growing interest as they offer various advantageous mechanical and electrochemical properties. Ability of such electrolytes in terms of fabrication as thin film devices, miniaturization to ensure high energy density, and proper electrode/electrolyte contacts due to flexibility is the most attractive part from the solid-state device point of view.

Polymer Electrolytes: Characterization Techniques and Energy Applications,
First Edition. Edited by Tan Winie, Abdul K. Arof, and Sabu Thomas.

This chapter deals with various aspects of polymer electrolyte-based supercapacitors that are possible portable power sources of present and future generations. To begin with, principles and working processes of different classes of supercapacitors have been described. The following section is devoted to polymer-based electrolytes, suitable for supercapacitor design. The classification of the polymer-based electrolytes is presented according to the types of host polymers and solvents employed including the most reported systems, namely, solvent-free (solid) polymer electrolytes, plasticized polymer electrolytes, and popularly known gel polymer electrolytes (GPEs). Performance characteristics of the polymer electrolyte-based supercapacitors have been presented in the subsequent section. Starting from the description of some basic characterization techniques, the performance characterization of different classes of supercapacitors (particularly, electrical double layer capacitors [EDLCs], pseudo-capacitors and hybrid supercapacitors) has been described, citing some important example studies. Finally, the challenges, possible improvements, and future trends of supercapacitors are presented in the subsequent sections.

9.2 Principle and Working Process of Supercapacitors

A capacitor is a passive electronic component, which consists of two parallel electrodes separated by a dielectric and has the ability to store charges, hence energy, electrostatically. For charging a capacitor, a potential difference (voltage) is applied across the electrodes, which results in induced charges (ions) of opposite polarity on the surface of electrodes. A charged capacitor, connected in a circuit, acts as a voltage source. The capacitance (C) of a capacitor is expressed as $C = Q/V$, where Q is electric charge on each electrode and V is the potential difference across them. The parallel plate capacitor is the simplest form of capacitor whose capacitance (C) can be expressed as $C = \varepsilon_0 \varepsilon_r A/d$, where ε_0 is the permittivity of free space and ε_r is the relative permittivity of the material between the plates. Three major factors that determine the capacitance of parallel plate capacitors are (i) plate area, (ii) separation distance between the electrodes, and (iii) properties of the dielectric (insulator) used. The important parameters of an energy/power storage device are its specific energy and specific power. The energy E (Wh kg^{-1}) stored in a capacitor is given by

$$E = \frac{1}{2M} CV^2 \tag{9.1}$$

The maximum energy can be achieved when capacitance of electrode material and electrostability or breakdown strength (V) of the dielectric are maximum. In general, power (P) is defined as the rate of energy delivery per unit time. The maximum power of the capacitor is estimated by matched impedance (i.e. the resistance of the load is same as the internal resistance of the cell), which is given as

$$P_{\max} = \frac{V^2}{4M \times R} \tag{9.2}$$

Broadly, supercapacitors are categorized into two classes, namely, (i) EDLCs and (ii) redox capacitor or pseudocapacitors, on the basis of charge storage

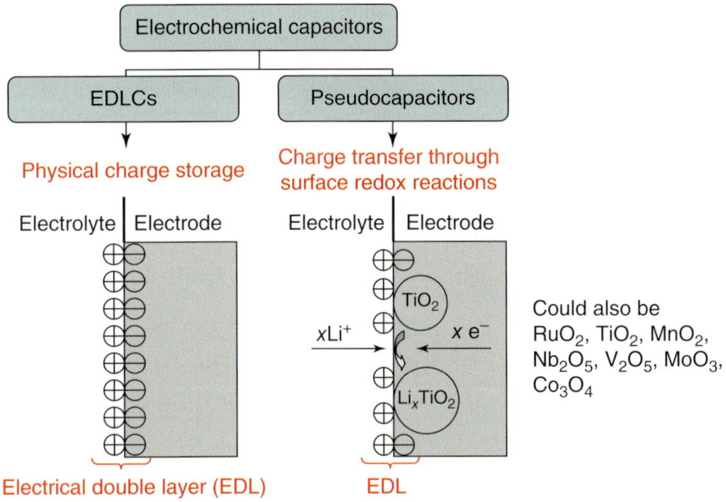

Figure 9.1 Classification of supercapacitors into EDLCs and pseudocapacitors. Source: Reproduced with permission of Laurent Pilon, University of California, Los Angeles. https://www.seas.ucla.edu/~pilon/EES.html.

mechanisms [2, 5]. In EDLCs, the translational charges are electrostatically stored/accumulated at the electrode/electrolyte interfaces, whereas the fast Faradaic redox reaction(s) is responsible for the pseudocapacitance [2, 5]. The classification of the supercapacitors is schematically shown in Figure 9.1 according to the charge storage mechanisms and types of electrode materials (https://www.seas.ucla.edu/~pilon/EES.html). Here, metal oxides like Nb_2O_5, V_2O_5, MoO_3, and Co_3O_4 give pseudocapacitive performance under specific conditions.

9.2.1 Charge Storage Mechanisms in EDLCs

An electric double layer appears when a charged object is placed into an electrolyte. The balancing counter charges are induced on the electrolyte side, concentrating near the surface. Few models are proposed for such interface between a solid electrode and an electrolyte, particularly liquid electrolyte. Figure 9.2 illustrates three important models, namely, the Helmholtz model, Gouy–Chapman model, and Stern model. In all these models, Ψ is referred as the potential profile at the interface, and Ψ_0 is the potential on the electrode surface. Helmholtz in 1879 first described and modeled the existence of a "double layer" in an electrochemical cell when he investigated the distribution of opposite charges at the interface of colloidal particles [7, 8]. Thereafter, various models have been developed to explain the formation of double layer at the electrode/electrolyte interfaces. These theoretical developments on the formation of electrical double layers are briefly described as follows.

The Helmholtz double layer is a simple approximation that the surface charges are balanced by counterions of opposite sign, placed at a definite separation "*d*" at the electrode/electrolyte interface (Figure 9.2a). It is assumed that the layer of opposite charges of counterions is perfectly rigid like parallel plate capacitor.

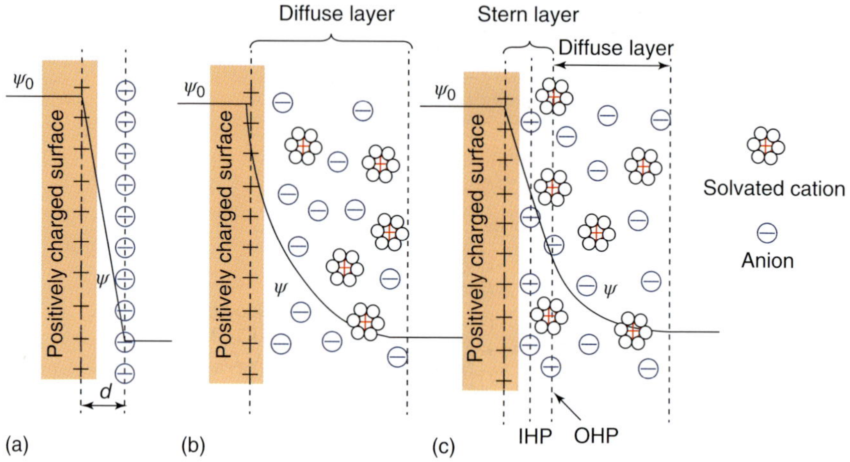

Figure 9.2 The electrical double layer formed at a positively charged electrode in aqueous electrolyte: (a) Helmholtz, (b) Gouy–Chapman, and (c) Stern models. Source: Frackowiak et al. 2013 [7]. Reproduced with permission of Elsevier.

The separation is approximately equal to the radius of ions and the Helmholtz layer pass through the center of ions. This model suggests that the profile of the surface charge potential (ψ), developed from the surface to the counterions, is linear (Figure 9.2a).

Gouy–Chapman extended the Helmholtz model, referred to as the "Gouy–Chapman double layer or diffuse model." According to this model, the counter electrolyte ions, which balance the rigid charged electrode surface, are not rigid; rather they form a cloud of opposite charged ions in the solution. The concentration of these ions decreases with distance from the surface, as shown in Figure 9.2b, and is referred to as diffuse double layer. Such diffuse layers are created due to thermal motion of cations and anions in the electrolyte [7, 8]. The surface charge potential (ψ) profile in such situations exponentially decreases from the rigid charge surface and into the bulk of the electrolyte. This model appears to be close to the real situation; however, it leads to an overestimation of the double layer capacitance, mainly due to the assumption that the electrolyte ions are point charges and close to the electrode surface.

Stern modified the Gouy–Chapman model of diffuse double layer by assuming ions to have finite size; hence they would have a separation of a few nanometers from the surface. It is also assumed that there is a possibility of some of the ions to be adsorbed by the surface, referred to as the Stern or compact layer. Thus, Stern combined the Helmholtz and Gouy–Chapman models and identified two regions of ion distribution, namely, (i) the inner region, known as the compact layer or Stern layer, strongly adsorbed by the electrode. This layer consists of specifically adsorbed ions and non-specifically adsorbed counterions known as the inner Helmholtz plane (IHP) and outer Helmholtz plane (OHP), respectively (Figure 9.2c). The surface charge potential (ψ) varies linearly in this compact layer, and (ii) the outer or second region is referred to as the diffuse layer region, which begins from OHP and has a thickness in the 10–100 nm range. After the diffuse layer, the bulk electrolyte begins, similar to the Gouy–Chapman model, mentioned earlier. The surface charge potential (ψ) varies exponentially in the diffusion region beginning at the OHP. Thus, the capacitance due to the electric double layer (C_{dl}) should be a combination of the capacitances from two regions, the Stern type of compact double layer capacitance (C_H) and the diffusion region capacitance (C_{diff}), expressed as

$$\frac{1}{C_{dl}} = \frac{1}{C_H} + \frac{1}{C_{diff}} \tag{9.3}$$

This model gives a satisfactory description of the electrical double layer on plane surfaces but fails to describe the real charge storage in porous electrodes.

In porous electrodes like activated carbon (AC), carbon nanotubes (CNTs), carbon nanofibers (CNFs), etc., not all the pores of the electrode are necessarily accessible to the electrolyte ions, and there is a nonlinear relation between the capacitance shown by the material and its specific surface area [9]. The relationship between the pore size and ion size is shown in few recent studies [10, 11]. It may be noted that the pores of size below 0.5 nm are not accessible to hydrated ions [10], whereas the pores less than 1 nm are not accessible to organic electrolytes [11]. However, Chmiola et al. recently reported an anomalous result that pores of sizes <1 nm significantly contribute to the specific capacitance [12]. In addition, the small amounts of mesopores (pore sizes from 2 to 50 nm) are necessary for a rapid supply of electrolyte ions to the micropore surface where main charge storage takes place, which leads to the enhanced power capability of supercapacitors [13]. The wider pores are not useful for capacitive charge storage; however, they contain free space that decreases the density of electrodes [14].

A number of attempts have been made to describe the capacitance of the carbon materials according to the pore size and shape. A schematic representation of a porous carbon containing micropores and mesopores is shown in Figure 9.3. For mesoporous carbons having pores of cylindrical shape, the conventional model of electrical double cylindrical capacitor (EDCC) is used and capacitance (C) can be expressed as [15, 16]

$$\frac{C}{A} = \frac{\varepsilon_r \varepsilon_o}{b \ln\left(\frac{b}{b-d}\right)} \tag{9.4}$$

where A is the surface area, d is the distance between the carbon surface and center of ion, and b is the pore radius. In the cases of micropores (size ≤2 nm), the ions line up in the center of a cylindrical pore, so the capacitance is estimated as [15, 16]

$$\frac{C}{A} = \frac{\varepsilon_r \varepsilon_o}{b \ln\left(\frac{b}{a_o}\right)} \tag{9.5}$$

where a_o is the effective size of the ion. A new slit-pore model is proposed recently [15, 17] in which the more realistic approximation to the pore shape in carbons is a slit, not a cylindrical shape. The specific capacitance for such pores can be expressed as

$$\frac{C}{A} = \frac{\varepsilon_r \varepsilon_o}{L/2 - a_o} = \frac{\varepsilon_r \varepsilon_o}{d_{eff}} \tag{9.6}$$

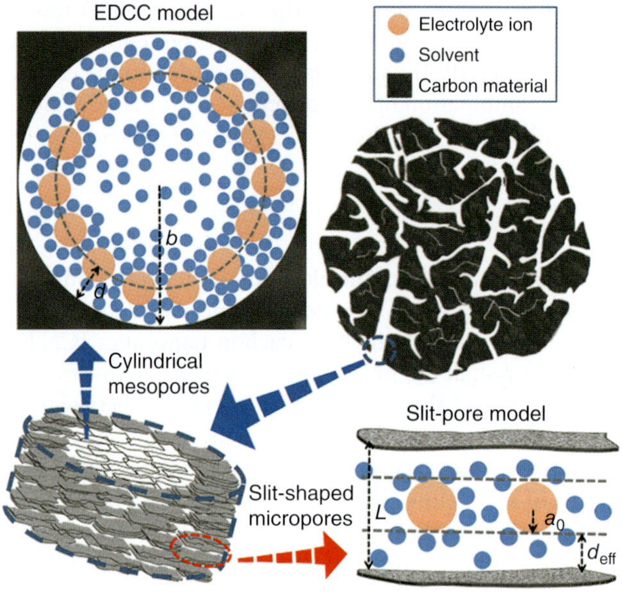

Figure 9.3 Schematic of a porous carbon particle containing micropores and mesopores, in which the double layers are formed based on EDCC model and slit-pore model. Source: Hsieh et al. 2015 [15]. Reproduced with permission of Royal Society of Chemistry.

where L is the width of slit pores and d_{eff} $(=L/2 - a_{\mathrm{o}})$ is the effective thickness of the double layer.

9.2.2 Charge Storage Mechanisms in Pseudocapacitors

The term "pseudocapacitance" was first introduced in 1971 for a specific electrode material RuO_2 as it involved Faradaic charge transfer reactions at the interface [18]. It may be noted that both the rechargeable batteries and redox capacitors/pseudocapacitors store charges via a redox process of the metal ions (e.g. Li^+, Na^+, Mg^{++}, etc.) in the electrode active materials. However, the difference is that, in batteries, the redox reaction is controlled by cationic diffusion within the crystalline framework of active electrode material, whereas it does not occur so in pseudocapacitors [19]. So, basically the pseudocapacitance is a Faradaic charge storage mechanism based on fast and highly reversible redox reactions at the surface or near surface of electrodes. In general, the materials offering pseudocapacitance can be classified as either intrinsic or extrinsic [20]. The intrinsic materials, showing pseudocapacitive behavior, have bigger particle sizes, while extrinsic materials have nanosized particles, though such materials do not show pseudocapacitive nature in the bulk [14]. The charge storage mechanisms for pseudocapacitive electrodes are identified in three categories, namely, (i) underpotential deposition, (ii) redox pseudocapacitance, and (iii) intercalation pseudocapacitance, as pictorially represented in Figure 9.4.

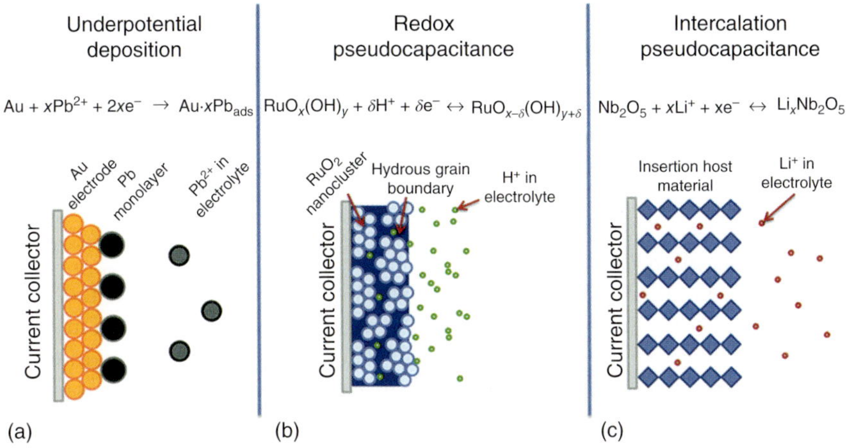

Figure 9.4 Different types of reversible redox mechanisms that give rise to pseudocapacitance: (a) underpotential deposition, (b) redox pseudocapacitance, and (c) intercalation pseudocapacitance. Source: Augustyn et al. 2014 [20]. Reproduced with permission of Royal Society of Chemistry.

9.2.2.1 Underpotential Deposition

In the underpotential deposition, monolayer of metal ions is formed at the surface of different metals above their redox potential [20]. One popular example of underpotential deposition is the deposition of lead (redox potential of Pb ~1.46 V) on the surface of gold (redox potential of Au ~1.50 V). The chemical reaction involved is as follows [21]:

$$2e^- + xPb^{++} + Au \leftrightarrow Au \cdot xPb$$

9.2.2.2 Redox Pseudocapacitance

In this case, the charge transfer process at the electrode/electrolyte interface is Faradaic in nature: i.e. it occurs by reduction–oxidation (redox) reactions. In this process, the oxidation state of the species increases during oxidation, while it lowers when reduction occurs. Trasatti et al. [18] observed the redox pseudocapacitive behavior for the first time in hydrous ruthenium oxide ($RuO_2 \cdot xH_2O$), which exhibits pseudocapacitance by accepting protons from the protic electrolyte and releasing them back into it (Figure 9.4b). The charge/discharge in hydrous RuO_2 follows the following redox reaction, which includes simultaneous proton and electron exchange:

$$RuO_x(OH)_y + ze^- + zH^+ \leftrightarrow RuO_{x-z}(OH)_{y+z} \tag{9.7}$$

The protons and electrons get inserted in RuO_2, resulting in the changes in oxidation state of Ru from +4 to +3 leading to the charge storage and giving rise to pseudocapacitance [18, 20]. The distinct current peaks in cyclic voltammograms corresponding to each reaction are the indication of the pseudocapacitance in $RuO_2 \cdot xH_2O$-based supercapacitors. Recently, Suzana et al. [22] demonstrated two different redox reactions in hydrous RuO_2 superimposed on the double layer capacitance. The first reaction is owing to double proton–electron exchange reaction, observed in acidic media at lower anodic potentials, whereas the second reaction is due to dissociative adsorption of water followed by the incorporation of oxygen atom into the bulk of the ruthenium oxide. The second reaction is observed in neutral media and at higher anodic potentials. In such reactions, ions are adsorbed on to the surface of the electrodes accompanied by the change in the oxidation state.

9.2.2.3 Intercalation Pseudocapacitance

There exist a number of insertion type oxides/sulfides that are potential electrode materials offering pseudocapacitive behavior. Insertion of cations into the bulk lattice of such solid electrode materials is termed as intercalation. In order to maintain the charge neutrality, the appropriate number of electrons is transferred to the host during intercalation of ions in the electrode, which leads to the pseudocapacitance. This process is schematically shown in Figure 9.4c, taking the example of Li ion insertion in an important host material Nb_2O_5. The conditions for ion intercalation are as follows:

(i) The lattice array must be intact during the process of intercalation or de-intercalation.
(ii) For the intercalation of the cations from the electrolyte and electrons from the external circuit, the host material should have a suitable reducing agent [23]. It may be noted that the intercalation of ions is limited by the ability of diffusion in to the electrode material.

Although all three pseudocapacitance mechanisms, mentioned earlier, occur in different types of electrode materials, they show, in general, similar electrochemical signatures, which are represented by the relation between potential and extent of charge [23], as expressed by

$$E \approx E^{o} - \frac{RT}{nF} \ln \left(\frac{X}{1 - X} \right) \tag{9.8}$$

where E is the potential, T is the temperature (K), n is the number of electrons, R is the ideal gas constant, and F is the Faraday constant (96 485 C mol^{-1}). The parameter X is different for different pseudocapacitance mechanisms – for example, X is two-dimensional site occupancy fraction in underpotential deposition, while $X = [\mathrm{Ox}]/([\mathrm{Ox}] + [\mathrm{Red}])$ is the extent of reaction for redox pseudocapacitance – and is occupancy fraction of layer lattice for the intercalation pseudocapacitance [24]. The real capacitance is defined in regions where the plot of E versus X or t (time) is linear and can be evaluated from

$$C = \frac{X}{V} \left(\frac{nF}{m} \right) \tag{9.9}$$

where m is the mass of the active electrode material on the current collector [20]. Despite the charge storage mechanism being fast Faradaic in nature in the situation where the redox and/or intercalation processes are the function of electrode potential, the derivative dq/dV has significance of "capacitance" and it is a measurable quantity experimentally [2, 23, 24]. Hence, it is termed as "pseudocapacitance" instead of electrostatic capacitance.

9.3 Electrolytes for Supercapacitors

Electrolyte is one of the important components of supercapacitors employed as separator/ion source. An electrolyte, employed as separator in supercapacitors, should have the following characteristics [7, 25]:

- High ionic conductivity.
- High concentration of free ions, high mobility, and minimum ion pairing tendency.
- High electrochemical stability and wide working voltage window.
- Low solvated ionic radius.
- Low viscosity, volatility, and toxicity.
- Cost effective.

The electrical or electrochemical behavior of the supercapacitors is determined by the electrolyte in the following ways [2]:

(i) High ionic conductivity of the electrolyte gives low equivalent series resistance (ESR) and, hence, high specific power of the supercapacitor.
(ii) Wide electrochemical stability window (ESW) leads to larger operating voltage range and, hence, higher specific energy delivered by the system.
(iii) The specific capacitance of the electrodes (carbon, oxides, or conducting polymers) is partly determined through ion adsorption from the electrolyte.

Two classes of electrolytes are primarily used as separators, namely, liquid electrolytes (aqueous, organic, or ionic liquids), and solid electrolytes, particularly polymer-based electrolytes. Most of the supercapacitors, reported in literature, are employed with liquid electrolytes; however, in second-generation supercapacitors, polymer-based electrolytes have been popularly employed. These two categories of electrolytes, their advantages and disadvantages, are discussed in the following sections. Particular emphasis has been given to the polymer-based electrolytes, which are used to develop quasi-solid-state supercapacitors.

9.3.1 Liquid Electrolytes

Liquid electrolytes, used in supercapacitors, are aqueous, nonaqueous, or molten salts (ionic liquids). All these types of electrolytes have their own advantages and disadvantages when used as separators/electrolytes. The aqueous electrolytes are generally strong electrolytes, which show a high value of dissociation constant (α) and possess higher ionic concentration. They also exhibit minimum ion pairing due to high dielectric constant and higher conductivities ($\sim$1 S cm^{-1}). Aqueous electrolytes have smaller solvated ions, which access even the smaller pores of carbonaceous electrodes, and hence such supercapacitors offer higher specific capacitance [7, 25]. The non-toxicity and environment-friendly nature make aqueous electrolytes attractive for supercapacitor fabrication. The aqueous solutions of KOH, H_2SO_4, Li_2SO_4, Na_2SO_4, and NH_4Cl are important examples of electrolytes frequently used in supercapacitors. However, they offer a relatively smaller voltage window ($\sim$1.2 V) due to water dissociation beyond this voltage [26]; therefore they limit the performance of the capacitors in terms of both the specific energy and specific power.

To meet a better supercapacitor performance in terms of the energy and power requirements, organic electrolytes are preferred. The operating voltage of organic electrolytes varies from ~2.5 to 3 V, which gives rise to the specific energy six to nine times as compared with aqueous electrolyte-based capacitors [26]. The liquid electrolyte 1.0 M tetra ethyl ammonium tetrafluoroborate (Et_4NBF_4) in propylene carbonate (PC) or acetonitrile has been demonstrated as one of the best candidate, although a number of such organic electrolytes are reported under this category. However, lower value of specific capacitance is observed in EDLCs or pseudocapacitors employing organic electrolytes, which is mainly due to larger size of solvated ions, which leads to limited accessibility at the interfaces and into pores of the electrodes [26]. Further, the low conductivity (~20 mS cm^{-1}) as compared with aqueous electrolyte (1 S cm^{-1}) results in higher ESR and, hence, power deterioration in the devices [7]. The other problems of organic electrolytes are low dielectric constant, complex purification procedure, and safety concerns due to flammability and toxicity.

From the last few decades, room temperature ionic liquids (RTILs) are popularly used as electrolytes in supercapacitors [27]. RTILs belong to environment-friendly and safe class of liquid electrolytes [7], popularly referred to as "green electrolytes" [8]. These are liquids at ambient or even lower temperatures and are composed of larger organic cations and relatively smaller inorganic anions. They have attracted research interest worldwide as electrolytes for electrochemical devices due to their various attractive properties including high thermal stability, high electrochemical stability over a wide potential window (>4 V), low vapor pressure, non-toxicity, nonflammability, sufficient ionic conductivity at elevated temperatures, etc. [7, 8, 25]. However, due to their relatively high viscosity and, hence, relatively low conductivity, the RTIL-based supercapacitors offer considerably higher value of ESR as compared with conventional organic electrolytes. However, the higher value of ESR can be compensated by increasing the operating potential window to obtain an acceptable specific energy [28].

In general, the organic cations, used to synthesize ionic liquids, include mainly imidazolium, pyrrolidinium, asymmetric quaternary ammonium ions, etc. [27]. The inorganic anions involved are tetrafluoroborate, trifluoromethanesulfonate, bis(trifluoromethanesulfonyl)imide, bis(fluorosulfonyl)imide, or hexafluorophosphate [28–30]. The nature of cations and anions determines the electrochemical and other physicochemical properties of the ILs, for example, aliphatic quaternary ammonium and pyrrolidinium-based ILs offer a wider potential window (>5 V) but lower conductivity as compared with that of 1-ethyl-3-methylimidazolium-based ILs, which offer conductivity of ~10^{-2} S cm^{-1} at room temperature [29]. Some important ionic liquids, employed as electrolytes in supercapacitor design, with their different properties are given in Table 9.1.

As described earlier, liquid electrolytes exhibit excellent performance in supercapacitor design, even on the commercial scale; however, they are associated with the well-known disadvantages of corrosion, leakage, self-discharge, bulky design, low energy density, etc. [25, 32]. Therefore, the development of high ion conducting, electrochemically and mechanically stable polymer-based electrolytes has been proposed as a good substitute for liquid electrolytes [33–42].

Table 9.1 Some examples of ionic liquids with their properties [31].

Ionic liquids	t_m (°C)	η (mm s^{-1} at 25 °C)	σ (mS cm^{-1})	ESW (V)
1-Ethyl-3-methylimidazolium bis(trifluoromethylsulfonyl) imide (EMITFSI)	−17	18	8.8	4.1
1-Ethyl-3-methylimidazolium trifluoromethanesulfonate (EMITf)	−9	43	9.2	4.1
1-Ethyl-3-methylimidazolium tetrafluoroborate ($EMIBF_4$)	15	32.4	11.5	4.3
1-Butyl-1-methyl pyrrolidinium bis(trifluoromethylsulfonyl) imide (BMPTFSI)	−50	71	2.2	5.5
1-Hexyl-3-methylimidazolium hexafluorophosphate ($HMIPF_6$)	−80	548	1	5.5
1-Ethyl-3-methylimidazolium dicyanamide (EMIDCA)	−21	17	27	5.9

t_m, melting point; η, viscosity; σ, conductivity; ESW, electrochemical stability window.

9.3.2 Polymer-Based Electrolytes

Polymer-based electrolytes are the class of ion-conducting solids, which are important due to their various favorable properties including high ionic conductivity (~10^{-4} to 10^{-2} S cm^{-1}) at room temperature, high chemical, thermal and electrochemical stability, ease of preparation with large area and small thickness, flexible nature, and excellent compatibility with the electrode materials [33, 34]. The preparation of polymer-based electrolytes in the form of films of desirable area and thickness provides an added advantage of the possibility to reduce the internal resistance of the device. A large number of polymer electrolytes with a variety of mobile ions, viz. H^+, Li^+, Na^+, Mg^{+2}, etc., have been studied in terms of their physicochemical and electrochemical properties for their application in various electrochemical devices like supercapacitors.

Poly(ethylene oxide) (PEO) complexed with sodium and potassium salts, introduced in 1973, were the first ion conducting polymers [43]. After that a large number of polymer electrolytes have been investigated. The polymer electrolytes exhibit reasonably good electrochemical properties, which are essential for electrochemical devices like supercapacitors. Based on various preparatory methods, adopted during the film cast and physical conditions, the polymer-based electrolytes are divided into following broad categories [44, 45]. These include (i) polymer-salt complexes or solvent-free polymer electrolytes, (ii) solvent-swollen polymer electrolytes, (iii) plasticized polymer electrolytes, (iv) polyelectrolytes, (v) rubbery electrolytes, (vi) porous polymer electrolytes (PPEs), (vii) GPEs, and (viii) composite polymer electrolytes. Initially, the majority of polymer electrolyte-based EDLCs/supercapacitors were reported with solvent-free or plasticized polymer electrolytes [46–49]. Subsequently, a large number of systems have been reported with different types of GPEs. PPEs belong to another important class of polymer-based electrolytes, introduced by Tarascon and coworkers [50]. These electrolytes are reported as promising separators/electrolytes mainly in LIBs. Few such systems are used in sodium ion batteries also [51, 52]. Despite their excellent electrochemical and mechanical properties, PPEs are hardly used as electrolytes in EDLCs/supercapacitors, except a few recent studies [53–55]. The aforementioned three classes of polymer-based electrolytes are described in the following sections along with their merits and demerits when employed in supercapacitors.

9.3.2.1 Solvent-Free Solid Polymer Electrolytes (SPEs)

Solid polymer electrolytes (SPEs) or solvent-free polymer–salt complexes comprise high molecular weight polar polymers, namely, PEO, poly(propylene oxide) (PPO), etc., complexed with ionic salts of larger anions (e.g. $LiClO_4$, $LiCF_3SO_3$, $LiN(SO_2CF_3)$, NaI, KI, etc.) [33, 56]. SPEs are attractive electrolytes/separators for the supercapacitors in view of the device safety as they perfectly prevent direct contact between the electrodes; however, they possess substantially lower ionic conductivity ($\sigma \sim 10^{-5}$ to $10^{-6}\ \mathrm{S\,cm^{-1}}$ at room temperature) than liquid electrolytes, and they also suffer from problems like poor interfacial contacts at electrode/electrolyte interfaces [2, 5]. In order to circumvent these problems, one of the initial approaches was the plasticization, i.e. addition of small amount of low molecular weight polar liquids/plasticizers, e.g. PC, ethylene carbonate (EC), polyethylene glycol (PEG), etc., in the SPEs [33, 34, 57]. Some of the plasticized SPEs have been reported as potential solid-like electrolytes in supercapacitors [48, 49]. The ionic conductivity profiles of a few important SPEs with respect to temperature are shown in Figure 9.5.

9.3.2.2 Gel Polymer Electrolytes (GPEs)

GPEs are quasi-solid-state form of electrolytes, in which large amounts of liquid electrolytes are entrapped in host polymers [36]. These electrolytes exhibit

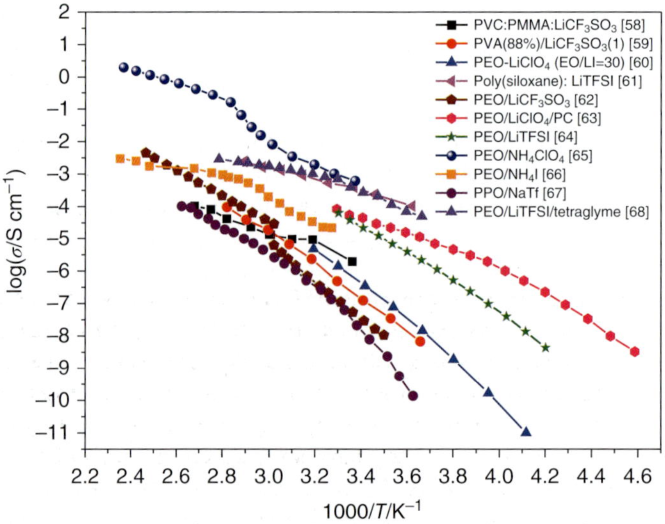

Figure 9.5 Temperature-dependent ionic conductivity profiles of some important SPEs.

intermediate properties possessing both the cohesive properties of solids and the diffusive property of liquids offering high ionic conductivities. The GPEs with different polymer hosts, namely, PEO [69], poly(vinylidene fluoride) (PVdF) [70, 71], poly(acrylonitrile) (PAN) [72, 73], poly(methyl methacrylate) (PMMA) [74, 75], poly(vinylidene fluoride-*co*-hexafluoropropylene) (PVdF-HFP) [41, 76], cellulose acetate (CA) [77], etc., have been reported. These electrolytes have ionic conductivities in the range of 10^{-4} to 10^{-3} S cm^{-1} at room temperature. With high ionic conductivity (almost comparable with liquid electrolytes) and moderately advantageous mechanical properties, such as their ability to form free-standing films of desirable area and thickness, flexible nature, and ability to have proper electrode/electrolyte contacts due to their softness, etc., GPEs are widely acceptable for their application as electrolytes/separators in various devices including supercapacitors [37].

The entrapment of liquid electrolytes increases the amorphous content of the polymer hosts with a single glass transition temperature as low as −40 °C, which, in turn, increases the ionic mobility in GPEs. Therefore, the increase in the overall conductivity of GPEs is observed to be mainly due to the enhancement in the ionic mobility on the account of diffusive transport property of the liquid phase. Further, the presence of liquid phase in excessive amount in GPEs leads to many disadvantageous characteristics, similar to those in liquid electrolytes. The other problem encountered, especially when Li^+ ion-conducting GPEs are employed in lithium batteries, is the reactivity of the electrolyte with the lithium metal surface. This affects the stability window of the electrolytes [78, 79]. Such reduction in the stability window may also occur in supercapacitors in specific situations, particularly when Li salt containing GPEs are employed.

Generally, organic liquids, namely, EC, PC, dimethylformamide (DMF), diethyl carbonate (DEC), dimethyl carbonate (DMC), etc., are used as solvents for various salts (e.g. $LiClO_4$, $LiCF_3SO_3$, etc.) in GPEs [80]. Although the ionic conductivities are significantly high due to the liquid-like conduction of ions in organic solvent-based GPEs, they always suffer from disadvantages like flammability, volatility, and electrochemical and thermal instabilities [37]. Therefore, it is of the prime consideration to develop GPEs that consist of nonvolatile solvents with a wider potential window and better thermal and electrochemical stabilities. Recently, new GPEs have been developed with nonvolatile solvents such as ionic liquids and solid-like plastic crystals. Redox-active electrolytes constitute another class of GPEs, recently introduced as excellent substitute due their extraordinary redox properties. These classes of GPEs are discussed in the following sections.

9.3.2.2.1 GPEs Incorporated with Organic Solvents Fenullade and Perche, in 1975, first demonstrated the idea of immobilizing an aprotic solution of alkali metal salt in organic solvent in the matrix of a polymer, which resulted in the formation of gels possessing high ionic conduction [81]. Since then, a large number of GPEs have been reported containing various types of salts and organic solvents, which exhibited the ionic conductivity values in the range of 10^{-4} to 10^{-3} S cm^{-1} at room/ambient temperature. Table 9.2 gives examples of few important GPEs incorporated with organic solvents, which have been mostly used in supercapacitors, along with their conductivities.

Table 9.2 Some important GPEs incorporated with organic solvents.

S. No.	GPEs	Ionic conductivity σ ($S\,cm^{-1}$)	Temperature (°C)	References
1.	PAN/PC:EC:DMF/$LiClO_4$	4×10^{-4}	22	[73]
2.	PAN–EC/DEC–$LiClO_4$	4×10^{-3}	RT	[82]
3.	PAN–EC/PC–$LiClO_4$	4×10^{-3}	25	[83]
4.	PVC–EC/PC–$LiClO_4$	1×10^{-3}	25	[83]
5.	PAN–EC/PC–$LiCF_3SO_3$	1×10^{-3}	20	[84]
6.	PMMA–EC–$LiCF_3SO_3$	3.4×10^{-5}	27	[85]
7.	PMMA–EC/PC–$LiClO_4$	1×10^{-3}	25	[86]
8.	PVdF–EC/PC–$LiBF_4$	6×10^{-3}	RT	[82]
9.	PVdF-HFP–EC/DEC–$LiN(CF_3SO_2)_2$	1×10^{-3}	RT	[87]
10.	PVdF-HFP–EC/PC–$Mg(ClO_4)_2$–MgO	8×10^{-3}	25	[88]
11.	PMMA–EC/PC–γ BL–$LiCF_3SO_3$	1×10^{-3}	RT	[89]
12.	PMMA–EC/DMC–$LiN(CF_3SO_2)_2$	1×10^{-3}	RT	[75]
13.	PMMA–EC/PC–$NaClO_4$ –SiO_2 (fumed)	3.4×10^{-3}	20	[90]

PEO-based GPEs, comprising EC and/or PC as solvents and salts, viz. $LiClO_4$, $LiCF_3SO_3$, $NaClO_4$, etc., form soft solids with very high room temperature conductivity $\sim10^{-3}\,S\,cm^{-1}$ [69, 91]. Mechanical strength of such GPEs is poor mainly due to the problem of solubility of PEO in the solvents [69]. The cross-linking of PEO minimizes this problem and, hence, the mechanical properties of the gels could be improved. The PAN- and PVdF-based GPEs are widely studied systems. PAN with dispersed salts and organic solvents forms homogeneous electrolyte films. Due to the absence of oxygen atoms in the PAN polymer matrix, a greater dissociation of lithium salts could be attained and could result in higher lithium ion transference number as high as 0.5 [92]. PVdF was chosen as another polymer host for the formation of gel electrolytes due to the fact that it possesses strong electron-withdrawing functional group (–C–F) as well as a high dielectric constant ($\varepsilon = 8.4$), which helps in greater dissociation of lithium salts. The mechanical strength of the as-casted GPE films varies with the PVdF content, while the viscosity of the medium and the concentration of the ionic salt control the magnitude of the ionic conductivity. Although PVdF-based GPEs offer excellent electrochemical properties, it has also been observed that their electrolyte properties could be further improved when PVdF is copolymerized with hexafluoropropylene (HFP). The PVdF-HFP copolymer exhibits greater solubility toward organic solvents and have lower crystallinity with reduced glass transition temperature as compared with pure PVdF polymer in the gel form [50]. Ionic conductivity profiles as a function of temperature of some common GPEs, which are reported as supercapacitor electrolytes, are given in Figure 9.6.

9.3.2.2.2 GPEs Incorporated with Ionic Liquids Many GPEs, synthesized by replacing organic solvents by ionic liquids, have recently been tested for their

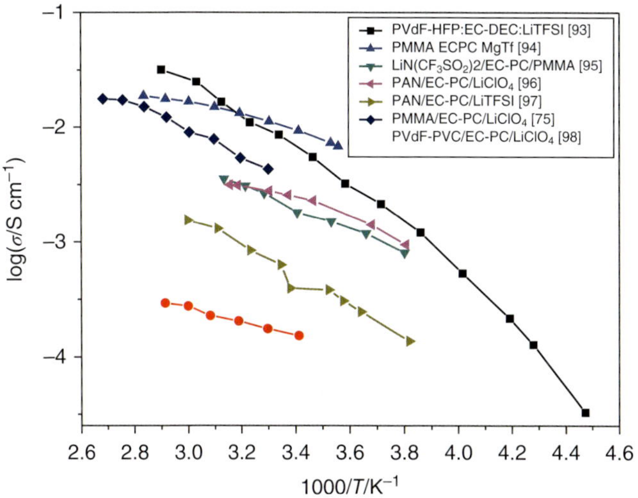

Figure 9.6 Temperature dependence of ionic conductivity of some important GPEs containing organic solvents.

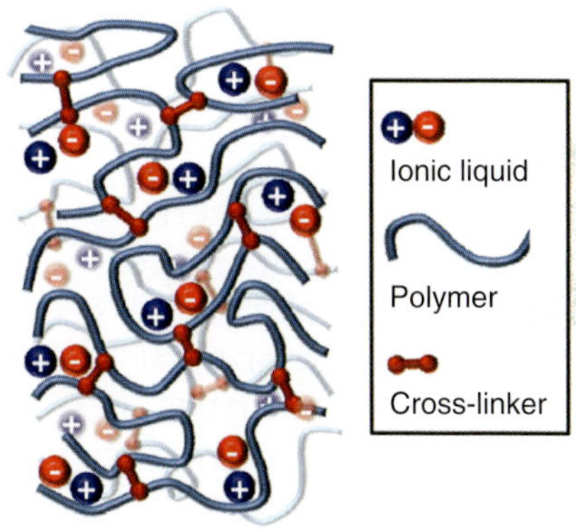

Figure 9.7 Schematic illustration of cross-linked IL-incorporated GPE. Source: Park et al. 2013 [99]. Reproduced with permission of John Wiley & Sons.

application as promising electrolytes in batteries and supercapacitors. The IL-based GPEs have specific limitation, particularly when employed in lithium/sodium batteries. The contribution of target (Li/Na) ions to the overall ion conduction becomes quite low due to substantial transport of component ions of ILs. However, IL-GPEs are suitable electrolytes for supercapacitors, where target ions are not much important (except in some specific cases like hybrid supercapacitors). In addition, the chemical, thermal, and environmental stability of ILs ensures safe operation and long life for the EDLCs/supercapacitors. Further, structure of host polymers for IL-GPEs is an important aspect. For example, cross-linking of IL-incorporated polymer electrolytes is generally implemented, as shown in Figure 9.7. Such cross-linked polymer (e.g. PEO chains) has the ability to increase the ionic liquid content to increase the ionic conductivity without any loss of mechanical integrity [99].

The results, available in literature, indicate that PVdF-based polymer electrolytes show highest conductivity at room temperature [100, 101]. PVdF-HFP is an important host polymer for IL-based GPEs also due to the reasons mentioned earlier, in addition to its properties such as excellent mechanical strength, chemical resistance, thermal stability, and high hydrophobicity [102]. In general, ionic conductivities of the gels increase with increasing temperature and weight fraction of IL [37]. The addition of room temperature ILs to PVdF-HFP improves the ionic conductivity and interfacial and thermal stability. Many IL-based GPEs with different salts were investigated by various research groups, as listed in Table 9.3.

The selection of suitable IL according to the applications is an important aspect. For example, RTIL, namely, 1-butyl-1-methyl pyrrolidinium bis(trifluoromethane sulfonyl)imide (BMPTFSI), has many desirable properties. In addition to the usual properties like low volatility, nonflammability, and non-toxicity, the other properties like low viscosity (~65 cP), high dielectric constant (~15) at room temperature, wide electrochemical window (~4–6 V), and high ionic

Table 9.3 Some ionic liquid-incorporated polymer-based electrolytes.

IL-based GPEs	Conductivity ($S\,cm^{-1}$)	Temperature (°C)	References
PVdF-HFP–EMITf–$EMIBF_4$	$\sim 5 \times 10^{-3}$	RT	[103]
PVdF-HFP–EMITf–$Zn(Tf)_2$	$\sim 1 \times 10^{-3}$	25	[104]
PVdF-HFP–EMITf–NaTf	$\sim 5.7 \times 10^{-3}$	27	[105]
PVdF-HFP–EMITf–$Mg(Tf)_2$	$\sim 4.8 \times 10^{-3}$	20	[41]
PVdF-HFP–EMIBETI–LiBETI	$\sim 10^{-3}$	RT	[106]
PVdF-HFP–DMOImTf–$LiCF_3SO_3$	$\sim 10^{-4}$	RT	[107]
PVdF-HFP–DMPIImTFA	$\sim 2 \times 10^{-3}$	RT	[108]
PVdF-HFP–BMMITFSI–LiTFSI	$\sim 3 \times 10^{-3}$	RT	[109]
PVdF-HFP–EImTFSI–HTFSI	$\sim 10^{-2}$	140	[110]
PVdF-HFP–PYR_{13}TFSI–LiTFSI	$\sim 10^{-3}$	RT	[111]
PYR_{13}TFSI–$P(EO)_{20}$LiTFSI	$\sim 3 \times 10^{-4}$	20	[112]
PVdF-HFP–P_{13}TFSI–LiTFSI	$\sim 3 \times 10^{-4}$	RT	[113]
PVdF-HFP–PY_{24}TFSI–LiTFSI–EC–PC	$\sim 10^{-2}$	110	[42]
PEO–BMITFSI–LiTFSI	$\sim 3.2 \times 10^{-4}$	25	[114]
PEO–LiTFSI–PYR_{1A}TFSI	$>10^{-4}$	20	[115]
$P(EO)_{20}$LiTFSI–BMPyTFSI	$\sim 7 \times 10^{-4}$	40	[116]
PVdF-HFP–LiTFSI–EMITFSI	2.11×10^{-3}	25	[117]
PVdF-HFP–EC/PC–LiTFSI–PDMITFSI	10^{-3}	30	[118]
P(EO)–BMPyTFSI–LiTFSI	6.9×10^{-4}	40	[116]
PVdF–PDPA–CFM–$LiClO_4$–PC	3.6×10^{-3}	25	[119]
PVdF-HFP–PEG–PEGDMA–$LiPF_6$–EC–DEC	1×10^{-3}	RT	[120]
PEGDA–PVdF–F127(PEO–PPO–PEO)	1.895×10^{-3}	RT	[121]
PVdF-HFP–BMmPBTFSI–PC	10^{-2}	60	[122]
PVdF–PEMA	1.5×10^{-4}	28	[123]

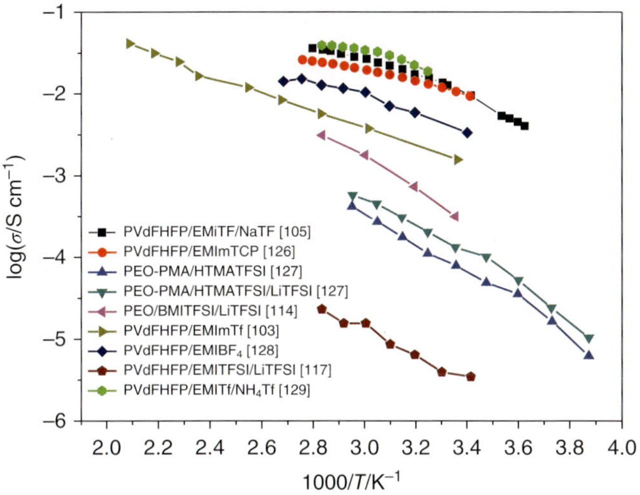

Figure 9.8 Temperature dependence of ionic conductivity of some important IL-incorporated GPEs.

conductivity ($\sim 2.4 \times 10^{-3}$ S cm^{-1} [124]) make it an excellent environmental-friendly electrolyte for supercapacitors [125]. The temperature dependent profiles of ionic conductivity of some important IL-incorporated GPEs are given in Figure 9.8.

9.3.2.2.3 Plastic Crystal-Incorporated GPEs Plastic crystals, a specific class of materials, have got recent attention due to their use in solid-like electrolytes for their applications in different electrochemical devices [130, 131]. Plastic crystals are basically soft matters, like liquid crystals, which have a transitional stage between real solids and real liquids (http://en.wikipedia.org/wiki/Plastic_crystal). Basically, plastic crystalline materials are composed of weakly interacting molecules, in which the centers of mass of the molecules form a crystalline lattice but the molecules show orientational disorder (http://en.wikipedia.org/wiki/Plastic_crystal, http://www.physics.uni-augsburg.de). There are three types of materials known to exhibit plastic crystalline nature, which offer ionic conductivity either themselves or after adding ionic salts in them [131]:

(i) Inorganic salts, e.g. Li_2SO_4 and α-Na_3PO_4, are reported long ago, which show rotator phases and plastic crystal nature at high temperatures (>500 °C). From experimental measurements, particularly, on Li_2SO_4, it was hypothesized that the rotator motions of the anion (SO_4^{2-}) support the motion of the Li^+ ion from one cation site to another site like a revolving door, as schematically represented in Figure 9.9 [132]. Some workers reported the conduction of cations in such materials via "paddle wheel" rotation of the anions [133, 134].
(ii) Organic ionic plastic crystals (OIPCs) are widely reported class of plastic crystals, which possess a relatively large and symmetric organic cation in combination with an inorganic anion that is normally either symmetrical or

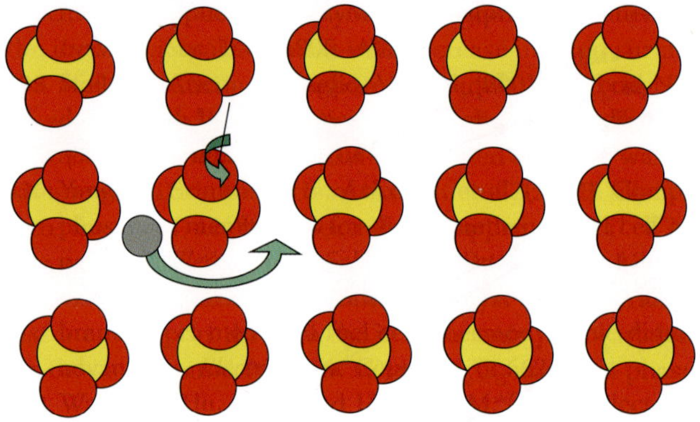

Figure 9.9 Hypothesized revolving door mechanism for the transport of Li^+ through a matrix of rotating SO_4^{2-} ions. Source: MacFarlane and Forsyth 2001 [132]. Reproduced with permission of John Wiley & Sons.

has a diffuse charge, just like in ionic liquids. These materials offer conductive and plastic phase around ambient temperature; hence they directly act as solid-like electrolytes [130, 132].

(iii) Molecular species such as succinonitrile (SN), referred to as nonionic plastic crystal, which allow good ionic conductivity on the addition of different ionic salts [135, 136].

Both the organic and non-organic plastic crystals show one or more solid–solid phase transitions before melting. These are associated with the onset of rotational or translational motions of the molecules/ions. A series of progressive transformations occur from an ordered crystalline phase to another crystalline phase and the increasingly disordered structure [132]. SN is a popular nonionic plastic crystal, which shows a stable plastic phase in the temperature range from −39 to 60 °C [137–139]. Below −39 °C, SN molecules exist in gauche conformation only with a rigid monoclinic structure in which all of the rotational motions are restricted [135, 140, 141]. Configuration of the SN molecule is shown in Figure 9.10, which has two $C_2H_4(CN)_2$ molecules per unit cell [131]. The plastic nature occurs due to the coexistence of trans- and gauche isomers of SN. The trans-isomers act as an impurity phase, resulting in the increase in lattice defects and, hence, decrease in activation energy for high ionic conduction [135, 140, 141]. This is one of the reasons for the high polarity and high dielectric constant (~55 at room temperature); hence SN is able to dissociate the ionic salts for high ion transport [135, 140–143]. The other attractive features are small vapor pressure and minimal flammability [131].

Figure 9.10 Configuration structure of succinonitrile. Source: Kumar and Rhee (2012). Reproduced from http://commons.wikimedia.org.

Table 9.4 SN-based electrolytes/gel polymer electrolytes along with their conductivity values.

Materials	σ (S cm^{-1})	Temperature (°C)	References
PEO–$LiCF_3SO_3$–nanochitosan–SN	$\sim 1 \times 10^{-2}$	60	[150]
PEO–SN–LiI–I_2	3×10^{-4}	25	[151]
PEO–LiBOB–SN	$(1.1–1.2) \times 10^{-3}$	80	[152]
PAN–$LiClO_4$–SN	6.2×10^{-3}	25	[138]
PEO–$NaCF_3SO_3$–SN	1.1×10^{-4}	25	[153]
PEO–$Mg(Tf)_2$–SN	6×10^{-4}	27	[154]
PAN–$LiClO_4$–SN	$(4.0–7.0) \times 10^{-3}$	25	[155]
PAN–LiTFSI–SN	1.3×10^{-3}	25	[156]
PEG–LiTFSI–SN	9.8×10^{-4}	25	[156]
PEO–LiTFSI–SN	3.0×10^{-4}	25	[156]
PEO–SN–LiTFSI	1.0×10^{-3}	30	[149]
PVdF-HFP–$LiClO_4$–SN	1.0×10^{-3}	10	[148]
PVdF-HFP–$LiPF_6$–SN	1.0×10^{-3}	20	[148]
PVdF-HFP–Li BETI–SN	1.0×10^{-3}	−10	[148]
PVdF-HFP–LiTFSI–SN	2×10^{-3}	−20	[157]
PVdF-HFP–BMPTFSI–SN	2.6×10^{-3}	25	[131]
PVdF-HFP–EMITf–SN	7.5×10^{-3}	25	[131]

Few binary systems of OIPCs or nonionic (molecular) plastic crystal SN with various ionic (including Li^+ and Na^+) salts have been reported as potential electrolytes, primarily for rechargeable batteries [139, 142, 144–147]. It may be noted that the OIPC or SN-based electrolytes are in liquid or molten phase and lack mechanical integrity. In order to obtain a defined shape, particularly, in the form of a flexible free-standing film, the immobilization of such liquid electrolyte in polymer hosts, for example, PEO, PAN, PVdF-HFP, etc., form polymer electrolytes [137–139, 148, 149]. Few important reports exist on polymer-based electrolytes incorporated with SN as solvent, as listed in Table 9.4.

Recently, a few studies have been reported on polymer-based electrolytes/GPEs incorporating nonionic plastic crystal SN, which plays the role of solid-state solvent [131, 138, 148]. Maier and coworkers reported the introduction of SN into PEO and PVdF-HFP based polymer electrolytes and demonstrated it as a versatile additive to improve electrochemical and mechanical properties of polymer electrolytes [137, 149]. The large orientational disorder of SN molecules helps in achieving high ionic mobility in the polymer–salt systems [131, 137, 143]. In fact, the pure SN-based electrolytes lack the mechanical integrity to provide shape like flexible films to the electrolytes, which are most appropriate for their application as separators in batteries/supercapacitors. The immobilization of such electrolytes in the polymer hosts, e.g. PEO, PAN, PVdF-HFP, etc., is a useful approach in this direction [137–139, 148, 149].

The ionic conductivity of SN-based polymer electrolytes/GPEs has been reported in the range of ~10^{-4} to 10^{-3} S cm^{-1} at room temperature [137–139, 142, 145, 148, 149, 158]. Yue et al. reported their studies on SN/$LiClO_4$ incorporated in low amounts of PEO and established a relationship between ionic conductivity with morphology, phase transition, and thermal properties [159]. Fan et al. reported comparative studies on SN/LiX (X = Cl^-, ClO_4^-, $CF_3SO_3^-$, PF_6^-, and $BETI^-$), i.e. bis(perfluoroethyl sulfonimide) added in PEO and PVdF-HFP, and demonstrated best ionic conductivity profiles for the SN/LiBETI-based polymer electrolytes [148, 149]. Bhattacharya and coworkers reported LiTFSI and SN immobilized in PAN, PEO, and PEG, which offer ionic conductivity of ~10^{-3} S cm^{-1} at 25 °C, suitable for lithium batteries [138, 139]. Effect of the addition of SN in PEO-$NaCF_3SO_3$ has been reported by the same group [153], which offers ~1.1×10^{-4} S cm^{-1} at 25 °C.

Recently our group reported GPE films comprising SN mixed with ionic liquids (EMITf, BMPTFSI, and SET_3·TFSI) immobilized in PVdF-HFP, as potential electrolytes for EDLCs/supercapacitors [160, 161]. We have recently presented comparative studies on SN-incorporated GPE and GPE with organic solvents (PVdF-HFP/SN/LiTFSI and PVdF-HFP/EC:PC/LiTFSI, respectively) [157]. It has been demonstrated that SN-incorporated GPEs not only give high ionic conductivity (~10^{-3} S cm^{-1} at room temperature) but also possess substantially high mechanical strength (in terms of tensile strength, maximum elongation at break, and toughness) as compared with organic solvent-based GPEs [157]. The temperature-dependent profiles of the ionic conductivity of some important SN-based polymer electrolytes/GPEs are shown in Figure 9.11.

Due to their plastic nature, PVdF-HFP/SN-based electrolytes possess the ability to accommodate mechanical stress, i.e. volume changes during the ions insertion/

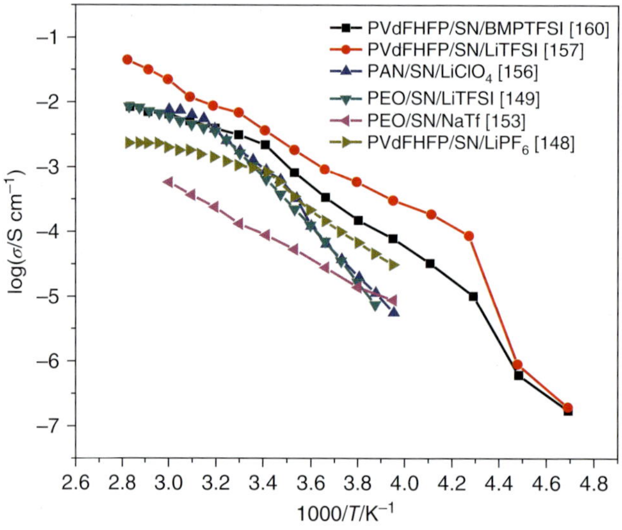

Figure 9.11 Temperature dependence of ionic conductivity of some important SN-based polymer electrolytes/GPEs.

extraction at the electrode/electrolyte interfaces in electrochemical operations, and hence provide better electrode/electrolyte contacts [142].

9.3.2.2.4 Redox-Active GPEs An interesting class of GPEs has recently been evolved to use as supercapacitor electrolytes, particularly for EDLCs with highly improved performance. This class comprises mainly aqueous GPEs, e.g. PVA/H_2SO_4, PVA/KOH, etc., incorporated with various redox additives, e.g. KI, hydroquinone (HQ), methylene blue (MB), etc. [162–173]. The introduction of redox additives first happened for liquid electrolytes, which led to improvement in the specific capacitance and specific energy of the EDLCs. Now, this approach has been extended to GPEs for quasi-solid-state EDLCs. Some important examples of redox-active GPEs along with their ionic conductivity are given in Table 9.5.

The addition of redox additives not only enhances the ionic conductivity but also improves the redox activity substantially at the electrode/electrolyte interfaces. This additional pseudocapacitance due to the redox reaction of the additives gives rise to enhancement in specific capacitance. Figure 9.12 shows the typical examples of ionic conductivity variations in PVA–H_2SO_4 incorporated with different concentrations of redox additives $VOSO_4$ and Na_2MoO_4 [171]. After attaining the maxima, the decrease in conductivity in redox-active GPEs

Table 9.5 Polymer-based electrolytes incorporated with redox additives.

S. No.	Host polymer electrolyte	Redox additives	Maximum room temperature ionic conductivity (mS cm^{-1})	References
1	PVA–KOH	KI	12.7	[162]
2	PVA–H_2SO_4	KI	—	[163]
3	PVA–H_2SO_4	$VOSO_4$	—	[163]
4	PVA–H_2SO_4	KI + $VOSO_4$	—	[163]
5	PVA–H_2SO_4	Hydroquinone (HQ)	29.3	[164, 168]
6	PVA–H_2SO_4	Methylene blue (MB)	29.6	[164]
7	PVA–H_2SO_4	p-benzenediol	34.8	[165]
8	PVA–H_2SO_4	Indigo carmine (IC)	20.3	[166]
9	PVA–KOH	KI	—	[167]
11	PVA–PVP–H_2SO_4	MB	36.3	[169]
12	PVA–H_2SO_4	Na_2MoO_4	—	[170]
13	PVA–H_2SO_4	$VOSO_4$	18.5	[171]
		Na_2MoO_4	31.1	
14	PVA–H_2SO_4	1-Anthraquinone sulfonic acid sodium (AQQS)	28.5	[172]
15	PVA–H_3PO_4	2-Mercapto pyridine (PySH)	22.5	[173]
16	PMMA–PC–$LiClO_4$	Hydroquinone (HQ)	—	[174]

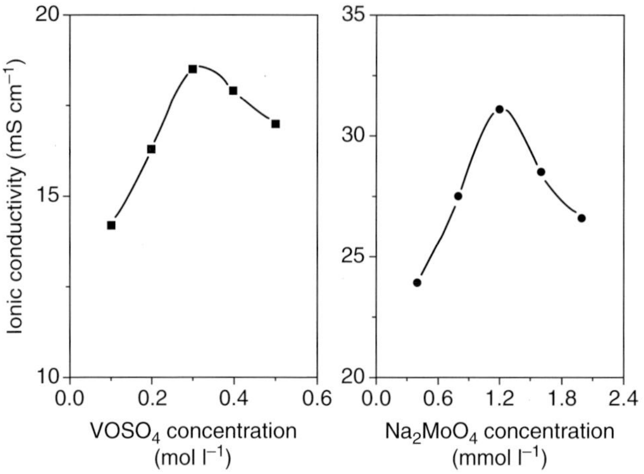

Figure 9.12 Ionic conductivity variations of GPEs PVA–H_2SO_4/$VOSO_4$ and PVA–H_2SO_4/Na_2MoO_4 as a function of concentrations of $VOSO_4$ and Na_2MoO_4. Source: Fan et al. 2016 [171]. Reproduced with permission of Elsevier.

is generally observed due to the aggregation and/or crystallization of redox-active species [171].

9.3.2.3 Porous Polymer Electrolytes

PPEs are another important class of polymer-based electrolytes that show promising results for their use in energy storage devices. They have been reported mostly for batteries and have also been found to be useful for supercapacitors [175]. In such electrolytes, the liquid electrolyte is absorbed and retained in the pores/cavities of the PPE film. The system remains solid after absorption of the electrolyte. The electrolyte may be retained in macropores ranging in size from 0.5 to 10 μm [175, 176] and/or in nanometer scale cages between individual polymer chains as in conventional GPE films, referred to as gelation. The gelation effect improves ion transfer between individual pores, and retention of the electrolyte in larger pores provides high conductivity as expected in its bulk liquid form. These systems also have excellent ESWs enabling higher energy densities. Hence, such PPEs not only show very good ionic conductivity and electrochemical stability but also provide desirable electrode/electrolyte interface for long cycling stability.

Tarascon et al. first reported electrochemical deployment of porous polymer films using liquid extraction and activation process from PVdF-HFP (host polymer), in which dibutyl phthalate (DBP) was extracted and 1 M $LiPF_6$ in EC-DMC was used for activation [50]. These porous polymer films, prepared using Bellcore technology, showed good mechanical stability and flexibility, but suffered from low ionic conductivity [50, 177]. An improvement was done by adding fumed silica in the Bellcore film; however, this too had the problem of low charge rates [178]. Stephan and Teeters [177] proposed that the porosity and pore diameter must be optimized to get a good porous separator for the type of liquid electrolyte to be used. Also, the conductivity values and transport numbers are affected by the type of polymer and salt interactions in the porous structure.

A number of reports have been published till date to optimize the properties of PPEs by preparing them in different ways using various polymers and activating with various liquid electrolytes. Park et al. [179] prepared porous films using coagulation of a polysulfone/*N*-methyl-2-pyrrolidone solution with water vapor-induced phase inversion at 20 °C and obtained well defined pores having an average diameter of ~10 μm. Stephan and Teeters [177] reported PVdF-HFP membranes prepared using phase-inversion technique and studied the effect of changing the non-solvent (pentane or 1-butanol). The membranes had nanopores having average size of 12 nm for pentane and 40 nm for 1-butanol non-solvent and the overall pore size ranged from 2 to 50 nm for both. Also, the pore size was uniform for pentane but nonuniform for 1-butanol. In another study [180], these films were compared with Celgard 2400, which has a mean pore size of 26 nm. The films showed room temperature ionic conductivity of 3.79 mS cm^{-1} for pentane and 2.17 mS cm^{-1} for 1-butanol when activated with a Li ion-containing liquid electrolyte. Wu et al. [181] reported porous films using a mixture of PVdF-HFP and polymeric surfactant P123 (polyethylene oxide-*co*-polypropylene oxide-*co*-polyethylene oxide), by steam-induced phase inversion. The mechanically stable films had pore size in the wide range of 2 nm to 5 μm. The room temperature conductivity, when activated with 1 M $LiClO_4$ in EC-PC, was found to be 2.6 mS cm^{-1} for 70 wt% P123 in PVdF-HFP and an excellent electrochemical stability up to 4.5 V. Another interesting work was reported by Xi et al. [182] when they made a porous membrane using a blend of two immiscible polymers – PVdF and PEO. Presence of PEO in the blend helps in improving the pore geometry, size, porosity and connectivity in the membrane. The pore size was found to be ~10 μm and the room temperature ionic conductivity of ~10^{-2} S cm^{-1} was obtained upon activation with 1 M $LiClO_4$ in PC. Li et al. [183] employed a phase inversion process using water as the non-solvent and pore inducer in PVdF-HFP dissolved in DMF. The resultant porous membrane had interconnected pores, as shown in Figure 9.13, and gave a

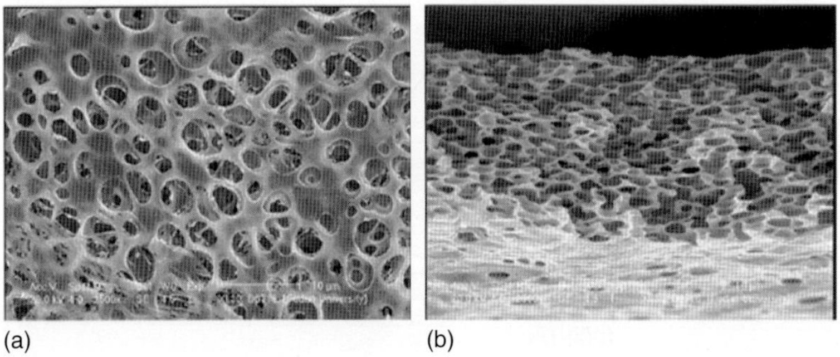

Figure 9.13 (a) Surface and (b) cross-section morphologies of the porous polymer membrane prepared by phase inversion process using water as non-solvent and pore inducer in PVdF-HFP dissolved in DMF. Source: Li et al. 2008 [183]. Reproduced with permission of Elsevier.

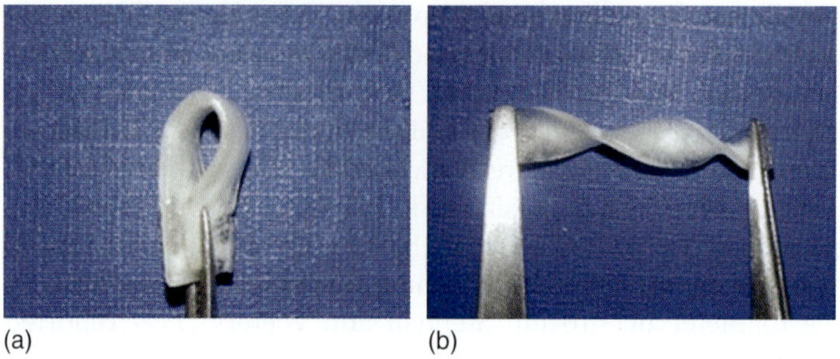

Figure 9.14 Photographs of flexible porous polymer electrolyte film (a) folded in closed loop and (b) twisted in spiral form. Source: Yadav et al. 2017 [175]. Reproduced with permission of Elsevier.

room temperature conductivity of 1.76 mS cm^{-1} upon activation with 1 M $LiPF_6$ in DMC:EMC:EC (1 : 1 : 1 w/w/w). The ESW of this film was reported to be 4.7 V versus Li/Li^+.

Recently our group demonstrated the use of PVdF-HFP-based PPE film as separator/electrolyte in an EDLC with activated-carbon electrodes [175]. The porous PVdF-HFP film was prepared via phase inversion method using steam (at the atmospheric pressure) as the non-solvent, and DMF as the solvent. The porous PVdF-HFP film was activated with 1 M $NaClO_4$ in EC:PC solution to obtain PPE film. The electrolyte retention capability was found to be ~400% with respect to the mass of un-activated film. The PPE film is found to be mechanically stable and shows flexible nature as depicted in Figure 9.14.

The PPE film showed the ionic conductivity of ~2 mS cm^{-1} at room temperature. The temperature dependence of ionic conductivity, as shown in Figure 9.15,

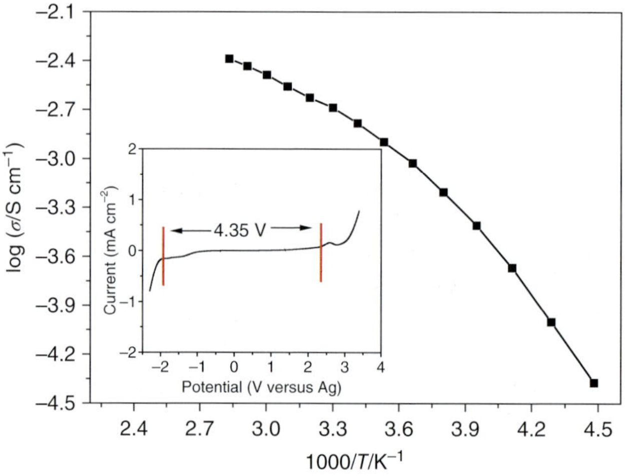

Figure 9.15 Temperature dependence of ionic conductivity of the PPE film, showing VTF fitting. LSV pattern of the film to estimate electrochemical stability window is shown as inset. Source: Yadav et al. 2017 [175]. Reproduced with permission of Elsevier.

indicates a typical curved plot, which is fitted with Vogel–Tamman–Fulcher (VTF) equation:

$$\sigma = AT^{-1/2} \exp\left(-\frac{B}{T - T_o}\right) \tag{9.10}$$

where B is pseudo-activation energy (in K) indicating the rate of change of conductivity with temperature, T_o is ideal glass transition temperature (in K), and A is the pre-exponential factor that indicates the conductivity at infinitely high temperature. The ESW of the electrolyte, estimated by linear sweep voltammetry (LSV), as shown in the inset of Figure 9.15, was found to be ~4.35 V. The mechanical stability, flexible nature, high ionic conductivity, and sufficient ESW make the PPE film suitable for EDLCs/supercapacitors.

9.4 Performance Characteristics

Performance characteristics of the supercapacitors mainly depend upon the characteristics of individual components, namely, electrodes, electrolytes, current collectors and lamination used to protect from undesirable environmental conditions. Emphasis has been given mainly on the performance characterization of capacitor electrodes, electrolytes, and the supercapacitors in the following sections.

9.4.1 Electrode Characterization

As discussed in previous section, the electrodes play important role in supercapacitive performance and they can be categorized according to charge storage mechanisms, and hence, two broad classes of supercapacitors namely EDLCs and pseudocapacitors (or redox supercapacitors) exist. There are various factors to be considered while selecting suitable electrode materials for high performance EDLCs or pseudocapacitors. These include their electrical (electronic + ionic) conductivity, chemical and electrochemical stability, compatibility with electrolytes (particularly with polymer-based electrolytes in the present context), wettability, flexibility, and most importantly types of porosity. Conductive additives, generally different types of carbons (e.g. carbon black, acetylene black, super-P, etc.) are incorporated in order to enhance electronic conductivity of electrodes. It may be noted that some electrode materials have inherent electronic conductivity; however many of them are electronically insulators (e.g. several oxides, viz. MnO_2, Mn_3O_4, TiO_2, Co_3O_4, V_2O_5, etc.). For such materials, conductive additives are essentially needed. The conductive additives not only improve the electronic conductivity but also have other advantages, for example, they act as electron-transfer catalysts, substrates for current leads and as an agent to control surface area, porosity, etc. [184].

All the physical, microscopic, spectroscopic, and electrochemical techniques, employed to characterize materials and electrochemical cells, are equally applicable to characterize supercapacitive electrode materials and their half cells. In order to investigate the morphological/structural properties, apart from conventional scanning electron microscopy/transmission electron microscopy (SEM/TEM) and X-ray diffraction (XRD) studies, Raman analysis is the most powerful tool for capacitor electrode materials. Raman spectroscopy is a nondestructive, fast, and high resolution technique, which gives maximum structural and electronic information, particularly for carbon electrode materials including carbon black, graphite, CNTs, and graphene. It is an invaluable method to understand many basic aspects of all sp^2 carbons. This includes the diameter of CNTs, presence of disorder/defects, and effect of nanotube-to-nanotube interactions [185]. The signature Raman bands are primarily D, G, D′, and G′ (2D) bands, which are mostly associated with defects and degree of disorder of the sp^2 carbons [186, 187] and are very sensitive to the morphology and any physical/chemical changes. Analysis related to these bands are well correlated with the supercapacitive performance of the carbon electrodes reported extensively [157, 161, 188].

Specific surface area and different types of porosity in the electrode materials (especially in carbon electrodes) play a very important role toward the performance characteristics of supercapacitors, particularly, the specific capacitance values (hence the specific energy) and rate capability, which affect the specific power and the cycle life of the devices. A general idea about theoretical aspects and quantitative estimation of specific surface area and porosity is briefly given as follows. When a gas is physically adsorbed in a solid, there is a weak van der Waals attraction between the gas molecules (adsorbates) and solid surface. Using this phenomenon, the porous materials can be characterized in terms of specific surface area, pore size distribution, and types of porosity [189].

According to Langmuir, the first theory of adsorption, the gas molecules adsorbed on the solid surface is monolayer. Subsequently Brunauer, Emmett, and Teller assumed in 1938 that physical adsorption results in the multilayer formation. The assumption of brunauer–emmett–teller (BET) theory is that "the solid surface possesses uniform sites and that adsorption at one site does not affect adsorption at neighboring sites and molecules can be adsorbed in second, third, …, and n^{th} layers, the surface area available for the nth layer being same to that of $(n\text{-}1)^{th}$ layer." Accordingly, the following equation is derived, known as BET equation [189]:

$$\frac{1}{W_{\text{total}}\left(\frac{1}{P/P_o}-1\right)}=\frac{1}{W_{\text{mono}}\times C}+\frac{C-1}{W_{\text{mono}}\times C}\left(\frac{P}{P_o}\right) \tag{9.11}$$

where W_{total} is total weight of gas adsorbed at relative pressure P/P_o, W_{mono} is the weight of gas adsorbed as monolayer which is constant for a given gas–solid system and C is BET constant depending upon the nature of gas. The plot of $1/[W_{\text{total}}\{1/(P/P_o)-1\}]$ versus P/P_o should be a straight line [189]. The slope of the linear plot gives the value of $(C-1)/(W_{\text{mono}}\times C)$, while the intercept gives the value of $1/(W_{\text{mono}}\times C)$. Thus, from the slope and intercept, W_{mono} and C can be evaluated. By knowing the W_{mono}, the surface area of the adsorbent material can be evaluated using the assumption that the molecules of the gas adsorbed in the first layer are closely packed on the surface. The most common adsorbate used is

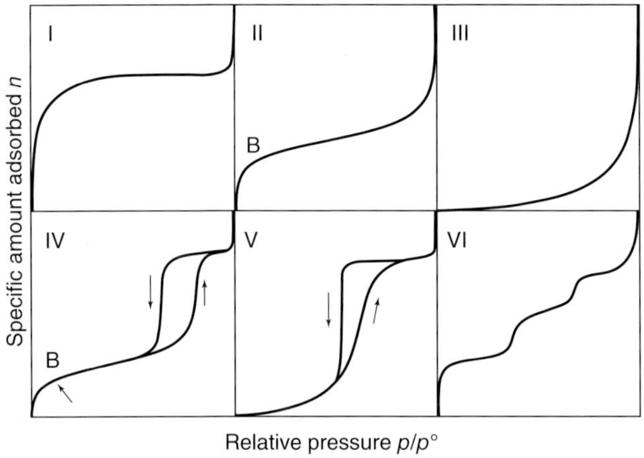

Figure 9.16 Types of physisorption isotherms. Source: Harry Marsh 2006 [190]. Reproduced with permission of Elsevier.

nitrogen gas; however, other adsorbates such as Ar, CO_2, CO, O_2, and C_4H_{10} are also used frequently [189].

The adsorption isotherm for a material under test is recorded in which an amount of gas is adsorbed across a wide range of relative pressures at a constant temperature (e.g. at liquid N_2 temperature, 77 K). Desorption isotherms are also recorded by measuring the gas removed from the pores as pressure is reduced. Six types of adsorption isotherms are classified according to the IUPAC classification, described by *Brunauer, Deming, Deming and Teller*, as shown in Figure 9.16. These classifications give different information about porous character of the materials, described as follows [189–191].

The type I isotherm is reversible, which is given by microporous solids having relatively small external surfaces (e.g. ACs). The type II isotherm is also reversible obtained with nonporous or macroporous powders. The type III isotherms are characterized by heats of adsorption less than the adsorbate heat of liquefaction. The adsorption proceeds as the adsorbate interaction with an adsorbed layer is greater than the interaction with the adsorbent surface. The behavior of type IV occurs on porous adsorbents with pores in the range of 1.5–100 nm. At higher pressures the slope shows increased uptake of adsorbate as pores become filled. Type V isotherm is observed where adsorbate–absorbent interaction is weak. It is related to the type III isotherm. Finally, the type VI isotherm is attributed to several possibilities, the most likely being if the temperature is below the adsorptive triple point, that the adsorbate is more like a solid forming a structured layer, i.e. epitaxial growth.

In addition, few more important pieces of information can be obtained from the types of hysteresis observed in the desorption isotherms, which are associated with the capillary condensation in mesopore interiors. The hysteresis loops are classified by IUPAC, in which the shapes of the hysteresis loops (types H1–H4) are correlated with the texture of the adsorbent [190, 191], as shown in Figure 9.17. Type H1 is associated with porous materials exhibiting a narrow

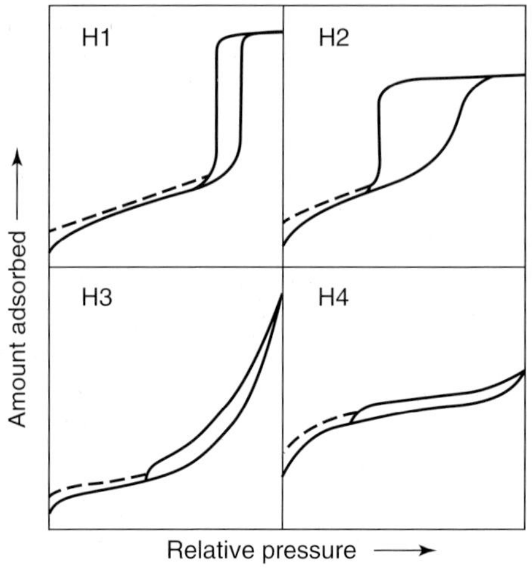

Figure 9.17 Types of hysteresis loops. Source: Harry Marsh 2006 [190]. Reproduced with permission of Elsevier.

distribution of relatively uniform (cylindrical-like) pores. Type H2 hysteresis is observed for materials in which the distribution of pore size and shape is not well defined. Such type of hysteresis loops is observed for narrow necks and wide bodies, often referred to as "ink bottle" pores [190]. The isotherms with H3-type hysteresis do not exhibit any limiting adsorption at high P/P_o. These are observed for non-rigid aggregates of plate-like particles giving rise to slit-shaped pores. Finally, the type H4 hysteresis is generally observed with complex materials containing both micropores and mesopores.

Barrett–Joyner–Halenda (BJH) method is adopted to estimate the pore size distribution and total pore volume, which are important characteristics of the capacitor electrode materials. Another important method, referred to as t-plot method, is frequently used to estimate parameters like micropore volume, micropore area, and external surface area, which are also important factors from the point of view of capacitor electrode materials. It may be noted that the external surface area calculated by t-method is equivalent to BET specific surface area. Details about these methods are available in literature [189–191].

9.4.2 Characterization of Supercapacitors

9.4.2.1 Electrochemical Characterization Techniques and Important Parameters

Basically, three important techniques, namely, electrochemical impedance spectroscopy (EIS), cyclic voltammetry (CV), and charge–discharge methods for prolonged cycles, are employed for electrochemical characterization of supercapacitors. These techniques are briefly described as follows.

9.4.2.1.1 Electrochemical Impedance Spectroscopy (EIS) EIS is a powerful technique to evaluate the performance of supercapacitors in frequency domain [2, 5, 192]. It is used to characterize the electrochemical behavior of supercapacitors because of their direct connectivity between the real system and its idealized equivalent circuit. The schematic representation of the impedance behavior for an ideal and

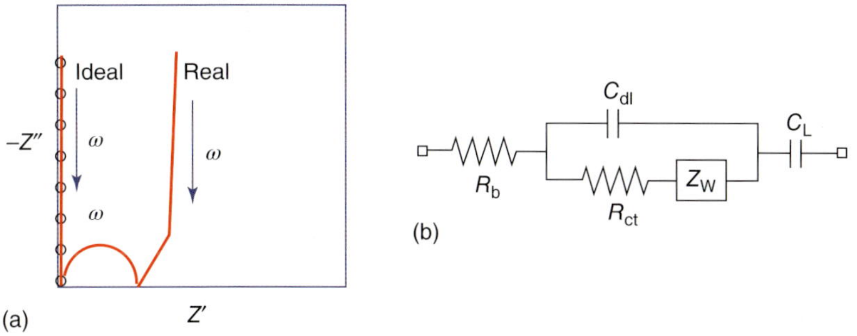

Figure 9.18 (a) Schematic representation of EIS plots for ideal and real supercapacitors and (b) possible equivalent circuit of a real supercapacitor.

a real capacitor cell along with the equivalent circuit is illustrated in Figure 9.18. The impedance pattern of an ideal capacitor shows a steep rising dispersion line overlapping with imaginary axis of the impedance plot. On the other hand, the real capacitor shows a semicircle representing a parallel combination of a bulk resistance R_b (resistance component mainly due to the ion migration through the bulk electrolyte) and a geometrical capacitance C_g due to dielectric polarization of electrolyte. The semicircular spur is followed by a linear pattern, making an angle of ~45° angle with real axis, which represents the characteristics of porous electrodes and diffusion-controlled Warburg-type impedance (Z_W) characterized by the constant phase element (CPE) (Figure 9.18b). In general, the CPE arises from the distribution of relaxation times as a result of inhomogeneity at the electrode/electrolyte interfaces, porosity of the electrodes, and dynamic disorder associated with diffusion [157]. The steeply rising curve, almost parallel to the Z''-axis, in the lower frequency region indicates capacitive behavior of the cells.

The overall capacitance (C) of the capacitor cells are evaluated using the relation:

$$C = \frac{-1}{\omega \cdot Z''} \tag{9.12}$$

where ω (= $2\pi f$) is the angular frequency and Z'' is the imaginary part of the total complex impedance at the respective frequencies. The single electrode specific capacitance values are evaluated by multiplying the overall capacitance by a factor of two and dividing by the active mass of the single electrode material.

9.4.2.1.2 Cyclic Voltammetry The electrochemical behavior of the capacitor cells can be observed by cyclic voltammetry (CV), performed either in three electrode configuration (generally for liquid electrolyte-based capacitors) or in two electrode configuration (usually for polymer electrolyte-based solid-state supercapacitors) [2, 5]. A typical voltammogram for a supercapacitor cell appears like a rectangle, as shown in Figure 9.19.

The current response of the supercapacitor cell is generally fast due to the electrostatic charge storage as in EDLCs or fast electrochemical charge transfer reactions as in pseudocapacitors. Changing the direction of the potential scan thus

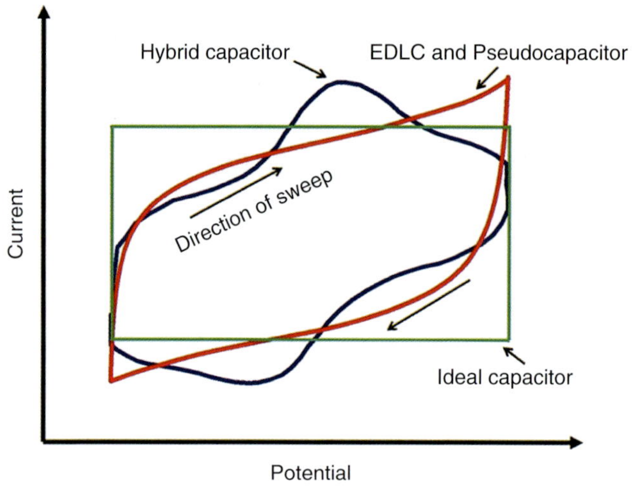

Figure 9.19 Schematic representation of a cyclic voltammograms of an ideal capacitor, real (resistive) capacitor (EDLC and pseudocapacitor), and a hybrid capacitor.

gives rise to a sharp drop or rise in the current responses. The current remains constant throughout the potential range of measurement (window) in the ideal case. The voltammogram adopts a more slanting shape depending on the resistance of the device and, hence, the current over the potential window is no longer constant. From the voltammogram, the capacitance can be calculated by the following equation:

$$C = \frac{i}{s} \tag{9.13}$$

where i is the constant voltammetric current and s is the scan rate ($\mathrm{d}V/\mathrm{d}t$). An ideal capacitor with no resistive component would display a rectangular shape, but real EDLC or pseudocapacitors show the tilted CV patterns, as shown in Figure 9.19. Prominent peaks that occur within narrow voltage ranges are usually due to faradaic reaction(s) in hybrid capacitors.

Faster scan rates correspond to charging and discharging at higher power levels. Multiple plots obtained at increasing scan rates are therefore often displayed to show the power levels on the charge–discharge characteristics. CV patterns also indicate the degree of reversibility of possible electrode reactions. A voltammogram that depicts a mirror image with respect to zero current line represents a reversible reaction. On the other hand, an irreversible process indicates two separate charge and discharge profiles.

9.4.2.1.3 Charge–Discharge In the charge–discharge method, a real capacitor/supercapacitor is represented by a series combination of an ideal capacitor C_e and the ESR (R) [2, 5]. The charge–discharge method is either potentiostatic or galvanostatic. In potentiostatic method, a potential step is applied across the capacitor cell and an exponentially decaying current is obtained. The behavior of the current i with respect to time on application of the potential step V_0 can be expressed as

$$i = \frac{V_0}{R} \exp(-t/\tau) \tag{9.14}$$

where τ $(=RC_e)$ is the time constant. The capacitor cells start discharging when the operation is reversed and the pattern of discharge curve is similar to the charging curve but with the reverse polarity of the current. The charge or discharge curves can be integrated numerically to obtain the total charge stored or released from the cell. The capacitance C_e can be evaluated by the expression (Eq. (9.15)):

$$C = \frac{\int i \cdot dt}{V_0} \tag{9.15}$$

In the galvanostatic method, a capacitor cell is charged by a constant current i. The total voltage V developed across the cell can be expressed as

$$V = i\left(R + \frac{t}{C_e}\right) \tag{9.16}$$

Thus, for a constant current step, the voltage across an ideal capacitor cell is expected to increase and decrease linearly with time while charging and discharging, respectively. A schematic V–t curve showing the linear variation is shown in Figure 9.20.

In the real capacitors, an initial sharp jump/fall in voltage while charging/discharging of the capacitor is observed (Figure 9.20), which is associated with the ohmic loss due to the ESR of the cell. The discharge capacitance (C_d) of the cell can be evaluated from the linear region of the curve following the relation:

$$\mathrm{Cd} = \frac{i \cdot \Delta t}{\Delta V} \tag{9.17}$$

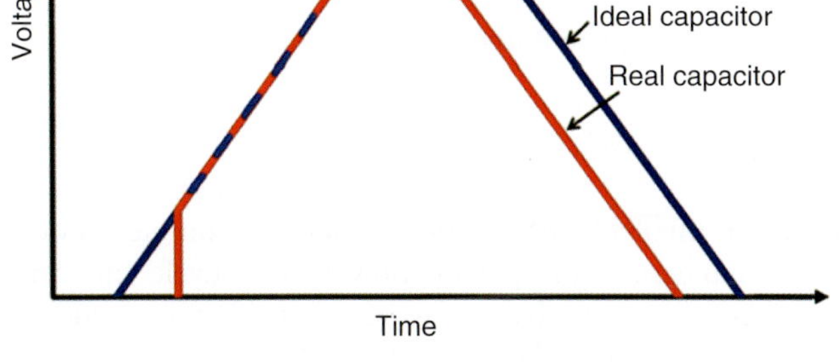

Figure 9.20 Galvanostatic charge–discharge behavior of an ideal and a real capacitor.

where ΔV is the change in voltage for Δt interval of time. The charge–discharge experiment can be carried out for several cycles to test cyclic efficiency of the cells.

Few more parameters can be extracted from galvanostatic charge–discharge curves, such as the columbic efficiency, specific energy, and specific power. Columbic efficiency η is the efficiency of a device to deliver charge on discharging, which was accumulated during charging at a constant current. It can be calculated easily from the galvanostatic (constant current) charge–discharge experiment using the expression:

$$\eta = \frac{t_D}{t_C} \times 100\% \tag{9.18}$$

where t_D and t_C are the duration of discharging and charging, respectively. This expression is basically applicable to ideal systems where galvanostatic charge–discharge variations are perfectly linear. However, for nonideal cases, where charge–discharge patterns are not perfectly linear, the efficiency should be evaluated by the ratio of discharge energy (E_D) and charge energy (E_C) of the cell [193], i.e.

$$\eta = \frac{E_D}{E_C} \tag{9.19}$$

where E_D and E_C are evaluated from the area under the galvanostatic discharging and charging curves, respectively [193].

The important parameters associated with the power sources, e.g. specific energy (E) and power density (P) of capacitor cell, can be evaluated using the following expressions:

$$E = \frac{1}{2m} C_d V^2 \tag{9.20}$$

$$P = \frac{V^2}{4m \times \text{ESR}} \tag{9.21}$$

where C is in Farad, V is voltage excluding the iR drop (observed at the beginning of the discharge curve), and m is the mass of active electrode material. The energy and power of the supercapacitors are often displayed in the form of Ragone plots, which are the indication of energy storage capability of supercapacitor at different power levels. Most devices exhibit a "knee"-shaped characteristic, with energy density rapidly decreasing at a critical power level.

9.4.2.2 Performance of Polymer Electrolyte-Based Supercapacitors: Some Case Studies

9.4.2.2.1 Carbon-Based Capacitors: EDLCs Supercapacitors based on carbon electrodes, also referred to as EDLCs, are the most popular devices because of various reasons including the involvement of almost inert carbon electrodes, simple charge–storage mechanism, high specific power, long cycle life, and safety. Principles and mechanisms of charge storage have been discussed earlier in Section 9.2.1. ACs (powder, fiber, or fabric) are the most widely used electrode materials for EDLCs [2–4, 6, 194–196]. However, many other forms of nanostructured carbons, such as CNTs [4, 126, 194, 197, 198], CNFs [26, 199, 200], carbon aerogels [201], carbide-derived carbons (CDCs) [25], onion-like carbons (OLCs) [202], and graphene [3, 161, 196, 203–205], have been explored extensively.

9.4.2.2.1.1 AC-Based EDLCs In general, activated carbon-based EDLCs offer a high specific capacitance (hence high specific energy) due to large surface area of AC electrodes (1000–2500 $m^2\ g^{-1}$). However, they suffer from many drawbacks, such as poor cyclic performance, relatively low power, and poor rate performance [184, 206]. Only about 10–20% of the maximum theoretical capacitance can be practically achieved primarily due to predominant presence of micropores, which are inaccessible to the solvated ions of electrolytes and wetting limitations of the electrolyte on the electrode surface [207] and inability of the proper formation of double layer in the pores [208]. In principle, the specific capacitance of carbon materials should increase with increasing surface area. However, Shi and coworker [209, 210] reported several ACs with different surface area, pore sizes, pore size distribution, and pore volumes and correlated these parameters with specific capacitance and demonstrated that the hypothesis is not practically true. Basically, the ionic mobility within the pores is different from the mobility in the bulk of the electrolyte solution, and it is highly influenced by pore size. Therefore, the pore size must be chosen to suit the electrolyte and thereby ensure that the pore size distribution is optimal based on the ionic sizes [211].

Most of the reported EDLCs are based on liquid electrolytes, which are soaked with different types of commercially available separators [4, 194, 201–204, 212]. However, they are associated with the well-known disadvantages of corrosion, self-discharge, low energy density, and bulky design. The solid-state form of EDLCs employing polymer-based electrolytes have attracted growing interest as they offer advantages such as the fabrication as thin film devices and miniaturization, despite anticipated difficulties in obtaining proper electrode/electrolyte contact. Some early reports on solid-state EDLCs include various polymer-based electrolytes such as PMMA-PEO/PC/tetraalkyl ammonium salts [213], PAN/PC/tetraalkyl ammonium salts [214], PEO/PC/$LiClO_4$[215], PAN/PC/$LiAsF_6$ [216], PVA–H_3PO_4 [217], PEO–PEG–$LiCF_3SO_3$[217], etc. In recent years, the latest generation of EDLCs employs GPEs as separators/electrolytes due to their advantageous properties including their flexible nature and high ionic conductivity comparable to liquid electrolytes [126, 161, 196–198, 200]. Few important EDLCs with polymer-based electrolytes (solvent-free polymer electrolytes, plasticized polymer electrolytes and GPEs) and various forms of carbon electrodes are listed in Table 9.6.

Ionic liquid-incorporated GPEs, due to their thermal and electrochemical stabilities, are the latest attraction as electrolytes/separators in EDLCs/supercapacitors [237]. The GPEs, incorporated with plastic crystals (primarily, SN as solid solvent) is another group of potential electrolytes for EDLCs.

Suleman et al. [160] recently reported solid-state EDLCs employing symmetrical AC electrodes (prepared from coconut shell), separated by GPE films comprising SN and ionic liquids BMPTFSI and 1-ethyl-3-methyl imidazolium trifluoromethanesulfonate (EMITf) entrapped in PVdF-HFP. Performance of

Table 9.6 Some important examples of polymer electrolyte-based EDLCs employing AC, CNTs, and graphene-based electrodes.

EDLC cell	Specific capacitance ($F g^{-1}$)	Specific energy ($Wh kg^{-1}$)	Specific power ($kW kg^{-1}$)	References
AC-based EDLCs				
AC|PEO–KOH–H_2O|AC	90 (1 $mA g^{-1}$)	—	—	[218]
ACF|PEO–PEG–$LiCF_3SO_3$|ACF	4	—	—	[5]
ACF|PMMA–PEO–$(C_2H_5)_4NBF_4$–PC|ACF	20 (0.11 $mA cm^{-2}$)	—	—	[219]
ACP|PAN–$EMIBF_4$|ACP	200	—	—	[220]
AC|PVDF–HFP–EMITFSI|AC	—	5	2.45	[221]
AC|PAN–EMITf|AC	130	—	—	[222]
ACF|PVA–H_3PO_4|ACF	550 $mF cm^{-2}$ (1 $mA cm^{-2}$)	—	—	[5]
AC|PVA–KOH|AC	136 (0.8 $A g^{-1}$)	3.98 (0.8 $A g^{-1}$)	4.88 (0.8 $A g^{-1}$)	[162]
AC|PMMA–$NaClO_4$–EC–PC|AC	78 (0.64 $mA cm^{-2}$)	10.8 (0.64 $mA cm^{-2}$)	0.19 (0.64 $mA cm^{-2}$)	[223]
CNTs-based EDLCS				
CNT-paper|EMIM–NTf_2–silica gel|CNT-paper	135 (2 $A g^{-1}$)	41 (2 $A g^{-1}$)	164 (2 $A g^{-1}$)	[224]
MWCNT|$LiPF_6$–EMImFAP–(PVdF-HFP)|MWCNT	127	—	—	[76]
f-MWCNT|EMIFSI–PVdF–SiO_2|f-MWCNT	80.7 (0.3 $A g^{-1}$)	32.2 (1 $A g^{-1}$)	0.9 (1 $A g^{-1}$)	[225]
MWCNT|Mg(Tf_2)–EMITf–PVdFHFP-EC:PC|MWCNT	41 (0.2 $A g^{-1}$)	17 (0.2 $A g^{-1}$)	13 (0.2 $A g^{-1}$)	[197]
MWCNT|EMImFAP–PVdFHFP|MWCNT	76 (0.5 $A g^{-1}$)	17.2 (0.5 $A g^{-1}$)	18.9 (0.5 $A g^{-1}$)	[226]
CNF–MWCNT|PVA–H_2SO_4|CNF-MWCNT	101 (3 $mA cm^{-2}$)	5.06 (0.5 $mA cm^{-2}$)	7.67 (0.5 $mA cm^{-2}$)	[227]
Aligned MWCNT–polyaniline|PVA–H_3PO_4|aligned MWCNT–polyaniline	233 (1 $A g^{-1}$)	—	—	[228]
CNT|PVA–H_3PO_4|CNT	20 (0.1 $A g^{-1}$)	0.515 (0.1 $A g^{-1}$)	19 (0.1 $A g^{-1}$)	[229]

(Continued)

Table 9.6 (Continued)

EDLC cell	Specific capacitance ($F\,g^{-1}$)	Specific energy ($Wh\,kg^{-1}$)	Specific power ($kW\,kg^{-1}$)	References
Graphene-based EDLCs				
PANI–graphene|PVA–H_3PO_4|PANI–graphene	261 (0.38 $A\,g^{-1}$)	23.2 (0.38 $A\,g^{-1}$)	399 (0.38 $A\,g^{-1}$)	[230]
rGO-cMWCNT|potassium polyacrylate-KCl|CFP–PPy	82 (0.5 $A\,g^{-1}$)	28.6 (0.5 $A\,g^{-1}$)	15.1 (0.5 $A\,g^{-1}$)	[231]
CNF–rGO|PVA–H_2SO_4|CNF–rGO	203 (0.5 $A\,g^{-1}$)	15.5 $mW\,cm^{-2}$	20 $mWh\,cm^{-2}$	[232]
GO|PVDFHFP-EMIMBF_4|GO	135 (1 $A\,g^{-1}$)	32.4 (10 $A\,g^{-1}$)	6.6 (10 $A\,g^{-1}$)	[233]
rGO|PVA–H_3PO_4|rGO	108 $F\,cm^{-3}$ (0.7 $A\,cm^{-3}$)	7.5 $Wh\,cm^{-3}$ (0.7 $A\,cm^{-3}$)	2.9 $W\,cm^{-3}$ (0.7 $A\,cm^{-3}$)	[234]
Graphene hydrogel|PVA–H_2SO_4|graphene hydrogel	186 (1 $A\,g^{-1}$)	0.12	0.67	[235]
rGO–MoO_2|PVA–H_2SO_4|rGO–MoO_2	404 (0.5 $A\,g^{-1}$)	14 $Wh\,kg^{-1}$	0.5 $kW\,kg^{-1}$	[236]

the capacitor cell with EMITf-based gel electrolyte offers high specific capacitance, specific energy, and specific power, which are, respectively, ~254 $F\,g^{-1}$ (1 $A\,g^{-1}$), 35 $Wh\,kg^{-1}$, and 12 $kW\,kg^{-1}$, and better than the EDLC fabricated with BMPTFSI-based gel electrolyte. The superior performance of the EMITf-based cell is correlated with smaller sizes of EMITf component ions, responsible for easy ion switching through the smaller pores of AC also.

Recently, our group reported PVdF-HFP-based PPE, prepared by phase inversion technique and activated with EC-PC-$NaClO_4$ electrolyte solution (discussed in Section 9.3.2.3) as a potential electrolyte for quasi-solid-state EDLC [175]. Performance characteristics (i.e. CV and charge–discharge tests) of the PPE-based EDLC, employing commercial AC electrodes (surface area: ~1123 $m^2\,g^{-1}$) are shown in Figure 9.21. The capacitor operates for maximum voltage of 2.0 V and offers specific capacitance of (~150 $F\,g^{-1}$, evaluated from EIS), and specific energy and power of ~17.7 $Wh\,kg^{-1}$ and ~14.3 $kW\,kg^{-1}$(0.95 $A\,g^{-1}$), respectively, evaluated from charge–discharge tests. The EDLC shows stable specific capacitance (after ~17% initial fading) and high Coulombic efficiency (~99%) for ~10 000 charge–discharge cycles.

A novel class of aqueous polymer-based electrolytes, referred to as "redox-active polymer electrolytes" (discussed earlier in Section 9.3.2.2), is frequently employed these days as electrolytes in EDLC fabrication [162–173]. Senthilkumar et al. [238] reported a hydroquinone mediated PVA–H_2SO_4-based EDLC with AC (extracted from a biowaste *Eichhornia crassipes*). The EDLC delivers improved specific capacitance, specific energy, and specific power, which are 941 $F\,g^{-1}$ (1.7 $A\,g^{-1}$), 20 $Wh\,kg^{-1}$, and 0.33 $W\,kg^{-1}$, respectively, with respect to the device without hydroquinone (425 $F\,g^{-1}$at 1.7 $A\,g^{-1}$and 9 $Wh\,kg^{-1}$ at

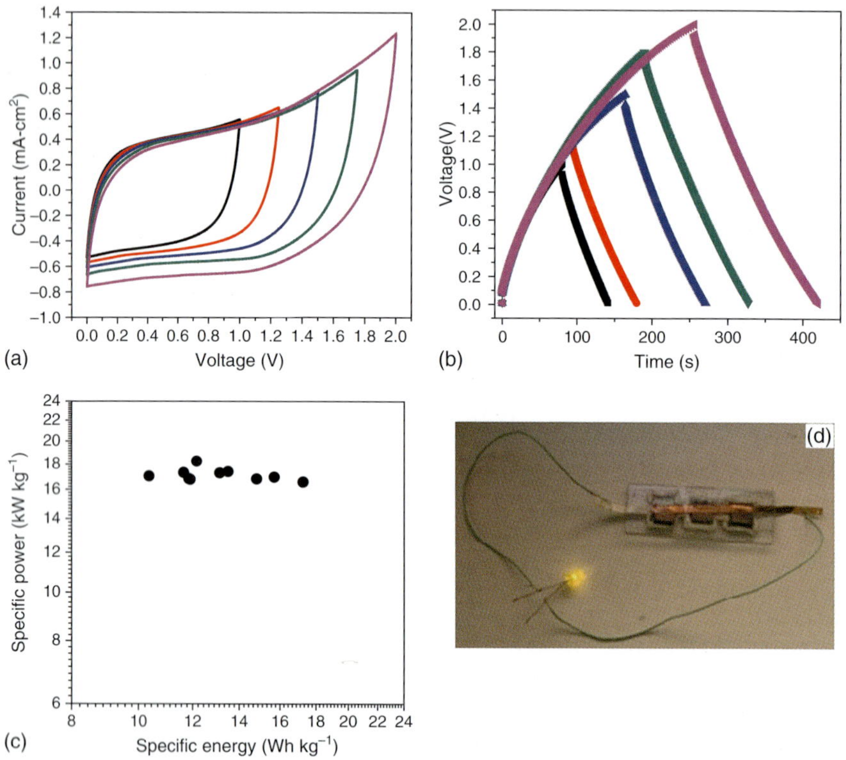

Figure 9.21 (a) Cyclic voltammograms at a scan rate of 10 mV s^{-1} for increasing voltages, (b) Galvanostatic charge–discharge characteristics, (c) Ragone plot for solid-state EDLC fabricated using AC electrodes and porous polymer electrolyte, and (d) photograph of LED, powered by the same EDLC. Source: Yadav et al. 2017 [175]. Reproduced with permission of Elsevier.

0.33 W kg^{-1}). Further improvement in performance characteristics of AC-based EDLC has been reported when PVA–H_2SO_4 has been employed, added with two different redox additives (e.g. KI and $VOSO_4$) [163]. The comparative data of specific capacitance values with PVA–H_2SO_4, PVA–H_2SO_4–KI, PVA–H_2SO_4–$VOSO_4$, and PVA–H_2SO_4–KI–$VOSO_4$ are 156.4, 404.6, 304.5, and 1232.8 F g^{-1}, respectively, at 0.5 A g^{-1}, which indicates the substantial improvement in performance [163]. Another novel approach of EDLC configuration with PVA–H_2SO_4 electrolytes containing two different additives, namely, $VOSO_4$ and Na_2MoO_4, has been reported by Fan et al. [171], as depicted in Figure 9.22. After introducing redox additives $VOSO_4$ and Na_2MoO_4, the specific capacitance and specific

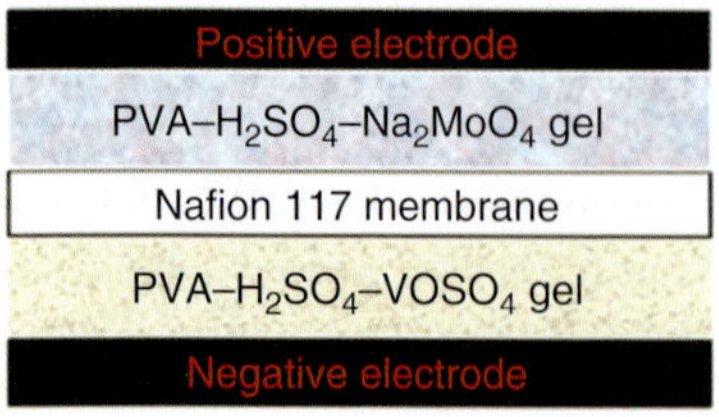

Figure 9.22 Schematic representation of EDLC configuration with PVA–H_2SO_4–$VOSO_4$/PVA–H_2SO_4–Na_2MoO_4 gel polymer electrolyte. Source: Fan et al. 2016 [171]. Reproduced with permission of Elsevier.

energy of the EDLC with PVA–H_2SO_4–$VOSO_4$/PVA–H_2SO_4–Na_2MoO_4 (separated by Nafion 117 membrane) reach to 543.4 F g^{-1} and 17.9 Wh kg^{-1} (0.5 A g^{-1}), respectively, which are about four times the values for the EDLC with a PVA–H_2SO_4 system.

9.4.2.2.1.2 CNTs-Based EDLCs An outstanding candidate for EDLC electrodes, alternative to ACs, is CNTs, first discovered by Iijima [239]. CNTs possess predominantly mesoporous interiors, and in general, they are available in the forms of single walled (SWCNTs) and multiwalled (MWCNTs). The mechanical strength, electronic conductivity, thermal conductivity, and optical properties of CNTs are attractive for various applications. The room temperature electrical conductivity of SWCNTs is of the order of 10^4 and 50 S cm^{-1} for MWCNTs films, aligned along the tube axis. Based on these properties, different CNT-based electrochemical devices have been reported, which include rechargeable batteries, fuel cells, and supercapacitors [240–243]. To be specific, CNTs are potential electrode materials for EDLCs due to their additional properties like chemical stability, low mass density, and reasonably high surface area (~400 m^2 g^{-1}) [239, 244]. CNTs are well known to have a cylindrical nanosized structure consisting of rolled graphene sheets built from sp^2 carbon atoms, with uniform mesopores (2–50 nm size) [245, 246]. Although the overall performance of SWCNTs is found to be better, MWCNTs are cost effective and favorable for practical applications, particularly in the fabrication of EDLCs. Few important polymer electrolyte-based EDLC configurations using SWCNTs or MWCNTs as electrodes are listed in Table 9.6.

Pandey and Hashmi recently reported the solid-state EDLCs, fabricated with nitric acid-treated MWCNTs using a novel ionic liquid-based GPE comprising 1-ethyl-3-methylimidazolium tris(pentafluoroethyl) trifluorophosphate (EMImFAP) immobilized in PVdF-HFP [226]. The acid-treated MWCNT electrodes offer the highest capacitance of ~76 F g^{-1}, the specific energy of ~17.2 Wh kg^{-1}, and specific power of ~18.9 kW kg^{-1} with high cycling stability up to ~10 000 cycles. Another solid-state flexible EDLC was reported, fabricated with CNTs (deposited on office paper) and ionic liquid-based GPE, comprising a mixture of fumed silica and IL 1-ethyl-3-methylimidazolium bis(trifluoromethyl-sulfonyl)imide (EMITFSI) [224]. The EDLC offered the specific capacitance of ~135 F g^{-1} at a current of 2 A g^{-1}. The optimum specific power and energy of the EDLC were 164 kW kg^{-1} and 41 Wh kg^{-1}, respectively. Gao et al. [227] reported an environment-friendly and cost-effective EDLC based on cellulose nanofibers (CNFs)/MWCNT hybrid aerogel electrodes with PVA–H_2SO_4 gel electrolyte. The one-dimensional CNFs effectively prevent the aggregation of MWCNTs and improve the utilization efficiency of the mesopores. The all-solid-state flexible EDLC exhibits the specific capacitance of ~178 F g^{-1} (measured from EIS) with excellent cyclic stability.

Yang et al. [229] recently reported an interesting work on the development of a highly stretchable, fiber-shaped EDLC, based on aligned CNT sheets and PVA–H_3PO_4 GPE. In this study, aligned CNT sheets and polymer electrolyte are sequentially wrapped on an elastic fiber to form coaxial EDLC, schematically shown in Figure 9.23A. Both rubber fiber substrate and GPE are stretchable so the

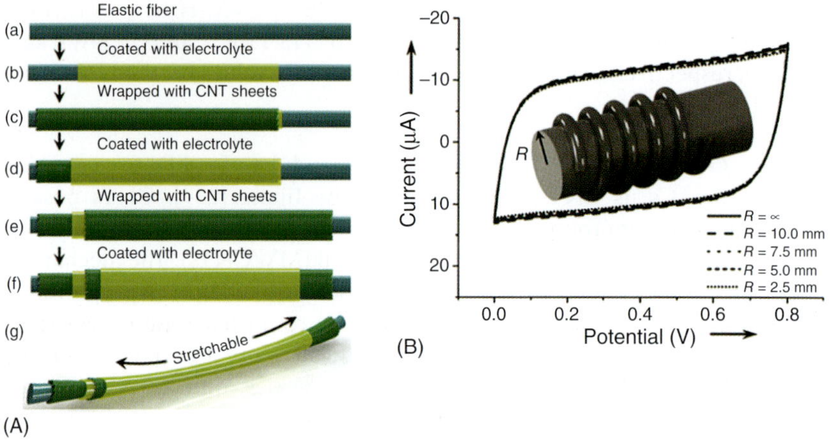

Figure 9.23 (A) Schematic representation of the fabrication of a highly stretchable, fiber-shaped EDLC with a coaxial structure and (B) CV curves of the fiber-shaped supercapacitor without bending and being bent with different radii *R* of curvatures. Source: Yang et al. 2013 [229]. Reproduced with permission of John Wiley & Sons.

EDLC is also stretchable. The specific capacitance varies from 11.0 to 19.2 $F\,g^{-1}$ (0.1 $A\,g^{-1}$) depending upon the thickness of CNT layers (from 110 to 330 nm).

The performance of the fiber capacitor has been tested under different bending and stretching conditions. The effect of bending on the performance has been tested by CV, as shown in Figure 9.23B. The shapes of CV curves are maintained to be almost the same under all the bending conditions, indicating the stability of the fiber EDLC.

Apart from many advantages, CNTs suffer from some drawbacks like poor wettability and dispersibility in different liquid media, which restrict their processing in various applications [247, 248]. Since the graphene layers are hydrophobic in nature and have attractive van der Waals forces between them, CNTs (the rolled form of graphene layers) have a tendency to be entangled in aqueous media and get agglomerated easily, thus showing poor dispersibility [249]. There exist various reports in which the CNTs are modified by functionalizing them with suitable functional groups to improve their dispersibility for easy processing [249, 250]. Ionic liquids (ILs) are recently reported as being used to modify the electrodes by several ways [251]. Fukushima et al. demonstrated the gel formation of SWCNTs with imidazolium cation-based IL, referred to as a "bucky gel of ionic liquids" [252]. Accordingly, the bundles of SWCNTs, which are heavily entangled in three-dimensional networks, are exfoliated during the process of gelation with the ILs [253–255]. EDLCs based on a gelled form of CNTs are rarely available in literature, except a few reports. An EDLC with a "bucky gel" of SWCNTs has recently been reported by Katakabe et al. [253], showing improvements in terms of high specific capacitance and lower electrode resistance as compared to pristine SWCNTs. Recently, our group [198] reported a "bucky gel" of MWCNTs prepared with an ionic liquid BMPTFSI, used as binderless electrodes in quasi-solid-state EDLCs. A GPE film consisting of

BMPTFSI, immobilized in PVdF-HFP, exhibiting a room temperature ionic conductivity of $1.5 \times 10^{-3}\,S\,cm^{-1}$, was used as electrolyte/separator in EDLCs. An improvement in specific capacitance (from 19.6 to $51.3\,F\,g^{-1}$) has been noted on replacing pristine MWCNTs by gelled MWCNT binderless electrodes. The gelled electrodes offer improvements in specific energy and power from 2.8 to $8.0\,Wh\,kg^{-1}$ and 2.0 to $4.7\,kW\,kg^{-1}$, respectively. Studies indicate that the gel formation of MWCNTs with ionic liquid is an excellent route to obtain high-performance EDLCs [198].

9.4.2.2.1.3 Graphene-Based EDLCs Interest has been raised in graphene-based materials due to their many advantageous properties that they are mechanically tough, highly elastic, nontoxic, and chemically and thermally tolerant [256, 257]. They exhibit superior electrical conductivity and a high charge carrier mobility ($\sim 20\,m^2\,V^{-1}\,s^{-1}$) [256, 258]. Very high theoretical specific surface area ($\sim 2630\,m^2\,g^{-1}$) and other properties make it attractive to be used in many promising applications including energy storage devices such as LIBs and supercapacitors [259, 260].

Different techniques including mechanical exfoliation of graphite, chemical vapor deposition, solvothermal synthesis and pyrolysis, wet chemical synthesis from graphite powder, liquid phase exfoliation, and oxidation of graphite are employed for the preparation of graphene [261, 262]. In general, the graphene powder is a mixture of monolayers, bilayers, and multilayers (3–10 monolayers) in the form of irregular structured flakes or flat sheets [256, 263]. The Hummer's method is most popular chemical method of graphene preparation, in which graphite is converted to graphene oxide (GO). For details, Refs [264] and [265] should be referred. The chemical reduction of GO sheets is performed using different reducing agents including hydrazine [266, 267], sodium borohydrate [267], and hydroquinone [268] to obtain reduced graphene oxide (rGO). The rGO contains a significant amount of oxygen and, possibly, significant numbers of defects. Recently, few workers produced graphene by electrochemical approach [256].

Another graphene-based structure, referred to as graphene nanoplatelets (GNPs), which is basically the multilayer graphene, has attracted significant interest. GNPs have drawn attention from the applications point of view because of their good mechanical properties ($\sim 1\,TPa$), excellent electrical conductivity ($\sim 10^6\,S\,cm^{-1}$), high thermal conductivity ($\sim 5000\,W\,m^{-1}\,K^{-1}$), and large surface area [250, 269]. Due to large surface area and high electrical conductivity, GNPs are suitable as electrodes for EDLCs/supercapacitors [250, 269–271]. Few important polymer electrolyte-based EDLCs, employing GO, rGO, and GNPs as electrodes, are listed in Table 9.6. To improve the performance of graphene-based materials as electrodes, their composites with conducting polymers and metal oxides have also been reported, which are also mentioned in Table 9.6.

Suleman et al. recently reported solid-state EDLCs fabricated with GO and rGO electrodes, and plastic crystal-based flexible GPE films [161]. The GO and rGO electrodes showed high-rate supercapacitive performance with SN-based GPEs. Due to the excessive oxygen functionalities in GO electrodes, additional pseudocapacitance results in higher specific capacitance and specific energy (~66 F g^{-1} and 18 Wh kg^{-1}, respectively) as compared to rGO electrodes (specific capacitance ~60 F g^{-1} and specific energy ~15.6 Wh kg^{-1}), measured at 1.7 A g^{-1}. The rGO electrodes offer about three times higher rate performance with respect to GO-electrodes, as indicated from the parameters namely: knee frequency f_k, response time τ_0, and pulse power P_0 (observed from EIS studies) and confirmed from the rectangular CV shapes up to scan rates of 5 V s^{-1} (for GO) and 10 V s^{-1} (for rGO). The rGO-based EDLC shows higher specific power (54.9 kW kg^{-1}) as compared to that of GO-based EDLC (33.3 kW kg^{-1}) [161].

In another study, Suleman et al. reported quasi-solid-state EDLCs based on activated GO/rGO electrodes and GPE film comprising plastic crystal (SN) and IL (triethylsulfonium bis(trifluoromethylsulfonyl)imide) mixture entrapped in polymer PVdF-HFP [272]. The 3D architecture of mesoporous activated rGO, in combination with SN-based GPE, shows a high rate performance with high value of knee frequency f_k (~966 Hz) and high pulse power P_0 (~111 kW kg^{-1}). The activated rGO-based EDLCs offer higher specific power (~118 kW kg^{-1}) than that for GO-based EDLC (~41 kW kg^{-1}), whereas the activated GO-based EDLCs show higher specific capacitance and specific energy (~88–96 F g^{-1} and 25 Wh kg^{-1}, respectively, at 1.1 A g^{-1}) as compared to the activated rGO-based EDLCs.

A solid-state, flexible EDLC has recently been reported by Singh et al. [157] employing GNPs as electrodes and free-standing GPE film containing 1 M LiTFSI in SN, immobilized in PVdF-HFP. The EDLC offers high rate capability, indicated from the EIS analysis in terms of high knee frequency (~58 Hz), low response time (~791 ms), and high pulse power (~16.4 kW kg^{-1}). The CV response further confirms the high rate performance as the EDLC shows rectangular pattern up to the scan rate of 1000 mV s^{-1} (Figure 9.24A). Galvanostatic charge–discharge patterns (Figure 9.24B) show the characteristics of EDLCs. Ragone plot of the EDLC is depicted in Figure 9.24C, which shows a typical behavior of a power source. The specific energy of the solid-state EDLC is low and not much attractive (E_{max} ~ 8.2 Wh kg^{-1}); however, it is comparable to few solid-state EDLCs, reported in literature.

9.4.2.2.2 Pseudocapacitors or Redox Supercapacitors As discussed earlier in Section 9.2.2, pseudocapacitors are an important class of supercapacitors that involve primarily fast redox reactions or intercalation–de-intercalation at the electrolyte/electrode interface. Generally, the pseudocapacitance per unit mass of active material is almost ten times higher than that of the double layer capacitance associated with carbon [273]. The main possible reasons for such observation are (a) high surface area, (b) contribution of the bulk of the material to charge storage, and (c) low resistance to electric current. The pseudocapacitive electrode materials are mainly conducting polymers and transition metal oxides (TMOs) or mixed transition metal oxides (MTMOs), presented as follows.

Among many available conducting polymers, polypyrrole, polythiophene and its derivatives (e.g. poly(3,4-ethylenedioxythiophene) (PEDOT), poly(3-methylthiophene) (PMeT), etc.), and polyaniline (PANI) are popular capacitor electrodes due to their various advantageous properties. Main advantages

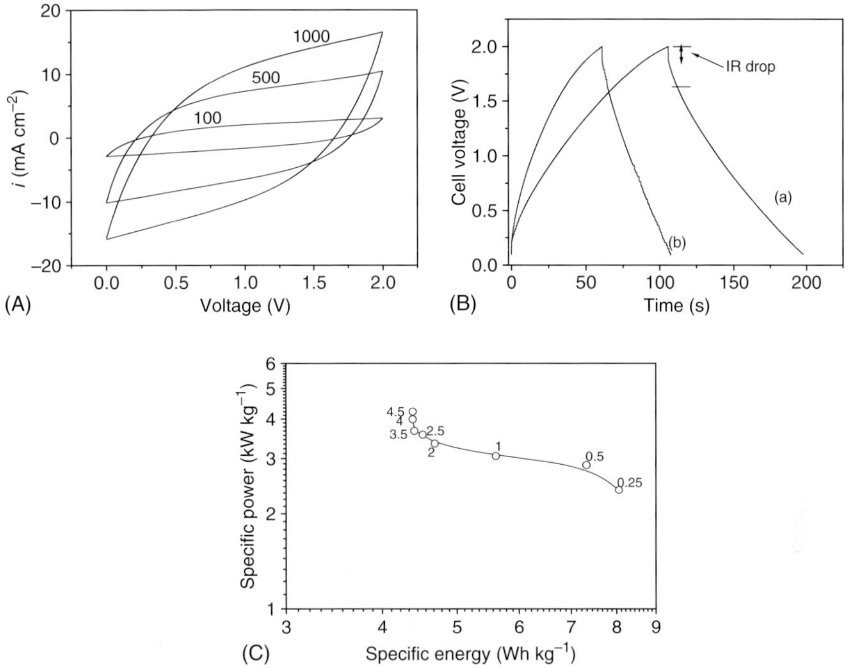

Figure 9.24 (A) CV responses of EDLC at different scan rates (the scan rates in mV s^{-1} are marked on each curve), (B) galvanostatic charge–discharge curves at current loads of (a) 0.25 A g^{-1} and (b) 0.5 A g^{-1}, and (C) Ragone plot where the current density (in A g^{-1}) are marked on each point. Source: Singh et al. 2015 [157]. Reproduced with permission of Elsevier.

of conducting polymers are their quick doping–de-doping, low density, low toxicity, low cost, sufficient voltage window, flexibility, and easy formability; however, their disadvantages include low cycling stability (<1000 cycles), particularly for n-doped polymers [25]. This disadvantage is primarily due to their swelling and deswelling upon charge and discharge. The observation of low capacitance compared to other redox active materials is another noted disadvantage. To circumvent these shortcomings, their composites with other redox materials, e.g. TMOs and carbonaceous materials, have been extensively reported [274–276].

TMOs are another class of electrode materials offering pseudocapacitance according to different mechanisms, namely, underpotential deposition at the surface, redox, and intercalation pseudocapacitance, as discussed in Section 9.2.2. TMO redox electrodes like RuO_x, IrO_x, Co_3O_4, NiO_2, Fe_2O_3, and MnO_2 [20, 277–280] undergo reversible reaction during charge–discharge accompanied with the movement of electrons and H^+ ions across electrode/electrolyte interface. A reaction for hydrous RuO_2 is given below [129]:

$$RuO_a(OH)_b + \delta H^+ + \delta e^- \leftrightarrow RuO_{a-\delta}(OH)_{b+\delta} \qquad (9.22)$$

In such reactions, the oxidation states of the transition metals change during the reduction or oxidation reactions, for example, $Ru^{+3} \leftrightarrow Ru^{+4}$, $Ru^{+4} \leftrightarrow Ru^{+5}$,

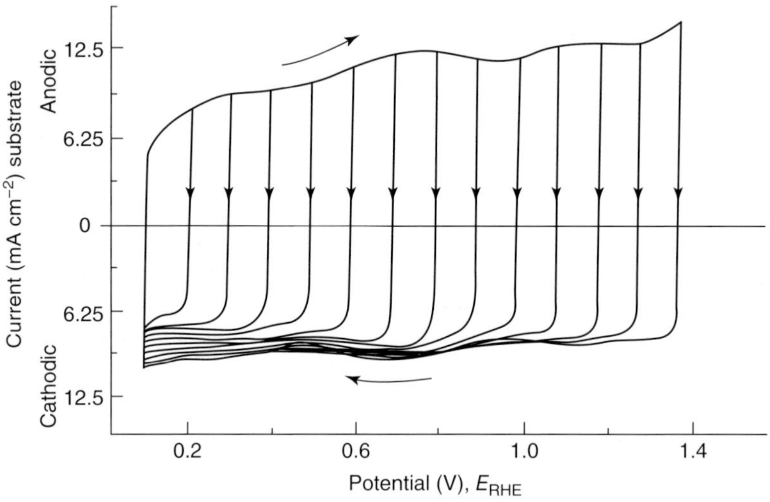

Figure 9.25 CV pattern of RuO_2 film electrode in 1 M H_2SO_4 showing reversible redox processes. Source: Conway 1999 [2]. Reproduced with permission of John Wiley & Sons.

$Mn^{+3} \leftrightarrow Mn^{+4}$, etc. Metals like ruthenium undergo multiple transitions in their oxidation states during a redox cycle and this can be seen from the nature of their CV curves, which are quasi-rectangular (Figure 9.25): i.e. clear redox peaks are not present due to overlap of one redox process with another.

A brief description of the conducting polymers and TMO-based capacitors with some references of solid- or quasi-solid-state pseudocapacitors are given as follows.

9.4.2.2.2.1 Conducting Polymer-Based Supercapacitors The doping in the conducting polymers is of p- or n-types [281], and depending on the type of conducting polymer being used, the pseudocapacitors are categorized in three different classes:

(i) Symmetric p–p type (which may be further divided on the basis of the p-type polymer used on either side, same or different polymer) [281]
(ii) Asymmetric p–n type [281]
(iii) Hybrid with p-type polymer as positive electrode and a carbon negative electrode [282].

The symmetric p–p type configuration offers low working voltage. From this point of view, the asymmetric p–n type configuration is attractive and has potential to offer high voltage range (~3 V); however, it was not very successful due to the unstable nature of n-type electrodes. The hybrid p-type polymer/carbon configuration replaced the p–n type due to this reason and is well reported.

A number of conducting polymer-based pseudocapacitors, employing mostly liquid electrolytes, have been reported [283]. Only a few reports exist on solid-state conducting polymer-based pseudocapacitors incorporating polymer-based electrolytes. Among such few important works, one of the early reports is that of Hashmi et al. [284] who studied polypyrrole-based supercapacitors (p–p type configuration) with proton and Li ion conducting polymer electrolytes. The electrodes were prepared electrochemically by depositing polypyrrole (PPy) on two substrates (ITO and high density flexible graphite sheets) and the polymer electrolytes were PVA–H_3PO_4 and PEO–PEG–$LiCF_3SO_3$. The polymer electrolytes showed room temperature conductivities of the order of 10^{-4} to 10^{-3} S cm^{-1}. The capacitor cells were characterized using EIS, CV, and galvanostatic charge-discharge (GCD) techniques. From the EIS plots, it has been observed that the ITO-coated PPy system with PVA–H_3PO_4 showed a highly capacitive nature [284]. The specific single electrode capacitance of this system was also highest among all the cells (84 F g^{-1}). Tripathi et al. [285] also studied capacitor cells using the same PPy electrodes and GPEs prepared by immobilizing solutions of $LiClO_4$, $NaClO_4$, or $(C_2H_5)_4NClO_4$ in EC-PC binary mixture in PVdF-HFP. The EIS, CV, and GCD plots showed typical capacitive behavior. The cells were tested for their cyclic durability and were found to have highly stable capacitance for ~5000 charge–discharge cycles after showing negligible initial fading. Interestingly, the type of cations in the electrolyte was not found to have significant effect on the electrochemical performance of the cells. Independent of the cations, all the cells offer almost same capacitance (~130 F g^{-1}), specific energy (19 Wh kg^{-1}), and specific power (0.44 kW kg^{-1}), measured at 0.34 A g^{-1}.

Prasad and Munichandraiah [286] tested PANI, deposited on stainless steel substrate, employing PAN/$LiClO_4$-based GPE film. The supercapacitor showed a high capacitance of 350 F g^{-1}(10 A g^{-1}), and high energy and power densities of 80 Wh kg^{-1} and 7.5 kW kg^{-1}, respectively. These values are important as they were obtained by enhancing the voltage limit of the cells by replacing their own previously reported aqueous electrolytes (0–0.75 V) with GPE (0–1 V).

In general, pure conducting polymer electrodes offer low rate capability and poor cyclic performance. Attempts have been made to address this issue by making composites of conducting polymers with carbonaceous materials (AC, CNTs, graphene, etc.). For example, Meng et al. [287] prepared PANI/CNT nanocomposite redox electrodes and utilized them in mechanically stable and flexible solid-state symmetric supercapacitors using GPE PVA/H_2SO_4. These devices show exceptional flexibility and display no change in electrical performance even on bending and twisting the cells. The high flexibility is one of the more attractive features due to which polymer electrolytes have gathered so much interest. The capacitor shows a high specific capacitance of 350 F g^{-1}(1 A g^{-1}). CV and charge–discharge profiles of the capacitor are shown in Figure 9.26a,b. It may be noted that corresponding to the redox peaks in CV pattern, plateau regions at the same voltages are seen in the charge–discharge curves (Figure 9.26b). The decay of capacitance with increasing current is not very significant indicating the high rate capability of the capacitors. Further, the fading in capacitance after 1000 cycles was only about 8% and Coulombic efficiency is about 100% (Figure 9.26c), highlighting the high cycling stability of the PANI/CNT composite-based cell.

Similarly, Yang et al. [288] obtained high capacitance and high cyclic stability for paper like electrodes made of PANI-PSS nanospheres stacked between

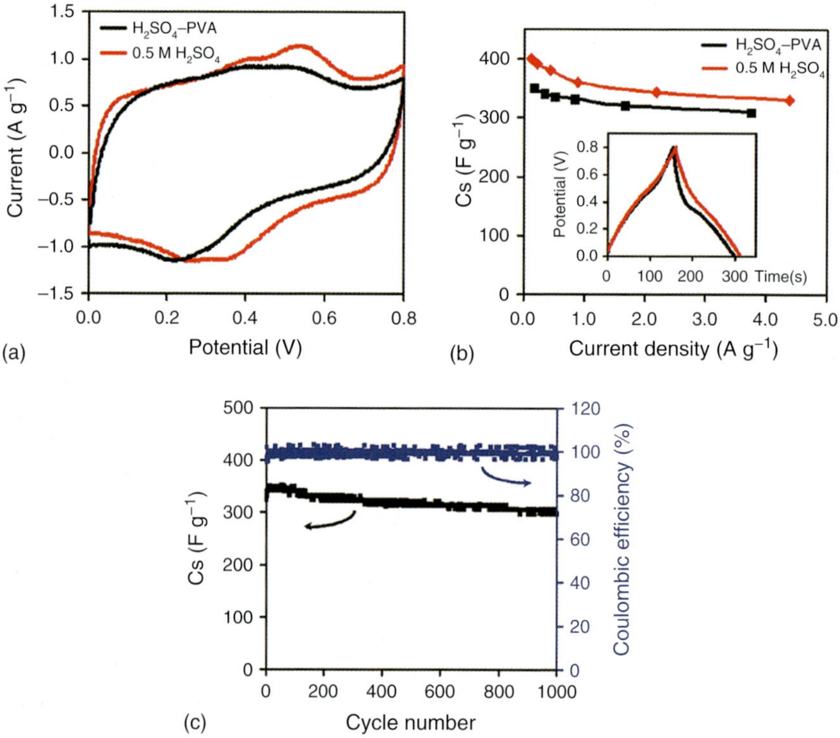

Figure 9.26 Comparison of (a) cyclic voltammograms and (b) charge–discharge performance of PANI/CNT pseudocapacitors with H_2SO_4–PVA gel electrolyte and 0.5 M aq. H_2SO_4 solution. (c) Cyclic stability of device with gel polymer electrolyte at 1 A g^{-1}. Source: Meng et al. 2010 [287]. Reproduced with permission of American Chemical Society.

layers of rGO sheets using the ability of PANI-PSS to stably disperse in a solution; the nanospheres have a very narrow size distribution, which enhances the stability of the stacking in the paper. Symmetric pseudocapacitor cells, fabricated using these electrodes and GPE PVA–H_2SO_4, are able to give a high areal capacitance of 217 F cm^{-2} and are able to retain up to 80% capacitance after 10 000 charge–discharge cycles, which is a higher value as compared with the values for other conducting polymer-based capacitors.

More recently, thiophene nanosheets and exfoliated graphene were reported in a stacked layer heterostructure geometry to make electrodes and used with GPE PVA–H_2SO_4 to produce symmetric supercapacitor cells, which offer high volumetric capacitance (375 F cm^{-3}) with high rate capability (scan rate: 1000 V s^{-1}) [289]. The stacked geometry of the cells offers rapid diffusion of ions leading to high rate capability and proper electron transfer through the entire bulk of the electrode leading to high capacitance.

9.4.2.2.2.2 Pseudocapacitors Based on Transition Metal Oxides (TMOs) and Mixed TMOs First pseudocapacitor was reported by Zheng et al. [273] who fabricated the cells consisting of $RuO_2 \cdot xH_2O$ powder electrodes and an aqueous electrolyte

0.5 M H_2SO_4. The ruthenium oxide was found to be electrically conducting ($\rho = 3.23\,m\Omega\,cm^2$), which is an advantageous property as electrode material. The best pseudocapacitor showed higher specific capacitance of $\sim$720 F g^{-1} with respect to the capacitance of similar system made with carbon electrodes and underlined the importance of redox processes for electrochemical charge storage. A lot of work has been reported on solid-state pseudocapacitors by varying the type of electrode materials, i.e. TMOs or conducting polymers, different combination of one oxide with the other oxide as high surface area support, composites of oxides with conducting polymers, composites of oxides and carbon materials, or that of conducting polymers with carbon materials. Carbon materials may serve as conducting additives and/or as structural supports in case of CNTs and fibers.

Ruthenium oxide is in general considered the best material for pseudocapacitors due to their high specific capacity, high electronic and protonic conductivity, and high cycling stability [290]. Many other oxides have also been tried due to the high cost involved with ruthenium oxide. The TMO-based redox active electrodes offer better capacitance value, electronic conductivity and electrochemical stability, and superior cycling durability than conducting polymer-based electrodes. Apart from reversible faradaic reactions, they also store part of their charge through the formation of a double layer [25]. TMO-based pseudocapacitors, utilizing SPE/GPE, are less reported. Nevertheless, they form a niche of their own in the supercapacitor family wall. Some examples of pseudocapacitors employing TMOs or their composites with carbon and conducting polymers are presented as follows.

Choudhury et al. [291] reported symmetric pseudocapacitors with PVA/PAA blend hydrogel electrolytes doped with $HClO_4$, NaOH, and NaCl, and RuO_2/C composite electrodes. CV studies, performed between −0.2 and 0.8 V, showed rectangular curves as expected for RuO_2 electrodes and a high capacitance of 1000 F g^{-1} was found with acidic hydrogel electrolytes. EIS study also revealed excellent capacitive behavior (phase angle of 82°) with lower charge transfer resistance (32 $\Omega\,cm^2$), indicating good compatibility of acidic PVA/PAA blend hydrogel electrolytes with the RuO_2 electrodes. Lian et al. [292] reported asymmetric pseudocapacitor cells with RuO_2 (cathode)/graphite (anode) separated by a PVA–HPA-based GPE. The cells offered a specific capacitance of 24 mF cm^{-2} (5 mA cm^{-2}) and stable performance till 5000 charge–discharge cycles in the potential window of 1–1.5 V.

Recently, Wang et al. [293] reported symmetric pseudocapacitors incorporating RuO_2/TiO_2 nanotube composite electrodes with an aqueous PVA–H_3PO_4 GPE. Four different weight percents (4, 10, 15, and 21 wt%) of $RuO_2{\cdot}xH_2O$, loaded in TiO_2 nanotubes, were tested in the cells. The CV studies were carried out at 5 mV s^{-1} scan rate between −0.2 and 0.85 V. The electrode containing 21 wt% $RuO_2{\cdot}xH_2O$ showed the maximum capacitance. The pseudocapacitance arising due to the RuO_2 component was evaluated to be between 1263 and 1095 F g^{-1} (0.7 A g^{-1}) of RuO_2 for each electrode in the cell. Metal oxide pseudocapacitors have good cycle life as there is no phase change involved and all the reactions are reversible with a simple movement of H^+ ions and this is seen in the cycling stability studies of cells based on $RuO_2{\cdot}xH_2O/TiO_2$ nanocomposite electrodes [293].

Sellam et al. [129] presented a symmetric, flexible pseudocapacitor using a RuO_2 composite with poly(3-methyl thiophene) (pMeT) as electrode and a solid gel polymer electrolyte comprising an ionic liquid EMITf and ammonium trifluoromethanesulfonate (NH_4Tf) trapped in a matrix of PVdF-HFP. The gel electrolyte film showed excellent room temperature conductivity of ~23 mS cm^{-1} and excellent electrochemical stability. The pseudocapacitor showed a high specific capacitance of 118 F g^{-1} and specific energy and power of 12.8 Wh kg^{-1} and 513 W kg^{-1}, respectively, measured at 0.37 A g^{-1}. The CV studies at different scan rates and different concentrations indicated the slow diffusion process of H^+ in RuO_2, mostly attributed to the high viscosity of the ionic liquid. The charge–discharge profiles were found fairly triangular and the increase in the maximum voltage with the RuO_2 to pMeT ratio indicating the dominant redox reactions at RuO_2 over the doping–de-doping of the conducting polymer. The cell showed excellent cyclic stability for as much as 5000 charge–discharge cycles after some initial fading, attributed to irreversible capacitance loss arising from the reaction of some loosely bound surface groups on the electrode surface. The same research group reported another symmetric all-solid-state pseudocapacitor with PEDOT-PSS/$RuO_2 \cdot x$HO composite electrodes and PVP–PVA–$EMIHSO_4$ GPE [294] offering very high rate performance with a maximum specific capacitance of 70 F g^{-1} (0.1 A g^{-1}), retaining that capacitance value for 1000 cycles.

MnO_2 is another low cost and environmentally nonhazardous material employable as electrodes in redox (pseudo-) capacitors [295]. One of the important features of MnO_2 is that it does not require specifically acidic electrolytes to show supercapacitive performance. Apart from its high specific capacitance, it also shows fast, reversible reactions allowing for high rate capability and hence high power density. A flexible solid-state pseudocapacitor utilizing PVP–$LiClO_4$ gel electrolyte with MnO_2 electrodes was recently reported by Chodankar et al. [296].

The electrodes were deposited on stainless steel substrates using the cost-effective chemical bath deposition method. The CV pattern of symmetrical pseudocapacitor cell shows a capacitance of 110 F g^{-1} at a scan rate of 5 mV s^{-1}. The decrease in capacitance with increasing scan rate has been reported, attributed to the diffusion-controlled charge transfer at the electrode/electrolyte interface, an important aspect to be considered from solid-state electrolyte point of view. Energy and power density evaluated from charge–discharge at ~1 A g^{-1} are ~23 Wh kg^{-1} and ~1.92 kW kg^{-1}, respectively, higher than similar reported systems [297, 298]. The system also showed excellent cycling stability with 92% retention of the initial capacity after 2000 cycles. Another highly flexible solid-state pseudocapacitor was fabricated by Yuan et al. [299] using MnO_2 nanorods and carbon nanoparticle hybrid electrodes with PVA–H_3PO_4 electrolyte. The composite electrodes were prepared using a flame synthesis procedure on a porous carbon fabric structure. The microporous texture, formed by the carbon nanoparticles on the fabric's surface, allows not only better mechanical properties but also better interaction with the gel electrolyte. The cells showed excellent specific capacitance of 800 F g^{-1} and ~97% retention after 20 000 charge–discharge cycles.

Few other pseudocapacitors with TMOs/hydroxides and their carbon composites as electrodes with GPEs also showed promising results in terms of specific capacitance and cyclic stability. For example, β-$Ni(OH)_2$-graphene nanohybrid thin film with alkaline PVA–KOH electrolyte, reported by Xie et al. [300], showed an areal capacitance of $3.2\,F\,cm^{-2}$ ($0.1\,A\,m^{-2}$) and significant stability for 2000 cycles. Liao et al. [301] reported electrodes of CoO_3 along with vertically aligned graphene nanosheets deposited on carbon fabric and fabricated symmetric supercapacitor with PVA–KOH polymer electrolyte. The highest capacitance achieved was $580\,F\,g^{-1}$ ($1\,A\,g^{-1}$) and the cells retained 86% of initial capacity even after 20 000 cycles.

Vanadium oxides are reported to be good electrode materials due to their higher availability and lower cost, so, despite the lower capacitance, these are attractive options for pseudocapacitors. However, vanadium oxides suffer from poor cyclic stability due to pulverization of the electrode material and tend to dissolve in acidic or basic environments [302]. Considering these merits and demerits, a novel solid-state pseudocapacitor based on vanadium oxide ($V_3O_7{\cdot}xH_2O$) nanowires was reported by Wang et al. [302] with a LiCl/PVA composite GPE. The gel electrolyte in the system minimizes the water content and maintains a neutral pH while providing the nanowires with a flexible and mechanically stable matrix that prevents pulverization during charge–discharge. Redox peaks are distinctly observed in CV patterns (Figure 9.27a) indicating the pseudocapacitive nature of the device (as claimed by the authors). The galvanostatic charge–discharge plots (Figure 9.27b) show nonlinear behavior, which also highlights the non-Faradaic processes happening in the system. The capacitance for a single electrode was found to be $262\,F\,g^{-1}$ ($0.22\,A\,g^{-1}$). The cyclic stability was also found to be high at 85% capacitance retention after 5000 cycles.

MTMOs are an emerging class of materials for use as pseudocapacitor electrodes [303]. Many such MTMOs have been tested for their electrochemical performance, reported in literature [295, 304–307]. The capacitance of these reported oxides varies from 200 to $1000\,F\,g^{-1}$ in neutral or basic aqueous

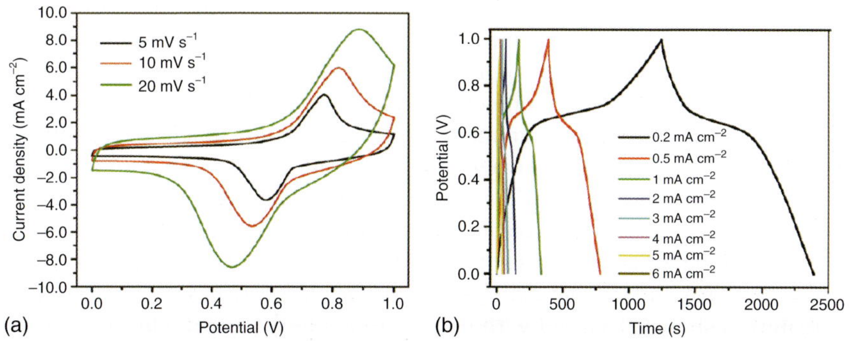

Figure 9.27 (a) Cyclic voltammogram for vanadium oxide pseudocapacitors with LiCl/PVA gel electrolyte at different scan rates in two-electrode geometry and (b) galvanostatic charge–discharge plots of the pseudocapacitor, measured in the gel electrolyte at various current densities. Source: Wang et al. 2012 [302]. Reproduced with permission of American Chemical Society.

electrolytes and almost all offer good cyclic stabilities. The introduction of another transition metal in TMO increases the available oxidation states and so the number of possible redox processes, whose synergetic effects increase the pseudocapacitance as well as the electrochemical potential window of the devices. However, MTMOs suffer from the drawback of low conductivity and ion transport [303]. Such drawbacks can be removed by making composites with carbon materials such as CNTs as reported by Basu et al. [308], where the CNTs drastically reduced the charge transfer resistance seen in electrodes. They reported micro-supercapacitor based on the composites of $CuCo_2O_4$ and CNTs and tested with PVA–H_3PO_4 GPE. This solid-state supercapacitor showed superior areal capacitance (2.29 mF cm^{-2}) compared with other similar capacitors reported elsewhere and was also highly stable for 9000 cycles without showing any significant fading. Wang et al. [309] tested $NiCo_2O_4$ nanoarrays on various conducting substrates with PVA–KOH gels and reported a high areal capacitance of 161 mF cm^{-2} with stability up to 3000 cycles. Most recently, another solid-state pseudocapacitor employing MTMO electrodes was reported by Raut et al. [310], who synthesized a $ZnFe_2O_4$ electrode using a simple ionic layer adsorption and reaction (SILAR) process. When used with a PVA–$LiClO_4$ gel electrolyte, the resulting pseudocapacitor yielded a specific capacitance of 471 F g^{-1}, and specific power and energy of 277 W kg^{-1} and 4.47 Wh kg^{-1}, respectively, measured at 1 A g^{-1}.

The TMO-based capacitors, discussed earlier, can be classified in two categories according to the electrochemical behavior at the electrode/electrolyte interfaces and nature of profiles observed in CV and galvanostatic charge–discharge patterns:

(i) Unlike carbon electrodes in EDLCs (whose charge storage mechanism is electrostatic in nature), the electrodes like $RuO_2 \cdot xH_2O$ and MnO_2, and their composites with carbonaceous materials or conducting polymers, show fast electrochemical (redox) charge transfer process. However, they show featureless rectangular shaped CV patterns and almost linear galvanostatic discharge profiles. According to Conway [2], in such materials "the pseudocapacitance arises at the electrode surfaces where a completely different charge-storage mechanism applies. It is faradaic in origin, involving the passage of charge across the double layer, as in battery charging or discharging, but capacitance arises in account of the special relation that can originate for thermodynamic reasons between the extent of charge acceptance (Δq) and the change of potential (ΔV), so that the derivative $d(\Delta q)/d(\Delta V)$ or dq/dV, which is equivalent to a capacitance, can be formulated and experimentally measured by dc, ac, or transient techniques."

(ii) On the other hand, the vanadium oxide or $Ni(OH)_2$ like active electrode materials show CV curves with distinct redox peaks and nonlinear charge–discharge profiles, generally containing short plateau regions, as typically shown in Figure 9.27. These materials do not satisfy the defining conditions for pseudocapacitors as discussed earlier; however, there exist many reports claiming such materials as pseudocapacitive electrodes. This is possibly due to their fast rate capabilities like any other supercapacitors, despite their battery like CV and charge–discharge behavior. According to recent reports addressing this issue, such devices should be categorized separately from pseudocapacitors [311–313]. The calculation of energy and power density for such devices should be performed as suggested by Akinwolemiwa et al. [311]. Also, the parameter "capacitance" should be replaced by a suitable parameter such as "capacity" for such devices, as the value of capacitance is different in different regions of the CV or charge–discharge profile.

9.4.2.2.3 Asymmetric/Hybrid Supercapacitors In order to improve the performance of supercapacitors, particularly their specific energy, various approaches have been adopted, discussed as follows. In general, the specific energy of a system can be improved either by improving specific capacitance and/or by expansion of operating voltage range [314]. According to a large number of reports, the asymmetric/hybrid supercapacitor configurations not only improve the values of specific capacitance (or capacity) but also widen the operating voltage range. As per the recent report of Brousse et al. [312], "the term "hybrid" supercapacitor should be used when pairing two electrodes with different charge storage behavior, i.e. one capacitive and one Faradaic, and the resulting device is in-between a supercapacitor and a battery. "Asymmetric" supercapacitor covers a wider range of electrode combinations because it can be used for supercapacitors using electrodes of the same nature but with different mass loading, or two electrodes using different materials. We suggest the term "asymmetric" should be used only when capacitive or pseudocapacitive electrodes are involved (such as AC//MnO_2 asymmetric supercapacitor) in order to avoid confusion with true "hybrid" devices. Various reports in literature suggest another type of devices, which have symmetric configuration, but electrodes, employed in the devices, are hybrid in nature comprising a composite of capacitive/battery electrode materials. These hybrid/composite electrodes have the ability of manifold enhancement in specific capacitance. Thus, three types of asymmetric/hybrid supercapacitors are under recent investigations, described in the following sections. Few important examples of these types of asymmetric/hybrid supercapacitors and their characteristics are given in Table 9.7.

9.4.2.2.3.1 Asymmetric Supercapacitors This type of capacitors utilize two carbon materials of different nature or two dissimilar conducting polymers/capacitive metal oxides or the configuration comprising one carbon electrode and the other conducting polymer/metal oxide electrode [316–318, 331–336]. Such asymmetric devices exhibit wider voltage range and hence improved specific energy. The charge storage mechanism of both the electrodes may be electrostatic or pseudocapacitive due to fast and near-surface redox reaction of different rates. In recent years, the coupling of an AC as negative electrode and a conducting polymer as positive electrode has received maximum attention to achieve higher energy, power, and better cycling stability than symmetric supercapacitors [285, 337, 338]. However, the issues related to instability in the conducting polymer-based devices cannot be ruled out. Lu et al. [319] recently reported a quasi-solid-state asymmetric supercapacitor based on vanadium nitride (VN) as anode and vanadium oxide (VO_x) nanowire as cathode separated

Table 9.7 Few important examples of asymmetric/hybrid and composite supercapacitors, and their characteristics.

S. No.	Cell configuration	Electrolyte	Voltage range (V)	Capacitance ($F\ g^{-1}$)	Specific energy ($Wh\ kg^{-1}$)	Specific power ($kW\ kg^{-1}$)	References
Asymmetric supercapacitors							
1.	PNDC//G-PANI	$BMIMPF_6$	—	194	114 ($2\ A\ g^{-1}$)	13.7 ($2\ A\ g^{-1}$)	[315]
2.	AC//NiO	PVA–KOH–H_2O	0–1.5	73.4 ($0.1\ A\ g^{-1}$)	26.1 ($0.1\ A\ g^{-1}$)	—	[316]
3.	AC//PMT	PEO-$(C_2H_5)_4NBF_4$	—	18 ($16\ A\ m^{-2}$)	10.2	0.096	[317]
4.	CNTs//MnO_2	PVA/Na_2SO_4/Silica	—	56	—	—	[318]
5.	VN//VO_x	PVA/LiCl	1.8	—	0.61 $mWh\ cm^{-3}$	0.85 $W\ cm^{-3}$	[319]
Hybrid supercapacitors							
1.	Fe_3O_4//CNTs	PVA/KOH Gel	0–1.6	—	1.56 $mWh\ cm^{-3}$	0.028 $W\ cm^{-3}$	[320]
2.	$Li_4Ti_5O_{12}$//LAC-25	LiTf/EC:PC/PVdFHFP	0–3.0	—	36	27	[321]
3.	MWCNTs// $LiFePO_4$@MWCNTs	LiTFSI/BMPTFSI/PVdF-HFP	0–2.5	—	19.8	524	[322]
4.	NPC//MPG	PVdF-HFP/$NaClO_4$ EC/DMC/DEC	4.2	—	168	501	[323]
Composite-type supercapacitors							
1.	PANI/CNT	PVA/H_2SO_4	0–2.5	350 ($1\ A\ g^{-1}$)	7.1 ($0.1\ A\ g^{-1}$)	2.19 ($4\ A\ g^{-1}$)	[287]
2.	GO/P(EO-*co*-PO)	PEO-PPO/$TEABF_4$/PC		130	—	—	[324]
3.	Graphene/ZnO	PVA/$NaNO_3$	0–1.0	156	21.7	2.6	[325]
4.	rGO–PMo_{12}	—	—	—	1.07 $mWh\ cm^{-3}$	12 $mW\ cm^{-3}$	[326]
5.	CNPs/MnO_2	PVA/H_3PO_4	0–0.8	6.1 $mF\ cm^{-2}$	4.8	14	[299]
6.	AC//CoS	PVA/KOH	0–1.8	47 ($2\ A\ g^{-1}$)	5.3 ($2\ A\ g^{-1}$)	1.8($2\ A\ g^{-1}$)	[327]
8.	rGO/Fe_3O_4/PANI	—	—	283 ($1\ A\ g^{-1}$)	47.7 ($0.5\ A\ g^{-1}$)	550 ($50\ A\ g^{-1}$)	[328]
9.	PTMA/$LiFePO_4$	—	2.5–4.0	130 $mAh\ g^{-1}$	—	—	[329]
10.	ZnO/AC	0.1 M Na_2SO_4 (aq.)	0–1.0	93 ($2\ mV\ s^{-1}$)	—	—	[330]

PANI, poly(aniline); PEO, poly(ethylene oxide); PPO, poly(propylene oxide); $TEABF_4$, tetraethylammonium tetrafluoroborate; CNP, carbon nanoparticle; PTMA, poly(2,2,6,6-tetramethyl-1-piperinidyloxy-4-yl methacrylate); PNDC, phosphorus and nitrogen co-doped carbon; PMT, poly(3-methyl thiopene); VN, vanadium nitride; NPC, nanoporous carbon; MPG, macroporous graphene.

by PVA-LiCl GPE. This assembly offered high energy density of 0.61 mWh cm^{-3} and high power density of 0.85 W cm^{-3} at current density of 5 mA cm^{-2}. Only 12.5% fading in capacitance was observed after 10 000 charge–discharge cycles.

9.4.2.2.3.2 Hybrid Supercapacitors The most popular and effective configuration of supercapacitors under investigation these days has two different technologies incorporated in the same device [339]. A capacitor electrode coupled with a battery electrode via a suitable electrolyte is referred to as hybrid supercapacitors, also popularly named as BatCap systems [340, 341]. This approach enables both Faradaic and non-Faradaic (i.e. electrostatic) processes to store charges in a single device. The capacitive electrode can either be EDLC-type or pseudocapacitive, and the other electrode can either be an anode or cathode [314]. The energy storage process in hybrid BatCaps is as follows. During charging or discharging, anions and cations conduct to the two respective electrodes. The bulk redox reactions occur at the battery-type electrode while ion accumulation (fast redox reaction) takes place at the capacitive electrode.

Different hybrid BatCap systems have been reported with different capacitor electrodes and electrodes derived from batteries: lead acid (lead oxides), alkaline (nickel oxides), and lithium (titanate, manganese oxide, and others) [320–323, 342–344]. Initial focus was on lead oxide, nickel hydroxide, and $Li_4Ti_5O_{12}$ (LTO) as one electrode, and AC, graphene, etc., as the other electrodes. Various other combinations of battery hybrid have also been studied and reported in literature. One of the important examples is the report by Pell and Conway [345], who proposed the hybrid capacitors comprising $PbO_2/PbSO_4$ positive electrodes (used in lead-acid batteries) and large surface area carbon as capacitive electrodes separated by aqueous sulfuric acid. Such system offered high energy efficiency (63–67%), high Coulombic efficiency (90–95%), and high cyclic performance. Another hybrid configuration of CNT/$LiPF_6$/TiO_2-B supercapacitor [346] showed an energy density of 12.5 Wh kg^{-1} at a rate of 10 C over a voltage region of 0–2.5 V. Cericola and Kötz [347] recently reviewed a number of hybrid BatCaps with different positive electrodes and capacitive AC as negative electrode with different aqueous and organic electrolytes. The performance characteristics suggest that this hybrid configuration has potential to bridge the gap between supercapacitors and batteries.

The hybrid BatCap devices are classified into different types such as acidic, alkaline, BatCaps with redox electrolytes, etc. [314]. Nowadays, ion-based capacitor are very popular after the successful performance of Li ion batteries. Different metal–ion hybrid BatCaps are presented as follows.

Li Ion Capacitors (LICs) From the configuration point of view, LIC is an analog to LIBs in which the carbonaceous material (mostly AC) replaces the cathode material of LiBs. It offers a promising avenue to bridge the gap between high-energy LIBs and high-power supercapacitors. In 2001, Amatucci et al. [55] first introduced an LIC comprising lithium titanate ($Li_4Ti_5O_{12}$) anode and an AC cathode, offering a high specific energy >10 Wh kg^{-1}. Thereafter, various such systems employing lithium insertion anodes and capacitive carbon/conducting polymer cathodes are widely reported. Various anode materials such as Ti-based oxides/compounds (TiO_2 [348], $Li_4Ti_5O_{12}$ [349] etc.), silicon [350] Fe_3O_4 [351], Nb_2O_5 [352, 353], etc., and their hybrids with carbons or conducting polymers are reported for LICs. Particularly, $Li_4Ti_5O_{12}$-based LICs are most widely reported because the Ti-based anodes completely avoid the electrolyte decomposition [314]. Recently, Naoi et al. [354] synthesized nano-$Li_4Ti_5O_{12}$, using the ultra-centrifuge technique, which offers a high capacity (78 mAh g^{-1}) even at 1200 C rate. The assembled full cell nano-$Li_4Ti_5O_{12}$//AC shows a maximum energy density of 28 Wh l^{-1} at the power density 10 kW l^{-1}.

LICs are also assembled employing lithium intercalation cathode and capacitive carbon anode, which have the advantage of high output voltage due to the high potential plateau of the cathode. Though the specific capacity of intercalation cathode is lower as compared to most of the battery anodes, even then it has higher charge storage ability than the capacitive cathodes. Nowadays, $LiCoO_2$, $LiMnO_2$, $Li_2Mn_4O_9$, and $LiCoPO_4$ are attracting attention as LIB cathodes due to their high energy density. The olivine-type cathode materials $LiMPO_4$ (M = Fe, Mn, Co, Ni) are well-known class of lithium insertion cathode compounds, which are widely reported in LIBs [314, 355]. Apart from all the merits, they have the limitations of extremely low electronic conductivity and low diffusion coefficient. Different approaches have been adopted to improve their performance [356, 357]. Conductive coating with carbonaceous materials such as sucrose, and micro-/mesoporous carbons have been found to be an effective way for significant improvement in the performance characteristics as battery cathode. For example, Wu et al. [358] obtained a cathode capacity of $Li(Ni_{0.5}Co_{0.2}Mn_{0.3})O_2$ with single-wall carbon as high as ~250 mA h g^{-1} with high voltage plateau >4.5 V. Brandt et al. [359] reported AC//$LiNi_{0.5}Mn_{1.5}O_4$ hybrid system, which offered high specific energy and power of ~50 Wh kg^{-1} and ~1100 W kg^{-1}, respectively, and retention capacity of 89% after 4000 cycles.

LICs are reported mostly with liquid electrolytes, except a few studies on their quasi-solid-state form using GPEs. Recently, Singh and Hashmi [321] reported a $Li_4Ti_5O_{12}$//AC LIC system using LiTf/EC:PC/PVdF-HFP GPE and obtained specific energy of ~15.5 Wh kg^{-1} at specific power of ~49.7 W kg^{-1}. Further improvement have been achieved by replacing pure AC cathode by $LiFePO_4$/AC composite to obtain remarkably improved specific energy of ~27.4 Wh kg^{-1} at ~75.7 W kg^{-1}. Another solid-state LIC has been studied by the same group [322], which is assembled with $LiFePO_4$@MWCNTs nanocomposite cathode and MWCNTs as anode. This hybrid system offers maximum energy density of 19.8 Wh kg^{-1} at power density of 524 W kg^{-1}. An LIC configuration with prelithiated carbon anode, coupled with AC cathode, has recently been commercialized with an energy density of 15 Wh kg^{-1} and a cell voltage of 3.8 V [360]. The advantage of the pre-doping process at low intercalation potential provides a higher output voltage and highest cycling stability among all Li ion anodes [314].

Na Ion Capacitors (NICs) Due to limited resources, uneven distribution on the Earth' crust, and very high reactivity of lithium, alternatives of Li ion technology have become the essential requirement. In this race, sodium ion energy storage technology has been found to be an ideal alternative as sodium has lower cost, natural and infinite abundance, low toxicity, low atomic mass (23.0), and high enough electrochemical reduction potential (−2.71 V versus standard hydrogen electrode (SHE)). In recent years, a large number of sodium ion battery electrode materials, employable in Na ion capacitors (NICs), have been reported, namely, $Na_3V_2(PO_4)_3$ [361], $Na_2Fe_2(SO_4)_3$ [362], $NaMnO_2$ [363], $Na_{0.85}Li_{0.17}Ni_{0.21}Mn_{0.64}O_2$ [364], $Na_x[Fe_{1/2}Mn_{1/2}]O_2$ [365], and $Na_{0.44}MnO_2$ [366]. They exhibit sloping plateaus like characteristic feature in charge–discharge curves. Few other battery electrodes such as V_2O_5, sodium titanate, Na_xMnO_2, K_xMnO_2, cobalt hexacynoferrate (CoHCF), $Na_4Mn_9O_{18}$, etc. have been used as battery-type electrode for Na ion BatCaps [314]. Wu and coworkers [367] investigated nanowires of $Na_{0.35}MnO_2$ showing better Na insertion peaks and higher capacity than rod-like $Na_{0.95}MnO_2$. The $Na_{0.35}MnO_2$//AC device exhibited 1.8 V output voltage and high energy density of 42.6 Wh kg^{-1}.

NICs with carbon//carbon configuration are most widely investigated. In such systems, the intercalation (hierarchical) carbons are employed as anodes, while ACs are used as cathodes. For example, Kuratani et al. [368] reported a predoped hard carbon and AC-based NIC. In general, micro-/nanostructures of carbon electrodes are prepared to improve the device performance. Fast sodium intercalation and the high surface area should be ensured for battery-type carbon anode and capacitive carbon cathode, respectively. Recently, Ding et al. [369] produced both types of carbon electrodes from peanut shells and assembled a carbon//carbon NIC. In this system, peanut shell ordered carbon (PSOC) was employed as an ion intercalation anode and 1 M $NaClO_4$ in EC:PC as electrolyte. The device offered excellent specific energy and power (201, 76, and 50 Wh kg^{-1} at 285, 6500, and 16 500 W kg^{-1}, respectively).

Recently, Wang et al. [323] reported a quasi-solid-state NIC with disordered carbon (DC) nanoparticles and macroporous graphene (MG) as the anode and cathode, respectively. The porous GPE (PVdF-HFP/1 mol l^{-1} $NaClO_4$ solution in EC/DMC/DEC) was used as electrolyte/separator. This hybrid capacitor cell showed ability to operate at a high voltage (4.2 V) with a specific energy of 168 Wh kg^{-1} at a specific power of 501 W kg^{-1}. A comparative Ragone plot shows the superiority of solid-state DC//MG NIC cell with respect to MG//MG symmetric cell in terms of energy and power density values (Figure 9.28).

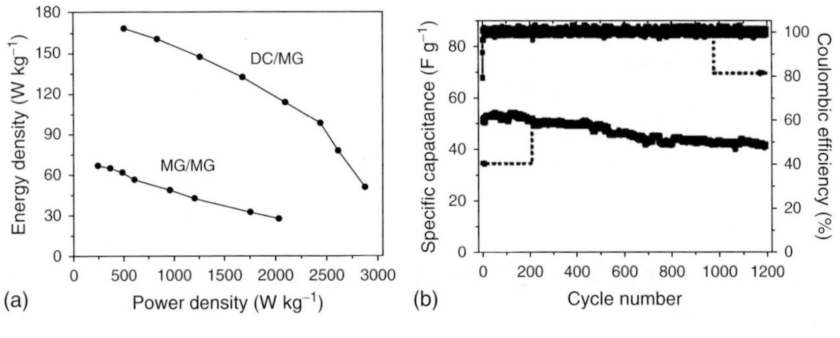

Figure 9.28 (a) Ragone plots of DC//MG NIC cell and MG//MG symmetric cell and (b) the cycling behavior of NIC at the current density of 1 A g^{-1}. Source: Wang et al. 2015 [323]. Reproduced with permission of John Wiley & Sons.

The NIC also showed stable cycling performance with 85% of the specific capacitance retention after 1200 cycles (Figure 9.28).

Al Ion Capacitors (AICs) Al ion capacitors (AICs) have been introduced as another class of supercapacitors and are becoming attractive due to the following reasons: (i) aluminum is very cheap and is the most abundantly available metal element on the Earth's crust and (ii) the Al-based energy storage devices have the ability of three-electron transportation during the electrochemical charge–discharge reactions, imparting larger storage capacity with respect to lithium- and sodium-based systems [370, 371]. Only limited reports on AICs are available due to the challenges faced in finding a suitable high performance anode with large capacity. It may also be noted that only liquid electrolyte-based AICs are studied and their solid-state counterparts do not exist. For example, the recently reported Ti_3C_2 delivers a capacitance of less than $30\,F\,g^{-1}$ in aqueous $Al_2(SO_4)_3$ electrolyte [372]. Wang et al. [373] reported high power AIC PPy@MoO_3//AC using aqueous $Al_2(SO_4)_3$ electrolyte, which offers an energy density of $30\,Wh\,kg^{-1}$ and an excellent cycling life with 93% capacitance retention after ~1800 cycles.

K ion [373] and Mg ion [374] capacitors have also been investigated recently with the same energy storage mechanism; however, at the time of writing, no reports were available about such systems based on polymer-based electrolytes.

9.4.2.2.3.3 Composite-Type Supercapacitors Such types of supercapacitors utilize composite electrode materials, fabricated in symmetric configuration. Both types of charge storage mechanisms (double layer and pseudocapacitive) occur in both the electrodes individually. For example, carbonaceous materials are added with pseudocapacitive materials, e.g. either conducting polymers or TMOs, to form the composite electrodes [375, 376]. The carbonaceous materials are responsible for capacitive double layer formation for charge storage and they provide a high surface area backbone that increases the contact between the pseudocapacitive materials and electrolyte. The pseudocapacitive materials are able to increase the capacitance of the electrode through Faradaic reactions [377, 378]. Such composite electrodes offer significant improvement in specific energy and power, along with their cycling performance [377, 378]. Ternary composite electrode materials have also been reported as potential supercapacitive electrodes. Most of such hybrid electrodes are tested in liquid electrolytes, except a very few reports on their solid-state counterparts. Recently, Mondal et al. [328] reported a ternary rGO/Fe_3O_4/PANI (rGFP) composite electrode for symmetric solid-state supercapacitor, which offered specific capacitance of $\sim 283.4\,F\,g^{-1}(1\,A\,g^{-1})$ and maximum energy density of $47.7\,Wh\,kg^{-1}$ at a power density of $550\,W\,kg^{-1}$. This system shows ~78% capacitance retention after 5000 charge–discharge cycles.

9.5 Challenges to Solid-State Supercapacitors and Future Scope of Improvement

Despite having advantages such as long cycle life (10^5–10^6 charge–discharge cycles), high Coulombic efficiency (>95%), and excellent performance characteristics with regard to power transfer, supercapacitors are still not as popular as primary and rechargeable batteries. Particularly, solid-state supercapacitors face two main challenges, such as their high cost and low energy density. The high cost arises from the choice of specific electrode materials (e.g. costly carbonaceous materials like CNTs, graphene, conducting polymers, rare TMOs) and solid electrolyte materials, especially costly organic solvents or ionic liquid incorporated polymer electrolytes. Further, unlike batteries, supercapacitors need to have fast movement of charge carriers across the electrolyte and into the pores of the electrodes. While maintaining low resistance and very long cycle life, both of these are difficult to attain, however, with the exception of ACs. In the case of solid-state supercapacitors, these problems become even more pronounced due to lesser penetration of electrolyte in the pores of the electrodes and due to the formation of solid electrolyte interface, which is not soluble, unlike liquid electrolytes. Research efforts are put to make better supercapacitor electrodes employing ACs from cheaper resources like biomasses/biowaste and by tailoring their pore shape and size and pore size distribution [379].

Another challenge is the low energy density, which limits the practical applications of supercapacitors, especially for high energy density applications like electric vehicles, escalators, etc. It may be noted that the energy is stored either inside the electrodes or at the electrode/electrolyte interfaces. Therefore, in order to achieve the high energy density, we need to focus on the active electrode materials and to identify suitable electrolytes (e.g. polymer-based electrolytes for solid/quasi-solid-state supercapacitors). For electrode materials, one of the methods is the densification by mechanical compression, which can be obtained by varying the synthesis process, for example, the case of CNT nanoforests, which are organized nanostructures and patterns by self-assembly or layer-by-layer assembly [380]. Woven carbon or metal oxide fibers are also promising electrode materials as they not only are dense and can contain capacitance-increasing additives in them but also make the device flexible [381].

As discussed in previous sections in detail, polymer-based electrolytes play an important role in supercapacitor performance. Apart from their cost, their associated technical parameters, namely, ionic conductivity and ESW, and the mechanical integrity of polymer electrolytes are important and essential considerations to fabricate high energy density, lightweight, flexible, safe, leak-proof, and durable solid-state supercapacitors. Many groups are working toward controlling the finer aspects of polymer electrolyte-based supercapacitors such as optimizing the properties of the liquid component in GPEs, viz. using better ionic liquids, salts, or host polymers; improving the interaction of GPEs with the electrodes; and improving mechanical and electrochemical properties of the polymer-based electrolytes.

References

1 Moseley, P.T. and Rand, D.A.J. (2010). *J. Power Sources* 195: 4423–4530.

2 Conway, B.E. (1999). *Electrochemical Supercapacitors—Scientific, Fundamentals and Technological Applications*. New York, NY: KluwerAcademic/Plenum Press.

3 Brousse, T., Bélanger, D., Chiba, K. et al. (2017). Materials for electrochemical capacitors. In: *Springer Handbook of Electrochemical Energy*, 1e (eds. C. Breitkopf and K. Swider-Lyons), 495–561. Berlin/Heidelberg: Springer-Verlag.
4 Shukla, A.K., Banerjee, A., Ravikumar, M.K., and Jalajakshi, A. (2012). *Electrochim. Acta* 84: 165–173.
5 Hashmi, S.A. (2004). *Natl. Acad. Sci. Lett.* 27: 27.
6 Béguin, F. and Frackowiak, E. (eds.) (2013). *Supercapacitors: Materials, Systems, and Applications.* Wiley-VCH Verlag GmbH.
7 Frackowiak, E., Abbas, Q., and Béguin, F. (2013). *J. Energy Chem.* 22: 226.
8 Zhang, L.L. and Zhao, X.S. (2009). *Chem. Soc. Rev.* 38: 2520.
9 Barbieri, O., Hahn, M., Herzog, A., and Kötz, R. (2005). *Carbon* 43 (6): 1303.
10 Qu, D. (2002). *J. Power Sources* 109: 403.
11 Kim, Y., Horie, Y., Ozaki, S. et al. (2004). *Carbon* 42: 1491.
12 Chmiola, J., Yushin, G., Gogotsi, Y. et al. (2006). *Science* 313: 1760.
13 Guterl, C.V., Frackowiak, E., Jurewicz, K. et al. (2005). *Carbon* 43: 1293.
14 González, A., Goikole, E., Barrena, J., and Mysyk, R. (2016). *Renewable Sustainable Energy* 58: 1189.
15 Hsieh, W., Horng, T.L.A., Huang, H.C., and Teng, H. (2015). *J. Mater. Chem. A* 3: 16535.
16 Huang, J., Sumpter, B., and Meunier, V. (2008). *Angew. Chem.* 120: 530.
17 Feng, G., Qiao, R., Huang, J. et al. (2010). *ACS Nano* 4: 2382.
18 Trasatti, S. and Buzzanca, G. (1971). *J. Electroanal. Chem. Interfacial Electrochem.* 29: A1.
19 Wang, Y., Song, Y., and Xia, Y. (2016). *Chem. Soc. Rev.* 45: 5925.
20 Augustyn, V., Simon, P., and Dunn, B. (2014). *Energy Environ. Sci.* 7 (5): 1597.
21 Sopcic, S., Mandic, Z., Inzelt, G. et al. (2011). *J. Power Sources* 196: 4849.
22 Max Planck Institute for Solid State Research. (2019) Intercalation/deintercalation. http://www.fkf.mpg.de/133111/De_Intercalation.
23 Conway, B.E. (1993). *Electrochim. Acta* 38: 1249.
24 Conway, B.E. and Pell, W.G. (2003). *J. Solid State Electrochem.* 7: 637.
25 Wang, G., Zhang, L., and Zhang, J. (2012). *Chem. Soc. Rev.* 41: 797.
26 Ghosh, A. and Lee, Y. (2012). *ChemSusChem* 5: 480–499.
27 Eftekhari, A. (2017). *Energy Storage Mater.* 9: 47–67.
28 Galiński, M., Lewandowski, A., and Stępniak, I. (2006). *Electrochim. Acta* 51: 5567.
29 Garche, J., Dyer, C.K., Moseley, P.T. et al. (2013). *Scrosati, Encyclopedia of electrochemical power sources.* Newnes.
30 Lewandowski, A. and Galiński, M. (2004). *J. Phys. Chem. Solids* 65: 281.
31 Stenger-Smith, J.D. and Irvin, J.A. (2009). *Mater. Matters* 4: 103.
32 Niu, J., Pell, W.G., and Conway, B.E. (2006). *J. Power Sources* 156: 725.
33 MacCallum, J.R. and Vincent, C.A. (eds.) (1987). *Polymer Electrolyte Reviews.* New York, NY: Elsevier Applied Sciences Publishers Ltd.
34 Gray, F.M. (1991). *Solid Polymer Electrolytes: Fundamentals and Technological Applications.* New York, NY: VCH Publishers.
35 Michot, T., Nishimoto, A., and Watanabe, M. (2000). *Electrochim. Acta* 45: 1347.

36 Stephan, A.M. (2006). *Eur. Polym. J.* 42: 21.
37 Agrawal, R.C. and Pandey, G.P. (2008). *J. Phys. D: Appl. Phys.* 41: 223001.
38 Jiang, J., Gao, D., Li, Z., and Su, G. (2006). *React. Funct. Polym.* 66: 1141.
39 Capiglia, C. (2000). *Solid State Ionics* 131: 291.
40 Lalia, B.S., Yoshimoto, N., Egashira, M., and Morita, M. (2009). *J. Power Sources* 194: 531.
41 Pandey, G.P. and Hashmi, S.A. (2009). *J. Power Sources* 187: 627.
42 Sirisopanaporn, C., Fernicola, A., and Scrosati, B. (2009). *J. Power Sources* 186: 490.
43 Fenton, D.E., Parker, J.M., and Wright, P.V. (1973). *Polymer* 14: 589.
44 Scrosati, B. and Vincent, C.A. (2000). *MRS Bull.* 25: 28.
45 Jacob, M.M.E., Hackett, E., and Giannelis, E.P. (2002). *J. Mater. Chem.* 13: 1.
46 Zhong, C., Deng, Y., Hu, W. et al. (eds.) (2016). *Electrolytes for ElectrochemicalSupercapacitors.* Boca Raton, FL: CRC Press.
47 Nègre, L., Daffos, B., Taberna, P.L., and Simon, P. (2015). *J. Electrochem. Soc.* 162: A5037–A5040.
48 Shuhaimi, N.E.A., Teo, L.P., Woo, H.J. et al. (2012). *Polym. Bull.* 69: 807–826.
49 Kadir, M.F.Z. and Arof, A.K. (2011). *Mater. Res. Innovations* 15: 217–220.
50 Tarascon, J.-M., Gozdz, A.S., Schmutz, C. et al. (1996). *Solid State Ionics* 86–88: 49–54.
51 Yang, Y.Q., Chang, Z., Li, M.X. et al. (2015). *Solid State Ionics* 269: 1–7.
52 Kim, J.I., Choi, Y., Chung, K.Y., and Park, J.H. (2017). *Adv. Funct. Mater.*: 1701768.
53 Li, H., Cheng, L., and Xia, Y. (2005). *Electrochem. Solid-State Lett.* 8: A433.
54 Pasquier, A.D., Plitz, I., Menocal, S., and Amatucci, G. (2003). *J. Power Sources* 115: 171.
55 Amatucci, G.G., Badway, F., Pasquier, A.D., and Zheng, T. (2001). *J. Electrochem. Soc.* 148: A930.
56 Armand, M. (1990). *Adv. Mater.* 1: 278.
57 Gray, F.M. (1997). *Polymer Electrolytes.* Cambridge: Royal Society of Chemistry.
58 Ramesh, S., Yahaya, A.H., and Arof, A.K. (2002). *Solid State Ionics* 152–153: 291–294.
59 Every, H.A., Zhou, F., Forsyth, M., and MacFarlane, D.R. (1998). *Electrochim. Acta* 43: 1465–1469.
60 Karan, N.K., Pradhan, D.K., Thomas, R. et al. (2008). *Solid State Ionics* 179: 689–696.
61 Hooper, R., Lyons, L.J., Moline, D.A., and West, R. (1999). *Organometallics* 18: 3249.
62 Robitaille, C.D. and Fauteux, D. (1986). *J. Electrochem. Soc.* 133: 315–325.
63 Das, S. and Ghosh, A. (2015). *AIP Adv.* 5: 027125.
64 Das, S. and Ghosh, A. (2015). *J. Appl. Phys.* 117: 174103.
65 Hashmi, S.A., Kumar, A., Maurya, K.K., and Chandra, S. (1990). *J. Phys. D: Appl. Phys.* 23: 1307–1314.
66 Murya, K.K., Srivastava, N., Hashmi, S.A., and Chandra, S. (1992). *J. Mater. Sci.* 27: 6357–6364.

67 Greenbaum, S.G., Pak, Y.S., Wintersgill, M.C., and Fonatanella, J.J. (1988). *NMR (Na Nuclear Magnetic Resonance) Studies of Na^+-Anion Association Effects in Polymer Electrolytes (TR-13).* New York: Hunter College, Department of Physics and Astronomy.

68 Porcarelli, L., Gerbaldi, C., Bella, F., and Nair, J.R. (2016). Super soft all-ethylene oxide polymer electrolyte for safe all-solid lithium batteries. *Scientific Reports* 6: 19892.

69 Chintapalli, S. and Frech, R. (1996). *Solid State Ionics* 86–88: 341.

70 Tsuchida, E., Ohno, H., and Tsunemi, K. (1983). *Electrochim. Acta* 28: 591.

71 Mohamed, N.S. and Arof, A.K. (2004). *J. Power Sources* 132: 229.

72 Wang, F.W.Z., Huang, B., Huang, H. et al. (1996). *J. Electrochem. Soc.* 143: 1510.

73 Watanabe, M., Kanba, M.O.T., and Nagaoka, K. (1982). *J. Appl. Polym. Sci.* 27: 4191.

74 Kumar, D. and Hashmi, S.A. (2010). *J. Power Sources* 195: 5101.

75 Appetecchi, G.B., Croce, F., and Scrosati, B. (1995). *Electrochim. Acta* 40: 991–997.

76 Pandey, G.P. and Hashmi, S.A. (2013). *J. Power Sources* 243: 211.

77 Guo, C., Zhou, L., and Lv, J. (2008). *J. Appl. Polym. Sci.* 1: 10, 594.

78 Ruch, P.W., Hardwick, L.J., Hahn, M. et al. (2009). *Carbon* 47: 38.

79 Honda, Y., Ono, T., Takeshige, M. et al. (2009). *Electrochem. Solid-State Lett.* 12: A45.

80 Abraham, K.M. and Alamgir, M. (1994). Room temperature polymer electrolytes, Chapter 3. In: *Lithium Batteries: New Materials, Developments and Perspectives*, vol. 5 (ed. G. Pistoia), 93–134. Amsterdam: Elsevier.

81 Feuillade, G. (1975). *J. Appl. Electrochem.* 5: 63.

82 Periasamy, P., Tatsumi, K., Shikano, M. et al. (2000). *J. Power Sources* 88: 269.

83 Alamgir, M. and Abraham, K.M. (1995). *J. Power Sources* 54: 40.

84 Watanabe, M., Togo, M., Sanui, K. et al. (1984). *Macromolecules* 17: 2908.

85 Othman, L., Chew, K.W., and Osman, Z. (2007). *Ionics* 13: 337.

86 Saito, Y., Capiglia, C., Kataoka, H. et al. (2000). *Solid State Ionics* 136–137: 1161.

87 Sekhon, S.S., Deepa, M., and Agnihotry, S.A. (2000). *Solid State Ionics* 136–137: 1189.

88 Pandey, G.P., Agrawal, R.C., and Hashmi, S.A. (2009). *J. Power Sources* 190: 563–572.

89 Quartarone, E., Tomasi, C., Mustarelli, P. et al. (1998). *Electrochim. Acta* 43: 1435.

90 Kumar, D., Suleman, M., and Hashmi, S.A. (2011). *Solid State Ionics* 202: 45–53.

91 Hooper, A. and North, J.M. (1983). The fabrication and performance of all solid state polymer-based rechargeable lithium cells. *Solid State Ionics* 9: 1161–1166.

92 DuPasquier, A., Sarrazin, C., Andrien, X., and Fauvarque, J.F. (1997). Poly(acrylonitrile)/ethylene carbonate-based gel electrolytes for ambient

temperature lithium and lithium-ion batteries. In: *Proceedings of the Symposium on Lithium Polymer Batteries*, vol. 96/17 (eds. J. Broadhead and B. Scrosati). Pennington, NJ: The Electrochemical Society.
93 Capiglia, C., Saito, Y., Yamamoto, H. et al. (2000). *Electrochim. Acta* 45: 1341.
94 Kumar, G.G. and Munichandraiah, N. (2002). *Electrochim. Acta* 47: 1013–1022.
95 Deepa, M., Sharma, N., Agnihotry, S.A. et al. (2002). *Solid State Ionics* 152–153: 253–258.
96 Abraham, K.M. and Alamgir, M. (1990). *J. Electrochem. Soc.* 137: 1657–1658.
97 Jayathilaka, P.A.R.D., Dissanayake, M.A.K.L., Albinsson, I., and Mellander, B.-E. (2003). *Solid State Ionics* 156: 179–195.
98 Rajendran, S. and Sivakumar, P. (2008). *Physica B* 403: 509–516.
99 Park, M.J., Choi, I., Hong, J., and Kim, O. (2013). *J. Appl. Polym. Sci.* 129: 2363.
100 Dong, Z., Zhang, Q., Yu, C. et al. (2013). *Ionics* 19: 1587.
101 Singh, B. and Sekhon, S.S. (2005). *Chem. Phys. Lett.* 414: 34.
102 Liu, F., Hashim, N.A., Liu, Y. et al. (2011). *J. Membr. Sci.* 375: 1.
103 Fuller, J., Breda, A.C., and Carlin, R.T. (1997). *J. Electrochem. Soc.* 144: L67.
104 Xu, J.J., Ye, H., and Huang, J. (2005). *Electrochem. Commun.* 7: 1309.
105 Kumar, D. and Hashmi, S.A. (2010). *Solid State Ionics* 181: 416–423.
106 Bansal, D., Cassel, F., Croce, F. et al. (2005). *J. Phys. Chem. B* 109: 4492.
107 Singh, B. and Sekhon, S.S. (2005). *J. Phys. Chem. B* 109: 16539.
108 Navarra, M.A., Panero, S., and Scrosati, B. (2005). *Electrochem. Solid-State Lett.* 8: A324.
109 Sutto, T.E., Ollinger, M., Kim, H. et al. (2006). *Electrochem. Solid-State Lett.* 9: A69.
110 Fernicola, A., Panero, S., Scrosati, B. et al. (2007). *ChemPhysChem* 8: 1103.
111 Tizzani, C., Appetecchi, G.B., Carewska, M. et al. (2007). *Aust. J. Chem.* 60: 47.
112 Shin, J.H., Henderson, W.A., and Passerini, S. (2005). *J. Electrochem. Soc.* 152: A978.
113 Ye, H., Huang, J., Xu, J.J. et al. (2007). *J. Electrochem. Soc.* 154: A1048.
114 Choi, J.W., Cheruvally, G., Kim, Y.H. et al. (2007). *Solid State Ionics* 178: 1235–1241.
115 Kim, G.T., Appetecchi, G.B., Alessandrini, F., and Passerini, S. (2007). *J. Power Sources* 171: 861.
116 Cheng, H., Zhu, C., Huang, B. et al. (2007). *Electrochim. Acta* 52: 5789.
117 Yang, P., Cui, W., Li, L. et al. (2012). *Solid State Sci.* 14: 598–606.
118 Choi, N.-S., Koo, B., Yeon, J.-T. et al. (2011). *Electrochim. Acta* 56: 7249.
119 Gopalan, A.I., Lee, K.-P., Manesh, K.M., and Santhosh, P. (2008). *J. Membr. Sci.* 318: 422.
120 Cheng, C.L., Wan, C.C., and Wang, Y.Y. (2004). *J. Power Sources* 134: 202.
121 Wang, Y.J. and Kim, D. (2008). *J. Membr. Sci.* 312: 76.
122 Kim, K.S., Park, S.-Y., Yeon, S.-H., and Lee, H. (2005). *Electrochim. Acta* 50: 5673.

123 Sivakumar, M., Subadevi, R., Rajendran, S. et al. (2007). *Eur. Polym. J.* 43: 4466.
124 Choi, J.A., Shim, E.G., Scrosati, B., and Kim, D.W. (2010). *Bull. Korean Chem. Soc.* 31: 3190.
125 Ohno, H. (ed.) (2005). *Electrochemical Aspects of Ionic Liquids*. Hoboken, NJ: Wiley.
126 Pandey, G.P. and Hashmi, S.A. (2013). *J. Mater. Chem. A* 1: 3372.
127 Egashira, M., Todo, H., Yoshimoto, N., and Morita, M. (2008). *J. Power Sources* 178: 729–735.
128 Fuller, J., Breda, A.C., and Carlin, R.T. (1998). *J. Electroanal. Chem.* 459: 29–34.
129 Sellam and Hashmi, S.A. (2014). *J. Solid State Electrochem.* 18: 465–475.
130 Pringle, J.M. (2013). *Phys. Chem. Chem. Phys.* 15: 1339.
131 Suleman, M., Kumar, Y., and Hashmi, S.A. (2013). *J. Phys. Chem. B* 117: 7436.
132 MacFarlane, D.R. and Forsyth, M. (2001). *Adv. Mater.* 13: 957.
133 Wilmer, D., Funke, K., Witschas, M. et al. (1999). *Physica B* 266: 60.
134 Aronsson, R., Jansson, B., Knape, H.E.G. et al. (1980). *Le J. Phys. Colloq.* 41: C6.
135 Timmermans, J. (1961). *J. Phys. Chem. Solids* 18: 1.
136 Sherwood, J.N. (1979). *Plastically Crystalline State: Orientationally Disordered Crystals*. London: Wiley.
137 Fan, L.-Z., Hu, Y.-S., Bhattacharyya, A.J., and Maier, J. (2007). *Adv. Funct. Mater.* 17: 2800.
138 Patel, M., Menezes, P.V., and Bhattacharyya, A.J. (2010). *J. Phys. Chem. B* 114: 5233.
139 Das, S., Prathapa, S.J., Menezes, P.V. et al. (2009). *J. Phys. Chem. B* 113: 5025.
140 Eijck, L.V., Best, A.S., Long, S. et al. (2009). *J. Phys. Chem. C* 113: 15007.
141 Derollez, P., Lefebvre, J., Descamps, M. et al. (1990). *J. Phys. Condens. Matter* 2: 6893–6903.
142 Alarco, P.-J., Abu-Lebdeh, Y., Abouimrane, A., and Armand, M. (2004). *Nat. Mater.* 3: 476–481.
143 Abu-Lebdeh, Y., Alarco, P.J., and Armand, M. (2004). *J. New Mater. Electrochem. Syst.* 7: 29.
144 MacFarlane, D.R., Huang, J., and Forsyth, M. (1999). *Nature* 402: 792–794.
145 Long, S., MacFarlane, D.R., and Forsyth, M. (2004). *Solid State Ionics* 175: 733–738.
146 Long, S., MacFarlane, D.R., and Forsyth, M. (2003). *Solid State Ionics* 161: 105–112.
147 Abouimrane, A., Whitfield, P.S., Niketic, S., and Davidson, I.J. (2007). *JPS* 174: 883–888.
148 Fan, L.-Z., Wang, X.L., and Long, F. (2009). *J. Power Sources* 189: 775–778.
149 Fan, L.-Z. and Maier, J. (2006). *Electrochem. Commun.* 8: 1753.
150 Karuppasamy, X.S.S.K., Thanikaikarasan, S., Antony, R., and Balakumar, S. (2012). *Ionics (Kiel)* 18: 737.
151 Kumar, R. and Rhee, H. (2012). *Electrochim. Acta* 76: 159.
152 Wu, X.-L., Xin, S., Seo, H.-H. et al. (2011). *Solid State Ionics* 186: 1.

153 Patel, M., Chandrappa, K.G., and Bhattacharyya, A.J. (2010). *Solid State Ionics* 181: 844–848.

154 Sharma, J. and Hashmi, S.A. (2013). *J. Solid State Electrochem.* 17: 2283.

155 Patel, M., Chandrappa, K.G., and Bhattacharyya, A.J. (2008). *Electrochim. Acta* 54: 209.

156 Patel, M. and Bhattacharyya, A.J. (2008). *Electrochem. Commun.* 10: 1912–1915.

157 Singh, M.K., Suleman, M., Kumar, Y., and Hashmi, S.A. (2015). *Energy* 80: 465.

158 Huang, J., Hill, A., Forsyth, M. et al. (2006). *Solid State Ionics* 177: 2569.

159 Yue, R., Niu, Y., Wang, Z. et al. (2009). *Polymer* 50: 1288–1296.

160 Suleman, M., Kumar, Y., and Hashmi, S.A. (2015). *J. Solid State Electrochem.* 19: 1347–1357.

161 Suleman, M., Kumar, Y., and Hashmi, S.A. (2015). *Electrochim. Acta* 182: 995–1007.

162 Yu, H., Wu, J., Fan, L. et al. (2011). *Electrochim. Acta* 56: 6881–6886.

163 Fan, L.-Q., Zhong, J., Wu, J.-H. et al. (2014). *Chem. A* 2: 9011–9014.

164 Zhong, J., Fan, L.-Q., Wu, X. et al. (2015). *Electrochim. Acta* 166: 150–156.

165 Yu, H., Wu, J., Fan, L. et al. (2012). *J. Power Sources* 198: 402–407.

166 Ma, G., Dong, M., Sun, K. et al. (2015). *J. Mater. Chem. A* 3: 4035.

167 Sankar, K.V. and Selvan, R.K. (2015). *Carbon* 90: 260–273.

168 Xu, R., Guo, F., Cui, X. et al. (2015). *J. Mater. Chem. A* 3: 22353–22360.

169 Yu, F., Huang, M., Wu, J. et al. (2014). *J. Appl. Polym. Sci.* 131 https://doi.org/10.1002/app.39784.

170 Senthilkumar, S.T., Kalai Selvan, R., Melo, J.S., and Sanjeeviraja, C. (2013). *ACS Appl. Mater. Interfaces* 5: 10541–10550.

171 Fan, L.-Q., Zhong, J., Zhang, C.-Y. et al. (2016). *Int. J. Hydrogen Energy* 41: 5725–5732.

172 Feng, E., Ma, G., Sun, K. et al. (2016). *RSC Adv.* 6: 75896.

173 Sun, K., Dong, M., Feng, E. et al. (2015). *RSC Adv.* 5: 22419.

174 Kim, D., Lee, G., Kim, D. et al. (2016). *Nanoscale.*

175 Yadav, N., Mishra, K., and Hashmi, S.A. (2017). *Electrochim. Acta* 235: 570–582.

176 Liu, Q., Tang, Z., Ou, B. et al. (2014). *Mater. Chem. Phys.* 144: 213–225.

177 Stephan, M. and Teeters, D. (2003). *Electrochim. Acta* 48: 2143.

178 Taranscon, J.M., Gozdz, A.S., Schmutz, C. et al. (1997). *Solid State Ionics* 86–88: 741.

179 Park, H.C., Kim, Y.P., Kim, H.Y., and Kang, Y.S. (1999). *J. Membr. Sci.* 156: 169.

180 Saito, Y., Kataoka, H., and Stephan, A.M. (2001). *Macromolecules* 34: 6955–6958.

181 Wu, C.-G., Lu, M.-I., and Chuang, H.-J. (2005). *Polymer* 46: 5929–5938.

182 Xi, J., Qiu, X., Li, J. et al. (2006). *J. Power Sources* 157: 501–506.

183 Li, G.C., Zhang, P., Zhang, H.P. et al. (2008). *Electrochem. Commun.* 10: 1883–1885.

184 Kumar, Y., Pandey, G.P., and Hashmi, S.A. (2012). *J. Phys. Chem. C* 116: 26118–26127.

185 Bokobza, L., Bruneel, J.-L., and Couzi, M. (2014). *Vib. Spectrosc.* 74: 57–63.
186 Dresselhaus, M.S., Jorio, A., and Saito, R. (2010). *Annu. Rev. Condens. Matter Phys.* 1: 89–108.
187 Bokobza, L. and Zhang, J. (2012). *eXPRESS Polym. Lett.* 6: 601–608.
188 Barsukov, I.V., Johnson, C.S., Doninger, J.E., and Barsukov, V.Z. (eds.) (2003). *New Carbon Based Materials for Electrochemical Energy Storage Systems.* Argonne, IL: Springer.
189 Rouquerol, J., Rouquerol, F., Llewellyn, P. et al. (2013). *Adsorption by Powders and Porous Solids: Principles, Methodology and Applications.* London: Academic Press.
190 Harry Marsh, F.R.R. (2006). *Activated Carbon.* Elsevier Science & Technology Books.
191 Condon, J.B. (2006). *Surface Area and Porosity Determinations by Physisorption Measurement and Theory.* Oxford, UK: Elsevier.
192 Bruce, P.G. (1987). Electrical measurements on polymer electrolytes. In: *Polymer Electrolyte Reviews-I* (eds. J.R. MacCullam and C.A. Vincent), 237–274. Elsevier Science.
193 Laheäär, A., Przygocki, P., Abbas, Q., and Béguin, F. (2015). *Electrochem. Commun.* 60: 21–25.
194 Sharma, P. and Bhatti, T.S. (2010). *Energy Convers. Manage.* 51: 2901–2912.
195 Choi, N.-S., Chen, Z., Freunberger, S.A. et al. (2012). *Angew. Chem. Int. Ed.* 51: 9994–10024.
196 Pandey, G.P., Hashmi, S.A., and Kumar, Y. (2010). *Energy Fuels* 24: 6644–6652.
197 Pandey, G.P., Hashmi, S.A., and Kumar, Y. (2010). *J. Electrochem. Soc.* 157: A105–A114.
198 Singh, M.K., Kumar, Y., and Hashmi, S.A. (2013). *Nanotechnology* 24: 465704–465713.
199 Lee, D., Jung, J.-Y., Jung, M.-J., and Lee, Y.-S. (2015). *Chem. Eng. J.* 263: 62–70.
200 Ren, J., Bai, W., Guan, G. et al. (2013). *Adv. Mater.* 25: 5965–5970.
201 Oschatz, M., Boukhalfa, S., Nickel, W. et al. (2017). *Carbon* 113: 283–291.
202 Pech, D., Brunet, M., Durou, H. et al. (2010). *Nat. Nanotechnol.* 5: 651–654.
203 Ervin, M.H., Pereira, C.M., Miller, J.R. et al. (2013). *ECS J. Solid State Sci. Technol.* 2: M3135–M3139.
204 Miller, J.R., Outlaw, R.A., and Holloway, B.C. (2011). *Electrochim. Acta* 56: 10443–10449.
205 Raccichini, R., Varzi, A., Passerini, S., and Scrosati, B. (2015). *Nat. Mater.* 14: 271–279.
206 Niu, C., Sichel, E.K., Hoch, R. et al. (1997). *Appl. Phys. Lett.* 70: 1997.
207 Sun, Y., Wu, Y.Q., and Shi, G. (2011). *Phys. Chem. Chem. Phys.* 13: 17249.
208 Liu, C., Yu, Z., Neff, D. et al. (2010). *Nano Lett.* 10: 4863.
209 Shi, H. (1996). *Electrochim. Acta* 40: 1633.
210 Qu, D. and Shi, H. (1998). *J. Power Sources* 74: 99–107.
211 Frackowiak, E. and Beguin, F. (2001). *Carbon* 39: 937.
212 Ye, S., Zhu, L., Kim, I.-J. et al. (2016). *J. Ind. Eng. Chem.* 43: 53–60.

213 Ishikawa, M., Morita, M., Ihara, M., and Matsuda, Y. (1994). *J. Electrochem. Soc.* 141: 1730.

214 Ishikawa, M., Morita, M., Ihara, M., and Matsuda, Y. (1995). *Electrochim. Acta* 40: 2217.

215 Liu, X. and Osaka, T. (1996). *J. Electrochem. Soc.* 143: 3982.

216 Xue, R., Huang, H., Chen, L., and Wang, K. (1993). *Solid State Ionics* 59: 1.

217 Hashmi, S.A., Latham, R.J., Linford, R.G., and Schlindwein, W.S. (1997). *J. Chem. Soc., Faraday Trans.* 93: 4177.

218 Lewandowski, A., Zajder, M., Frackowiack, E., and Beguin, F. (2001). *Electrochim. Acta* 46: 2777–2780.

219 Lassegues, J.C., Grondin, J., Becker, T. et al. (1995). *Solid State Ionics* 77: 311.

220 Pernaut, J.M. and Goulart, G. (1995). *J. Power Sources* 55: 93.

221 Frackowiak, E., Lota, G., and Pernak, J. (2005). *Appl. Phys. Lett.* 86: 164104.

222 Osaka, T., Liu, X., Nojima, M., and Momma, T. (1999). *J. Electrochem. Soc.* 146: 1724.

223 Hashmi, S.A., Kumar, A., and Tripathi, S.K. (2007). *J. Phys. D: Appl. Phys.* 40: 6527.

224 Kang, Y.J., Chung, H., Han, C.-H., and Kim, W. (2012). *Nanotechnology* 23: 065401. (6pp).

225 Ortega, P.F.R., Trigueiro, J.P.C., Silva, G.G., and Lavall, R.L. (2016). *Electrochim. Acta* 188: 809–817.

226 Pandey, G.P. and Hashmi, S.A. (2013). *Electrochim. Acta* 105: 333.

227 Gao, K., Shao, Z., Wang, X. et al. (2013). *RSC Adv.* 3: 15058–15064.

228 Lin, H., Li, L., Ren, J. et al. (2013). *Sci. Rep.* 3.

229 Yang, Z., Deng, J., Chen, X. et al. (2013). *Angew. Chem. Int. Ed.* 52: 13453.

230 Xie, Y., Liu, Y., Zhao, Y. et al. (2014). *J. Mater. Chem. A* 2: 9142–9149.

231 Yang, C., Shen, J., Wang, C. et al. (2014). *J. Mater. Chem. A* 2: 1458.

232 Gao, K., Shao, Z., Li, J. et al. (2013). *J. Mater. Chem. A* 1: 63–67.

233 Yang, X., Zhang, F., Zhang, L. et al. (2013). *Adv. Funct. Mater.* 23: 3353–3360.

234 Wang, M., Duong, L.D., Mai, N.T. et al. (2015). *ACS Appl. Mater. Interfaces* 7: 1348–1354.

235 Xu, Y., Lin, Z., Huang, X. et al. (2013). *ACS Nano* 7: 4042–4049.

236 Cao, X., Zheng, B., Shi, W. et al. (2015). *Adv. Mater.* 27: 4695–4701.

237 Pandey, G.P., Kumar, Y., and Hashmi, S.A. (2010). *Ind. J. Chem.* 49A: 743–751.

238 Senthilkumar, S.T., Kalai Selvan, R., Ponpandian, N., and Melo, J.S. (2012). *RSC Adv.* 2: 8937–8940.

239 Iijima, S. (1991). *Nature* 354: 56.

240 Baibarac, M., Lira-Cantú, M., Oró-Solé, J. et al. (2006). *Small* 2: 1075.

241 Claye, A.S., Fischer, J.E., Huffman, C.B. et al. (2000). *J. Electrochem. Soc.* 147: 2845.

242 Kaempgen, M., Lebert, M., Nicoloso, N., and Roth, S. (2008). *Appl. Phys. Lett.* 92: 1.

243 Suehiro, J., Zhou, G., and Hara, M. (2003). *J. Phys. D: Appl. Phys.* 36: L109.

244 Frackowiak, E. (2007). *Phys. Chem. Chem. Phys.* 9: 1774.

245 Ajayan, P.M. (1999). *Chem. Rev.* 99: 14.
246 Lu, W., Hartman, R., Qu, L., and Dai, L. (2011). *J. Phys. Chem. Lett.* 2: 655.
247 Chen, D.T.N., Chen, K., Hough, L.A. et al. (2010). *Macromolecules* 43: 2048.
248 Hu, L., Pasta, M., La Mantia, F. et al. (2010). *Nano Lett.* 10: 708.
249 Datsyuk, V., Kalyva, M., Papagelis, K. et al. (2008). *Carbon* 46: 833.
250 Ma, P.C., Siddiqui, N.A., Marom, G., and Kim, J.K. (2010). *Composites Part A* 41: 1345.
251 Wu, H. and Drzal, L.T. (2012). *Carbon* 50: 1135.
252 Fukushima, T., Kosaka, A., Ishimura, Y. et al. (2003). *Science* 300: 2072.
253 Katakabe, T., Kaneko, T., Watanabe, M. et al. (2005). *Electrochem. Soc.* 152: A1913.
254 Lee, J. and Aida, T. (2011). *Chem. Commun.* 47: 6757.
255 Mukai, K., Asaka, K., Kiyohara, K. et al. (2008). *Electrochim. Acta* 53: 5555.
256 Low, C.T.J., Walsh, F.C., Chakrabarti, M.H. et al. (2013). *Carbon* 54: 1.
257 Guo, H.-L., Wang, X.-F., Qian, Q.-Y. et al. (2009). *ACS Nano* 3: 2653.
258 Lee, J., Tao, L., Hao, Y. et al. (2012). *Appl. Phys. Lett.* 100: 1.
259 Pumera, M. (2011). *Energy Environ. Sci.* 4: 668.
260 Huang, X., Zeng, Z., Fan, Z. et al. (2012). *Adv. Mater.* 24: 5979.
261 Bolotin, K.I., Sikes, K.J., Jiang, Z. et al. (2008). *Solid State Commun.* 146: 351.
262 Berger, C., Song, Z., Li, X. et al. (2006). *Science* 312: 1191.
263 Yan, L., Zheng, Y.B., Zhao, F. et al. (2012). *Chem. Soc. Rev.* 41: 97.
264 Singh, V., Joung, D., Zhai, L. et al. (2011). *Prog. Mater Sci.* 56: 1178.
265 William, J., Hummers, S., and Offeman, R.E. (1958). *J. Am. Chem. Soc.* 80: 1339.
266 Gómez-Navarro, C., Weitz, R.T., Bittner, A.M. et al. (2007). *Nano Lett.* 7: 3499.
267 Becerril, H.A., Mao, J., Liu, Z. et al. (2008). *ACS Nano* 2: 463.
268 Wang, S., Chia, P.J., Chua, L.L. et al. (2008). *Adv. Mater.* 20: 3440.
269 Yu, A., Sy, A., and Davies, A. (2011). *Synth. Met.* 161: 2049.
270 Bonso, J.S., Rahy, A., Perera, S.D. et al. (2012). *J. Power Sources* 203: 227.
271 Wang, G., Yang, J., Park, J. et al. (2008). *J. Phys. Chem. C* 112: 8192.
272 Suleman, M., Othman, M.A.R., Hashmi, S.A. et al. (2017). *J. Alloys Compd.* 695: 3376–3392.
273 Zheng, J.P., Cygan, P.J., and Jow, T.R. (1995). *J. Electrochem. Soc.* 142: 2699.
274 Crispin, X., Jakobsson, F.L.E., Crispin, A. et al. (2006). *Chem. Mater.* 18: 4354.
275 Muthulakshmi, B., Kalpana, D., Pitchumani, S., and Renganathan, N.G. (2006). *J. Power Sources* 158: 1533.
276 Prasad, K.R. and Miura, N. (2004). *Electrochem. Solid-State Lett.* 7: A425.
277 Wu, N.-L. (2002). *Mater. Chem. Phys.* 75: 6.
278 Wu, M.-S., Huang, Y.-A., Yang, C.-H., and Jow, J.-J. (2007). *Int. J. Hydrogen Energy* 32: 4153.
279 Brousse, T., Toupin, M., Dugas, R. et al. (2006). *J. Electrochem. Soc.* 153: A2171.
280 Castro, E.B., Real, S.G., and Dick, L.F.P. (2004). *Int. J. Hydrogen Energy* 29: 255.

281 Rudge, A., Davey, J., Raistrick, I., and Gottesfeld, S. (1994). *J. Power Sources* 47: 89.

282 Laforgue, A., Simon, P., Fauvarque, J.F. et al. (2003). *J. Electrochem. Soc.* 150: A645–A651.

283 Ramya, R., Sivasubramanian, R., and Sangaranarayanan, M.V. (2013). *Electrochim. Acta* 101: 109–129.

284 Hashmi, S.A., Latham, R.J., Linford, R.G., and Schlindwein, W.S. (1998). *Polym. Int.* 47: 28.

285 Tripathi, S.K., Kumar, A., and Hashmi, S.A. (2006). *Solid State Ionics* 177: 2979.

286 Prasad, K.R. and Munichandraiah, N. (2002). *Electrochem. Solid-State Lett.* 5: A271.

287 Meng, C., Liu, C., Chen, L. et al. (2010). *Nano Lett.* 10: 4025.

288 Yang, C., Zhang, L., Hu, N. et al. (2016). *Appl. Surf. Sci.* 379: 206.

289 Wu, Z.-S., Zheng, Y., Zheng, S. et al. (2016). *Adv. Mater.* 29: 1521.

290 Bi, R.R., Wu, X.L., Cao, F.F. et al. (2010). *J. Phys. Chem. C* 114: 2448–2451.

291 Choudhury, N.A., Shukla, A.K., Sampath, S., and Pitchumanic, S. (2006). *J. Electrochem. Soc.* 153: A614.

292 Lian, K. and Tian, Q. (2010). *Electrochem. Commun.* 12: 517.

293 Wang, Y. and Zhang, X. (2004). *Electrochim. Acta* 49: 1957.

294 Sellam, S.A.H. (2013). *ACS Appl. Mater. Interfaces* 5 (9): 3875–3883.

295 Nakayama, M., Tanaka, A., Sato, Y. et al. (2005). *Langmuir* 21: 5907.

296 Chodankar, N.R., Dubal, D.P., Gund, G.S., and Lokhande, C.D. (2016). *J. Energy Chem.* 25: 463.

297 Wu, S., Chen, W., and Yan, L. (2014). *J. Mater. Chem. A* 2: 2765.

298 Yang, P., Ding, Y., Lin, Z. et al. (2014). *Nano Lett.* 14: 731.

299 Yuan, L., Lu, X.-H., Xiao, X. et al. (2011). *ACS Nano* 6: 656.

300 Xie, J., Sun, X., Zhang, N. et al. (2013). *Nano Energy* 2: 65.

301 Liao, Q., Li, N., Jin, S. et al. (2015). *ACS Nano* 9: 5310.

302 Wang, G., Lu, X., Ling, Y. et al. (2012). *ACS Nano* 6: 10296.

303 Yuan, C., Wu, H.B., Xie, Y., and Lou, X.W.D. (2014). *Angew. Chem. Int. Ed.* 53: 1488–1504.

304 Kim, H. and Popov, B.N. (2003). *J. Electrochem. Soc.* 150: D56.

305 Zhu, D., Wang, W., Zhu, J. et al. (2017). *J. Solid State Electrochem.* 21: 403.

306 Wang, Y. and Chen, Q. (2014). *ACS Appl. Mater. Interfaces* 6: 6196.

307 Prasad, K.R. and Miura, N. (2004). *Electrochem. Commun.* 6: 1004.

308 Basu, A., Bhardwaj, M., Gawli, Y. et al. (2016). *ChemistrySelect* 1: 5159.

309 Wang, Q., Wang, X., Liu, B. et al. (2013). *Mater. Chem. A* 1: 2468.

310 Raut, S.S. and Sankapal, B.R. (2016). *Electrochim. Acta* 198: 203.

311 Akinwolemiwa, B., Peng, C., and Chena, G.Z. (2015). *J. Electrochem. Soc.* 162 (5): A5054–A5059.

312 Brousse, T., Belanger, D., and Long, J.W. (2015). *J. Electrochem. Soc.* 162 (5): A5185–A5189.

313 Simon, P., Gogotsi, Y., and Dunn, B. (2014). *Science* 343: 1210–1212.

314 Zuo, W., Li, R., Zhou, C. et al. (2017). *Adv. Sci.*: 1600539.

315 Basnayaka, P.A., Kumar, A., Ram, M.K. et al. (2017). *ECS J. Solid State Sci. Technol.* 6 (6): M3168.

316 Yuan, C., Zhang, X., Wu, Q., and Gao, B. (2006). *Solid State Ionics* 177: 1237.

317 Sivaraman, P., Thakur, A., Kushwaha, R.K. et al. (2006). *Electrochem. Solid-State Lett.* 9: A435.

318 Kang, Y.J., Chung, H., and Kim, W. (2013). *Synth. Met.* 166: 40.

319 Lu, X., Yu, M., Zhai, T. et al. (2013). *Nano Lett.* 13: 2628.

320 Li, R.Z., Wang, Y.M., Zhou, C. et al. (2015). *Adv. Funct. Mater.* 25: 5384.

321 Singh, M.K. and Hashmi, S.A. (2017). Performance of solid-state hybrid supercapacitor with $LiFePO_4$/AC composite cathode and $Li_4Ti_5O_{12}$ as anode. *Ionics* 23: 2931–2942.

322 Singh, M.K. and Hashmi, S.A. (2017). Performance of solid-state hybrid supercapacitor with $LiFePO_4$/AC composite cathode and $Li_4Ti_5O_{12}$ as anode. *Ionics* 23: 2931–2942.

323 Wang, F., Wang, X., Chang, Z. et al. (2015). *Adv. Mater.* 27: 6962.

324 Tien, C.P. and Teng, H. (2010). *J. Power Sources* 195: 2414.

325 Li, Z., Zhou, Z., Yun, G. et al. (2013). *Nanoscale Res. Lett.* 8 (8).

326 Dubal, D.P., Guevara, J.S., Tonti, D. et al. (2015). *J. Mater. Chem. A* 3: 23483.

327 Subramani, K., Sudhan, N., Divya, R., and Sathish, M. (2017). *RSC Adv.* 7: 6648.

328 Mondal, S., Rana, U., and Malik, S. (2017). *J. Phys. Chem. C* 121 (14): 7573.

329 Vlad, A., Singh, N., Rolland, J. et al. (2014). *Sci. Rep.* 4: 4315.

330 Selvakumar, M., Bhat, D.K., Aggarwal, A.M. et al. (2010). *Physica B* 405: 2286–2289.

331 Roldán, S., Barreda, D., Granda, M. et al. (2015). *Phys. Chem. Chem. Phys.* 17: 1084.

332 Khomenko, V., Piñero, E.R., and Béguin, F. (2010). *J. Power Sources* 195: 4234.

333 Wang, H., Yoshio, M., Thapa, A.K., and Nakamura, H. (2007). *J. Power Sources* 169: 375.

334 Khomenko, V., Piñero, E.R., and Béguin, F. (2008). *J. Power Sources* 177: 643.

335 Snook, G.A., Kao, P., and Best, A.S. (2011). *J. Power Sources* 196: 1.

336 Arbizzani, C., Mastragostino, M., and Meneghello, L. (1996). *Electrochim. Acta* 41: 21.

337 Arbizzani, C., Mastragostino, M., and Soavi, F. (2001). *J. Power Sources* 100: 164.

338 Mastragostino, M., Arbizzani, C., and Soavi, F. (2002). *Solid State Ionics* 148: 493.

339 Dubal, D.P., Ayyad, O., Ruiz, V., and Romero, P.G. (2015). *Chem. Soc. Rev.* 44: 1777.

340 Fkovich, Y.M.V. and Serdyuk, T.M. (2002). *Russ. J. Electrochem.* 38: 935.

341 Choi, H.S., Im, J.H., Kim, T. et al. (2012). *J. Mater. Chem.* 22: 16986.

342 Zhang, Y., Feng, H., Wu, X. et al. (2009). *Int. J. Hydrogen Energy* 34: 4889.

343 Ganesh, V., Pitchumani, S., and Lakshminarayanan, V. (2006). *J. Power Sources* 158: 1523.

344 Marumo, C., Ando, N., and Taguchi, M. (2008). Design and performance of lithium-ion capacitors. In: *ESSCAP'08, 3rd European Symposium on Supercapacitors and Applications.*

345 Pell, W.G. and Conway, B.E. (2004). *J. Power Sources* 136: 334.
346 Wang, Q., Wen, Z.H., and Li, J.H. (2006). *Adv. Funct. Mater.* 16: 2141.
347 Cericola, D. and Kötz, R. (2012). *Electrochim. Acta* 72: 1.
348 Kim, H., Cho, M.Y., Kim, M.H. et al. (2013). *Adv. Energy Mater.* 3: 1500.
349 Fan, Q., Yang, M., Meng, Q. et al. (2016). *J. Electrochem. Soc.* 163: A1736.
350 Liu, X., Jung, H.G., Kim, S.O. et al. (2013). *Sci. Rep.* 3: 3183.
351 Zhang, F., Zhang, T.F., Yang, X. et al. (2013). *Energy Environ. Sci.* 6: 1623.
352 Lim, E., Kim, H., Jo, C. et al. (2014). *ACS Nano* 8: 8968.
353 Lim, E., Jo, C., Kim, H. et al. (2015). *ACS Nano* 9: 7497.
354 Naoi, K., Naoi, W., Aoyagi, S. et al. (2013). *Acc. Chem. Res.* 46: 1075.
355 Goodenough, J.B. and Park, K.S. (2013). *J. Am. Chem. Soc.* 135: 1167.
356 Kang, B. and Ceder, G. (2009). *Nature* 458: 190.
357 Zane, D., Carewska, M., Scaccia, S. et al. (2004). *Electrochim. Acta* 49: 4259.
358 Wu, Z., Ji, S., Zheng, J. et al. (2015). *Nano Lett.* 15 (8): 5590.
359 Brandt, A., Balducci, A., Rodehorst, U. et al. (2014). *J. Electrochem. Soc.* 161: A1139.
360 Simon, P. and Gogotsi, Y. (2008). *Nat. Mater.* 7: 845.
361 Jian, Z., Han, W., Lu, X. et al. (2013). *Adv. Energy Mater.* 3: 156.
362 Barpanda, P., Oyama, G., Nishimura, S. et al. (2014). *Nat. Commun.* 5: 4358.
363 Ma, X., Chen, H., and Ceder, G. (2011). *J. Electrochem. Soc.* 158: A1307.
364 Kim, D., Kang, S., Slater, M. et al. (2011). *Adv. Energy Mater.* 1: 333.
365 Yabuuchi, N., Kajiyama, M., Iwatate, J. et al. (2012). *Nat. Mater.* 11: 512.
366 Wang, Y., Liu, J., Lee, B. et al. (2015). *Nat. Commun.* 6: 6401.
367 Zhang, B.H., Liu, Y., Chang, Z. et al. (2014). *J. Power Sources* 253: 98.
368 Kuratani, K., Yao, M., Senoh, H. et al. (2012). *Electrochim. Acta* 76: 320.
369 Ding, J., Wang, H., Li, Z. et al. (2015). *Energy Environ. Sci.* 8: 941.
370 Wang, W., Jiang, B., Xiong, W. et al. (2013). *Sci. Rep.* 3: 3383.
371 Li, Z., Xiang, K., Xing, W. et al. (2015). *Adv. Energy Mater.* 5: 1401410.
372 Lukatskaya, M.R., Mashtalir, O., Ren, C.E. et al. (2013). *Science* 341: 1502.
373 Wang, F., Liu, Z., Wang, X. et al. (2016). *J. Mater. Chem. A* 4: 5115.
374 Yoo, H.D., Shterenberg, I., Gofer, Y. et al. (2014). *J. Electrochem. Soc.* 161: A410.
375 Pasquier, A.D., Plitz, I., Gural, J. et al. (2003). *J. Power Sources* 113: 62.
376 Mastragostino, M., Arbizzani, C., and Soavi, F. (2001). *J. Power Sources* 97–98: 812.
377 Jurewicz, K., Delpeux, S., Bertagna, V. et al. (2001). *Chem. Phys. Lett.* 347: 36.
378 Frackowiak, E., Khomenko, V., Jurewicz, K. et al. (2006). *J. Power Sources* 153: 413.
379 Kleszyk, P., Ratajczak, P., Skowron, P. et al. (2015). *Carbon* 81: 148–157.
380 Yoon, Y., Lee, K., Kwon, S. et al. (2014). *ACS Nano* 8: 4580–4590.
381 Dalton, A.B., Collins, S., Muñoz, E. et al. (2003). *Nature* 423: 703.

10

Polymer Electrolytes for Quantum Dot-Sensitized Solar Cells (QDSSCs) and Challenges

T.M.W.J. Bandara[1] *and J.L. Ratnasekera*[2]

[1] *University of Peradeniya, Department of Physics, Faculty of Science, Peradeniya, 20400, Sri Lanka*
[2] *Rajarata University of Sri Lanka, Department of Physical Science, Faculty of Applied Sciences, Mihinthale 50300, Sri Lanka*

10.1 Demand and Supply of Energy

The world is becoming increasingly energy hungry. We are utilizing more and more energy to make our lives more comfortable and safer. According to the International Energy Outlook 2016 (IEO2016), there will be a significant increase in worldwide energy demand from 2012 to 2040 [1]. As it is forecasted, the total world consumption of marketed energy will increase from 5.79226×10^{20} J (549 quadrillion British thermal units [Btu]) in 2012 to 8.59871×10^{20} J (815 quadrillion Btu) in 2040 [1], which is almost a 48% increase and could be attributed to the pure dependency of the modern life on the prevailing energy resources. While the global demand for energy is springing up, the growth of energy resources to meet the demand side is indeed questionable. Thus, amid this disruption of equilibrium of energy demand and supply, the world is also consuming nonrenewable energy resources, which leads to endangering the entire life on Earth. Consequently, there is a growing consensus among researchers regarding the increasing global-level detrimental consequences of the overexploitation of extant energy resources and crafting sustainable solutions to create a demand and supply equilibrium for energy.

Currently, the fossil fuel supply is about 87% (oil, 33%; natural gas, 24%; and coal, 30%) of the total world energy [2]. The predictions on peak oil and oil depletion may change however, it is obvious that the fossil fuel resources will be depleted in the near future due to large consumptions [3]. In addition, environmental and ecological damages due to oil drilling, transportation, and combustion have become an acute problem today [4, 5], especially the emissions of greenhouse gases, carbon, sulfur dioxide, nitrogen oxides, etc. The situation with the dirty fuel, coal, is worse than that of oil and natural gas.

Nuclear power is another energy source that is also potentially life-threatening, mainly due to accidents in the nuclear power plants and issues created by radioactive waste disposal. Ninety-nine nuclear accidents, resulted in either the loss of human life or property damages worth of more than US$ 50 000, have been

Polymer Electrolytes: Characterization Techniques and Energy Applications,
First Edition. Edited by Tan Winie, Abdul K. Arof, and Sabu Thomas.

reported in nuclear power plants from 1952 to 2009 [6]. Further, the possibility of disasters such as those happened in Chernobyl (1986) and Fukushima (2011), the risk of acquiring highly dangerous nuclear weapons by terrorist groups, problems associated with nuclear waste disposal, possible damages to plants due to earthquakes and tsunami, and the contamination of the environment by radioactive elements due to imperfect constructions of power plants and inappropriate waste disposal are the main problems associated with nuclear power.

On the other hand, the utilization of hydropower is limited by the availability of resources and can only be used in certain locations and climates, for example, it is difficult to utilize in deserts or during drought seasons.

In summary, there are plentiful energy resources on Earth. However, as discussed, the irresponsible utilization of existing energy resources has led to numerous problems such as environmental damages and rapid depletion of resources. When we utilize energy resources, which are limited on Earth, or their regeneration process is very slow, then we should ensure that some portion of those resources is sustained for future generations. In other words, it is very important to convert abundantly available energy resources on Earth to usable forms of energy to fulfill present and future energy requirements.

Electricity and chemical energy are the most usable forms of energy. In particular, electricity is more advantageous as it provides facile conversion to other forms of energy that are being widely used every day.

For example, electricity can be converted to following forms of energies very easily:

(1) Heat energy (e.g. hotplates, ovens)
(2) Sound energy (e.g. radios, mobile phones, TV, etc.)
(3) Light energy (e.g. tungsten filament or light-emitting diode [LED])
(4) Mechanical energy (e.g. electric motor)
(5) Chemical energy (e.g. batteries)

In general, a solar cell is a device that directly converts sunlight into highly usable forms of energy, i.e. electricity or chemical energy.

10.2 The Sun as a Potential Energy Resource

In the sight of rapid depletion of fossil fuels, there is a growing consensus among researchers on converting highly available and long-term existing sources of energy to usable forms. Solar energy is recognized as a promising energy source due to many reasons. The annual average solar energy receives on Earth is enormous, about 3×10^{24} J. Significantly, this amount exceeds the present global energy consumption by a factor of 10^4 [7], which means that the amount of energy received on Earth in one hour is more than enough to fulfill the energy needs of entire the world for a year. To illustrate, the total annual energy consumption in the year 2014 was about 3×10^{20} J [8] and the entire energy requirements of the world can be fulfilled if 0.1% of the Earth's surface is covered by solar cells with 10% efficiency.

10.3 Advantages of Solar Cells

As discussed previously, the sun is considered as one of the best energy resources. Significantly, sunlight is freely available in spite of the diverse political, religious, national, ethnic, or economic grounds of the nations. Hence, due to the zero cost for the input energy source (sunlight), one has to pay only for the installation of solar cells for power generation. The additional advantage is that the maintenance cost of solar cells is negligible since there are no moving or mechanical parts in them. Further, they run on free fuels (sun light) come to every doorstep.

The solar cells are zero emission devices and, thus, they are considered as green energy sources. Notably, solar cells do not cause any detrimental environmental consequences unlike the widely utilized energy resource, i.e. fossil fuel [4, 5]. Consequently, the usage of solar cells is increasingly receiving global attention due to the minimal environmental damage they cause. The solar cells are suitable to be installed even in urban or crowded areas due to zero emission, silent functioning, safety, and free and abundant supply of sunlight anywhere. Also, the solar panels can be installed on rooftops and thus they do not need extra space. Additionally, they are safe and less hazardous, compared to many other resources. The solar cells do not produce any harmful waste products, whereas the waste disposal is the main problem associated with nuclear power generation. Unlike nuclear energy, there is no threat of using solar cells for war activities or threat of acquiring the technology or waste products by terrorists.

The solar cells convert light energy directly to electricity, which is the most useful form of energy as mentioned earlier. They can directly be associated with low power applications such as mobile phones, calculators, garden lamps, toys, torches, watches, charges, etc.

10.4 Photo-Electrochemical Solar Cells

Up to now, the solar cells have been evolved through three generations. Single crystal and multi-crystal solar cells based on silicon wafers belong to the first generation [9], and the most frequently used solar cells today are these first-generation ones. Significantly, though their cost of production is relatively high, they exhibit relatively high efficiencies of about 15–20%.

The second-generation solar cells are based on thin films of amorphous silicon, cadmium telluride/cadmium sulfide (CdTe/CdS), copper indium selenide (CIS), and copper indium gallium selenide (CIGS), etc. Their efficiencies are lower and lie between 10% and 15%, but the production cost is lower compared to the first-generation cells. These two factors, i.e. the efficiency and the production cost, do compete with each other, and therefore the cost per watt is a more appropriate parameter to assess the performance of solar cells since surface coverage is not a big issue. It is notable that the cost per watt produced is lower in second-generation cells compared with the first-generation ones [9]. In addition, the thin films used for the second-generation cells can also be grown on flexible substrates to produce more flexible solar cells. However, most of the materials used to prepare second-generation cells are toxic.

The third-generation consists of a variety of solar cells that are showing their potential to fulfill the future energy needs with prospected higher efficiencies and cost reduction. Some cells have already shown better performances; however, long-term stability and scaling up are found to be challenging.

The third-generation solar cells are photo-electrochemical (PEC) solar cells that include:

i. Dye-sensitized solar cells (DSSCs)
ii. Nanocrystalline or nanostructured semiconductor thin film-based solar cells
iii. Quantum dot-sensitized solar cells (QDSSCs)
iv. Organic–inorganic hybrid perovskite-based solar cells.

There is a growing body of research on improving third-generation solar cells, specifically, to reach higher energy efficiencies at lower cost. There are a number of novel technologies such as use of organic–inorganic hybrid perovskites and quantum dots (QDs) exhibiting better performances. However the commercialization process of those technologies has been slowed down due to the particular issues related to chemical and physical stability of the cells.

Among all the types of solar cells, the DSSC is the widely studied one within the last two decades due to the wider choice of improving different components in the cell and high possibility of using facile and low cost preparation methods. One such study is by O'Regan and Grätzel (1991) [10] who reported a DSSC prepared by using dye-sensitized nanocrystalline TiO_2 films and recoded the energy conversion efficiency of 7.1%. Up to now, a variety of DSSCs has been prepared by using different materials. Accordingly, there is a mounting interest among researchers on a wide-bandgap nanostructured semiconductors such as TiO_2, ZnO, NiO, and MgO, with a variety of morphologies, nanostructures, and arrays, as well as with different dopants as the photo-anode [11]. Sensitizers, in particular, ruthenium metal-based dyes [12], organic metal-free dyes [13], and natural pigments [14–16] have also attracted wider attention of contemporary investigators as light-harvesting species in DSSCs. Also, an increasing number of research initiatives could be observed to find promising electrolyte for DSSCs [17–19]. Iodide tri-iodide redox couple is the widely studied redox couple to be used in DSSCs. In addition, the electrolytes based on cobalt have also shown the potential to be used in DSSCs [20].

The records report a higher energy conversion efficiency value of 11–13% for DSSCs using sensitizer ruthenium polypyridyl complexes, zinc porphyrin complexes, and metal-free organic dyes containing carboxy-anchor groups [20–22]. The efficiency in DSSCs has enhanced to about 14% by capitalizing the advantage of collaborative sensitization of multiple dyes [23].

A large number of sensitizers and electrolytes have been studied for DSSCs. However, the QD is recognized as a reliable alternative sensitizer to harvest visible light efficiently. In addition, quasi-solid-state electrolytes are showing their potential to be a stable and efficient redox mediator of QDSSCs. Prior to the discussion of polymer electrolytes-based QDSSCs, it would be useful to understand the mechanisms of different PEC solar cells.

10.4.1 General Mechanism of a Photo-Electrochemical Solar Cell

In general, solar cells can be categorized as photovoltaic devices and PEC devices. The photovoltaic cell is a Schottky barrier type solar cell, which uses n- or p-type semiconductor as well as n- and p-type junctions. The action of a solar cell is the conversion of energy in light into electricity. The mechanism of all the solar cells is similar since they all are to harvest light energy. However, different cells use different materials to absorb light energy to excite electrons (charge generation) and transport them to external circuit (charge separation and transport).

In general, the solar cell mechanism can be divided into three processes:

1. Charge generation
2. Charge separation
3. Charge transport

The photons whose energy is greater than E_g can be absorbed by electrons in the valence band (VB) and these electrons jump to the conduction band (CB) leaving out holes in the VB, and thus creating conduction electrons and holes. This process is the charge generation. The charge generation alone is not sufficient for the solar cell action and there should be a mechanism for these excited electrons to move away before they relax in order to prevent the excited electrons from falling back into their ground state or recombine.

There are many ways to interpret a charge generation and separation mechanism depending on the materials and the type of the solar cell; the basic idea is depicted in Figure 10.1. The charge separation can be achieved by assisting the

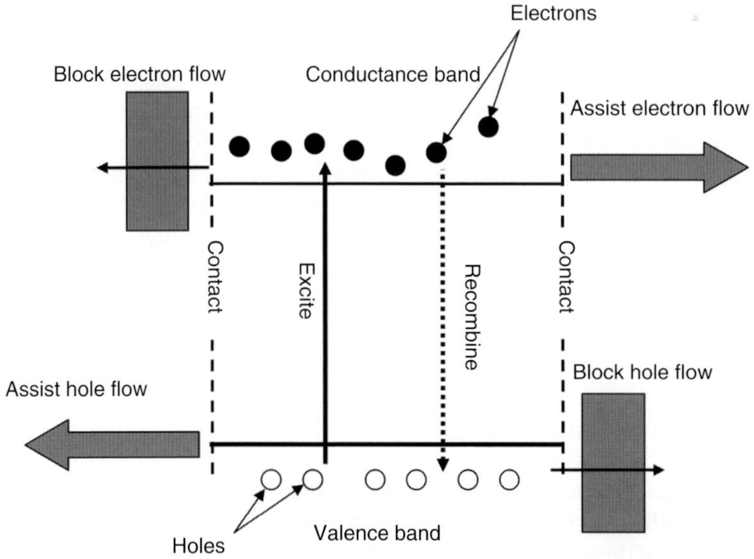

Figure 10.1 A schematic diagram of the desired mechanism for the charge separation in a solar cell. One contact should assist electron flow, but block holes and vice versa in the other.

electron flow by blocking holes to one contact and vice versa in the other side as in Figure 10.1. The charge separation can be understood by band bending at interfaces. The excited electrons can generate a potential difference to drive these CB electrons into the external circuit.

The driving force for photocurrent in a solar cell is the charge separation. In PEC solar cells, the chemical potentials and ion transport in the electrolyte play a significant role to achieve charge separation and transportation.

10.4.2 Mechanism of a Photo-Electrochemical Solar Cell

In a conventional PEC solar cell, the charge generation is done by n-type or p-type narrow bandgap semiconductors. The charge separation is facilitated by band bending at interfaces. Figure 10.2a,b shows the mechanism of a PEC solar cell. As demonstrated in Figure 10.2b, the electrons in the VB of the semiconductor jump to CB by absorbing photons in the visible light. The charge transport is governed by the separation of the Fermi level of the semiconductor and redox potential of the electrolyte when the cell is irradiated. This can be understood by band bending at interfaces as shown in Figure 10.2. The path of the photo-generated electrons is indicated by dotted arrows in Figure 10.2. In a PEC solar cell, the electrochemical junction plays an important role in achieving charge separation and transportation.

The contact between a semiconductor and an electrolyte represents an electrochemical junction. The charge transportation to the semiconductor surface is performed by a redox couple in the electrolyte, and the behavior of the charge transport at the interface depends on the positions of Fermi level of the semiconductor and redox potential (level) of the electrolytes.

Figure 10.3 shows the schematic diagram of energy levels of n-type semiconductor and electrolyte before contact. After a semiconductor is kept in contact with an electrolyte, initially, the charges travel across the junction and reach an equilibrium. At the equilibrium, the Fermi level (E_F) of the semiconductor and redox potential of the electrolyte (E_{redox}) reach the same level as shown in Figure 10.4 [24]. Therefore, at the equilibrium band bents at the interface, as illustrated in Figure 10.4 for n-type semiconductor and electrolyte where the work function of the semiconductor, is higher than the redox potential of the electrolyte.

Therefore, when an n-type semiconductor, such as TiO_2, is kept in contact with an electrolyte containing appropriate redox couple (e.g. iodide/tri-iodide), a small potential barrier is created at the interface. This barrier prevents the recombination of excited electrons with ionic species in the electrolytes. In other words, this barrier blocks the flowing of the CB electron to the electrolytes. Further, the bending of the VB at the interface, as illustrated in Figure 10.4, facilitates the flow of VB electrons to the electrolyte through the barrier.

The energy levels of the oxidized and reduced species can be associated with the CB and VB energies, respectively, for interfacial charge transport. The behavior of the electrochemical junction is almost similar to that of the Schottky barrier. However, the disadvantage of this junction is that the semiconductor surface

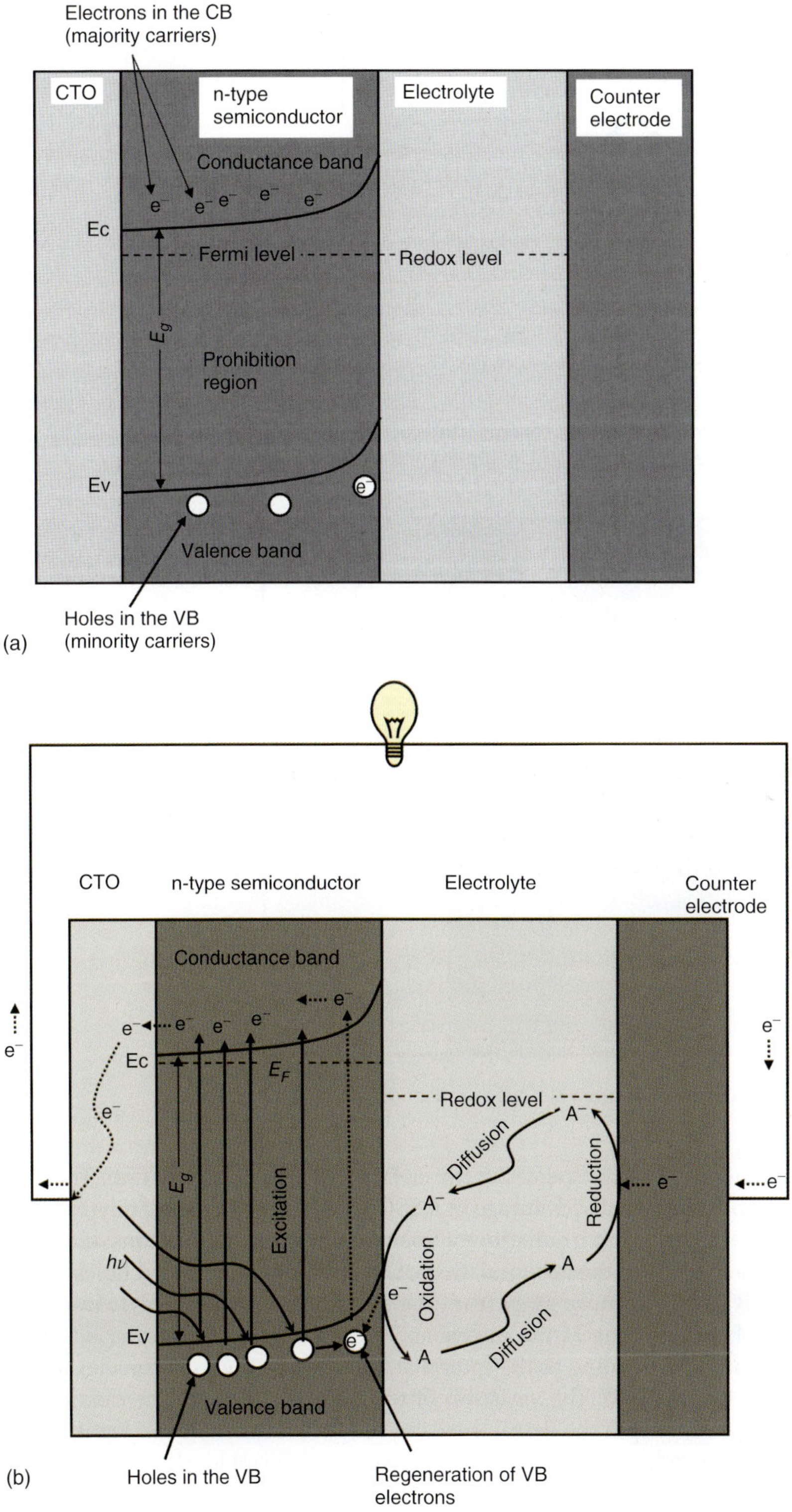

Figure 10.2 The mechanism of a PEC cell, containing an n-type semiconductor (a) in dark and (b) in light.

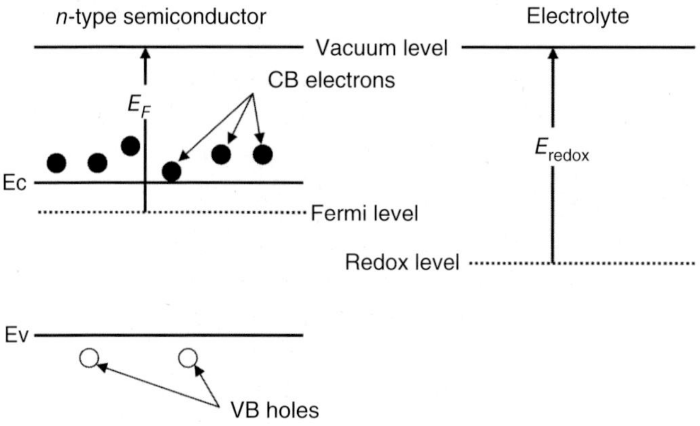

Figure 10.3 Schematic diagram of energy bands of n-type semiconductor and conductor before making the contact. Filled circles represent the conduction electrons.

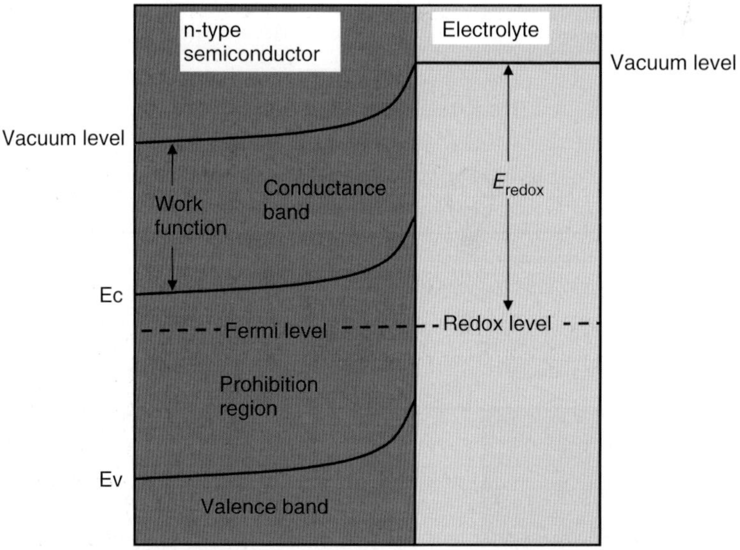

Figure 10.4 Schematic diagram of n-type semiconductor/electrolyte junction.

can react chemically with species in the electrolyte under illumination. This has become a considerable disadvantage of DSSCs based on liquid electrolytes.

As discussed earlier, an irradiation of a semiconductor with photons of energy higher than E_g leads to the accumulation of electrons in the CB and holes in the VB. The CB electrons move away from the semiconductor/electrolyte interface due to the band bending at the interface. The VB holes move to the interface as a result of band bending at the interface and oxidize the ionic species in the electrolyte (Figure 10.5). The variation of the number of charge carriers in the CB and VB leads to split the Fermi level and redox level as shown in Figure 10.5. Therefore, the irradiated semiconductors have quasi-Fermi levels (Figure 10.5).

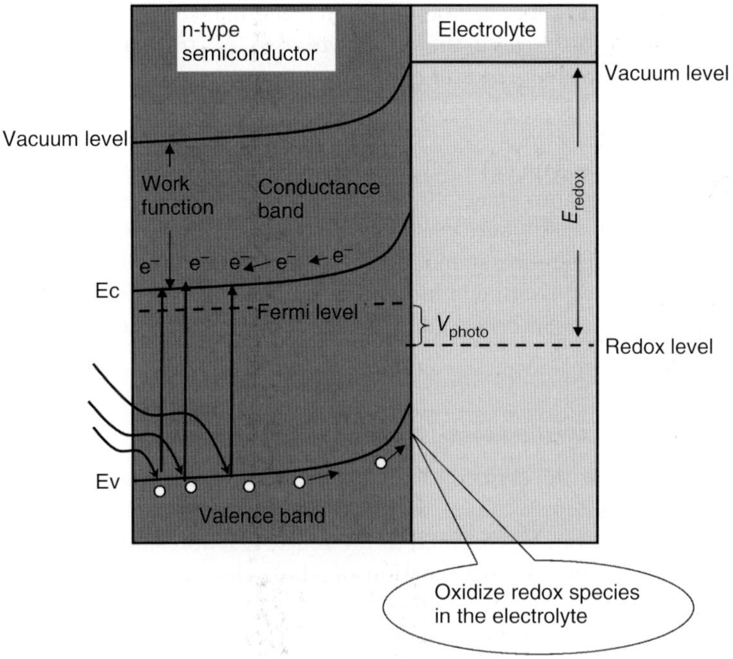

Figure 10.5 Under illumination, photo-generated electrons accumulate in the n-type semiconductor. This process raises the electron Fermi level and generates a photovoltage (V_{photo}).

The difference between the Fermi level and the redox level (E_F and E_{redox}) at the equilibrium under the illumination is the photovoltage V_{photo} of the cell (Figures 10.2b and 10.5). This V_{photo} is the open circuit voltage (V_{oc}) of the solar cell when there is no current in the external circuit (external circuit is open; Figure 10.5). As a result of potential deference created, CB electrons travel through the external circuit and reduce the ions in the electrolyte at the electrolyte/counter electrode interface (the external circuit is closed; Figure 10.2b). Therefore, a photo-generated current can be observed through the external circuit. The photocurrent depends on the flow of excited electrons through the external circuit. The maximum photocurrent flows through the external circuit when the external resistance is zero (short circuit). This maximum photocurrent is the short circuit current (I_{sc}) of the cell.

A certain number of holes created are directed toward the *n*-type semiconductor/electrolyte interface due to band bending, thereby oxidizing the reduced species in the electrolyte solution (Figures 10.2b and 10.5). If the redox species in the electrolyte are A/A^-, reaction (10.1) will take place at the electrolyte/semiconductor interface:

$$A^- + h^+ \rightarrow A \tag{10.1}$$

The accumulated photo-generated electrons in the CB are directed toward the external circuit due to the band bending, and thus they flow to the external circuit through an ohmic contact. These electrons return to the electrolyte in the PEC cell through a counter electrode, where a reduction reaction takes place as shown in Figure 10.2b. If the redox species in the electrolyte are A/A^-, the reaction (10.2) will take place at the electrolyte/counter electrode interface:

$$A + e^- \rightarrow A^- \tag{10.2}$$

At the equilibrium, there is no net reaction in the electrolyte. The redox species will shuttle between the photo-electrode and the counter electrode. This photocurrent is absent in the dark where the concentration of holes is very low, and hence no reaction with the redox species in the electrolyte is possible. On the contrary, in the case of p-type semiconductors under irradiation, electrons assist in a reduction process at the interface and a current is produced by the holes in the semiconductor.

10.4.3 Semiconductor/Polymer Electrolyte Junction

The chemical reactions at the semiconductor/electrolyte junction can be suppressed by replacing aqueous medium by a polymer medium. In this way, the reactions at the interface can be confined to the redox reaction since polymers are chemically inert. Hence, the degradation of PEC solar cells can be minimized by using polymer electrolytes. Polymer electrolytes are more reliable for dye-sensitized solar cells where photo-electrons are produced with the help of a photo-sensitive dye and the charge separation is produced by a semiconductor and a polymer electrolyte [25, 26].

10.4.4 Photo-sensitization of Wide Bandgap Semiconductors

As discussed earlier, in order to achieve the charge generation, the bandgap energy of the semiconductor should be less than the energy of photons incident on the semiconductor. However, wide bandgap semiconductors (semiconductors with bandgap higher than incident photons) can also be used to prepare PEC solar cells. So as to the wide bandgap semiconductor should be photosensitized by using dyes, QDs, nanoparticles, etc. Here, these sensitizers get excited by absorbing photons in the visible light and the excited sensitizers inject electrons to the CB of the semiconductor as illustrated in Figure 10.6. In this case, the semiconductor is acting as a mediator for charge transport. A wide range of dyes and metal nanoparticles, such as Au, Ag, etc., have been studied to be used as photo-sensitizers.

However, the main focus of this chapter is the discussion of QDSSCs. PdS, CdS, and CdSe have widely been used as sensitizer for QDSSCs. In addition, the QDs based on CdTe, $CuInS_2$, Cu_2S, PbSe, InP, InAs, Ag_2S, Bi_2S_3, and Sb_2S_3 are also considered as suitable sensitizers for solar cells.

A schematic diagram of a dye-sensitized solar cell is given in Figure 10.6, which also illustrates the mechanism of the photo-sensitized wide bandgap semiconductor-based solar cells.

In DSSCs, the dye (sensitizer) absorbs the photons of the light, and the electrons in the dye get excited (some electrons in the dye move from a ground state to an excited state). This is the photoexcitation of sensitizer S to S*:

$$S + h\nu \rightarrow S^*$$

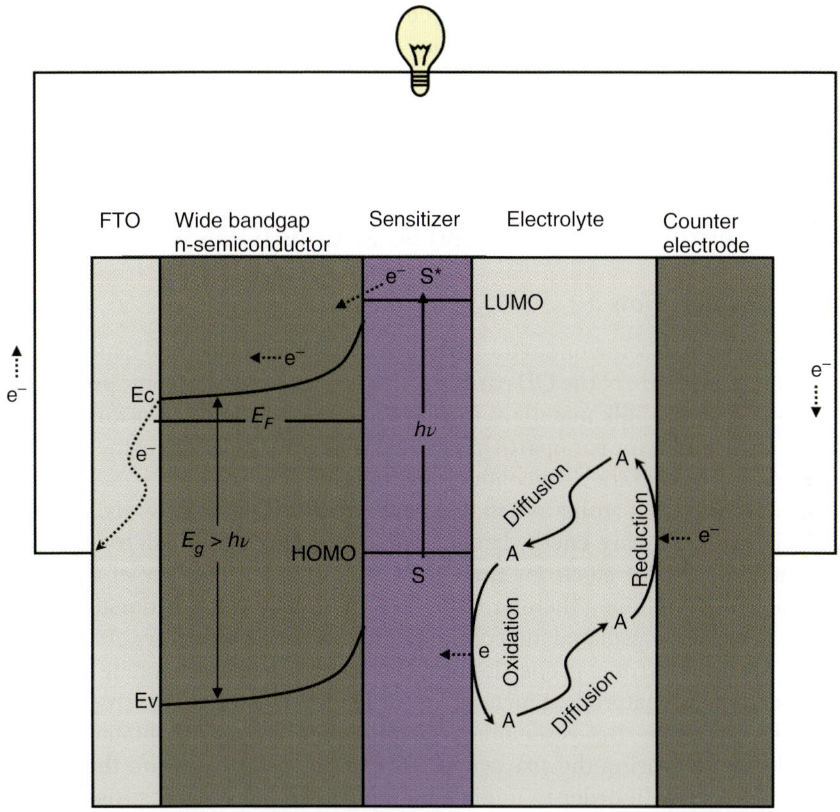

Figure 10.6 A schematic diagram to show the mechanism of a photosensitized electrochemical solar cell. This illustration is for an n-type semiconductor.

The excited dye injects electrons to the wide bandgap semiconductor (e.g. TiO_2) layer:

$$S^* \rightarrow S^+ + e^-_{\text{(Semiconductor)}}$$

The injected election to the semiconductor percolates to the conductive glass substrate. At that point, this electron travels through the external circuit and reaches the counter electrode (Figure 10.6). The redox species in the electrolyte reduce by taking electrons from the counter electrode. If symbol A/A^- is used for the redox couple, then the reaction at the electrolyte counter electrode interface can be written as

$$A + e^-_{\text{(Counter electrode)}} \rightarrow A^-$$

The oxidized sensitizer molecules (S^+) obtain electrons from the redox species in the electrolyte and regenerate the sensitizer (S). Therefore, the reduced ions in the electrolyte are oxidized at the photo-electrode/electrolyte interface:

$$A^- + S^+ \rightarrow A + S$$

Then, this regenerated sensitizer molecules (S) are excited again by absorbing photons under the continued illumination. The whole process is repeated as far as the cell is illuminated. Therefore, a steady electrical current can be observed in the external circuit till the cell is illuminated. QDs can also be used to photo-sensitize wideband semiconductors instead of dye sensitizers.

10.5 Quantum Dot-Sensitized Solar Cells (QDSSCs)

10.5.1 Quantum Dots

QDs are very small semiconductor particles, and their size is a few nanometers. Due to the smaller size of the QD particles, their optical and electronic properties do deviate from the bulk materials. In addition, the optical and electronic properties of semiconductor nanoparticle can be engineered by changing the size, the shape, and the type of the semiconductor material.

It is well known that atoms and molecules have discrete energy levels, whereas solid bulk materials have energy bands. The QDs are small enough to have discrete energy levels for electrons since there are a limited number of atoms or molecules in the vicinity. Therefore, the QDs demonstrate transitional properties between those of isolated molecules and of bulk semiconductors. As a result, the optical and electronic properties of QDs can be tuned by changing the size, shape, and type of material, as mentioned earlier. For instance, by increasing the size of the QD properties resembling the bulk material can be obtained. Consequently, by decreasing the size properties can be directed toward that of the isolated molecules. In order to understand optical and electronic properties of QDs, it is necessary to study the energy band structures in QDs.

As illustrated in Figure 10.7, the individual atoms have discrete energy levels for electrons. These energy levels are specified as K, L, M, N and s, p d, f. Some of these levels are filled and some are empty, depending on the number of electrons in the atom. However, when two atoms are brought to proximity, each atomic level splits into two levels depending on the distance between two atoms

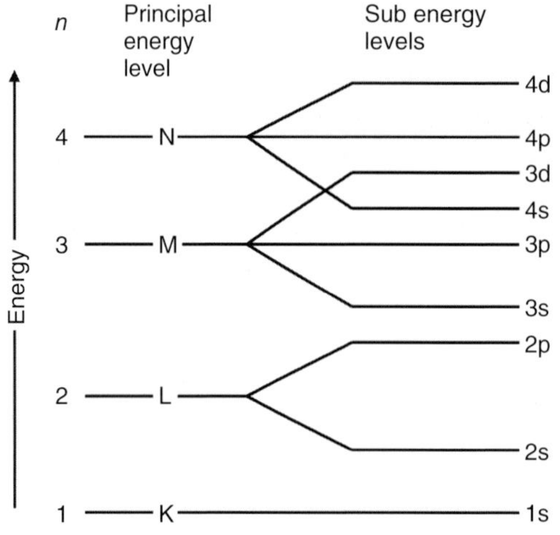

Figure 10.7 Schematic diagram to show energy levels in an isolated atom (the scale is arbitrary).

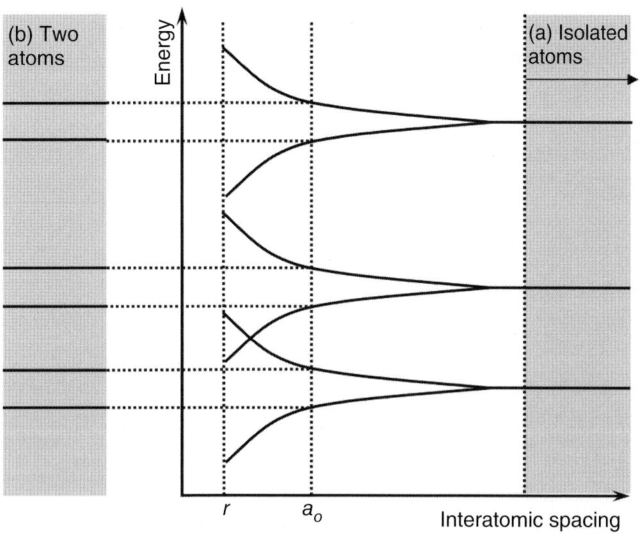

Figure 10.8 Schematic diagram to show splitting of energy levels of isolated atoms when two atoms brought into proximity. (a) Energy levels of an isolated atom. (b) Split atomic energy levels of when the interatomic spacing is a_o. r is the atomic radius (the scale is arbitrary).

as illustrated schematically in Figure 10.8. The right side of Figure 10.8a shows some atomic energy levels (energy levels of isolated atoms), while the left side of Figure 10.8b shows the split atomic energy levels of two-atom system at the atomic separation a_o.

Consequently, when many atoms are brought into proximity, each atomic level splits into many levels depending on the distance between atoms as shown in Figure 10.9. However, a very large number of atoms should be brought close to each other to form a bulk material. Therefore, in bulk materials the atomic energy levels split into so many energy levels, which are very close to each other. These very close energy levels in bulk material can be considered as quasi continues regions for electrons or energy bands. For instance, a schematic diagram of energy bands in a semiconductor is shown in Figure 10.10. Here, the lowest unfilled band is the CB, while the highest filled band is the VB. As Figure 10.10 illustrates, Ec and Ev are the edges of CB and VB, respectively. E_g is energy difference between Ec and Ev. For a semiconductor, E_g is about 1 eV.

QDs are tiny semiconductor particles of size of few nanometers, which can be assumed as a zero-dimensional crystal. They can also be considered as artificial atoms. A bound state of an electron and a hole, which are attracted to each other by the Coulomb force, is an exciton. The exciton can be considered as an electrically neutral quasi-particle that exists in insulators, semiconductors, etc. The size of the bulk material is very much greater than that of the exciton. However, when the exciton is in QDs, the space for exciton is restricted by the size of the QD. The size of the exciton can be geometrically constricted in all three directions by limiting the size of QD (quantum confinement), when the QD size is smaller than that of exciton [27].

Generally, the smaller the size of the crystal, the higher the separation between VB and CB. Therefore, the bandgap energy can be increased by decreasing the

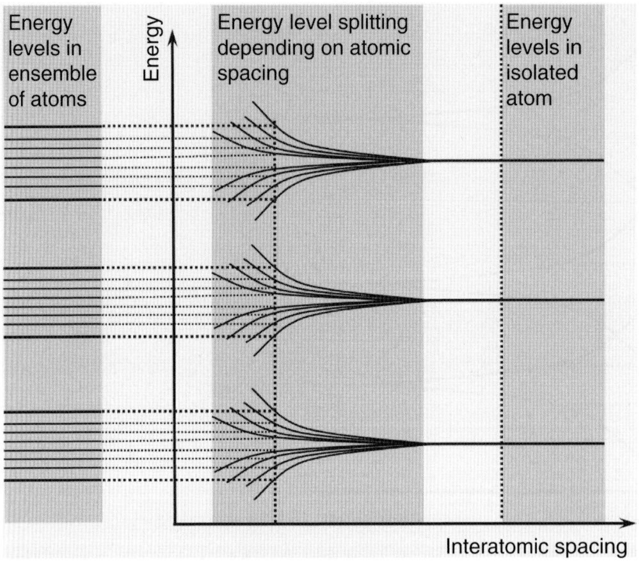

Figure 10.9 Schematic diagram to show splitting of energy levels in isolated atoms when many atoms are in proximity (arbitrary scale).

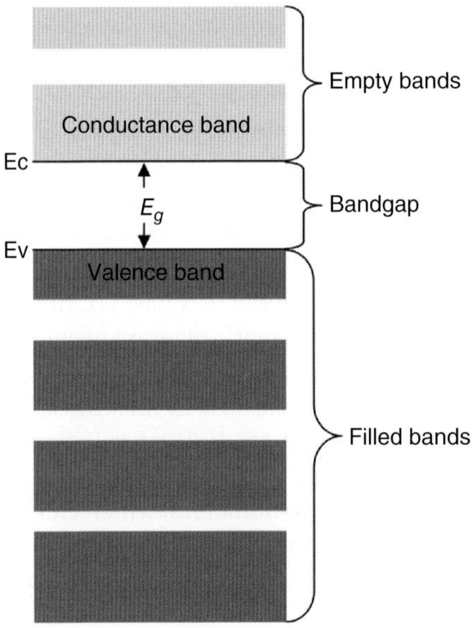

Figure 10.10 Schematic diagram to show energy bands in a semiconductor.

size of the dot. As a result, more energy is needed to excite the electron in the QD than that of the bulk material. The bandgap (E_g) energy of QD particles can be calculated by the modified Brus equation [27, 28]:

$$E_{g(QD)} = E_{\text{Bulk}} + \frac{h^2}{8R^2}\left(\frac{1}{m_e^*} + \frac{1}{m_h^*}\right) - \frac{1.786e^2}{4\pi\varepsilon_0\varepsilon_r R^2} - 0.248(E \times Ry)$$

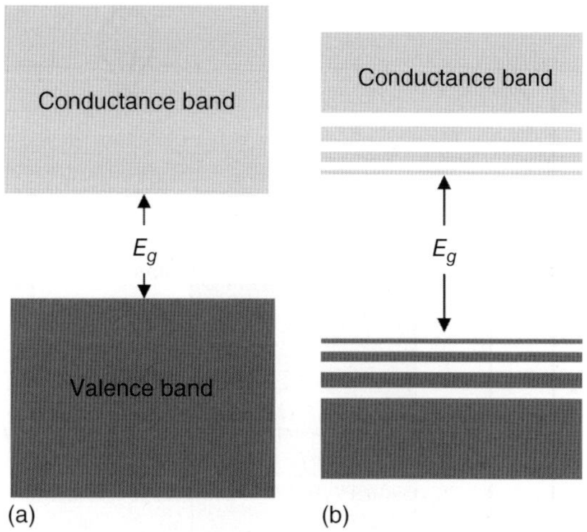

Figure 10.11 A schematic diagram to illustrate valence and conductance band positions and increase in E_g. (a) Bulk semiconductor and (b) quantum dot.

where

$E_{g(\mathrm{QD})}$ – bandgap energy of the quantum dot
E_{bulk} – bandgap energy of bulk semiconductor
R – radius of the quantum dot
m_e^* – effective mass of the excited electron
m_h^* – effective mass of the excited hole
h – Planck's constant
ε_0 – permittivity of vacuum
ε_r – relative permittivity
$E \times Ry$ – Rydberg energy

The energy bands or levels of a QD are illustrated schematically in Figure 10.11. As discussed earlier, the QDs show intermediate properties between bulk metals and individual atoms. The QDs have neither the discrete energy levels like atoms nor the energy bands like bulk materials. The number of atoms in a QD is not large enough to show smooth energy bands. Therefore, in QDs the CB and VB split or show smaller bands as illustrated in Figure 10.11. The bandgap of QDs is larger than that of the bulk material.

As mentioned earlier, the bandgap energy increases with the decreasing size of the QD. Therefore, larger dots can excite low energy photons (longer wavelengths), whereas smaller dots can excite by high energy photons (short wavelengths). The size dependent of bandgap energy of QDs can be used to prepare solar cells with wide absorption spectrum.

10.5.2 Mechanism of a QDSSC

The mechanism of the QDSSC is different from that of the other PEC solar cells. In QDSSCs, the photons in light are absorbed (harvested by) QDs in the photo-electrode. The excited electrons in QDs by absorbing photons transferred to a wide bandgap semiconductor substrate such as TiO_2, SnO_2, ZnO, etc. The mechanism of a QDSSC is illustrated schematically in Figure 10.12. The

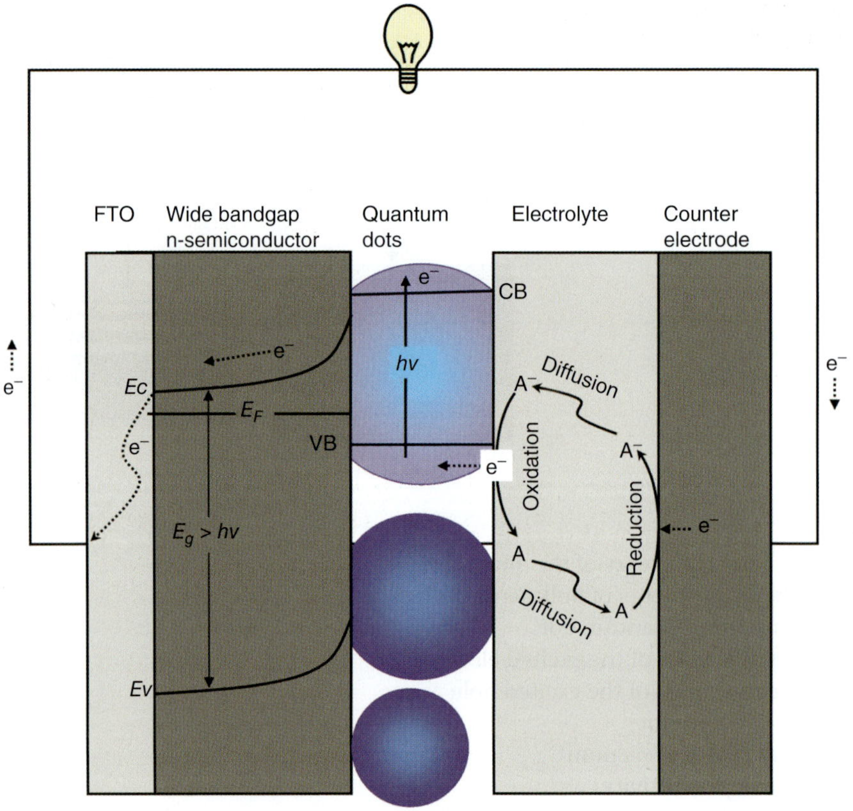

Figure 10.12 A schematic diagram to show the mechanism of a QDSSC.

electrons transferred to the wide bandgap semiconductor reach the QD after traveling through transparent conducting oxide (TCO), external circuit counter electrode, and electrolyte, as shown in Figure 10.12.

10.5.3 Quantum Dot-Sensitized Solar Cells (QDSSCs)

A schematic diagram showing the configuration of a QDSSC is given in Figure 10.13. In general, QDSSC contains a QD-sensitized photo-electrode, an electrolyte, and a counter electrode. In a QDSSC, a porous, nanostructured layer of a wide bandgap semiconductor prepared on TCO substrate is photo-sensitized by QDs, similar to the DSSC sensitized by the dye. The commonly used wide bandgap semiconductor is TiO_2 films prepared on fluorine-doped tin oxide (FTO) substrate. Thin nanostructured TiO_2 films are prepared mainly by using a TiO_2 nanopowder containing colloidal suspension. The widely used nanostructured film preparation methods used for QDSSC preparations are as follows:

(i) Doctor blade
(ii) Spin coating
(iii) Screen printing
(iv) Spray pyrolysis method

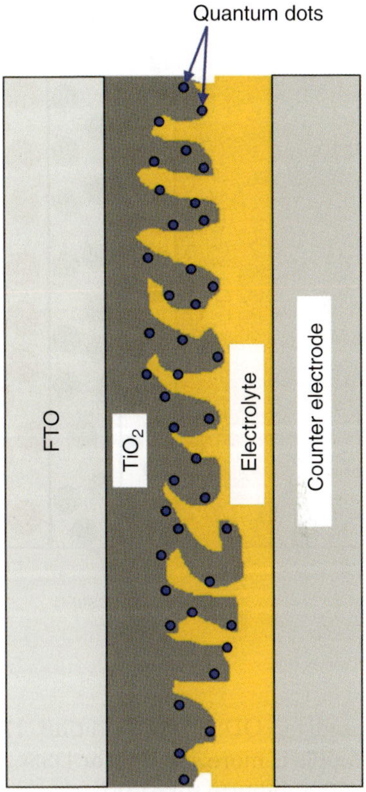

Figure 10.13 Schematic diagram to show the configuration of a QDSSC.

The film can then be made photoactive (photo-sensitized) by coating semiconductor QDs. The QD layers are commonly deposited by using following methods:

(i) Chemical bath deposition (CBD)
(ii) Electrophoretic deposition (EPD)
(iii) Electro-depositing (ED)
(iv) Successive ionic layer adsorption and reaction (SILAR)
(v) Spin coating (SP)

After preparing the photo-anode, the QDSSC assembly is then completed by sandwiching a liquid or solid electrolyte containing appropriate redox couple between the photo-anode and counter electrode as shown in Figure 10.13. The commonly used counter electrode materials are as follows:

(i) Platinum (Pt)
(ii) Copper(I) sulfide (Cu_2S)
(iii) Gold (Au)
(iv) Carbon (C)

Cu_2S-based counter electrodes are more reliable for QDSSCs, while Pt electrodes are more suitable for DSSCs. Further, the polysulfide electrolytes are widely used in QDSSCs, and they have shown relatively higher efficiencies [29]. In addition, the electrolytes containing iodide/tri-iodide redox couple are also

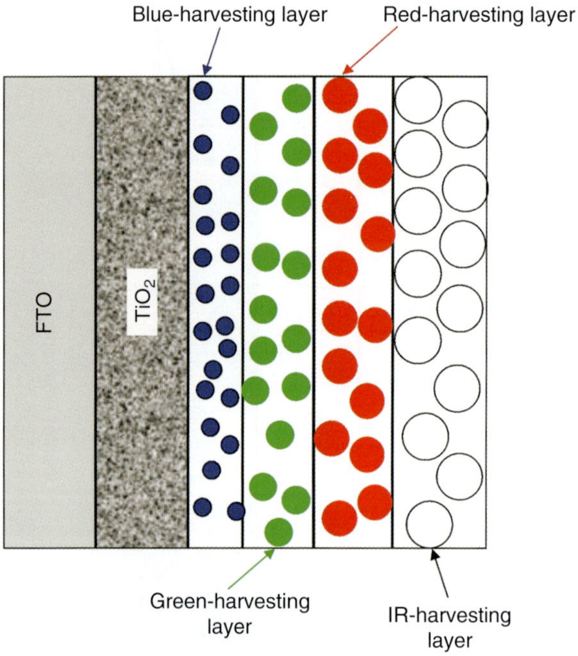

Figure 10.14 Multiple layers of QDs can be used to improve the light-harvesting efficiency in QDSSCs.

used for QDSSC preparation. However, it seems that iodide/tri-iodide redox couple is more suitable for DSSCs than for QDSSCs.

The ability to tune the bandgap makes QDs more desirable for solar cells, which give the possibility of preparation photo-electrodes for solar cells containing different layers of QDs, which can harvest different wavelength ranges in the sunlight as schematically illustrated in Figure 10.14. For instance, smaller dots can harvest shorter wavelengths such as green and blue, while bigger dots can harvest longer wavelengths such as red. The size of the QDs can be further increased to harvest infrared (IR) region of the sun light. In that way, most of the photons in visible light can be harvested and then improve the light to electricity conversion efficiency.

The QDSSCs are reliable devices to fulfill future energy demand due to their prospective higher efficiency, stability, facile, low cost fabrication techniques, and environmentally friendly nature in operations. In order to get higher efficiencies, the bandgap of wide bandgap semiconductor can be tuned by changing the size and type of QDs and, thus, very high efficiencies are expected from QDSSCs [30, 31]. However, the energy conversion efficiency of QDSSC is still less than that of DSSC, and therefore more future research initiatives on developing QDSSCs are vitally important.

The following section reviews the current status of polymer electrolyte-based QDSSCs. The key focus areas of the section are fabrication method and material and cell efficiency. The section also discusses the limitations, strengths, and possibilities for further improvements of QDSSCs.

10.5.4 Polymer Electrolytes for QDSSCs

The studies on solid-state ionic conductors have led to a new discipline commonly termed as solid polymer electrolytes (SPEs) specifically with the discovery of the ability to dissolve ionic salts in poly(ethylene oxide) (PEO) polymer disrupting the crystallinity of the material and ion conducting properties of the PEO complexes by P.V. Wright and coworkers in the 1970s [32, 33]. Subsequently, Armand et al. [34] proposed the suitability of such electrolytes used in solid-state batteries, and the discovery was extended to other polymers such as poly(propylene oxide) (PPO) and various salts that have broadened this field of study. Since then, polymer electrolytes have received increased research interest as a distinct research field. This trend is further intensified with the discovery that the polymer electrolytes as a sustainable solution for many applications such as secondary batteries, fuel cells, electrochromic devices, and supercapacitors and PEC solar cells.

It is apparent that the practical application of polymer electrolytes can bring about numerous advantages due to their low cost, easy fabrication, possibility to prepare as thin films, and mechanical flexibility. The flexible nature of polymer electrolytes brings the advantage in fabrication of space-efficient battery designs with variable dimensions, flexible or bendable solar cells, etc. Additionally, their low reactivity and the chemical inertness of the host polymer is remarkable. Therefore, the electron transfer is confined to the solute and the redox reaction is limited to the electrode/electrolyte interface. Since polymers do not react with QDs, the polymer electrolytes enhance the stability of QDSSCs as well. Furthermore, polymer electrolytes appear to be the best compromise between liquid and solid inorganic electrolytes [35]. The device scaling up and layer by layer (LBL) assembly is ensured by the possibility of preparation of thin polymer electrolyte films with large area.

In the recent past, the gel polymer electrolytes (or the quasi-sold-state electrolytes) have attracted wide researcher attention due to their proven capability to offer better chemical and physical stability for PEC solar cells. In general, the gel polymer electrolytes have higher ionic conductivities compared to that of SPEs. However, their conductivity is less than that of the respective liquid electrolytes. The gel polymer electrolytes are increasingly getting the popularity among many researchers since they offer facile cell preparation and higher long-term stability and efficiencies close to that of liquid electrolyte-based devices. The long-term stability of the device is a result of suppression of evaporation of solvents in the electrolyte by the host polymer.

In general, the quasi-solid-state electrolytes are prepared by blending ingredients of the liquid electrolyte with organic polymer matrix or gelator. However, there are many methods to fabricate gel polymer electrolytes. Commonly used polymer matrixes are based on PEO (or polyethylene glycol [PEG]), poly(methyl methacrylate) (PMMA), poly(ethylene glycol) methyl ether methacrylate (PEGMA), polyacrylonitrile (PAN), poly(vinyl chloride) (PVC) and poly(vinylidene fluoride-*co*-hexafluoropropylene) (PVdF-HFP), poly(vinyl alcohol) (PVA), and chitosan. These host polymers are more suitable for iodide ion-conducting electrolytes intended for DSSCs. Nevertheless, according to reports published so far, the polysulfide ion-conducting hydrogel electrolytes have given better results for QDDSCs.

10.6 Performances of Different QDSSCs Assemblies Based on Polymer Electrolytes

This section discusses the advantages of using polymer electrolytes and QDs in PEC solar cells, their mechanism, and possibilities for further development. It is clear that the polymer electrolyte-based QDSSCs are increasingly accepted as a very reliable candidate to address the future energy demands. Though the results obtained for QDSSCs so far are not very satisfactory, higher efficiencies are anticipated [30]. Furthermore, the stability of QDSSCs can be improved by using polymer electrolytes. Hence, the world needs more research initiatives on preparing stable and efficient QDSSC since this field of study still is in the infancy stage. The following section reviews the research works on preparing polymer electrolyte-based QDSSC. The section particularly discusses the current status of the research, performances of fabricated cells, practical imitations, and the possible alternatives for future improvements.

10.6.1 Quasi-Solid-State QDSSCs Based on Polyacrylamide Hydrogel Electrolytes

For the first time, Zhexun Yu et al. [36] developed a polysulfide hydrogel electrolyte based on chemically cross-linked polyacrylamide hydrogel polymer matrix for quasi-solid-state solar cells. Here, the solar cell is co-sensitized by sensitized CdS and CdSe QDs.

10.6.1.1 Hydrogel Electrolyte with Polyacrylamide

A detailed description of the hydrogel electrolyte preparation was given by Zhexun Yu et al. [36]. To brief out their process, the electrolyte has been prepared by dissolving 0.1 g acrylamide monomer in 0.9 ml of distilled water and adding bis-acrylamide (MBA) cross-linker [36]. The mixture has been degassed for 10 minutes, and then ammonium persulfate (0.4% weight of the monomer) has been added to the solution. To complete the polymerization process, the mixture has been heated to 70 °C for one hour. The reaction is shown in Figure 10.15. Subsequently, xerogel has been prepared by removing the water by heating the hydrogel sample to 80 °C until a constant weight is reached [36]. The gel polymer electrolyte has been prepared by immersing the xerogel into the polysulfide electrolyte (1 M Na_2S and 1 M S in aqueous solution) for 12 hours. The room temperature conductivity of resulted chemically cross-linked polyacrylamide-based polysulfide electrolyte is 0.093 S cm^{-1} [36].

According to Zhexun Yu et al. [36], the QDSSC has been assembled using the following procedure. For the photo-anode preparation, 10-μm-thick TiO_2 film has been prepared on FTO glass substrate. The sequential chemical bath deposition (SCBD) technique has been used to attach CdS and CdSe QDs onto TiO_2

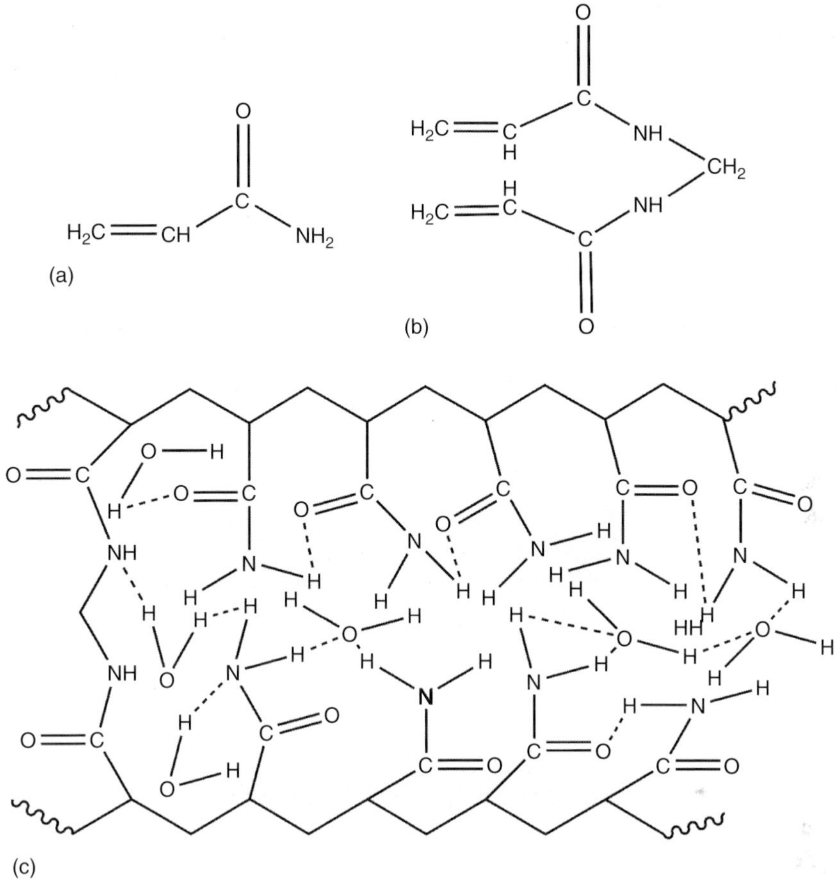

Figure 10.15 Structures of (a) acrylamide monomer (b) bis-acrylamide (MBA) cross-linker, and (c) chemically cross-linked polyacrylamide-based hydrogel. Reproduced from [36] with kind permission from Elsevier.

porous layer. CdS and CdSe have been deposited for 0.5 and 5.5 hours, respectively, at 10 °C [36]. Subsequently, ZnS passivation layer has been prepared twice on CdS/CdSe co-sensitized TiO_2 photo-anode. The QDSSC is assembled by sandwiching a slice of polymer gel electrolyte between a CdS/CdSe co-sensitized TiO_2 photo-anode and a Cu_2S counter electrode [36].

The quasi-solid-state QDSSC based on the hydrogel electrolyte has shown short circuit current density (J_{sc}) of 12.4 mA cm^{-2} and open circuit voltage (V_{oc}) of 534 mV. The energy conversion efficiency and the fill factor (FF) of the QDSSC are 60.1% and 4.0%, respectively [36].

10.6.2 CdS-Sensitized Cell with PAN and PVDF Electrolytes

To the best of our knowledge, the first work on polymer electrolytes based on nonaqueous solvents intended for QDSSC was reported by Kumaraarachchi and coworkers in 2011 [37] where cadmium sulfide (CdS) QDs have been

Figure 10.16 (a–c) SEM images of compact TiO_2 film at different scales.

sequentially assembled onto a nanocrystalline TiO_2 films to prepare a TiO_2/CdS photo-electrode. In order to prepare photo-anode, two layers of TiO_2 dense (compact) and mesoporous layers are coated on fluorine-doped conducting tin oxide (FTO) glasses [37]. For the dense layer preparation, an ethanolic solution containing titanium-tetra-isopropoxide ($Ti[OCH(CH_3)_3]_4$), few drops of HNO_3 (pH 2), and 5 ml of acetic acid has been spin coated on FTO at 3000 rpm for one minute [37]. The scanning electron microscopy (SEM) images of TiO_2 dense layer is given in Figure 10.16a–c, at different scales. Later, films have been sintered at 500 °C for 30 minutes and the process has been repeated to obtain ~100 nm thick compact layer of TiO_2. To prepare a porous layer, a paste of TiO_2 is prepared by grinding 200 mg of TiO_2 (P-25 powder) with 11 drops of glacial acetic acid, one drop of Triton X-100, and about 2.00 ml of ethanol. This paste is coated on the TiO_2 compact layer using the "doctor blade" method and then the films are sintered again at 450 °C for 45 minutes. The thickness of the resulted mesoporous TiO_2 film is about 6 μm [37]. The SEM images of TiO_2 porous layer is given in Figure 10.17a–c at different scales.

The CdS QDs have been deposited onto the mesoporous TiO_2 layer using the SCBD method [37]. The prepared TiO_2 electrode, with compact and mesoporous layers, has been immersed in an ethanolic solution of 0.5 M $Cd(NO_3)_2$ for five

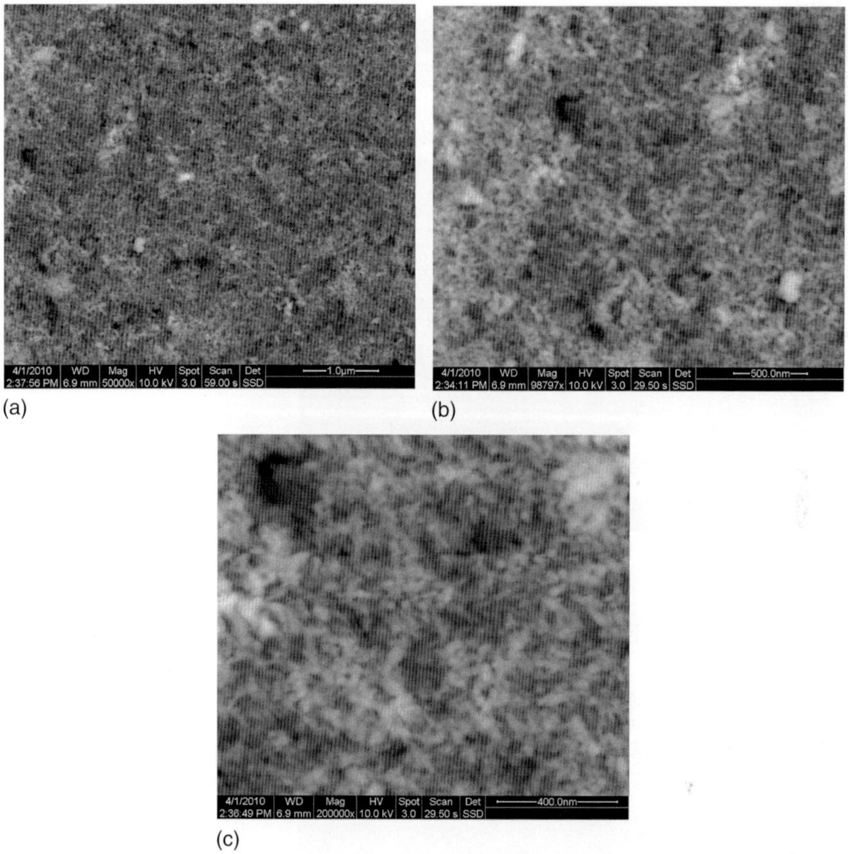

Figure 10.17 (a–c) SEM images mesoporous TiO_2 film at different scales.

minutes, and then rinsed with ethanol, and subsequently immersed in a methanolic solution of 0.5 M Na_2S for five minutes, and then rinsed with methanol. This two-step dipping process is considered as one SCBD cycle. This process has been repeated four times (i.e. 4 SCBD cycles) to get the desired number of QDs on TiO_2 substrate. After the synthesis of CdS QDs, the electrodes have been annealed at 400 °C for 10 minutes under N_2 atmosphere. The SEM images of CdS QDs sensitized photo-anode are shown in Figure 10.18.

In the search for good gel polymer electrolyte for CdS QDSSCs, the authors have tried out several redox couples and polymers polyvinylenefluoride (PVdF) and PAN together with nonaqueous solvents [37]. Figure 10.19 shows the structures of polymers PAN and PVDF and nonaqueous solvents, ethylene carbonate (EC) and propylenecarbonate (PC). The QDSSCs have tested with the mentioned TiO_2/CdS electrode and gel type polymeric electrolytes containing polymers and redox couple given below:

(1) PVdF-based polymer electrolyte with polysulfide ($S^{2-}/S_x^{\,2-}$) redox couple
(2) PVdF-based polymer electrolyte with Fe^{2+}/Fe^{3+} redox couple
(3) PAN-based polymer electrolyte with polysulfide ($S^{2-}/S_x^{\,2-}$) redox couple
(4) PAN-based polymer electrolyte with iodide tri-iodide ($I^-/I_3^{\,-}$) redox couple

(a)

(b)

Figure 10.18 (a, b) SEM images of CdS QD-sensitized photo-anode at different scales.

Here, a limited amount of ethylene carbonate and propylene carbonate aprotic co-solvents have been added to gellify the polymer electrolyte. The gel polymer electrolyte system having the composition of 19.6% PVdF: 15.7% EC: 15.7% PC: 47% $Na_2S{\cdot}8H_2O$: 2% S (by weight ratio) has given a conductivity of 1.75 S m^{-1} at room temperature. The QDSSC containing gel polymer electrolyte with this composition has shown a power conversion efficiency of 0.78% with a short circuit current density of 2.6 mA cm^{-2} and 410 mV open circuit voltage under the irradiance of 100 mW cm^{-1} (AM 1.5) [37]. The fill factor of the cell is 0.73.

The other electrolyte systems studied by Kumaraarachchi and coworkers have given lower efficiencies [37]. The efficiencies with PAN-based (I^-/I_3^-) electrolyte,

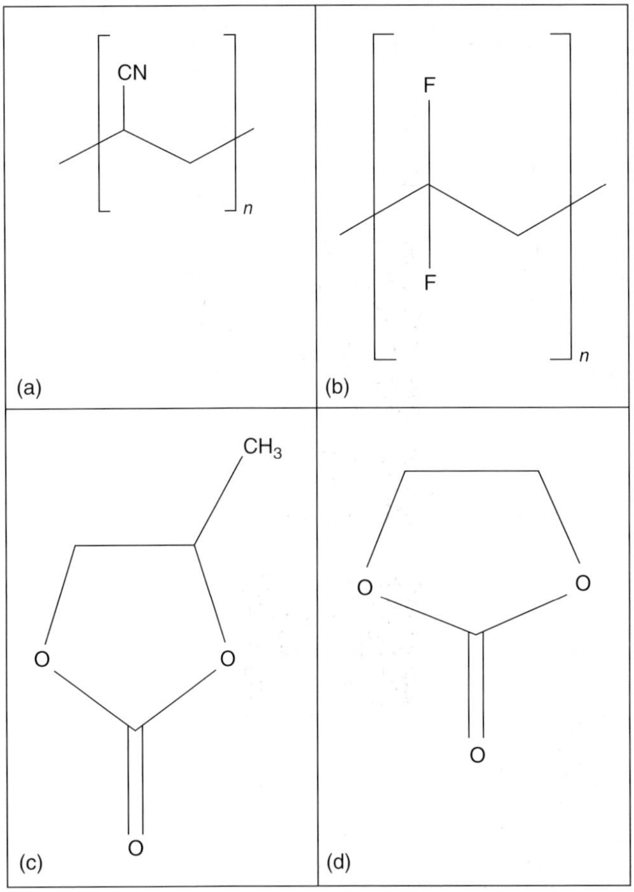

Figure 10.19 The structures of polymers, (a) polyacrylonitrile (PAN) and (b) polyvinylenefluoride (PVDF), and nonaqueous solvents, (c) ethylene carbonate (EC) and (d) propylenecarbonate (PC).

PAN-based (S^{2-}/S_x^{2-}) electrolyte, and PVdF-based Fe^{2+}/Fe^{3+} electrolytes were 0.24%, 0.16%, and 0.01%, respectively [37]. Thus, this is a strong indication on the necessity of future research activities to find the suitable redox couples, aprotic solvents, host polymers, photo-electrodes with a variety of wide band gap semiconductor, and QDs, which are needed to improve the performance of such devices.

10.6.3 ZnO-Based Quasi-Solid QDSSCs Sensitized with CdS and CdSe

Karageorgopoulos et al. [38] have introduced a new method to fabricate thin, transparent ZnO nanocrystalline films for QDSSCs. For the ZnO film preparation, poly(propylene glycol) bis(2-aminopropyl) has been added to the Zinc acetate solution under vigorous steering and then the mixture has been heated to 80 °C in a closed bottle. The poly(propylene glycol) bis(2-aminopropyl) has been used as a template for the ZnO formation. In their work, they have synthesized ZnO films by using the dip coating method where they prepared ZnO thin films with different particles of sizes from 10 to 30 nm changing the amount of poly(propylene glycol) bis(2-aminopropyl) in the starting solution.

The SILAR method has been used to prepare CdS QDs on ZnO thin films [38]. For CdS QDs preparation, the ZnO film has been immersed in two solutions 0.1 M aqueous $Cd(NO_3)_2$ and then in 0.1 M aqueous Na_2S solutions for 10 minutes, respectively [38]. The SILAR cycle was followed by heating to 100 °C and cycles has been repeated several times. Subsequently, authors have deposited CdSe QDs on ZnO/CdS substrate by CBD technique. Finally, a ZnS passivation layer has been deposited on CdSe QDs to complete the photo-anode preparation. The ZnO films co-sensitized by CdS and CdSe QD and passivated by ZnS have been used to assemble quasi-solid-state QDSSCs [38].

10.6.3.1 Quasi-Solid-State Electrolyte Preparation

Hybrid organic–inorganic material named ICS-PPG230 has been prepared by using poly(propylene glycol)bis(2-aminopropylether) (molecular weight 230) and 3-isocyanatopropyltriethoxysilane according to the procedure described in detail [39]. The gel electrolyte has been synthesized by dissolving 0.7 g of the functionalized ICS-PPG230 in 2.4 g of methanol under vigorous stirring, followed by the addition of 0.25 M sulfur and 1 M Na_2S. Finally, the quasi-solid-state solar cell has been assembled by sandwiching the gel electrolyte between platinized FTO glass and the ZnO electrode co-sensitized by CdS and CdSe [38, 39].

The quasi-solid-state QDSSC prepared with monodispersed ZnO particles have shown maximum efficiency of 1.5% for 12 CdS coating cycles (SILAR) [38]. However, the efficiency increases to 4.5% for the photo-electrode having ZnO-aggregated particles. The higher performance shown for ZnO-aggregated particles resulted from increased light scattering and photon localization within the ZnO film [38].

10.6.4 Natural Polysaccharide Thin Film-Based Electrolyte for Quasi-Solid State QDSSCs

Natural polysaccharide, Konjac glucomannan (KGM) has been studied to be use as the polymer matrix for gel electrolyte in CdS/CdSe QD-sensitized solar cells by Shen Wang et al. [40]. The in situ preparation of the gel electrolyte and Cu_2S counter electrode in one step has provided facile cell preparation method.

Shen Wang et al. have fabricated the electrolyte and counter electrode using the following procedure [40]. In order to prepare the electrolyte, 0.5 g of KGM powder has been dissolved in 100 ml of deionized water by stirring at 80 °C for one hour. Then, after filtering out the impurities, the KGM concentration has been doubled by evaporating the solvent. The KGM sol has been doctor-bladed on HCl-treated brass plate and then about 25-μm-thick film has been prepared by evaporating the water at 80 °C. An aqueous polysulfide electrolyte, containing 1 M Na_2S and 1 M S in water has been added drop-wise onto the KGM after it was cooled down [40]. Both hydrogel electrolyte and counter electrode are ready to use after removing the surface liquid. The conductivity in the quasi-solid-state electrolyte is 0.075 S cm^{-1}, whereas that in liquid electrolyte is 0.078 S cm^{-1}. More than 90% intake of liquid electrolyte into KGM polymer matrix has led to higher conductivity [40].

For the photo-anode preparation, the TiO_2 has been coated on FTO substrate by doctor blade method [40]. The CdS/CdSe QDs have been deposited on TiO_2 layer by the CBD [40] and have been assembled by immersing the TiO_2 film in their respective chemical baths at 10 °C for 50 minutes and 5.5 minutes. Finally, the ZnS passivation layer also has been prepared on QDs to complete photo-anode preparation.

The cells prepared with KGM-based polysulfide electrolyte, TiO_2/CdS/CdSe/ZnS photo-anode, and Cu_2S counter electrode have exhibited energy conversion efficiency of 4.0% under AM 1.5 illumination of 100 mW cm^{-2} [40]. The quasi-solid-state QDSSC has shown excellent stability compared to that of liquid electrolyte-based cell. In order to study cell stability, sealed QDSSCs with liquid and gel electrolytes has been stored in similar condition for 1000 hours. The QDSSC with gel electrolyte has given 73% of its original efficiency after 1000 hours, while that of liquid electrolyte-based cell dropped to 10% [40]. The higher stability in quasi-solid-state QDSSC is due to the strong affinity between polymer matrix and solvent, which can hinder the leakage and evaporation of electrolyte components [40]. The work is a clear example of higher stability offered by quasi-solid-state electrolytes compared with that of liquid electrolytes for device applications.

10.6.5 Dextran-Based Hydrogel Polysulfide Electrolyte for Quasi-Solid-State QDSSCs

Chen et al. [41] have prepared dextran gelator-based hydrogel polysulfide electrolyte for quasi-solid-state QDSSCs. In the reported work [41], the hydrogel electrolyte with gelator concentration of 15 wt% has been used for solar cell preparation and to study the conductivity in gel electrolyte. The conductivity in the gel electrolyte has shown values close to that of the liquid electrolyte [41]. In the study, the heated electrolyte solution (before cooling) has been injected to the space between the preheated QD sensitized photoanodes and Pt counter electrode in order get better pore filling [41].

In order to prepare photo-anodes, screen printed TiO_2 films have been co-sensitized by CdS and CdSe. The CdS and CdSe QDs have been directly electrodeposited successively onto the TiO_2 substrate [41].

Chen et al. have shown that QDSSC containing hydrogel polysulfide electrolyte with dextran gelator gives power conversion efficiency of 3.23% under AG 1.5 G 1 sun under 100 mW cm^{-2} irradiation. The QDSSC prepared with liquid electrolyte (without gelator) has shown efficiency of 3.69% [41]. Additionally, an increase in power conversion efficiency with decreasing light intensity has also reported for this system. For example, an efficiency of 4.58% has been obtained for dextran gelator-based polysulfide electrolyte and CdS/CdSe QD-based solar cells under 0.12 sun irradiation [41]. Therefore, it can be concluded that efficiencies in QDSSCs as well as DSSCs [42] increase with decreasing light

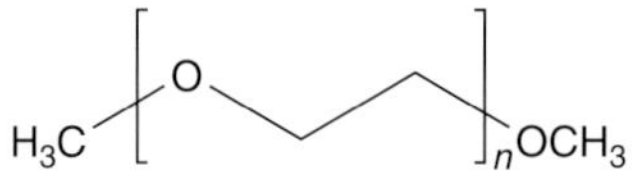

Figure 10.20 The structure of poly(ethylene glycol) dimethyl-ether (PEGDME) host polymer.

intensity. The study provides examples for electrodeposited QD-based solar cells and the performance variation with change of light intensity in such cells.

10.6.6 Carbon Dots Enhance Light Harvesting in a Solid-State QDSSC

Narayanan et al. [43] have investigated a QDSSC with cadmium sulfide (CdS) QDs in zinc sulfide (ZnS) monolayers. Successive ZnS, CdS, and ZnS layers have been prepared on the current collector using ionic layer adsorption and desorption (SILAR) technique. The carbon dots (C-dots) have been deposited using electrophoretic method. The sulfide ion-conducting gel polymer electrolyte is based on copper phthalocyanine (CuPc). The photo-electrodes are prepared in the following configuration for quasi-solid-state QDSSCs [43]:

1. ZnS/CdS/ZnS/
2. ZnS/CdS/ZnS/CuPc
3. ZnS/CdS/ZnS/C-dot/CuPc

The luminescent and conducting C-dots) have been incorporated in the anode to improve the electron transfer and transport characteristics of the solar cell. The studies on fluorescence quenching and lifetime have shown a dominant energy transfer from the blue/green absorbing ZnS/CdS/ZnS donor QDs to the red absorbing CuPc acceptor. The work reveals that the charge transfer rate from CuPc to the current collector through C-dots in the cell is faster and the electron recombination lifetime (with the oxidized species in the electrolyte) is longer [43].

The best solar cell performance has given by the ZnS/CdS/ZnS/C-dot/CuPc configuration. This configuration has given J_{sc}, V_{oc}, FF, and efficiency values of 1.88 mA cm^{-2}, 605 mV, 31%, and 0.35%, respectively. The reported energy conversion efficiency is poor; however the work [43] reveals the increase of charge transport by incorporating C-dots.

10.6.7 Quantum Dot-Sensitized Solar Cells Based on Oligomer Gel Electrolytes

Kim et al. [44] have investigated solar cells based on three-dimensional nanostructured ZnO films sensitized by CdSe/CdS QDs. To improve the stability of this QDSSC, methanol-based polysulfide electrolyte has gellified by adding poly(ethylene glycol) dimethyl-ether (PEGDME) and fumed silica [44]. The structure of PEGDME host polymer is shown in Figure 10.20.

To improve the cell efficiency, ZnO nanowires of about 10 μm long have been grown on FTO substrate using a hydrothermal method [44]. Then, CdS and CdSe QDs have been prepared using SILAR method and CBD method, respectively, on ZnO nanowires. The CdS QDs have been deposited using 20 SILAR cycles. One SILAR cycle involves dipping in a 200 mM $CdSO_4$ aqueous solution for 30 seconds and then rinsing with deionized water for 30 seconds followed by dipping in aqueous 200 mM Na_2S solution for 30 seconds [44]. After preparing CdS dots on the ZnO electrode, it has been immersed in 2.5 mM $Cd(CH_3COO)_2$, 0.25 M Na_2SeSO_3, and 45 mM NH_4OH for three hours at 95 °C to prepare CdSe QDs. The desired thickness of QD layer has been obtained by repeating the process three times [44]. The counter electrode prepared using Pt and carbon nanotubes has increased the electrocatalytic activity at the counter electrode/electrolyte interface.

The optimized cell composed of ZnO nanowires co-sensitized by CdSe and CdS QDs and gellified polysulfide electrolyte has shown 5.45% power conversion efficiency and good stability over 5000 seconds operation time [44]. The study has significantly contributed to the extant knowledge base on the fronts of using nanowire-based semiconductor substrate for QDs and experimenting a higher energy conversion efficiency in QDSSC [44].

10.6.8 QDSSCs with Thiolate/Disulfide Redox Couple and Succinonitrile-Based Electrolyte

A quasi-solid electrolyte based on thiolate/disulfide redox couple has been prepared using organic plastic crystal succinonitrile ($NCCH_2CH_2CN$) as a matrix by Meng et al. [45]. In the study, guanidinium thiocyanate (GuSCN), lithium perchlorate ($LiClO_4$), 4-*tert*-butylpyridine (TBP), or *N*-methylbenzimidazole (NMBI) have been used as additives to improve the cell performance. The optimized electrolyte contains 5-mercapto-1-methyltetrazole *N*-tetramethylammonium:disulfide:$LiClO_4$:NMBI molar ratio of 8 : 8 : 1 : 1 [45]. The chemical structure of 5-mercapto-1-methyltetrazole *N*-tetramethylammonium and the disulfide are shown in Figure 10.21a,b, respectively.

Here, the optimized quasi-solid-state electrolyte has been used to assemble quasi-solid-state QDSSCs with CoS counter electrode. The power conversion efficiency of the quasi-solid-state QDSSCs is 0.94% [45]. The poor efficiency has been attributed to limitation of visible light absorption by CdS QDs.

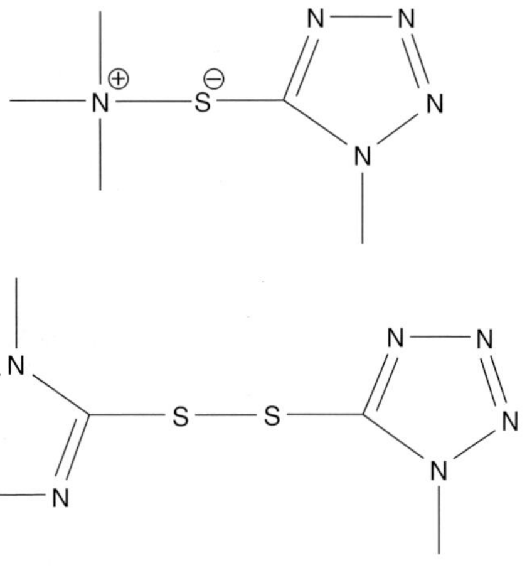

Figure 10.21 The chemical structures of (a) 5-mercapto-1-methyltetrazole N-tetramethylammonium and (b) the disulfide used to prepare thiolate/disulfide-based electrolyte.

Additionally, Duan et al. [46] have studied solid-state electrolyte composed of plastic crystal succinonitrile and sodium sulfide (Na_2S) for solar cells. The ionic conductivity and photovoltaic performance have been optimized by varying the amounts of succinonitrile and Na_2S in the electrolyte. The electrolyte based on succinonitrile plastic crystal has been prepared by mixing succinonitrile and Na_2S different molar ratios (succinonitrile:Na_2S = 1 : 1, 2 : 1, 4 : 1, 6 : 1, or 8 : 1) with vigorous stirring for five hours at 65 °C. The maximum power conversion efficiency of 1.29% has given for the QDSSC with succinonitrile to Na_2S molar ratio of 2 : 1 [46]. The results reported by Duan et al. [46] and Meng et al. [45] show that polysulfide electrolyte-based succinonitrile are suitable to prepare electrolytes intended for QDSSCs and further investigation of such electrolytes would help to improve performance of solid/quasi-solid-state QDSSCs.

10.6.9 Graphene-Implanted Polyacrylamide Gel Electrolytes for QDSSCs

Graphene-implanted gel electrolytes based on polyacrylamide (PAAm-G) have been prepared by Duan et al. [47] for quasi-solid-state QDSSCs. The chemical structure of polyacrylamide is shown in Figure 10.22. Dense or microporous polyacrylamide matrices with graphene and without graphene have been immersed in a liquid electrolyte consisting of 1 M sulfur and 1 M Na_2S aqueous solution for five days to prepare electrolytes. The room temperature ionic conductivity of pristine PAAm-based gel electrolyte is ~115 mS cm^{-2} and the conductivity has been enhanced to 125 mS cm^{-2} for the graphene-implanted gel electrolyte (PAAm–7 wt‰ G) [47].

A TiO_2 film with an average thickness of about 10 μm has been screen printed on FTO glass and it has been calcined in a muffle furnace at 450 °C for 30 minutes before QD deposition. The screen printed TiO_2 films have been sensitized by CdS QDs using SILAR method [47]. In order to synthesize QDs, the TiO_2 film has been immersed into 0.1 M $Cd(NO_3)_2$ ethanol solution for 1.5 minutes, rinsed with anhydrous ethanol and dried by N_2 gas, then soaked in 0.1 M Na_2S methanol solution for 1.5 minutes, rinsed with anhydrous methanol, and dried by N_2 gas [47]. The process is repeated for 12 times (SILAR cycles) to prepare the desired number of QDs on TiO_2 film.

The quasi-solid-state QDSSC has been assembled by sandwiching a gel electrolyte film of 0.5 mm thickness between a CoSe counter electrode and QD-sensitized TiO_2 film. It has shown a power conversion efficiency of 2.34% for the quasi-soli-state QDSSC with microporous graphene-implanted gel electrolytes [47]. The cell, assembled with gel electrolyte containing microporous polyacrylamide (without graphene), has given 1.64% power conversion efficiency

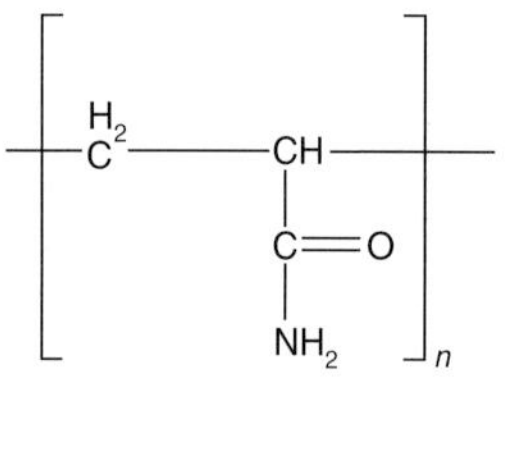

Figure 10.22 The structure of polyacrylamide host polymer.

[47]. The incorporated graphene has improved the charge transport in the bulk of the electrolyte and at the electrolyte/counter electrode interface, and thus cell performance is improved.

10.6.10 PEO and PVDF-Based Electrolyte for Solid-State Electrolytes for QDSSCs

SPE based on PEO-PVDF polymer blends with S/tetramethylammonium sulfate (S/TMAS) redox species for QDSSCs has been investigated by Ying Yang et al. [48]. The structures of PEO and PVDF are given in Figure 10.23. The polysulfide electrolyte has been prepared using aprotic co-solvents dimethyl-ether (DME)/PC. The TiO_2 film for photo-anode has been prepared on FTO by the doctor blade method using a paste containing TiO_2 (P25) powder. The CdSe/ZnS QD layer was deposited on the TiO_2 film by the SILAR method. A solution of 30 mM $Cd(NO_3)_2$ in ethanol, 100 mM $Zn(Ac)_2$ in water has been used for CdSe QDs preparation [48]. ZnS has been deposited by immersing in a Zn^{2+} and S^{2-} ion solutions for one minute each. A platinum plate has been used as the counter electrode. The study has shown some improvement of photon-to-current conversion efficiency and stability in QDSSCs with S/TMAS redox additive, although the results are not highly completive [48].

10.6.11 Hydroxystearic Acid-Based Polysulfide Hydrogel Electrolyte for CdS/CdSe QDSSCs

A quasi-solid-state QDSSC has been fabricated adding 12-hydroxystearic acid ($CH_3(CH_2)_5CH(OH)(CH_2)_{10}COOH$) as a gelator for the polysulfide electrolyte by Huo et al. [49]. The chemical structure of 12-hydroxystearic acid is given in Figure 10.24. The polysulfide aqueous electrolyte has been prepared by dissolving Na_2S (1.0 mol), sulfur (1.0 mol), and sodium hydroxide (0.1 mol) in one liter of deionized water. The hydrogel electrolyte has been prepared by mixing 3 wt% of 12-hydroxystearic acid into liquid electrolyte. The mixer has been heated and stirred to get polysulfide hydrogel electrolyte. The gel-to-liquid transition temperature of this polysulfide hydrogel electrolyte is as high as 96 °C.

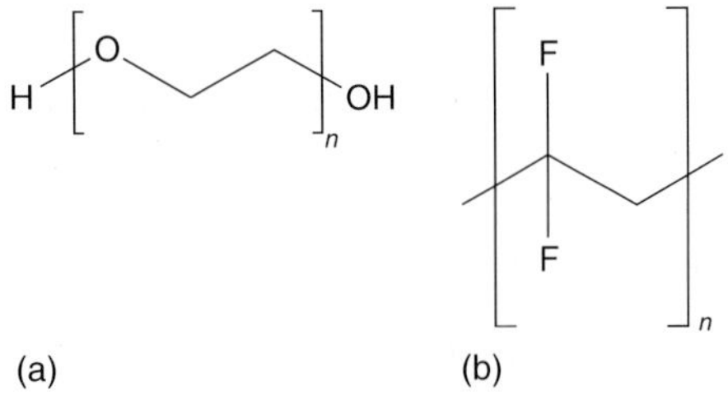

Figure 10.23 Structures of (a) PEO and (b) PVDF.

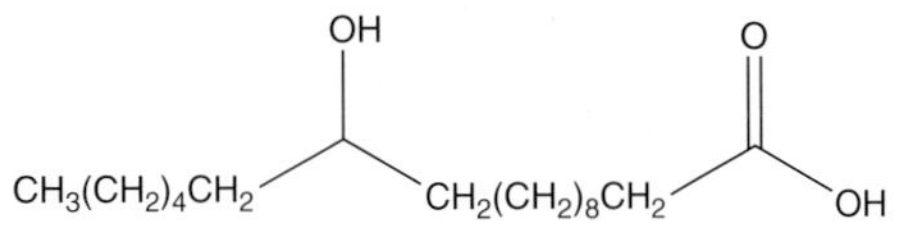

Figure 10.24 The structure of 12-hydroxystearic acid.

The colloidal TiO_2 suspension has been prepared by hydrolysis of titanium tetra-isopropoxide. The mesoporous TiO_2 film of about 10 μm thickness has been coated on FTO glass substrate using screen-printing method followed by sintering the samples at 500 °C for 30 minutes in air. The fabricated TiO_2 films have co-sensitized by preparing CdS and CdSe QDs using SILAR method. In order to synthesize, CdS QDs, the TiO_2 films have been immersed in a solution of 0.5 M cadmium nitrate tetrahydrade in ethanol solution and 0.2 M sodium sulfide nonahydrate ($Na_2S{\cdot}9H_2O$) in methanol solution for five minutes. To complete SILAR cycle, the TiO_2 photo-anodes were rinsed with ethanol and methanol consecutively, and dried with N_2 gas.

In order to prepare CdSe QDs, the prefabricated TiO_2/CdS electrodes have been dipped in 0.5 M $Cd(NO_3)_2$ ethanol solution for five minutes at room temperature and then dipped in sodium seleno-sulfate (Na_2SeSO_3) aqueous solution for one hour at 50 °C [49]. The QDs of CdS and CdSe have been deposited using 7 and 4 SILAR cycles, respectively.

The quasi-solid-state QDSSCs sensitized by CdS and CdSe have given an energy conversion efficiency of 2.40% at AM 1.5 (100 mW cm^{-2}) with polysulfide hydrogel electrolyte. Here, the liquid electrolyte based QDSSC has given an efficiency of 2.88%. However, the quasi-solid-state QDSSCs have shown impressive stability compared with that of liquid electrolyte-based cells during the accelerated thermal test [49]. The energy conversion efficiency of the liquid electrolyte-based QDSSC has decreased to 29% of its initial value during 220 hours of the accelerated aging test [49]. Nevertheless, efficiency of quasi-solid-state QDSSCs has dropped only to 92% of its initial value during 220 hours of aging test. The results highlight that gel polymer electrolytes offer better stability for QDSSCs [49].

10.6.12 QDSSCs Based on a Sodium Polyacrylate Polyelectrolyte

Feng et al. [50] prepared a polysulfide gel electrolyte with sodium polyacrylate for quasi-solid-state QDSSCs. The sodium polyacrylate added electrolytes have shown better conductivities. The conductivity values obtained for sodium polyacrylate-added polysulfide electrolytes are comparable with that of respective liquid electrolytes [50].

The liquid electrolyte has been prepared by dissolving 4.8 g of $Na_2S{\cdot}9H_2O$ and 0.64 g of sulfur in 10 ml of deionized water [50]. According to the authors, the gel electrolyte has been prepared by adding a different amount of sodium polyacrylate into polysulfide liquid electrolyte under stirring at room temperature for 10 minutes. Different weight ratios of sodium polyacrylate (5%, 10%, 15%, and 20%) have been added to the liquid electrolyte in order to gellify the liquid electrolyte [50]. However, the compositions of gellified and non-gellified electrolytes are not clear [50].

Further, the counter electrode has been prepared by screen printing $Cu_{2-x}S$ nanoparticles on FTO substrate [50] and the photo-electrode has been prepared by using two layers of TiO_2. These two layers have been prepared by screen printing a transparent layer and a light scattering layer on FTO glass and the thicknesses of the layers are about 9.0 and 6.0 μm, respectively [50]. The QDs CdSeTe and CdSe have been prepared to construct quasi-solid-state QDSSCs based on gel electrolyte containing sodium polyacrylate [50].

Water-soluble and oil-soluble QDs of CdSe and CdSeTe have been used to prepare QDSSCs. Then QD sensitizers were absorbed on the double layered TiO_2 substrate by using drop casting method [50].

A power conversion efficiency of 8.54% in one sun irradiation has been achieved for QDSSCs with CdSeTe [50]. The short circuit current densities around 20 mA cm^{-2} and open circuit voltage around 650 mV are given by the CdSeTe-sensitized QDSSCs [50]. The light-soaking stability of the quasi-solid-state cells containing CdSeTe and CdSe QDs have improved significantly compared with that of the cell prepared with polysulfide-based liquid electrolyte. The improvement of stability is attributed to the improvement of the water-holding capacity of sodium polyacrylate that can hinder the volatilization and leakage of liquid electrolytes [50]. Unprecedentedly high energy conversion efficiency reported for a quasi-solid-state QDSSC is significative.

10.7 Summary

The demand for energy is at a rapid expansion. However, the contemporary unsustainable ways of utilizing the current energy resources, specifically nonrenewable energy resources, have endangered life on Earth. In crafting sustainable solution for the issue by developing alternative energy sources, solar cells are increasingly getting attention of contemporary scientists and scholars. In general, a solar cell is a device that directly converts sunlight into most usable forms of energy, electricity or chemical energy, in an environmentally friendly manner. The additional advantage of using these devices is that no moving or mechanical parts are used, which leads to minimal maintenance cost and wear and tear. Further, input cost of generating energy is almost zero for solar cells, since sunlight is the input of energy production. On these grounds, the solar cell is recognized as a green and zero emission energy conversion device.

As discussed in this chapter, solar cells have evolved through three generations. The third generation consists of a variety of solar cells that have potentials to cater for the future demand of energy needs since most of the third-generation cells are oriented at reaching higher efficiencies with low cost. There are a number of novel technologies that have already shown high potential and exhibited better performance. However, many of the third-generation cells are in the stage of experimental level. This raises the need of further research initiatives on the area to push the development of cells from the developmental stage to commercialization stage.

The solar cell mechanism can be divided into three processes: charge generation, charge separation, and charge transportation. Based on work reported so far, we observe that charge generation, charge separation, and charge transportation efficiencies need to be improved in order to enhance the solar cell performances.

Wide bandgap semiconductors (semiconductors with bandgap higher than incident photons) can be used to prepare PEC solar cells by means of photo-sensitization. A wide range of dyes and metal nanoparticles, such as Au, Ag, etc., have received much attention from scientists and scholars as photo-sensitizer. The QDSSCs are reliable devices to fulfill future energy demand due to prospected higher efficiency and stability, facile and low cost fabrication techniques, and environmentally friendly operation. QDs of PdS, CdS, or CdSe have widely been used to photo-sensitize wideband gap semiconductors. In addition, the QDs based on CdTe, $CuInS_2$, Cu_2S, PbSe, InP, InAs, Ag_2S, Bi_2S_3, and Sb_2S_3 are also considered as suitable sensitizers for solar cells.

The QDs are very small semiconductor particles and the size of a QD is a few nanometers. Due to the smaller size of the QD particles, their optical and electronic properties do deviate from the bulk materials. In addition, the optical and electronic properties of semiconductor nanoparticle can be engineered by changing the size, shape, and type of the semiconductor material. As a result, the optical and electronic properties of QDs can be tuned by changing the size, shape, and the type of material in order to prepare efficient solar cells. In particular, the size dependent of bandgap energy of QDs can be used to prepare solar cells with wide absorption spectrum.

The photo-electrode is a key component in a QDSSC prepared by photo-sensitizing a nanostructured wide bandgap semiconductor. In general, the nanostructured film is prepared by using *doctor blade, spin coating, screen printing,* or *spray pyrolysis* methods.

The nanostructured film is photo-sensitized by coating semiconductor QDs on it. The QD layers are commonly deposited by using *CBD, EPD, ED, SILAR,* or *SP* methods. All these methods have been used to prepare conventional liquid electrolyte-based QDSSCs [29, 30]. However, only a limited number of research woks have reported on polymer electrolyte-based QDSSCs. In particular, we could not find any report on polymer electrolyte-based QDSSCs that were prepared using EPD and SP methods, which strongly prompts the future research on the area.

In general, after preparing the photo-anode, the QDSSC assembly is then completed by sandwiching a liquid or solid electrolyte containing appropriate redox couple between the photo-anode and counter electrode. The commonly used counter electrode materials are *platinum (Pt), copper(I) sulfide* (*Cu_2S*), *gold (Au), or carbon (C).*

Cu_2S-based counter electrodes are more reliable for QDSSCs, while Pt electrodes are more suitable for DSSCs. Further, the polysulfide electrolytes are widely used in QDSSCs and they have shown relatively higher efficiencies, whereas iodide/tri-iodide redox couple is more suitable for DSSCs than for QDSSCs.

In general, the stability in PEC solar cells can be increased by substituting liquid electrolytes by polymer electrolytes. The chemical reactions at the semiconductor/electrolyte junction can be suppressed by replacing aqueous medium by a polymer medium. Hence, the degradation of PEC solar cells can be controlled using polymer electrolytes. The gel polymer electrolytes ensure the long-term stability since the polymer suppresses the evaporation of solvent in the electrolyte. According to reports published so far, the polysulfide ion-conducting hydrogel electrolytes have given better results for QDDSCs.

Table 10.1 The efficiencies of different QDSSCs prepared with different wide bandgap semiconductor substrates, QDs, redox couples, host polymers, QD preparation methods, and the publication year.

Substrate	Sensitizer	Redox couple	Polymer/gelator	QD preparation	Efficiency (%)	Year	References
TiO_2	CdS/CdSe	S^{2-}/S_x^{2-}	Poly-acrylamide	SCBD	4.0	2010	[36]
TiO_2	CdS	S^{2-}/S_x^{2-}	PVdF	SCBD	0.78	2011	[37]
TiO_2	CdS	I^-/I_3^-	PAN	CBD	0.24	2011	[37]
ZnO	CdS/CdSe	S^{2-}/S_x^{2-}	ICS-PPG230	SILAR	4.5	2012	[38]
TiO_2	CdS/CdSe	S^{2-}/S_x^{2-}	KGM	CBD	4.0	2013	[40]
TiO_2	CdS/CdSe	S^{2-}/S_x^{2-}	Dextran	ED	3.23	2013	[41]
ZnS	CdS	S^{2-}/S_x^{2-}	Poly(acrylamide)	SILAR/CBD	0.35	2013	[43]
ZnO	CdS/CdSe	S^{2-}/S_x^{2-}	PEGDME	SILAR/CBD	5.45	2014	[44]
TiO_2	CdS	Thiolate/disulfide	Succinonitrile	SILAR	0.94	2014	[45]
TiO_2	CdS	S^{2-}/S_x^{2-}	Succinonitrile	SILAR	1.29	2015	[46]
TiO_2	CdS	S^{2-}/S_x^{2-}	PAAm	SILAR	1.64	2015	[47]
TiO_2	CdS	S^{2-}/S_x^{2-}	PAAm-G	SILAR	2.34	2015	[47]
TiO_2	CdSe/ZnS	S^{2-}/S_x^{2-}	PEO-PVDF	SILAR		2015	[48]
TiO_2	CdS/CdSe	S^{2-}/S_x^{2-}	12-Hydroxystearic	SILAR	2.40	2015	[49]
TiO_2	CdSeTe	S^{2-}/S_x^{2-}	Sodium polyacrylate	CBD/drop cast	8.54	2016	[50]

Though very high efficiencies are expected from QDSSCs, the energy conversion efficiency of QDSSC is still less than that of DSSC, and therefore more researches would be needed on developing polymer electrolyte-based QDSSCs are required.

Finally, the efficiencies of different QDSSCs prepared with different wide bandgap semiconductor substrates, QDs, redox couples, host polymers, and QD preparation methods are summarized in the Table 10.1.

Though very high efficiencies are expected from QDSSCs, the energy conversion efficiency of the QDSSC is still less than that of the DSSC and, thus, this provides a clear indication on the need for future research on developing polymer electrolyte-based QDSSCs.

References

1 International Energy Outlook 2016. U.S. Energy Information Administration, Washington, DC 20585.

2 Mearns, E. (2015). Global Energy Trends – BP Statistical Review 2015. Posted on 12 June 2015.

3 Höök, M. and Tang, X. (2013). Depletion of fossil fuels and anthropogenic climate change—A review. *Energy Policy* 52: 797–809.

4 Höök, M., Sivertsson, A., and Aleklett, K. (2010). Validity of the fossil fuel production outlooks in the IPCC Emission Scenarios. *Nat. Resour. Res.* 19 (2): 63–81.

5 Green, S.J., Demes, K., Arbeider, M. et al. (2017). Oil sands and the marine environment: current knowledge and future challenges. *Front. Ecol. Environ.* 15 (2): 74–83.

6 Sovacool, B.K. (2010). A critical evaluation of nuclear power and renewable electricity in Asia. *J. Contemp. Asia* 40 (3): 369–400.

7 Bella, F., Gerbaldi, C., Barolo, C., and Grätzel, M. (2015). Aqueous dye-sensitized solar cells. *Chem. Soc. Rev.* 44 (11): 3431–3473.

8 International Energy Agency (IEA) (2016). Key world energy statistics. Head of Communication and Information Office 9, rue de la Fédération, 75739 Paris Cedex 15, France

9 Kibria, M.T., Ahammed, A., Sony, S.M., and Hossain, F. (2014). A review: comparative studies on different generation solar cells technology. In: *International Conference on Environmental Aspects of Bangladesh*, 51–53.

10 O'Regan, B. and Grfitzeli, M. (1991). A low-cost, high-efficiency solar cell based on dye-sensitized. *Nature* 353 (6346): 737–740.

11 Cavallo, C., Di Pascasio, F., Latini, A. et al. (2017). Nanostructured semiconductor materials for dye-sensitized solar cells. *J. Nanomater.* 2017: 1–31.

12 Qin, Y. and Peng, Q. (2012). Ruthenium sensitizers and their applications in dye-sensitized solar cells. *Int. J. Photoenergy* 2012: 1–21.

13 Mishra, A., Fischer, M.K., and Bäuerle, P. (2009). Metal-free organic dyes for dye-sensitized solar cells: from structure: property relationships to design rules. *Angew. Chem. Int. Ed.* 48 (14): 2474–2499.

14 Shalini, S., Prasanna, S., Mallick, T.K., and Senthilarasu, S. (2015). Review on natural dye sensitized solar cells: operation, materials and methods. *Renewable Sustainable Energy Rev.* 51: 1306–1325.
15 Gong, J., Liang, J., and Sumathy, K. (2012). Review on dye-sensitized solar cells (DSSCs): fundamental concepts and novel materials. *Renewable Sustainable Energy Rev.* 16 (8): 5848–5860.
16 Ludin, N.A., Mahmoud, A.A.A., Mohamad, A.B. et al. (2014). Review on the development of natural dye photosensitizer for dye-sensitized solar cells. *Renewable Sustainable Energy Rev.* 31: 386–396.
17 Wu, J., Lan, Z., Lin, J. et al. (2015). Electrolytes in dye-sensitized solar cells. *Chem. Rev.* 115 (5): 2136–2173.
18 Wu, J., Lan, Z., Hao, S. et al. (2008). Progress on the electrolytes for dye-sensitized solar cells. *Pure Appl. Chem.* 80 (11): 2241–2258.
19 Gorlov, M. and Kloo, L. (2008). Ionic liquid electrolytes for dye-sensitized solar cells. *Dalton Trans.* (20): 2655–2666.
20 Yella, A., Lee, H.W., Tsao, H.N. et al. (2011). Porphyrin-sensitized solar cells with cobalt (II/III)-based redox electrolyte exceed 12 percent efficiency. *Science* 334 (6056): 629–634.
21 McGehee, M.D. (2011). Paradigm shifts in dye-sensitized solar cells. *Science* 334 (6056): 607–608.
22 Mathew, S., Yella, A., Gao, P. et al. (2014). Dye-sensitized solar cells with 13% efficiency achieved through the molecular engineering of porphyrin sensitizers. *Nat. Chem.* 6 (3): 242–247.
23 Kakiage, K., Aoyama, Y., Yano, T. et al. (2015). Highly-efficient dye-sensitized solar cells with collaborative sensitization by silyl-anchor and carboxy-anchor dyes. *Chem. Commun.* 51 (88): 15894–15897.
24 Grätzel, M. (2001). Photoelectrochemical cells. *Nature* 414 (6861): 338–344.
25 Pagliaro, M., Ciriminna, R., and Palmisano, G. (2008). Flexible solar cells. *ChemSusChem* 1 (11): 880–891.
26 Ileperuma, O.A., Dissanayake, M.A.K.L., Somasunderam, S., and Bandara, L.R.A.K. (2004). Photoelectrochemical solar cells with polyacrylonitrile-based and polyethylene oxide-based polymer electrolytes. *Sol. Energy Mater. Sol. Cells* 84 (1): 117–124.
27 Chukwuocha, E.O., Onyeaju, M.C., and Harry, T.S. (2012). Theoretical studies on the effect of confinement on quantum dots using the Brus equation. *World J. Condens. Matter Phys.* 2 (02): 96.
28 Brus, L.E. (1984). Electron–electron and electron–hole interactions in small semiconductor crystallites: The size dependence of the lowest excited electronic state. *J. Chem. Phys.* 80 (9): 4403–4409.
29 Jun, H.K., Careem, M.A., and Arof, A.K. (2013). Quantum dot-sensitized solar cells—perspective and recent developments: a review of Cd chalcogenide quantum dots as sensitizers. *Renewable Sustainable Energy Rev.* 22: 148–167.
30 Nozik, A.J. (2002). Quantum dot solar cells. *Physica E* 14 (1): 115–120.
31 Beard, M.C., Luther, J.M., and Nozik, A.J. (2014). The promise and challenge of nanostructured solar cells. *Nat. Nanotechnol.* 9 (12): 951–954.

32 Fenton, D.E., Parker, J.M., and Wright, P.V. (1973). Complexes of alkali metal ions with poly(ethylene oxide). *Polymer* 14 (11): 589.

33 Wright, P.V. (1975). Electrical conductivity in ionic complexes of poly(ethylene oxide). *Polym. Int.* 7 (5): 319–327.

34 Armand, M.B., Chabagno, J.M., and Duclot, M.J. (1979). *Fast Ion Transport in Solids* (eds. P. Vashishta, J.N. Mundy and G.K. Shenoy), 131. North Holland.

35 Armand, M. (1989). Charge transfer at polymer electrolytes. *Faraday Discuss. Chem. Soc.* 88: 65–76.

36 Yu, Z., Zhang, Q., Qin, D. et al. (2010). Highly efficient quasi-solid-state quantum-dot-sensitized solar cell based on hydrogel electrolytes. *Electrochem. Commun.* 12 (12): 1776–1779.

37 Careem, M.A., Senadeera, G.K.R., Mellander, B.E. et al. (2011). *Polymer Electrolyte Based CdS Quantum Dot Sensitized Solar Cells*. Kandy: Institute of Fundamental Studies.

38 Karageorgopoulos, D., Stathatos, E., and Vitoratos, E. (2012). Thin ZnO nanocrystalline films for efficient quasi-solid state electrolyte quantum dot sensitized solar cells. *J. Power Sources* 219: 9–15.

39 Stathatos, E., Chen, Y., and Dionysiou, D.D. (2008). Quasi-solid-state dye-sensitized solar cells employing nanocrystalline TiO_2 films made at low temperature. *Sol. Energy Mater. Sol. Cells* 92 (11): 1358–1365.

40 Wang, S., Zhang, Q.X., Xu, Y.Z. et al. (2013). Single-step in-situ preparation of thin film electrolyte for quasi-solid state quantum dot-sensitized solar cells. *J. Power Sources* 224: 152–157.

41 Chen, H.Y., Lin, L., Yu, X.Y. et al. (2013). Dextran based highly conductive hydrogel polysulfide electrolyte for efficient quasi-solid-state quantum dot-sensitized solar cells. *Electrochim. Acta* 92: 117–123.

42 Bandara, T.M.W.J., Jayasundara, W.J.M.J.S.R., Fernado, H.D.N.S. et al. (2015). Efficiency of 10% for quasi-solid state dye-sensitized solar cells under low light irradiance. *J. Appl. Electrochem.* 45 (4): 289–298.

43 Narayanan, R., Deepa, M., and Srivastava, A.K. (2013). Förster resonance energy transfer and carbon dots enhance light harvesting in a solid-state quantum dot solar cell. *J. Mater. Chem. A* 1 (12): 3907–3918.

44 Kim, H., Hwang, I., and Yong, K. (2014). Highly durable and efficient quantum dot-sensitized solar cells based on oligomer gel electrolytes. *ACS Appl. Mater. Interfaces* 6 (14): 11245–11253.

45 Meng, K. and Thampi, K.R. (2014). Efficient quasisolid dye-and quantum-dot-sensitized solar cells using thiolate/disulfide redox couple and CoS counter electrode. *ACS Appl. Mater. Interfaces* 6 (23): 20768–20775.

46 Duan, J., Tang, Q., He, B., and Chen, H. (2015). All-solid-state quantum dot-sensitized solar cell from plastic crystal electrolyte. *RSC Adv.* 5 (42): 33463–33467.

47 Duan, J., Tang, Q., Li, R. et al. (2015). Multifunctional graphene incorporated polyacrylamide conducting gel electrolytes for efficient quasi-solid-state quantum dot-sensitized solar cells. *J. Power Sources* 284: 369–376.

48 Yang, Y. and Wang, W. (2015). A new polymer electrolyte for solid-state quantum dot sensitized solar cells. *J. Power Sources* 285: 70–75.

49 Huo, Z., Tao, L., Wang, S. et al. (2015). A novel polysulfide hydrogel electrolyte based on low molecular mass organogelator for quasi-solid-state quantum dot-sensitized solar cells. *J. Power Sources* 284: 582–587.

50 Feng, W., Li, Y., Du, J. et al. (2016). Highly efficient and stable quasi-solid-state quantum dot-sensitized solar cells based on a superabsorbent polyelectrolyte. *J. Mater. Chem. A* 4 (4): 1461–1468.

11

Polymer Electrolytes for Perovskite Solar Cell and Challenges

Rahul Singh[1], *Hee-Woo Rhee*[1], *Bhaskar Bhattacharya*[2], *and Pramod K. Singh*[3]

[1] *Sogang University, Polymer Materials Laboratory, Department of Chemical and Biomolecular Engineering, 35 Baekbeom-Ro, Mapo-Gu, Seoul 121-742, Republic of Korea*
[2] *Banaras Hindu University, Department of Physics (MMV), Varanasi, 229205, India*
[3] *Sharda University, Material Research Laboratory, School of Basic Sciences and Research, Department of Physics, Knowledge Park-III, Greater Noida 201310, India*

11.1 Introduction

Organometallic halide perovskites are promising next-generation materials for high efficient and low-cost photovoltaics. Initially research was focused on the use of organometallic halide perovskite layers in mesoscopic solar cells by replacing the organic dyes as a sensitizer material (N3, N719, N749, etc.) [1, 2]. Hybrid perovskite methylammonium lead iodide ($MAPbI_3$) is the most prominent choice owing to its outstanding properties for a solar cell absorber, including a high extinction coefficient, a medium band gap, a small exciton binding energy, and long exciton and charge diffusion lengths [3–5]. Apart from having a suitable bandgap, it also has excellent electronic properties like high carrier mobility, shallow defect levels, high carrier diffusion lengths, etc. [5, 6]. The power conversion efficiency (PCE) of the first perovskite-based solar cell device reported in 2009 was 3.8% with liquid electrolyte [7], but the certified efficiency has recently reached above 23.6% for solid state device [8, 9]. Perovskite solar cell (PSC) can be fabricated as n-i-p and p-i-n type of architectures, to fabricate n-i-p type devices or solid-state meso-super structured PSCs. Spiro-MeOTAD is used as the hole transport and mesoporousTiO_2 as an electron transport material, which reported a PCE of 9.7% perovskite based solid-state mesoscopic heterojunction solar cell [10]. The most commonly used organic hole transport medium (HTM) are 2,2′,7,7′-tetrakis-(*N*,*N*-di-p-methoxyphenylamine)-9,90-spiro-biuorene (spiro-MeOTAD) [11, 12]; poly-[3-hexylthiophene-2,5-diyl] (P3HT) [13, 14]; poly-[[9-(1-octylnonyl)-9H-carbazole-2,7-diyl]-2,5-thiophenediyl-2,1,3-benzothiadiazole-4,7-diyl-2,5-thiophenediyl] (PCDTBT) [15, 16]; poly-triarylamine (PTAA) [17]; 4-(diethylamino)-benzaldehyde diphenylhydrazone (DEH) [18]; *N*,*N*′-dialkylperylenediimide (PDI) [19]; polypyrrole (PPy) and polyaniline (PANI) [17]; poly(3,4-ethylenedioxythiophene):poly(styrene sulfonate) (PEDOT:PSS); and poly(3,4-ethylenedioxythiophene) (PEDOT) [20, 21]. On the other hand, inorganic (p-type

Polymer Electrolytes: Characterization Techniques and Energy Applications,
First Edition. Edited by Tan Winie, Abdul K. Arof, and Sabu Thomas.

semiconductors) HTM is copper-based materials (CuSCN, CuI, Cu_2O, CuO) [22], NiOx [23], MoOx [24], polymer electrolyte [25], and carbon materials (including graphene and carbon nanotube [CNT]) [26].

Semiconductor Al_2O_3 as meso-structured scaffold is also implied in PSC instead of the semiconductor TiO_2 [2, 3]. The fact that this type of solar cell (incorporating either a semiconducting or an insulating scaffold layer) could deliver very high device efficiency indicates excellent ambipolar charge transport efficiency within the perovskite photoactive layer. A characteristic of mesoporous structured solar cells is that the perovskite layer is deposited on a mesoporous layer, e.g. TiO_2 or Al_2O_3. A subsequent report by Snaith et al. showed that planar heterojunction PSCs incorporating vapor-deposited perovskite layer could be used to generate devices with a PCE over 15% without the need for any mesoporous layers [27]. Since then, planar solar cells using a solution-processed photoactive layer have reached PCEs close to 20% using compositional [28] or interfacial engineering.

Lots of improvements in solar cell architecture and molecular and compositional engineering to form uniform morphology of perovskite materials over the last six years have led to enhancement in the PCE of PSCs. The most studied perovskite material achieving high cell performance has been methyl ammonium lead iodide ($CH_3NH_3PbI_3$). Lead halide perovskites have attracted considerable interest as photo absorbers in photovoltaic applications over the last few years. Developing technological advancement further to achieve PCEs near theoretical values continues to be among the most important challenges in the solar cell community to achieve easily. Till now, various types of perovskite materials have been studied such as $MAPbX_3$, MA = $CH_3NH_3^+$, FA = $CH_3(NH_2)_2^+$, Cs^+, and Rb^+; X = Cl^-, Br, I^-, polyanion:BF_4^-; PF_6^-; SCN^- [29–35]. The prime advantage of the organo-lead trihalide hybrid semiconducting perovskite absorbers is their direct bandgap with large absorption coefficients over a wide range [27], which enables efficient light absorption in ultrathin films, high efficiency and low-cost alternative to conventional silicon-based solar cell, long diffusion length [28, 36], high carrier mobility [37, 38], low exciton binding energy [17, 39], and simple easy preparation. At present, stability of these devices is the main concern for the researches. Known to be stabilized depending on the ionic radii of A and B cations in relation with tolerance factor, the perovskite ABX_3 (X = halogens) structure consists of organic components in cuboctahedral A site and an inorganic component in octahedral B site, and the chemistry of the organic and inorganic components can be tailored to tune the optical, electronic, magnetic, and mechanical properties of hybrid materials [40–44].

In this chapter, we are focusing on the challenges faced in formation of efficient and long-term stable device. Great efforts have been made by researchers in synthesizing planar and inverted perovskite thin film cells with high uniformity. Pinhole free surface using different techniques has been reported such as hot casting technique [45], spin coating [46, 47], vapor deposition [48], and thermal evaporation [49]. The spin coating method is the conventional method for preparing highly efficient devices using one-step and two-step processes. PSCs are very sensitive to moisture and air; generally the high efficient devices reported are very small scale and fabricated in controlled conditions. PSC fabricated in ambient atmosphere is less expensive and thus more attractive.

11.2 Principle and Working Process of Perovskite Solar Cell

In a general principle, a perovskite molecule absorbs light and an electron is excited from the highest occupied molecular orbital (HOMO) to the lowest unoccupied molecular orbital (LUMO). This process resembles the promotion of electron from the valence bond to the conduction band in TiO_2 semiconductors, within which they are transported out of the solar cell. The holes remaining on the perovskite crystals are then transferred to a redox-active electrolyte, or a solid-state hole conductor, and subsequently transported to the external circuit [50]. It soon became apparent that the perovskites can function as both hole and electron conductors and that the bulk of a few hundred-nanometer-thick solid perovskite thin film can even sustain charge generation and transport [50, 51]. To produce photocurrent, the exciton needs to overcome the binding energy and dissociate into free charge electrons and holes. Working principle of a PSC involves photon absorption, charge separation, charge transport, and charge collection.

Photoexcitation:

$$\mathrm{PVK} + h\nu \rightarrow (e^-/h^+)_{\mathrm{PVK}} \tag{11.1}$$

Electron injection:

$$(e^-/h^+)_{\mathrm{PVK}} \rightarrow e^-{}_{\mathrm{cb(TiO_2)}} + h^+{}_{\mathrm{PVK}} \tag{11.2a}$$

$$h^+{}_{\mathrm{PVK}} \rightarrow h^+{}_{\mathrm{(HTM)}} \tag{11.2b}$$

Hole injection:

$$(e^-/h^+)_{\mathrm{PVK}} \rightarrow h^+{}_{\mathrm{(HTM)}} + e^-{}_{\mathrm{PVK}} \tag{11.3a}$$

$$e^-{}_{\mathrm{PVK}} \rightarrow e^-{}_{\mathrm{cb(TiO_2)}} \tag{11.3b}$$

The overall photovoltaic conversion efficiency of mesoporous structure perovskite cell is controlled through the aforementioned processes. In order to achieve high PCE, charge generation and transport (Eqs. (11.1)–(11.3)) should have faster rate comparing with undesired recombination. On the other hand, the significant high efficiencies in PSCs without the use of electron transporting material [52, 53] and without hole transporting material [54] indicate that PSC can work in a sensibly different configuration than dye sensitized solar cell (DSSC). These results constitute a strong indication that photogenerated electrons and holes coexist in $CH_3NH_3PbX_3$ absorber material and travel to the selective contacts where they are separately collected. Organometallic perovskites such as $CH_3NH_3PbI_3$ have showed both electron and hole transport properties [4]. Therefore, PSCs could be constructed both as p-n junction and p-i-n junction.

11.2.1 Perovskite Materials

The basic building component of the organic–inorganic perovskite family is the ABX_3 perovskite structure (Figure 11.1). This simple structure consists of a 3D network of corner-sharing BX_6 octahedra, where the B atom is typically a metal cation and X is an anion, with the appropriate charge to balance the A and B cations (where A > B) [55]. The A cations fill the large 12-fold coordinated holes between the octahedra. Generally, a perovskite structure consists of corner-sharing BX_6 octahedra with the A ion placed in the cuboctahedral interstices that belong to the cubic Pm3m crystal structure [56]. In the hybrid organometal halide perovskites, A is an organic cation (i.e. MA^+ or FA^+) or

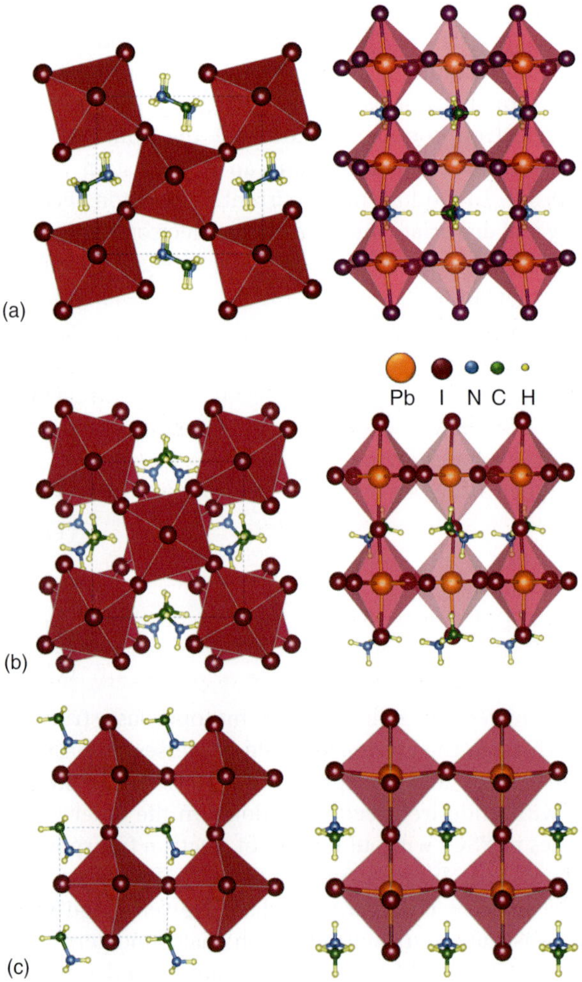

Figure 11.1 ABX_3 perovskite structure showing BX_6 octahedral and larger A (Pb) cation occupied in cubo-octahedral site (CH_3NH_3I). The crystal structures of the (a) orthorhombic, (b) tetragonal, and (c) cubic phases of $CH_3NH_3PbI_3$ at different orientations (100) and (001), respectively.

inorganic cation (i.e. K^+, Rb^+, Cs^+), B is a metal cation (i.e. $Sn_2{}^+$ or $Pb_2{}^+$), and X is a halide anion (i.e. Cl^-, Br^-, or I^-) [57, 58]. In particular, hybrid organometal halide perovskites $ABX_{3-x}Y_x$ with mixed halides, for example, $MAPbI_{3-x}Cl_x$ and $MAPbI_{3-x}Br_x$, have also attracted special attention due to their tunable optical properties that lead to improved performance in PSCs [44, 59, 60].

Stability of perovskite device is a major challenge, and one can overcome this issue in layered perovskites by producing thin perovskite films free from grain boundaries and high-quality single crystalline material, in which the crystallographic planes of the inorganic perovskite component have a strongly preferential out-of-plane alignment with respect to the contacts in planar solar cells to facilitate efficient charge transport. For a perfectly packed cubic perovskite structure, the conditioned A, B, and X ions have to satisfy Eq. (11.4):

$$t = \frac{(R_A - R_X)}{\sqrt{2}(R_B + R_X)} \tag{11.4}$$

where R_A, R_B, and R_X are the corresponding ionic radii and the tolerance factor, $t = 1$. It can be used to evaluate which mismatches in size of the A, B, and X ions are tolerated to form perovskite-like structures. Empirically it is found that $0.8 \leq t \leq 1$ for most cubic 3D perovskite structures. Below ~0.80, other structures such as the ilmenite type ($FeTiO_3$) are more stable due to the similar sizes R_A and R_B, where the cations A and B become similar in size; therefore, the A cation comes in too close contact to the halides. Thus, mainly covalent bonds are formed with new structure. Values larger than 1 lead to hexagonal structures where layers of face-sharing octahedra are introduced into the structure [61, 62]. If the value of t is smaller, it leads to a lower symmetry tetragonal or orthorhombic structure; however, if the value of t is larger, it can destabilize the 3D network and perovskite structure get distorted, creating a 2D structure [61, 62]. The methylammonium cation is expected to be an appropriate choice for the 3D perovskite structure, but the methylammonium cation undergoes quick isotropic reorientation.

Estimating the exact ionic radius of the molecular cation for the stability of 3D perovskite from tolerance factor is very critical and challenging. The reason behind this is van der Waals interactions and hydrogen bonding that makes slight variation in bond lengths, the change in steric size with respect to the anionic counterpart; therefore it is difficult to estimate an effective radius, even for symmetric cations such as NH^{4+} or $(CH_3)_4N^+$. However, charge balance is very essential for any perovskite so that fully occupied with "A" is monovalent cation, "B" is divalent cation, and "X" is a halogen. On the other hand, materials for divalent "B" cations include Ge^{2+}, Sn^{2+}, Pb^{2+}, Ca^{2+}, Eu^{2+}, and Yb^{2+} [63]. Depending on the nature of the anionic species (X), oxide (O^{2-}) and non-oxide perovskites such as chalcogenide (S^{2-}, Se^{2-}, Te^{2-}) and metal halide perovskites are distinguished. Moreover, molecular anions such as $HCOO^-$ [64], BF^{4-} [31, 34], PF^{6-} [31], and SCN^- [33] were successfully used as counterion.

The formability of metal halide perovskites depends on three basic requirements: (i) charge neutrality between cations and anions, i.e. $Z(A) + Z(B) = 3Z(X)$, whereby Z represents the valence of the respective A, B, or X ions [65]; (ii) the stability of the BX_6 octahedra, which can be estimated by the octahedral factor

Table 11.1 Effective ionic radii of various anions.

S. No.	Anion	X	Effective radius r_x (pm)	References
1	Fluoride	F^-	129	[63]
2	Formate	$HCOO^-$	136	[63]
3	Chloride	Cl^-	181	[63]
4	Bromide	Br^-	196	[63]
5	Iodide	I^-	220	[70]

(μ); (iii) the ionic radii of A, B, and X that need to meet the requirements for the Goldschmidt tolerance factor (t) [66]:

$$\mu = \frac{R_B}{R_X} \tag{11.5}$$

The octahedral factor l, which is the ratio of the radii of the B-site cation (R_B) and the halide counterion (R_X), can be used to estimate the stability of the BX_6 octahedra (Eq. (11.5)) [67, 68]. The incorporation of the B-site cation is limited by ionic size restrictions defined by the X_6 octahedron. For μ values between 0.442 and 0.895, metal halide perovskites have been found to be stable [69]. Some of the ionic radii of anion and cations were tabulated in Tables 11.1 and 11.2.

11.2.2 Perovskite Structure

Varying the temperature can generate phase transition in $MAPbI_3$ such as orthorhombic phase (*Pnma*) and can be formed below temperature 162 K, in which the PbI_6 octahedra are strongly deformed assuming a rectangular basal plane. Such deformation restricts the rotational degrees of freedom of MA in the rhombus-shaped interstitial region, thus imposing a spatial ordering to CH_3NH_3. In this case, the organic cation is pinned and can only rotate along the C–N axis. The molecular cation is in a fixed position as shown in Figure 11.1a and remains stable at this temperature. On the other hand, tetragonal phase can be formed at temperature 165 K and above that cubic phase around 327 K [62, 63]. The molecular cations are no longer in a fixed position as in the orthorhombic phase. The molecules are disordered between two nonequivalent positions in every structure. In the case of tetragonal formation, the MA cations are aligned as in the orthorhombic phase, toward the face of the perovskite structure, i.e. (100) in the cubic basis. The MA in different (001) planes are approximately orthogonal to one another as shown in Figure 11.1b. In the tetragonal phase (*I4/mcm*), the CH_3NH_3 molecules become free to rotate between the octahedral cages. Above room temperature, such rotation is fast, to a point where both crystallographic analysis [72] and NMR measurements [73] have shown that

Table 11.2 Effective ionic radii of organic or inorganic molecular cations A and B.

Cation A	Chemical structure	Effective radius $r_{A,eff}$ (pm)	References	Cation B	Effective radius $r_{B,eff}$ (pm)	References
Acetamidinium	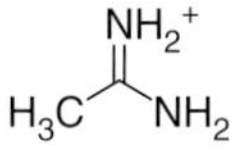	—	—	Pb^{2+}	119	[71]
5-Azaspiro [4.4] nonan-5-ium	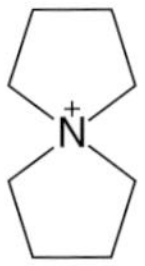	—	—	Sn^{2+}	110	[71]
1,4-Benzene diammonium	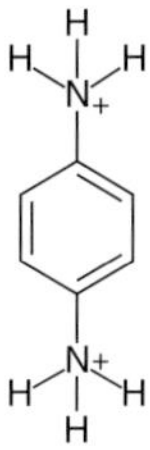	—	—	Sn^{4+}	69	[71]
Benzylammonium	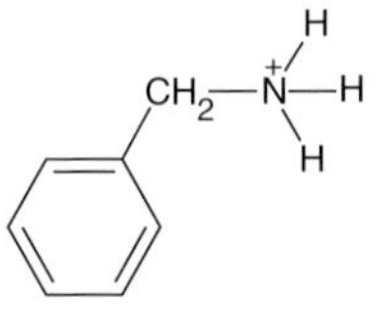	—	—	Ge^{2+}	73	[71]
iso-Butylammonium	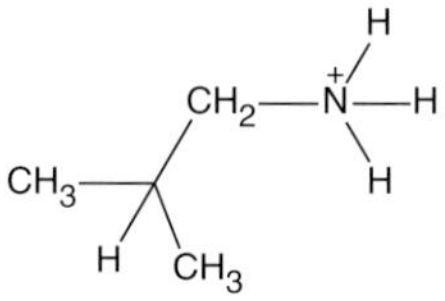	—	—	Mg^{2+}	72	[71]
n-Butylammonium	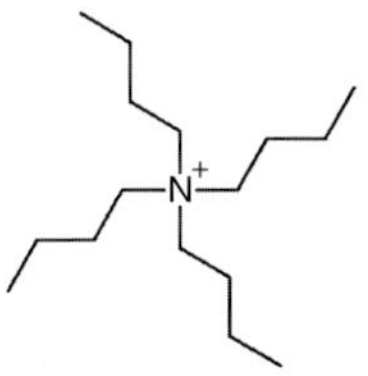	—	—	Ca^{2+}	100	[71]
t-Butylammonium	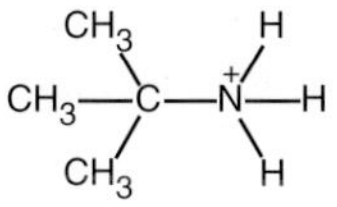	—	—	Sr^{2+}	118	[71]
Cyclohexylammonium	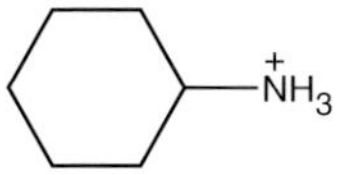	—	—	Ba^{2+}	135	[71]

(Continued)

Table 11.2 (Continued)

Cation A	Chemical structure	Effective radius $r_{A,eff}$ (pm)	References	Cation B	Effective radius $r_{B,eff}$ (pm)	References
1,4-Diazabicyclo [2,2,2] octane-1,4-diium	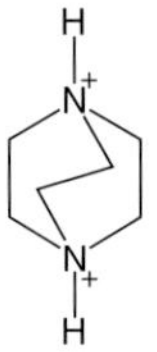	—	—	Cu^{2+}	73	[71]
Diethylammonium	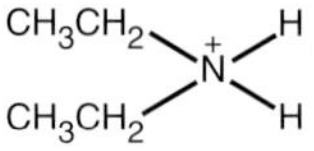	272	[70]	Fe^{2+}	78	[71]
Dimethylammonium	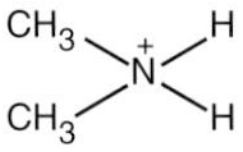	—	—	Pd^{2+}	86	[71]
Ethane-1,2-diammonium	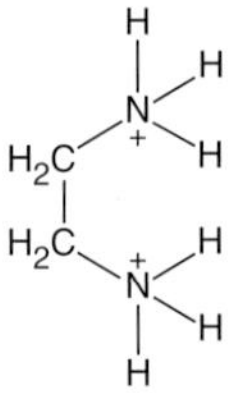	—	—	Eu^{2+}	117	[71]
Ethylammonium	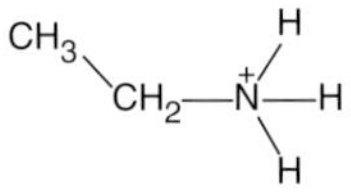	274	[70]	Tm^{2+}	103	[71]
Formamidinium	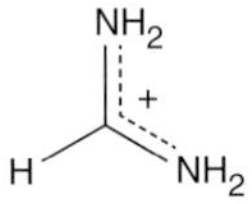	253	[70]	Yb^{2+}	102	[71]
Guanidinium	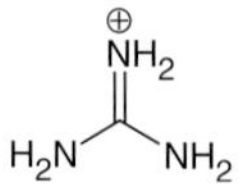	278	[70]	TI^{+}	150	[71]
n-Hexylammonium	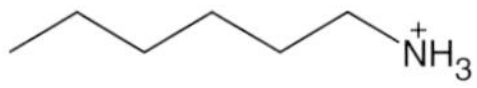	—	—	Au^{+}	137	[71]
Imidazolium	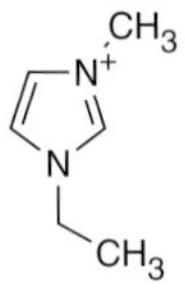	258	[70]	Au^{3+}	85	[71]

(Continued)

Table 11.2 (Continued)

Cation A	Chemical structure	Effective radius $r_{A,eff}$ (pm)	References	Cation B	Effective radius $r_{B,eff}$ (pm)	References
Methylammonium	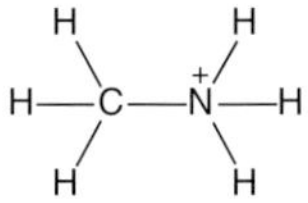	217	[70]	Sb^{3+}	76	[71]
n-Octylammonium	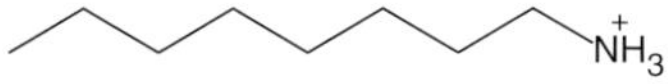	—	—	Bi^{3+}	103	[71]
Phenethylammonium	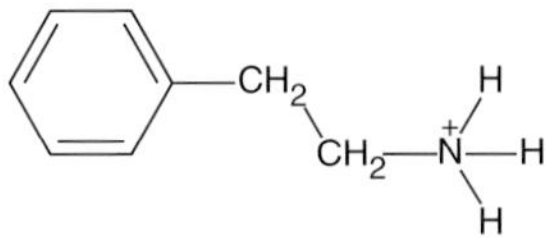	—	—	Te^{4+}	97	[71]
Phenylammonium	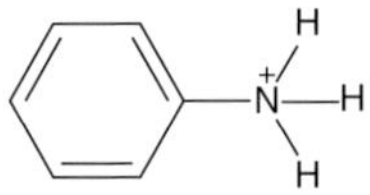	—	—	La^{3+}	103	[71]
Piperazine-1,4-diium	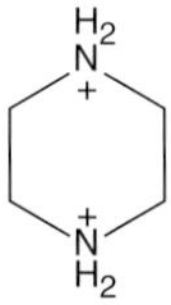	—	—	Ce^{3+}	101	[71]
Piperidinium	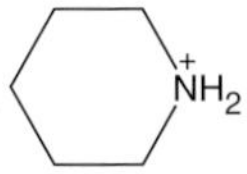	—	—	Pr^{3+}	99	[71]
Propane-1,3-diammonium	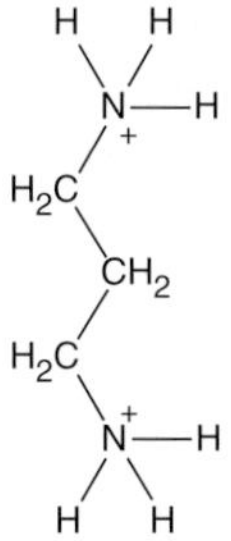	—	—	Nd^{3+}	98	[71]
iso-Propylammonium	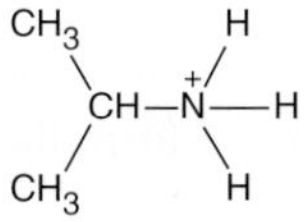	—	—	Sm^{3+}	96	[71]
n-Propylammonium	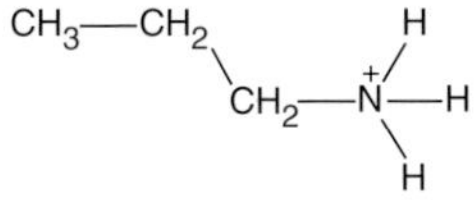	—	—	Eu^{3+}	95	[71]

(Continued)

Table 11.2 (Continued)

Cation A	Chemical structure	Effective radius $r_{A,eff}$ (pm)	References	Cation B	Effective radius $r_{B,eff}$ (pm)	References
Pyrrolidinium	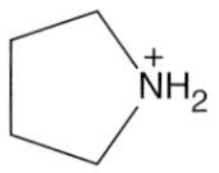	—	—	Gd^{3+}	94	[71]
Quinuclidin-1-ium	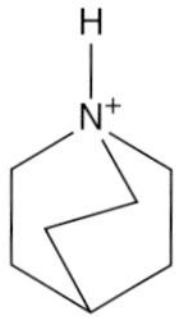	—	—	Dy^{3+}	91	[71]
Potassium	K^+	164	[71]	Er^{3+}	89	[71]
Rubidium	Rb^+	172	[71]	Tm^{3+}	88	[71]
Cesium	Cs^+	188	[71]	Lu^{3+}	86	[71]
				Pu^{3+}	100	[71]
				Am^{3+}	98	[71]
				Bk^{3+}	96	[71]

the exact location of the MA groups cannot be determined. The cubic phase at high temperature has space group $Pm\bar{3}m$ (octahedral symmetry), whereas the methylammonium ions have C_{3v} symmetry and the orientational disorder gives rise to the effective higher symmetry on average. Temperature dependence structural phase transformation is shown in Table 11.3.

$MAPbI_3$ undergoes a tetragonal phase at ambient room temperature. Therefore, 2D perovskite material with methylammonium cation can be the material of interest for good device performance. A recent study revealed that the motion of MA^+ in $MAPbX_3$ is too fast to be associated with hysteresis and hypothesized that it is due to polarization of ionic charges in the perovskite layer under the influence of the applied as well as the built-in voltage [75]. These results indicate that the motion of the organic cation is not giving significant contribution to the observed hysteresis. However, more information about the causes and possible strategies, how to avoid them, are needed, since hysteresis behavior is causing a low stability of halide perovskites and complicates the determination of the "real" performance of a solar cell. A list of organic–inorganic perovskite families ABI_3, $ABBr_3$, and $ABCl_3$, with their chemical formula and molecular weight, is tabulated in Tables 11.4–11.6.

Table 11.3 Temperature-dependent structural data of $CH_3NH_3PbI_3$, $CH_3NH_3PbBr_3$, and $CH_3NH_3Cl_3$ perovskite.

Perovskite	Phase	Temperature (K)	Structure	Space group	Lattice (Å) a	b	c	Volume (10^6 pm^3)	References
$MAPbI_3$	α	>327.4	Cubic	*Pm*3*m*	6.328	6.328	6.328	253.5	[74]
	β	162.2-327.4	Tetragonal	I4/m*cm*	8.849	8.849	12.642	992.6	[74]
	γ	<162.2	Orthorhombic	P*na*2$_1$	8.861	8.581	12.620	959.5	[74]
$MAPbBr_3$	α	>236.9	Cubic	P*m*3*m*	5.901			206.3	[74]
	β	155.1–236.9	Tetragonal	I4/*mcm*	8.322	11.833		819.4	[74]
	γ	149.5–155.1	Tetragonal	P4/*mmm*	5.8942	5.8612			[74]
	δ	<144.5	Orthorhombic	P*na*2$_1$	7.979	8.580	11.849	811.1	[74]
$MAPbCl_3$	α	>178.8	Cubic	P*m*3*m*	5.675			182.2	[74]
	β	172.9–178.9	Tetragonal	P4/*mmm*	5.655	5.630		180.1	[74]
	γ	<172.9	Orthorhombic	P222$_1$	5.673	5.628	11.182	375	[74]

11.2.3 Synthesis of Perovskite

The methylammonium lead iodide ($CH_3NH_3PbI_3$) perovskite was prepared by solution cast method. The precursor was dissolved in γ-butyrolactone solvent with equimolar CH_3NH_3I and PbI_2 on a nanocrystalline 18NRT TiO_2 surface. The concentration of the coating solution was 40 wt%, which contains 0.395 g of CH_3NH_3I (1.23 mmol) and 1.157 g of PbI_2 (1.23 mmol) in 2 ml of γ-butyrolactone kept for overnight stirring at 60 °C. The perovskite coating solution was spread on the annealed TiO_2 film (50 μl cm^{-2}) and was spun for 30 seconds at a speed of 2000 rpm in ambient atmosphere. The perovskite ($CH_3NH_3PbI_3$) formed on the TiO_2 surface was dried at 100 °C for 15 minutes. Likewise, other perovskite layers should also be prepared such as $CH_3NH_3PbBr_3$ [76], $CH_3NH_3Pb(I_{1-x}Br_x)_3$ [44], $CH_3NH_3PbI_2Cl$ [3], $CH_3NH_3SnI_3$ [77], etc.

11.2.3.1 Solution-Processed Method

Spin coating technique is one of the simplest and low-cost methods for synthesizing the perovskite layer. One-step and two-step sequential deposition method falls under solution process spin coating method and seems to be better in terms of minimizing the processing step. Both methods have their advantages and drawbacks. Therefore, one-step method is easiest and fastest method for synthesizing the thin film, whereas good morphology of the film can be obtained from two-step sequential deposition method. Here, PbI_2 was deposited on a substrate prior to spin coating of CH_3NH_3I. Thus, uniform morphology control is crucial to obtain in one-step method. Crystal growth can be controlled by anti-solvent process [46, 47].

Table 11.4 Shows list of cations "A" from organic–inorganic perovskite family ABI_3 with its chemical formula and its molecular weight.

S. No.	Cation A	Iodides based perovskite	Chemical formula	Molecular weight (g mol^{-1})
1	Acetamidinium	Acetamidinium iodide	$C_2H_7IN_2$	185.99
2	5-Azaspiro [4.4] nonan-5-ium	5-Azaspiro [4.4] nonan-5-ium iodide	$C_8H_{16}IN$	253.12
3	1,4-Benzene diammonium	1,4-Benzene diammonium iodide	$C_6H_{10}I_2N_2$	363.97
4	Benzylammonium	Benzylammonium iodide	$C_7H_{10}IN$	235.07
5	*iso*-Butylammonium	*iso*-Butylammonium iodide	$C_4H_{12}IN$	201.05
6	*n*-Butylammonium	*n*-Butylammonium iodide	$C_4H_{12}IN$	201.05
7	*t*-Butylammonium	*t*-Butylammonium iodide	$C_4H_{12}IN$	201.05
8	Cyclohexylammonium	Cyclohexylammonium iodide	$C_6H_{14}IN$	227.09
9	1,4-Diazabicyclo [2,2,2] octane-1,4-diium	1,4-Diazabicyclo [2,2,2] octane-1,4-diium iodide	$C_6H_{14}I_2N_2$	368.00
10	Diethylammonium	Diethylammonium iodide	$C_4H_{12}IN$	201.05
11	Dimethylammonium	Dimethylammonium iodide	C_2H_8IN	173.00
12	Ethane-1,2-diammonium	Ethane-1,2-diammonium iodide	$C_2H_{10}I_2N_2$	315.92
13	Ethylammonium	Ethylammonium iodide	C_2H_8IN	173.00
14	Formamidinium	Formamidinium iodide	CH_5IN_2	171.97
15	Guanidinium	Guanidinium iodide	CH_6IN_3	186.98
16	*n*-Hexylammonium	*n*-Hexylammonium iodide	C_6H_1IN	229.10
17	Imidazolium	Imidazolium iodide	$C_3H_5IN_2$	195.99
18	Methylammonium	Methylammonium iodide	CH_6IN	158.97
19	*n*-Octylammonium	*n*-Octylammonium Iodide	$C_8H_{20}IN$	257.16
20	Phenethylammonium	Phenethylammonium iodide	$C_8H_{12}IN$	249.09
21	Phenylammonium	Phenylammonium iodide	C_6H_8IN	221.04
22	Piperazine-1,4-diium	Piperazine-1,4-diium iodide	$C_4H_{12}I_2N_2$	341.96
23	Piperidinium	Piperidinium iodide	$C_5H_{12}IN$	213.06
24	Propane-1,3-diammonium	Propane-1,3-diammonium iodide	$C_3H_{12}I_2N_2$	329.95
25	*iso*-Propylammonium	*iso*-Propylammonium iodide	$C_3H_{10}IN$	187.02
26	*n*-Propylammonium	*n*-Propylammonium iodide	$C_3H_{10}IN$	187.02
27	Pyrrolidinium	Pyrrolidinium iodide	$C_4H_{10}IN$	199.03
28	Quinuclidin-1-ium	Quinuclidin-1-ium iodide	$C_7H_{14}IN$	239.10
29	Potassium	Potassium iodide	KI	166.00
30	Rubidium	Rubidium iodide	RbI	212.37
31	Cesium	Cesium iodide	CsI	259.81

Table 11.5 Shows list of cations "A" from organic–inorganic perovskite family $ABBr_3$ with its chemical formula and its molecular weight.

S. No.	Cation A	Bromides based perovskite	Chemical formula	Molecular weight
1	Acetamidinium	—	—	—
2	5-Azaspiro [4.4] nonan-5-ium	5-Azaspiro [4.4]nonan-5-ium bromide	$C_8H_{16}BrN$	206.12
3	1,4-Benzene diammonium	—	—	—
4	Benzylammonium	Benzylammonium bromide	$C_7H_{10}BrN$	188.06
5	*iso*-Butylammonium	*iso*-Butylammonium bromide	$C_4H_{12}BrN$	154.05
6	*n*-Butylammonium	*n*-Butylammonium bromide	$C_4H_{12}BrN$	154.05
7	*t*-Butylammonium	*t*-Butylammonium bromide	$C_4H_{12}BrN$	154.05
8	Cyclohexylammonium	Cyclohexylammonium bromide	$C_6H_{14}BrN$	180.09
9	1,4-Diazabicyclo [2,2,2] octane-1,4-diium	1,4-Diazabicyclo [2,2,2] octane-1,4-diium bromide	$C_6H_{14}Br_2N_2$	274.00
10	Diethylammonium	Diethylammonium bromide	$C_4H_{12}BrN$	154.05
11	Dimethylammonium	Dimethylammonium bromide	C_2H_8BrN	126.00
12	Ethane-1,2-diammonium	Ethane-1,2-diammonium bromide	$C_2H_{10}Br_2N_2$	221.92
13	Ethylammonium	Ethylammonium bromide	C_2H_8BrN	125.99
14	Formamidinium	Formamidinium bromide	CH_5BrN_2	124.97
15	Guanidinium	Guanidinium bromide	CH_6BrN_3	139.98
16	*n*-Hexylammonium	*n*-Hexylammonium bromide	$C_6H_{16}BrN$	182.10
17	Imidazolium	Imidazolium bromide	$C_3H_5BrN_2$	148.99
18	Methylammonium	Methylammonium bromide	CH_6BrN	111.97
19	*n*-Octylammonium	*n*-Octylammonium Bromide	$C_8H_{20}BrN$	210.16
20	Phenethylammonium	Phenethylammonium bromide	$C_8H_{12}BrN$	202.09
21	Phenylammonium	Phenylammonium bromide	C_6H_8BrN	174.04
22	Piperazine-1,4-diium	Piperazine-1,4-diium bromide	$C_4H_{12}Br_2N_2$	247.96
23	Piperidinium	Piperidinium bromide	$C_5H_{12}BrN$	166.06
24	Propane-1,3-diammonium	Propane-1,3-diammonium bromide	$C_3H_{12}Br_2N_2$	235.95

(Continued)

Table 11.5 (Continued)

S. No.	Cation A	Bromides based perovskite	Chemical formula	Molecular weight
25	*iso*-Propylammonium	*iso*-Propylammonium bromide	$C_3H_{10}BrN$	140.02
26	*n*-Propylammonium	*n*-Propylammonium bromide	$C_3H_{10}BrN$	140.02
27	Pyrrolidinium	Pyrrolidinium bromide	$C_4H_{10}BrN$	152.03
28	Quinuclidin-1-ium	Quinuclidin-1-ium bromide	$C_7H_{14}BrN$	192.10
29	Potassium	Potassium bromide	KBr	119.00
30	Rubidium	Rubidium bromide	RbBr	165.37
31	Cesium	Cesium bromide	CsBr	212.81

11.2.3.2 Hot Casting Technique

In hot casting process, the substrate was preheated at desired temperature before spin coating the solution on the substrate. The as-prepared solution of perovskite is used for spin coating by one-step method. The thickness of the perovskite layer was controlled by optimizing the molar ratio of the solution or by speed of spin coating, whereas the performance of the device is controlled by the substrate annealing temperature and time [45].

11.2.3.3 Vapor Deposition Method

In vapor deposition method, firstly the prepared solution of PbI_2 was spin coated on the fluorine-doped tin oxide (FTO)/c-TiO_2 substrates and further dried. Whereas, CH_3NH_3I powder was spread out around the PbI_2-coated substrates with a petri dish covering on the top, followed by annealing at 150 °C for a fixed time. Both the deposition of PbI_2 film and vapor treatment of PbI_2 film are carried out in a glove box. After cooling down, the as-prepared substrates were washed with isopropanol, dried, and annealed. The active layer-coated electrode was further used for fabricating solar cell [48].

11.2.3.4 Thermal Evaporation Technique

Organic and inorganic precursors are thermally evaporated from separate crucible sources at 10^{-5} mbar, and both sources were heated at deposition temperature for fixed time. The uniformity of the film was controlled by a rotating substrate holder. The thickness of the thin film was optimized by controlling the deposition rate and time using sensor above crucible. Deposition composition can be also varied at fixed annealing to achieve high performance device. The change in color of the sample was the indication for phase transformation. Hence, a dark brown color can be obtained from the conventional PbI_2 and CH_3NH_3I perovskite material. Afterward thermal evaporation substrate was kept for annealing at 100 °C [49, 56].

Table 11.6 Shows list of cations "A" from organic–inorganic perovskite family $ABCl_3$ with its chemical formula and its molecular weight.

S. No.	Cation A	Chloride based perovskite	Chemical formula	Molecular weight
1	Acetamidinium			
2	5-Azaspiro [4.4] nonan-5-ium	5-Azaspiro [4.4] nonan-5-ium chloride	$C_8H_{16}ClN$	161.67
3	1,4-Benzene diammonium			
4	Benzylammonium	Benzylammonium chloride	$C_7H_{10}ClN$	143.61
5	*iso*-Butylammonium			
6	*n*-Butylammonium			
7	*t*-Butylammonium			
8	Cyclohexylammonium			
9	1,4-Diazabicyclo [2,2,2] octane-1,4-diium	1,4-Diazabicyclo [2,2,2] octane-1,4-diium chloride	$C_6H_{14}Cl_2N_2$	185.09
10	Diethylammonium			
11	Dimethylammonium			
12	Ethane-1,2-diammonium			
13	Ethylammonium	Ethylammonium chloride	C_2H_8ClN	81.545
14	Formamidinium			
15	Guanidinium			
16	*n*-Hexylammonium			
17	Imidazolium			
18	Methylammonium	Methylammonium chloride	CH_6ClN	67.518
19	*n*-Octylammonium	*n*-Octylammonium chloride	$C_8H_{20}ClN$	165.70
20	Phenethylammonium			
21	Phenylammonium			
22	Piperazine-1,4-diium	Piperazine-1,4-diium chloride	$C_4H_{12}Cl_2N_2$	159.06
23	Piperidinium	Piperidinium chloride	$C_5H_{12}ClN$	121.61
24	Propane-1,3-diammonium	Propane-1,3-diammonium chloride	$C_3H_{12}Cl_2N_2$	147.05
25	*iso*-Propylammonium			
26	*n*-Propylammonium			
27	Pyrrolidinium	Pyrrolidinium chloride	$C_4H_{10}ClN$	107.58
28	Quinuclidin-1-ium	Quinuclidin-1-ium chloride	$C_7H_{14}ClN$	147.65
29	Potassium	Potassium chloride	KCl	74.55
30	Rubidium	Rubidium chloride	RbCl	120.92
31	Cesium	Cesium chloride	CsCl	168.36

11.3 Polymer Electrolyte for Perovskite Solar Cell

Electrolytic materials are in demand since they possess high ionic conductivity, which is essential for electrochemical devices (like batteries, supercapacitors, solar cell, fuel cells, etc.). Inorganic hole conductors, organic conducting polymers, and organic hole conductors show an excellent candidate for device application. Many efforts have been made to make less volatile electrolyte with high conduction. Redox couple is a main component to make an electrolyte. There are several redox couples reported for electrochemical devices such as iodide/tri-iodide (I^-/I^{3-}), ferrocene/ferricenium/(Fc/Fc^+), copper (I/II), and series of cobalt (II/III) and nickel (III/IV) complexes, Br^-/Br^{3-}, $SCN^-/(SCN)^{3-}$, $SeCN^-/(SeCN)^{3-}$, etc. As this chapter is basically on the inorganic hole transport layer (HTL) for PSC, it is restricted to discuss in great detail.

The redox electrolyte was prepared by dissolving 0.9 M LiI, 0.45 M I_2, 0.5 M *tert*-butyl pyridine *t*BP, and 0.05 M urea in ethyl acetate, which was introduced into the space of the sealed electrodes [7, 78]. Pt counter electrode was prepared by spreading a droplet of 7 mM H_2PtCl_6-$x$$H_2O$ in 2-propanol on an FTO substrate and heated at 400 °C for 20 minutes in air. In another system 0.4 M LiBr and 0.04 M Br_2 were dissolved in acetonitrile, while the $CH_3NH_3PbI_3/TiO_2$-based cell employed 0.15 M LiI and 0.075 M I_2 dissolved in methoxy acetonitrile [78]. In addition, a Li salt solution was composed of 0.032 M Li-TFSI, 0.032 M LiI, and 0.2 M *t*BP in the mixed solvent of chlorobenzene and acetonitrile (chlorobenzene:acetonitrile ¼ 1 : 0.1 v/v) [1].

In a common preparation method of polymer electrolyte, a fixed amount of poly(ethylene oxide) (Mw ~ 5 000 000, Sigma-Aldrich) and poly(ethylene glycol) (Mw ~ 200, Sigma-Aldrich) was dissolved in acetonitrile (A2) (20 ml) in a beaker with continuous stirring at room temperature. This ratio was used as a stock solution for the overall experiment (written here as B1). Stoichiometry ratio of potassium iodide (KI) and Iodine (I^2) was then dissolved in A2 acetonitrile (5ml) in another beaker (written here as B2). Adding B2 (drop by drop) in B1 formed the solid polymer electrolyte solution. During this process, the polymer salt solution was continuously stirred. Finally, this solution of the mixture is ready to use in PSC.

11.3.1 Device Fabrication

For laboratory scale PSC fabrication, we cut FTO (from Sigma-Aldrich) glass plate ($1 \times 1\,cm^2$). FTO glasses were etched with zinc powder and HCl (2 M) to obtain the required electrode pattern. The sheets were then washed with soap (water), then FTO-coated glass substrates were cleaned for 15 minutes with acetone, ethanol, and deionised (DI) water in an ultrasonic bath, and finally treated with isopropyl alcohol (IPA) for 10 minutes to remove the last traces of all organic residues. A compact hole-blocking layer of TiO_2 was spin-coated (at 4000 rpm for 60 seconds) on the substrate, using 1 : 10 ratio of titanium diisopropoxide bis(acetylacetonate) (75 wt% in isopropanol) in anhydrous ethanol. This layer was then baked at 500 °C for 30 minutes. A ~200–500-nm-thick layer of porous mp-TiO_2 was prepared on top of the TiO_2 layer using a diluted

TiO_2 paste (EtOH:Dyesol 18NRT TiO_2 paste in a ratio of 4.5 : 1) and calcining at 500 °C for 30 minutes. The resulting mp-TiO_2/FTO substrate was immersed for 15 minutes in a 0.04 M $TiCl_4$ solution at 70 °C, rinsed with DI water, and then annealed at 500 °C for 30 minutes.

11.3.2 Hole Transport Layer

30 μl of an HTL solution (72 mg of spiro-MeOTAD, 28.8 μl of 4-tert-butylpyridine, and 17.6 μl of LiI solution (720 mg of LiI in acetonitrile) in 1 ml of chlorobenzene) was formed via spin coating at 3000 rpm for 30 seconds. Transfer in the sputtering chamber for depositing of 100 nm thick Au.

11.4 Performance Characteristics

Optical absorption of hybrid perovskites shows very high absorbance for low thickness active layer with tunable bandgap [3, 54, 79]. Tuning the bandgap of inorganic metal halide by incorporating different halide I^-, Cl^-, or Br^- so that it becomes $CH_3NH_3PbI_{3-x}Cl_x$, $CH_3NH_3PbBr_{3-x}Cl_x$, or $CH_3NH_3PbI_{3-x}Br_x$ [3, 80, 81]. 3D perovskites behave very similar to bulk semiconductor. However, since only the metal-halide framework is responsible for the long charge carrier lifetimes, while the organic cation is not contributing energetically to relevant density of states near the *bandgap* [82, 83], it can be attempted to modify the optoelectronic properties without changing the BX_3^- network. The absorption for $CH_3NH_3PbI_3$ and $CH_3NH_3PbI_{3-x}Cl_x$ perovskite thin film shows absorption around 800 nm, whereas $CH_3NH_3PbI_{3-x}Br_x$ shows absorption around 700 nm [3, 7, 84, 85]. $CH_3CH_2NH_3PbI_3$ perovskite thin film shows absorption around 850 nm with bandgap of ~1.48 eV, but adding Br can enable to tune the bandgap of perovskite such as $CH_3CH_2NH_3PbI_{3-x}Br_x$ to show absorption around 550 nm and reach a bandgap to 2.23 eV for 100% Br. However, $CH_3NH_3Sn_xPb_{1-x}I_3$ based thin film can absorb above 1000 nm [42] $CH_3NH_3PbI_3$ perovskite that has carrier diffusion lengths for both electrons and holes of around 100 nm, whereas for $CH_3NH_3PbI_{3-x}Cl_x$ is around 1 μm [4, 28].

The spectral response of the PSC depends on the absorption properties of the perovskite phase. Electrical characterization of the solar cell depends on various electrical parameters that include the photocurrent and photo-potentials measured under different conditions (open and closed circuit, under monochromatic light or sunlight illumination: I_{oc}, V_{oc}, I_{sc}, and V_{sc}). The incident photon-to-electrical conversion efficiency (IPCE) is a quantum-yield term for the overall charge-injection collection process measured using monochromatic light (single wavelength source). The photocurrent measured under closed circuit I_{sc} is the integrated sum of IPCE measured over the entire solar spectrum:

$$I_{SC} = \int_0^{\infty} \mathrm{IPCE}(\lambda) \cdot I_{sun}(\lambda)\, d\lambda \tag{11.6}$$

$$\mathrm{IPCE}(\lambda) = 1240 \left(\frac{I_{SC}}{\lambda_{\varphi}} \right) \tag{11.7}$$

Table 11.7 The methods adopted for achieving high efficient cell parameter using redox electrolyte layer for perovskite solar cell.

S.No.	Electrolyte	Type	Method	Cell structure	V_{oc} (V)	J_{sc} (mA cm^{-2})	FF	Efficiency (%)	References
1	LiI	Mesoscopic n-i-p	Spin coating	FTO/TiO_2/PVK/ LiI/H_2PtCl_6	0.629	15.99	0.61	6.20	[78]
2	PEO:KI/I_2	Mesoscopic n-i-p	Spin coating	FTO/TiO_2/PVK/ PEO:KI/I_2/H_2PtCl_6	0.70	1.16	0.67	0.75	[25]
3	LiBr	Mesoscopic n-i-p	Spin coating	FTO/TiO_2/PVK/ LiBr/H_2PtCl_6	0.96	5.57	0.59	3.13	[7]
4	LiI	Mesoscopic n-i-p	Spin coating	FTO/TiO_2/PVK/ LiI/H_2PtCl_6	0.61	11.0	0.57	3.81	[7]
5	PANI-LiI	Mesoscopic n-i-p	Spin coating	FTO/TiO_2/PANI-$CH_3NH_3PbI_3$/ Li-salt	0.78	14.48	0.65	7.34	[17]
6	LiI	Mesoscopic n-i-p	Spin coating	FTO/TiO_2/PVK/ LiI/H_2PtCl_6	0.66	5.2	0.70	2.4	[86]

where λ is the wavelength, I_{sc} the current at short circuit (mA cm^{-2}), and φ is the incident radiative flux (W m^{-2}). The overall sunlight-to-electric PCE of a PSC is given by the following expression:

$$\eta = \frac{P_{max}}{P_{in}} = \frac{I_{sc}V_{oc}FF}{P_{in}} \tag{11.8}$$

The product of the two I_{max} and V_{max} is the maximum power of the solar cell. The fill factor (FF) is defined as the ratio of $(I_{max} \cdot V_{max}/I_{sc} \cdot V_{oc})$. I_{sc}, V_{oc}, FF, and η are the key performance parameters of the solar cell. In perovskite, solar cell J–V hysteresis is one of the major issues that limit the device performance that is observed by varying the direction and the rate of sweep voltage (Table 11.7).

11.5 Challenges and Improvement

Long-term device stability during operation under stressed conditions such as humidity, air, and elevated temperature and under ultraviolet light has yet to be well studied. PSCs have proven for producing high efficiency, but are not commercialized till now due to the poor stability. Real understanding of the physics behind these devices has to be examined in depth to know how structural and electronic properties determine the short-circuit current (J_{sc}) and open-circuit voltage (V_{oc}) are still lacking. However, the important issues and challenges that limited the commercialization of PSCs still exist. Still the role of charge transfer states in the PSC is a topic of debate. Still J–V hysteresis is observed in the J–V characteristic of PSC that limits the standardized characterization of device performance.

There are still lots of concern about the environment impact of these devices. The toxicity of lead-based perovskite is the main concern for commercialization. There are a handful of reports on lead-free PSC, but this class of solar cell does not show to be an efficient device. Still the manufacturing, operational, and disposal phases of PSCs are needed to be taken into consideration. Contamination is also related with the perovskite, which leads to the poor performance of the materials in terms of producing reproducible solar devices. Although the complexity of the diverse material preparation methods and device architectures make it more difficult to address these issues, recent progress has provided insights into these issues and the corresponding material properties.

11.6 Future Trends

Much progress has been made in PSC area over the past few years. Researchers are looking for cost-effective, clean, and reliable energy devices. From a commercialization point of view, the large-scale implementation of PSCs requires toxicity and stability issues to be resolved. Recently, work on lead free PSC has been started but still results obtained are expected to improve in terms of stability and toxicity. Instability under air and moisture conditions are the challenges for researchers to work on. Recently, few studies have shown good moisture and air stable solar devices based on 2D/3D layered PSC. These materials are truly remarkable absorbers and should surely be considered for commercialization. However, the perovskite material has inherent stability and toxicity problem. Material design and interface engineering is required for achieving more stable devices. Stability is a major challenge for these solar cells, improving stability of PSC with sealant such as glass and photocurable fluoropolymer on the top of perovskite devices also needed to focus. Due to tunable bandgap from 1.5 to 2.3 eV and high V_{OC}, which is great interest in implementing them in perovskite tandem devices with silicon or CIGS or perovskite cells. The Shockley–Queisser limit in terms of efficiency of the monolithic perovskite tandem cells can exceed 35% in the coming future. There are a handful of articles that talk about the sealant materials for perovskite layer.

A novel approach and method is needed to the design the device architecture for PSCs that should be motivated from low-cost, light weight, and flexible substrates. These devices are of commercial interest for cheap, large-scale roll-to-roll processing and for technological advancement applications. Lots of efforts have been made on perovskite-based solar devices on flexible, conductive substrates, such as poly(ethylene terephthalate), polyethylene naphthalate, and Ti foils. Additionally, high-transparency and colorful perovskite PV for building integration have also been studied with great interest and have a lot of scopes in future. The main advantage for fabrication of PSC is its simple and easy processing such as solution cast technique, vapor deposition, thermal evaporation, hot casting technique, and large area processing. In a very short span of time, other technologies that have reached efficiencies took decades to achieve. Now researchers are proving that PSC is an emerging type of technology that soon will be the face of the future.

11.7 Conclusion

Significant progress has been shown in the device architecture, synthesis of material, thin film formation, etc., of hybrid perovskite materials. We have tried to focus on the low-cost, simple design, and easy synthesis of hole transport material for PSC. It has been found that the most efficient PSCs follow a p-i-n type of architecture with a lot of possible p- and n-type materials for efficient carrier extraction. In this chapter, different types of perovskite materials have been discussed and the unique electrical, chemical, and physical properties with compatibility on other materials are also described, although these materials are of great interest to the scientific community working in the field of new generation photovoltaic. Comparison tables of the materials, challenges, improvements, and future trends have been discussed in this chapter so that a clear picture can be created that will definitely help to open up the new way in the photovoltaic field.

Competing Interests

The authors declare that they have no competing interests.

Acknowledgments

The authors are thankful to the Material Research Laboratory, Sharda University, India, and Centre for Ionics University of Malaya (CIUM), Universiti Malaya, Malaysia. Also thanks to the Department of Science and Technology, Government of India, and IUSSTF.

References

1 Park, N.G. (2016). Methodologies for high efficiency perovskite solar cells. *Nano Converg.* 3: 15.

2 Park, N.G. (2016). High efficiency mesoscopic organometal halide perovskite solar cells. In: *Unconventional Thin Film Photovoltaics*, Chapter 1 (eds. E. Da Como, F. De Angelis, H. Snaith and A. Walker), 1–31. Royal Society of Chemistry.

3 Lee, M.M., Teuscher, J., Miyasaka, T. et al. (2012). Efficient hybrid solar cells based on meso-superstructured organometal halide perovskites. *Science* 338: 643–647.

4 Xing, G., Mathews, N., Sun, S. et al. (2013). Long-range balanced electron- and hole-transport lengths in organic-inorganic $CH_3NH_3PbI_3$. *Science* 342: 344–347.

5 Miyata, A., Mitioglu, A., Plochocka, P. et al. (2015). Direct measurement of the exciton binding energy and effective masses for charge carriers in organic–inorganic tri-halide perovskites. *Nat. Phys.* 11: U582–U594.

6 Song, Z., Watthage, S.C., Phillips, A.B., and Heben, M.J. (2016). Pathways toward high-performance perovskite solar cells: review of recent advances in organo-metal halide perovskites for photovoltaic applications. *J. Photonics Energy* 6 (2): 022001.

7 Kojima, A., Teshima, K., Shirai, Y., and Miyasaka, T. (2009). Organometal halide perovskites as visible-light sensitizers for photovoltaic cells. *J. Am. Chem. Soc.* 131 (17): 6050–6051.

8 NREL chart. http://www.nrel.gov/ncpv/images/efficiency_chart.jpg (accessed 13 March 2016).

9 Bush, K.A., Palmstrom, A.F., Yu, Z.J. et al. (2017). 23.6%-efficient monolithic perovskite/silicon tandem solar cells with improved stability. *Nature Energy* 2: 17009.

10 Kim, H.S., Lee, C.R., Im, J.H. et al. (2012). Lead iodide perovskite sensitized all-solid-state submicron thin filmmesoscopic solar cell with efficiency exceeding 9%. *Sci. Rep.* 2: 591.

11 Wang, W., Zhang, Z., Cai, Y. et al. (2016). Enhanced performance of $CH_3NH_3PbI_{3-x}Cl_x$ perovskite solar cells by CH_3NH_3I modification of TiO_2-perovskite layer interface. *Nanoscale Res. Lett.* 11: 316.

12 Wang, S., Yuan, W., and Meng, Y.S. (2015). Spectrum-dependent spiro-OMeTAD oxidization mechanism in perovskite solar cells. *ACS Appl. Mater. Interfaces* 7 (44): 24791–24798.

13 Chen, H.-W., Huang, T.-Y., Chang, T.-H. et al. (2016). Efficiency enhancement of hybrid perovskite solar cells with MEH-PPV hole-transporting layers. *Sci. Rep.* 6: 34319.

14 Xiao, J., Shi, J., Liu, H. et al. (2015). Efficient $CH_3NH_3PbI_3$ perovskite solar cells based on graphdiyne (GD)-modified P_3HT hole-transporting material. *Adv. Energy Mater* 5: 1401943.

15 Heo, J.H., Im, S.H., Noh, J.H. et al. (2013). Efficient inorganic–organic hybrid heterojunction solar cells containing perovskite compound and polymeric hole conductors. *Nat. Photonics* 7: 486–491.

16 Seo, J., Noh, J.H., and Seok, S.I. (2016). Rational strategies for efficient perovskite solar cells. *Acc. Chem. Res.* 49 (3): 562–572.

17 Xiao, Y., Han, G., Chang, Y. et al. (2014). An all-solid-state perovskite-sensitized solar cell based on the dual function polyaniline as the sensitizer and p-type hole-transporting material. *J. Power Sources* 267: 1–8.

18 Bi, D., Yang, L., Boschloo, G. et al. (2013). Effect of different hole transport materials on recombination in $CH_3NH_3PbI_3$ perovskite-sensitized mesoscopic solar cells. *J. Phys. Chem. Lett.* 4 (9): 1532–1536.

19 Zhang, H., Xue, L., Han, J. et al. (2016). New generation perovskite solar cells with solution-processed amino-substituted perylene diimide derivative as electron-transport layer. *J. Mater. Chem. A* 4: 8724–8733.

20 Li, Y., Meng, L., Yang, Y. et al. (2016). High-efficiency robust perovskite solar cells on ultrathin flexible substrates. *Nat. Commun.* 7: 10214.

21 Chiang, C.H. and Wu, C.G. (2016). Bulk heterojunction perovskite–PCBM solar cells with high fill factor. *Nat. Photonics* 10: 196–200.
22 Li, M., Wang, Z.K., Yang, Y.G. et al. (2016). Copper salts doped spiro-OMeTAD for high-performance perovskite solar cells. *Adv. Energy Mater.* 6: 1601156.
23 Jung, J.W., Chueh, C.C., and Jen, A.K.Y. (2015). A low-temperature, solution-processable, Cu-doped nickel oxide hole-transporting layer via the combustion method for high-performance thin-film perovskite solar cells. *Adv. Mater.* 27 (47): 7874–7880.
24 Sung, H., Ahn, N., Jang, M.S. et al. (2016). Transparent conductive oxide-free graphene-based perovskite solar cells with over 17% efficiency. *Adv. Energy Mater.* 6: 1501873.
25 Bhattacharya, R.B., Singh, P.K., Singh, R., and Khan, Z.H. (2016). Perovskite sensitized solar cell using solid polymer electrolyte. *J. Hydrogen Energy* (4, 1): 2847–2852.
26 Aitola, K., Sveinbjörnsson, K., Correa-Baena, J.P. et al. (2016). Carbon nanotube-based hybrid hole-transporting material and selective contact for high efficiency perovskite solar cells. *Energy Environ. Sci.* 9: 461–466.
27 Hao, F., Stoumpos, C.C., Cao, D.H. et al. (2014). Lead-free solid-state organic–inorganic halide perovskite solar cells. *Nat. Photonics* 8: 489–494.
28 Stranks, S.D., Eperon, G.E., Grancini, G. et al. (2013). Electron-hole diffusion lengths exceeding 1 micrometer in an organometal trihalide perovskite absorber. *Science* 342: 341–344.
29 Saliba, M., Matsui, T., Seo, J.Y. et al. (2016). Cesium-containing triple cation perovskite solar cells: improved stability, reproducibility and high efficiency. *Energy Environ. Sci.* 9: 1989–1997.
30 Koh, T.M., Fu, K.W., Fang, Y.N. et al. (2014). Formamidinium-containing metal-halide: an alternative material for near-IR absorption perovskite solar cells. *J. Phys. Chem. C* 118: 16458–16462.
31 Hendon, C.H., Yang, R.X., Burton, L.A., and Walsh, A. (2015). Assessment of polyanion (BF^{4-} and PF^{6-}) substitutions in hybrid halide perovskites. *J. Mater. Chem. A* 3: 9067–9070.
32 Heo, J.H., Song, D.H., and Im, S.H. (2014). Planar $CH_3NH_3PbBr_3$ hybrid solar cells with 10.4% power conversion efficiency, fabricated by controlled crystallization in the spin-coating process. *Adv. Mater.* 26: 8179–8183.
33 Jiang, Q., Rebollar, D., Gong, J. et al. (2015). Pseudohalide-induced moisture tolerance in perovskite $CH_3NH_3Pb(SCN)_2I$ thin films. *Angew. Chem. Int. Ed.* 54: 7617–7620.
34 Nagane, S., Bansode, U., Game, O. et al. (2014). $CH_3NH_3PbI_{(3-x)}(BF_4)_x$: molecular ion substituted hybrid perovskite. *Chem. Commun.* 50: 9741–9744.
35 Prajongtat, P. and Dittricht, T. (2015). Precipitation of $CH_3NH_3PbCl_3$ in $CH_3NH_3PbI_3$ and impact on modulated charge separation. *J. Phys. Chem. C* 119: 9926–9933.
36 Dong, Q., Fang, Y., Shao, Y. et al. (2015). Electron-hole diffusion lengths > 175 μm in solution-grown $CH_3NH_3PbI_3$ single crystals. *Science* 347: 967–970.
37 Motta, C., Mellouhi, F.E., and Sanvito, S. (2015). Charge carrier mobility in hybrid halide perovskites. *Sci. Rep.* 5: 12746.

38 Brenner, T.M., Egger, D.A., Rappe, A.M. et al. (2015). Are mobilities in hybrid organic-inorganic halide perovskites actually 'high'? *J. Phys. Chem. Lett.* 6 (23): 4754–4757.

39 Lin, Q.Q., Armin, A., Nagiri, R.C.R. et al. (2015). Electro-optics of perovskite solar cellsNat. *Photonics* 9: 106–112.

40 Pellet, N., Gao, P., Gregori, G. et al. (2014). Mixed-organic-cation perovskite photovoltaics for enhanced solar-light harvesting. *Angew. Chem.* 53: 3151–3157.

41 Yang, W.S., Noh, J.H., Jeon, N.J. et al. (2015). High-performance photovoltaic perovskite layers fabricated through intramolecular exchange. *Science* 348: 1234–1237.

42 Ogomi, Y., Morita, A., Tsukamoto, S. et al. (2014). $CH_3NH_3Sn_xPb_{(1-x)}I_3$ perovskite solar cells covering up to 1060 nm. *J. Phys. Chem. Lett.* 5: 1004–1011.

43 Stoumpos, C.C., Malliakas, C.D., and Kanatzidis, M.G. (2013). Semiconducting tin and lead iodide perovskites with organic cations: phase transitions, high mobilities, and near-infrared photoluminescent properties. *Inorg. Chem.* 52: 9019–9038.

44 Noh, J.H., Im, S.H., Heo, J.H. et al. (2013). Chemical management for colorful, efficient, and stable inorganic–organic hybrid nanostructured solar cells. *Nano Lett.* 13: 1764–1769.

45 Nie, W., Tsai, H., Asadpour, R. et al. (2015). High-efficiency solution-processed perovskite solar cells with millimeter-scale grains. *Science* 347 (6221): 522–525.

46 Xiao, M., Huang, F., Huang, W. et al. (2014). A fast deposition-crystallization procedure for highly efficient lead iodide perovskite thin-film solar cells. *Angew. Chem.* 126 (37): 10056–10061.

47 Jeon, N.J., Noh, J.H., Kim, Y.C. et al. (2014). Solvent engineering for high-performance inorganic–organic hybrid perovskite solar cells. *Nat. Mater.* 13 (9): 897–903.

48 Chen, Q., Zhou, H., Hong, Z. et al. (2013). Planar heterojunction perovskite solar cells via vapor-assisted solution process. *J. Am. Chem. Soc.* 136 (2): 622–625.

49 Huang, F., Dkhissi, Y., Huang, W. et al. (2014). Gas-assisted preparation of lead iodide perovskite films consisting of a monolayer of single crystalline grains for high efficiency planar solar cells. *Nano Energy* 10: 10–18.

50 Stranks, S.D. and Snaith, H.J. (2015). Metal-halide perovskites for photovoltaic and light-emitting devices. *Nat. Nanotechnol.* 10: 391–402.

51 Gonzalez-Pedro, V., Juarez-Perez, E.J., Arsyad, W.S. et al. (2014). General working principles of $CH_3NH_3PbX_3$ perovskite solar cells. *Nano Lett.* 14 (2): 888–893.

52 Zhang, H., Shi, Y., Yan, F. et al. (2014). A dual functional additive for the HTM layer in perovskite solar cells. *Chem. Commun.* 50: 5020–5022.

53 Ball, J.M., Lee, M.M., Hey, A., and Snaith, H. (2013). Low-temperature processed meso-superstructured to thin-film perovskite solar cells. *J. Energy Environ. Sci.* 6: 1739–1743.

54 Etgar, L., Gao, P., Xue, Z. et al. (2012). Mesoscopic $CH_3NH_3PbI_3/TiO_2$ heterojunction solar cells. *J. Am. Chem. Soc.* 134: 17396–17399.

55 Lotsch, B.V. (2014). New light on an old story: perovskites go solar. *Angew. Chem. Int. Ed.* 53: 635–637.
56 Liu, M., Johnston, M.B., and Snaith, H.J. (2013). Efficient planar heterojunction perovskite solar cells by vapour deposition. *Nature* 501: 395–398.
57 Hu, M., Liu, L., Mei, A. et al. (2014). Efficient hole-conductor-free, fully printablemesoscopic perovskite solar cells with a broad light harvester $NH_2CH{=}NH_2PbI_3$. *J. Mater. Chem. A* 2: 17115–17121.
58 Hoefler, S.F., Trimmel, G., and Rath, T. (2017). Progress on lead-free metal halide perovskites for photovoltaic applications: a review. *MonatshChem* 148: 795–826.
59 Zhao, Y., Nardes, A.M., and Zhu, K. (2014). Mesoporous perovskite solar cells: material composition, charge-carrier dynamics, and device characteristics. *Faraday Discuss.* 176: 301–312.
60 Jiang, M., Wu, J., Lan, F. et al. (2015). Enhancing the performance of planar organo-lead halide perovskite solar cells by using a mixed halide source. *J. Mater. Chem. A* 3: 963–967.
61 Brehm, J.A., Bennett, J.W., Schoenberg, M.R. et al. (2014). The structural diversity of ABS_3 compounds with d 0 electronic configuration for the B cation. *J. Chem. Phys.* 140: 224703.
62 Mitzi, D. (1999). Synthesis, structure, and properties of organic-inorganic perovskites and related materials. In: *Progress in Inorganic Chemistry* (ed. K.D. Karlin), 48. Canada: Wiley.
63 Saparov, B. and Mitzi, D.B. (2016). Organic–inorganic perovskites: structural versatility for functional materials design. *Chem. Rev.* 116: 4558–4596.
64 Hu, K.L., Kurmoo, M., Wang, Z., and Gao, S. (2009). Metal–organic perovskites: synthesis, structures, and magnetic properties of $[C(NH_2)_3][M^{II}(HCOO)_3]$ (M=Mn, Fe, Co, Ni, Cu, and Zn; $C(NH_2)_3$= Guanidinium). *Chem. Eur. J.* 15: 12050.
65 Qian, J., Xu, B., and Tian, W. (2016). A comprehensive theoretical study of halide perovskites ABX_3. *Org. Electron.* 37: 61.
66 Goldschmidt, V.M. (1926). Die Gesetze der Krystallochemie. *Naturwissenschaften* 14: 477.
67 Travis, W., Glover, E.N.K., Bronstein, H. et al. (2016). On the application of the tolerance factor to inorganic and hybrid halide perovskites: a revised system. *Chem. Sci.* 7: 4548.
68 Huang, T.J., Thiang, Z.X., Yin, X. et al. (2016). $(CH_3NH_3)_2PdCl_4$: a compound with two-dimensional organic–inorganic layered perovskite structure. *Chem. Eur. J.* 22: 2146.
69 Li, C., Lu, X., Ding, W. et al. (2008). Formability of ABX_3 (X = F, Cl, Br, I) halide perovskites. *Acta Crystallogr., Sect. B: Struct. Sci.* 64: 702.
70 Kieslich, G., Sun, S., and Cheetham, A.K. (2014). Solid-state principles applied to organic–inorganic perovskites: new tricks for an old dog. *Chem. Sci.* 5: 4712.
71 Shannon, R.D. (1976). Revised effective ionic radii and systematic studies of interatomie distances in halides and chaleogenides. *Acta Crystallogr., Sect. A* 32: 751.

72 Baikie, T., Fang, Y., Kadro, J.M. et al. (2013). Synthesis and crystal chemistry of the hybrid perovskite $(CH_3NH_3)PbI_3$ for solid-state sensitised solar cell applications. *J. Mater. Chem. A* 1: 5628–5641.

73 Knop, O., Wasylishen, R.E., White, M.A. et al. (1990). Alkylammonium lead halides. Part 2. $CH_3NH_3PbX_3$ (X=Cl, Br, I) perovskites: cuboctahedral halide cages with isotropic cation reorientation. *Can. J. Chem.* 68 (3): 412–422.

74 Poglitsch, A. and Weber, D. (1987). Dynamic disorder in methylammoniumtrihalogenoplumbates (II) observed by millimeter-wave spectroscopy. *J. Chem. Phys.* 87: 6373–6378.

75 Meloni, S., Moehl, T., Tress, W. et al. (2016). Ionic polarization-induced current–voltage hysteresis in $CH_3NH_3PbX_3$ perovskite solar cells. *Nat. Commun.* 7: 10334.

76 Edri, E., Kirmayer, S., Cahen, D., and Hodes, G. (2013). High open-circuit voltage solar cells based on organic–inorganic lead bromide perovskite. *J. Phys. Chem. Lett.* 4 (6): 897–902.

77 Mitzi, D.B., Feild, C.A., Schlesinger, Z., and Laibowitz, R.B. (1995). Transport, optical, and magnetic properties of the conducting halide perovskite $CH_3NH_3SnI_3$. *J. Solid State Chem.* 114: 159–163.

78 Im, J.H., Lee, C.R., Lee, J.W. et al. (2011). 6.5% efficient perovskite quantum-dot-sensitized solar cell. *Nanoscale* 3: 4088–4093.

79 Chen, Q., Marco, N.D., Yang (Michael), Y. et al. (2015). Under the spotlight: the organic—inorganic hybrid halide perovskite for optoelectronic applications. *Nano Today* 10: 355–396.

80 Kulkarni, S.A., Baikie, T., Boix, P.P. et al. (2014). Band-gap tuning of lead halide perovskites using a sequential deposition process. *J. Mater. Chem. A* 2: 9221–9225.

81 Kumawat, N.K., Dey, A., Kumar, A. et al. (2015). Band gap tuning of $CH_3NH_3Pb(Br_{1-x}Cl_x)_3$ hybrid perovskite for blue electroluminescence. *ACS Appl. Mater. Interfaces* 7: 13119–13124.

82 Filip, M.R., Eperon, G.E., Snaith, H.J., and Giustino, F. (2014). Steric engineering of metal-halide perovskites with tunable optical band gaps. *Nat. Commun.* 5: 5757.

83 Menendez-Proupin, E., Palacios, P., Wahnon, P., and Conesa, J.C. (2014). Self-consistent relativistic band structure of the $CH_3NH_3PbI_3$ perovskite. *Phys. Rev. B* 90: 045207.

84 Mosconi, E., Amat, A., Nazeeruddin, M.K. et al. (2013). First-principles modeling of mixed halide organometal perovskites for photovoltaic applications. *J. Phys. Chem. C* 117: 13902–13913.

85 Qiu, J., Qiu, Y., Yan, K. et al. (2013). All-solid-state hybrid solar cells based on a new organometal halide perovskite sensitizer and one-dimensional TiO_2 nanowire arrays. *Nanoscale* 5: 3245–3248.

86 Im, J.H., Chung, J., Kim, S.J., and Park, N.G. (2012). Synthesis, structure, and photovoltaic property of a nanocrystalline 2H perovskite-type novel sensitizer $(CH_3CH_2NH_3)PbI_3$. *Nanoscale Res. Lett.* 7: 353.

12

Polymer Electrolytes for Electrochromic Windows

Li Na Sim and Agnieszka Pawlicka

Universidade de São Paulo, Departamento de Físico Química, Avenida Trabalhador Sancarlense, 400, Parque Arnold Schimidt, 13566-590 São Carlos - SP, Brazil

12.1 Introduction

Research on electrochromic materials began to gain significant attention from academics and industries since Deb's report on electrochromism in tungsten oxide (WO_3) [1]. Electrochromism is a phenomenon whereby a material's optical property alters in response to electrical charge. Electrochromic windows (ECWs), also known as smart windows, are glasses that can control the transmission of light and solar radiation into buildings and vehicles. ECWs work by changing their optical state by darkening and bleaching with the application of low electrical potential. This allows users to adjust the amount of visible light and/or heat that enter the premise according to their comfort and preferences. ECWs are just one of the applications that utilize electrochromic technology. Other electrochromic devices (ECDs) that are available commercially are automotive mirrors, eyewear, information displays, etc.

ECWs application is most useful to enhance indoor comfort and help users to generate energy savings. For example, the use of ECWs in tropical countries and during summer in seasonal countries can block heat from entering buildings and vehicles and reduce the use of air-conditioning. Consequently, the result is a long-term cost saving. Due to easy switching of ECDs, from colored to bleached states and vice versa upon application of a small potential, these windows can be colored to attenuate light and heat entrance into the room in tropical countries or during the summer in others. This results in energy saving with air conditioning and cost saving to users. The light and heat entrance controlled by ECWs provides comfort to occupants from the blazing sun. On the other hand, there are already well known and well established in the market electrochromic rear-view mirrors for cars [2]. These mirrors are devices that darken upon the reflection of strong light from cars behind, providing comfort for the drivers during their night travels. ECWs can also be seen in some airplanes, attenuating sunlight and letting passengers to rest during long intercontinental trips [3]. More recently, there were introduced electrochromic sunglasses, whose goal is to provide light comfort and protection from harmful ultraviolet (UV) rays. Overall, the use of

Polymer Electrolytes: Characterization Techniques and Energy Applications,
First Edition. Edited by Tan Winie, Abdul K. Arof, and Sabu Thomas.

ECWs increases the comfort level of users and promotes energy savings when applied in building windows.

In commercial ECWs, the ion conductor used is made up of gel or liquid electrolyte, e.g. lithium perchlorate dissolved in propylene carbonate (PC) solution. However, electrolyte in the liquid state poses several safety hazards to users such as possible leakage and flammability of the solvent and low chemical stability of the solution. Hence, research on solid-state and gel-like polymer electrolytes for ECWs is rising. This chapter discusses some polymer electrolyte systems for application in ECDs based on transition metal oxides that have been reported in literature over the past five years.

12.2 Principles and Working Process of Electrochromic Window

An ECW is made up of a multilayer structure, usually five layers of components, with the configuration of substrate|working electrode|electrolyte|counter electrode|substrate. The ECW is an electrochemical cell with two electrodes and electrolyte between them and each of the component has its own role.

The most outer layer of an ECW is the substrate that is usually made of glass or in some cases of flexible polyester such as poly(ethylene terephthalate) (PET) [4]. A substrate is coated with conductive oxide of either indium tin oxide (ITO) or fluorine tin oxide (FTO). Both ITO and FTO have good optical and electrical properties. The combination of the substrate and conductive oxide makes the transparent conductor. Besides serving as the conductive oxide, ITO or FTO has also been reported to act as the electrochromic electrode, which will be mentioned later in this chapter. The other possibility is that one of the substrates is coated with conductive reflective material, so the ECW can operate as mirror.

There are three possibilities of electrochromic materials that can be used in ECWs: the ones that always are in solution, the ones that change from solution to solid state, and the ones that always are in solid-state form [5]. The most interesting for practical device application are the ones that always are in solid-state form as the thin films of transition metal oxides. Therefore, to get good contact between conducting substrate and electrochromic material, the thin film of electrochromic layer, also called as working electrode, is coated with a conducting substrate. Depending on the type of material, the working electrode, e.g. WO_3, undergoes cathodic coloration, whereby it colors upon the simultaneous insertion of metal ions (e.g. Li^+, H^+, K^+) and electrons (e^-), and becomes transparent upon their removal. The same procedure is applied to counter electrode, which is also made from transition metal oxides. The counter electrode acts as ion storage and is either optically passive, e.g. cerium oxide (CeO_2), cerium oxide-doped titanium dioxide (CeO_2–TiO_2), and tin oxide (SnO_2), which remain colorless in both oxidized and reduced states. The counter electrode can also act as complementary electrochromic material, and examples of such coatings are vanadium pentoxide (V_2O_5), nickel oxide (NiO), iridium oxide (IrO_2), etc. These materials color under opposite way than the working electrode, providing additional color to be combined with working electrode. These coatings protect ITO from coloring too.

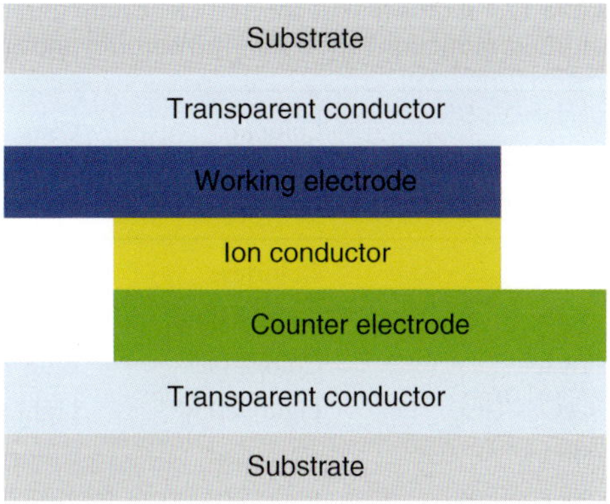

Figure 12.1 A schematic diagram of an ECW.

Simplified examples of cathodic coloration of WO_3 and anodic coloration of vanadium pentoxide (V_2O_5) in nonaqueous solutions are shown in Eqs. (12.1) and (12.2) [5]:

$$\underset{\text{Pale yellow}}{WO_3} + x(M^+ + e^-) \rightarrow \underset{\text{Blue}}{M_xWO_3} \quad (12.1)$$

$$\underset{\text{Very pale blue}}{M_xV_2O_5} \rightarrow \underset{\text{Brown-yellow}}{V_2O_5} + x(M^+ + e^-) \quad (12.2)$$

where M^+ refers to the metal ion and x is the number of metal ions and electrons.

The ECWs are assembled by placing the ion conductor (e.g. polymer electrolyte) in between working and storage electrodes that have coatings facing each other. The whole device is then sealed together, and additional copper-conducting tape can be glued on the free edge of each substrate to improve electrical connections. To avoid external humidity influence on electrolyte and/or electrolyte, leakage cells could be sealed using a Teflon® or some other sealing tape or glue. Several small ECWs were assembled, characterized, and described in the literature. These laboratorial devices usually have sizes of 1, 2, and 5 cm^2 [6] and are made of tungsten trioxide (WO_3) as electrochromic and cerium–titanium dioxide (CeO_2–TiO_2) as counter electrode coatings. Figure 12.1 illustrates the schematic diagram of an ECW.

12.3 Types of Electrochromic Electrodes

Various combinations of working and counter electrodes have been reported in ECDs employing polymer electrolytes. Examples of working electrodes are tungsten trioxide (WO_3), niobium pentoxide (Nb_2O_5), molybdenum dioxide (MoO_2), or Prussian blue (PB), while counter electrodes include cerium dioxide (CeO_2) and cerium–titanium dioxide (CeO_2–TiO_2).

Among solid-state electrochromic coatings, WO_3 is the most widely used inorganic electrochromic electrode due to its high coloration efficiency and possibility to obtain in several ways. Besides inorganic transition metal oxide

Table 12.1 Some electrochromic electrodes and their color at original state and during cathodic reduction (upon charge insertion) and anodic oxidation (upon charge removal).

Electrochromic electrode	Color at original state	Color upon cathodic reduction	Color upon anodic oxidation	References
WO_3	Colorless	Blue	Colorless	[5]
V_2O_5	Light yellow	—	Blue	[14]
TiO_2	Colorless	Blue	Colorless	[15]
CeO_2–TiO_2	Light yellow	Light yellow	Light yellow	[16]
NiO	Colorless	Colorless	Brown	[17]
Prussian Blue	Blue	Transparent	Blue	[5]
Nb_2O_5	Colorless	Blue	Colorless	[18]

coatings, there are also organic materials such as conducting polymers such as poly(3,4-ethylenedioxythiophene) (PEDOT) and its soluble form poly(3,4-ethylenedioxythiophene):poly(styrene sulfonate) (PEDOT:PSS). Another well-known and used as electrochromic coating is Prussian blue (PB), which is an inorganic potassium hexacyanoferrate compound. The difference between PB and WO_3 is that they color under opposite electrochemical conditions. WO_3 natural state is transparent, while PB's is blue, so WO_3 becomes blue colored under insertion of cations and electrons and PB becomes transparent under extraction of cations and electrons [7]. The reactions that occur during electrochemical conditions when PB is subjected to electrical stimuli are described in Eqs. (12.3) and (12.4) [8].

$$\underset{\substack{\text{PB} \\ \text{Blue}}}{KFe^{3+}Fe^{2+}(CN)_6} + e^- + K^+ \leftrightarrow \underset{\substack{\text{PW} \\ \text{Transparent}}}{K_2Fe^{2+}Fe^{2+}(CN)_6} \tag{12.3}$$

$$\underset{\text{Soluble PB}}{Fe_4{}^{3+}[Fe^{2+}(CN)_6]_3} + K^+ + e^- \leftrightarrow \underset{\text{Everitt's salt (PG)}}{K_2Fe^{2+}[Fe(CN)_6]} \tag{12.4}$$

ITO is also used as either the cathodic working electrode [9] or anodic counter electrode [10–13]. Table 12.1 lists the color of some electrochromic electrodes at the original state and during cathodic reduction and anodic oxidation. Table 12.2 shows some ECWs and the potentials at which their change the color.

12.4 Mechanism of ECW

Electrochromic effect is caused by simultaneous insertion of ions and electrons into the electrochromic materials [1]. When a small potential (between 1 and 5 V) is applied between the transparent electrical conductors of ECW, small and mobile ions such as Li^+, K^+, and H^+ are shuttled between the ion storage film

Table 12.2 Some electrochromic devices configurations and their reduction and oxidation potentials.

ECW configuration	Reduction potential (V)	Oxidation potential (V)	References
PB|HPC–electrolyte|CeO_2–TiO_2	−2.6	2.0	[19]
PB|PVB–electrolyte|CeO_2–TiO_2	−1.5	1.5	[8]
PB|chitosan–electrolytes|CeO_2–TiO_2	−2.8	2.8	[20]
WO_3|chitosan–electrolyte|CeO_2–TiO_2	−2.8	2.8	[20]
WO_3|agar–electrolyte|CeO_2–TiO_2	−2.5	2.0	[21]
WO_3|PAN–electrolyte|CeO_2–TiO_2	−1.25	−0.40	[22]
WO_3|PMMA–electrolyte|PB	−2.0	1.8	[23]
WO_3|gelatin–electrolyte|NiO	−1.1	1.8	[24]
PANI:DBSA|PVdF-HFP electrolyte|ITO	−2.9	2.9	[11]

PB, Prussian blue; HPC, hydroxypropyl cellulose; PVB, poly(vinyl butyrate); PAN, poly(acrylonitrile); PMMA, poly(methyl methacrylate); PANI, polyaniline; DBSA, dodecylbenzenesulfonic acid; ITO, indium tin oxide.

(counter electrode) and the electrochromic film (working electrode). At the same time the electrons from the electrical circuit are injected, through transparent conductors, into or from the coatings to promote a change of their optical properties. Hence, different processes occur in the electrochromic cell during coloration and bleaching, but almost always they are related to the redox reactions.

The original optical state of an ECW can be recovered by reversing the potential. In this technology, potential is required only during switching between the states of the device, and the optical state remains for a long time even after the removal of the applied potential due to memory effect.

A simple representation of the electrochromic process in WO_3 is given in Eq. (12.1).

12.5 Polymer Electrolytes for Electrochromic Windows

12.5.1 Background

Over the past four decades, research has been on the rise on polymer electrolytes (PEs), which can be used as ion conductors in the application of ECDs such as ECWs, car sunroofs, visors, antiglare mirrors, displays, and memory devices. Aiming to develop ECDs with high cycling number, large contrast between colored and bleached states, fast switching time, and high current density, there is a need to develop PEs that have conductivity and thermal, structural, optical, and electrochemical performance that provide ECDs the required characteristics.

12.5.2 Criteria of Polymer Electrolytes and Electrochromic Device

Usually the performance of an ECD is linked to the ionic conductivity of the electrolyte. High conductivity is desired as it facilitates rapid switching between colored and bleached states. Practical applications require conductivities of at least 10^{-5} S cm^{-1} to perform well, which is less than required for lithium batteries [25]. Moreover, the ions should be small, mobile, and easily inserted and removed from electrochromic and counter electrode layers.

The PEs employed in ECWs should possess high transparency (usually characterized by high bandgap of PE). Electrolytes which are slightly opaque exhibit lower optical transparency [26]. PEs should also be homogeneous in order to prevent light scattering, which reduces their transparency [27].

PE should also exhibit good mechanical and thermal stability as well as a wide electrochemical stability window over a wide range of potentials and temperatures. A sticky PE is more easily spread on the electrochromic layers and provides better adhesion and/or contact between the electrolyte and electrodes.

12.5.3 Types of Polymer Electrolytes Used in ECWs

An electrolyte or ion conductor that is used in ECWs can be in the form of either liquid, gel, or solid, and it is sandwiched between two electrochromic electrodes. This electrolyte contains ions that under applied potential intercalate into and out of the electrochromic electrodes, providing conditions to occur an electrochemical reaction. Therefore, it is a crucial component of an ECW.

Various kinds of PE systems have been reported in literature. In the past years, researchers are turning to biopolymers and their derivatives such as gellan gum [28], hydroxypropyl cellulose (HPC) [19], and deoxyribonucleic acid (DNA) [29] as polymer hosts in PEs. Polymer blend such as poly(ethyl methacrylate) (PEMA)/poly(vinylidenefluoride-*co*-hexafluoropropylene) (PVdF-HFP) [30] has also been reported. Commercial polymers, which have been employed as polymer hosts in ECDs, include poly(methyl methacrylate) (PMMA), poly(ethylene oxide) (PEO), poly(vinylidene fluoride) (PVdF), etc.

12.5.3.1 Solid Polymer Electrolytes (SPEs)

Assis et al. [19] developed an ECD employing transparent HPC incorporated with 10 wt% of acetic acid as proton source. The authors tested the performance of the electrolyte in a 1 cm^2-sized ECD using PB and CeO_2-TiO_2 as the electrochromic layer and counter electrode, respectively. With this configuration, the natural state of the ECD is blue because of PB and becomes transparent upon positive charge insertion. The color change is due to the transition from Prussian blue to Prussian white, as shown in Eq. (12.3), causing the change from deep blue to colorless.

The largest transmittance variation between the transparent and colored states were found between 600 and 800 nm. Bleaching and coloration of the ECD occurred upon application of −2.6 and 2.0 V, respectively. Transmittance variation of 32% was obtained after 15 seconds of potential application, while a higher transmittance variation was achieved after 30 and 60 seconds. The transmittance value at both bleaching and coloration processes was recorded at 686 nm. This result was comparable to the transmittance variation of ECD made with PE based on agar and acetic acid (CH_3COOH) [21] in the same configuration. Charge density of −6.0 mC cm^{-2} was obtained upon application of −2.6 V after 15 seconds, which increased to $-8.4\,mC\,cm^{-2}$ after 60 seconds. Reverse potential of 2.0 V applied for 15 and 60 seconds resulted in the charge densities of 1.6 and $3.3\,mC\,cm^{-2}$, respectively.

The CV diagrams displayed two reduction peaks located at −0.5 and −0.1 V and three oxidation peaks found at 0.5, 1.4, and 1.7 V. The authors attributed the additional peaks found between −2.6 and 2.0 V to the transition of PB to Prussian green (PG) as shown in Eq. (12.4). After 2000 cycles, the current density of the cathodic reaction dropped from about 6.0 to $-1.9\,mC\,cm^{-2}$, which was accompanied by the reduction of transmittance variation from 32% to 6%. Changes in both peaks position and shape with increasing bleaching/coloring cycles indicated that the PB redox reactions are not completely reversible. The ECD demonstrated memory effect whereby the transmittance during the transparent state was retained with only a slight drop of 2% in open circuit condition over the duration of three hours.

An all solid-state green-yellow reflective ECD was presented by [8] using poly(vinyl butyrate) (PVB) as the polymer host in the configuration of glass|ITO|PB|electrolyte|CeO_2–TiO_2|ITO|glass. Figure 12.2 illustrates the electrodeposition curves of PB and the CV curve of PB film in $1\,mol\,l^{-1}$ KCl solution.

The authors utilized the subtractive color mixing method of blue color from PB electrochromic layer and the yellow color from the lithium iodide (LiI)/iodine (I_2) dispersed in the electrolyte, and this mix produced green color. By applying a potential to the ECD, there was observed a color change from deep green (when the PB layer is blue in the original state) to yellow (when the PB layer is reduced and become transparent). Moreover, different shades of green could be obtained by adjusting the applied potential (Figure 12.3). The electrolyte of PVB:γ-butyrolactone (GBL):LiI/I_2, in proportion to 41.65 : 41.65 : 16.7, exhibited an ionic conductivity of $6.6\times10^{-7}\,S\,cm^{-1}$ at 23 °C and followed the Arrhenius rule, whereby the ions are transported along the PE chains by hopping mechanism. The authors also studied the effect of exposing the electrolyte to air humidity for 24 hours and reported an increase in ionic conductivity by three orders. The enhancement in ionic conductivity of the electrolyte exposed to air

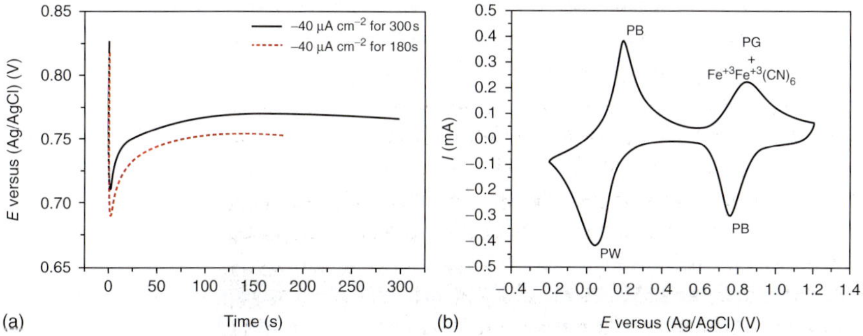

Figure 12.2 (a) Electrodeposition curves of PB at $-40\,\mu A\,cm^{-2}$ for different durations and (b) CV curve of cycle 3 of PB film in $1\,mol\,l^{-1}$ KCl solution recorded at $20\,mV\,s^{-1}$ (PB, Prussian blue; PG, Prussian green; and PW, Prussian white). Source: Assis et al. 2013 [8]. Reproduced with permission of Elsevier.

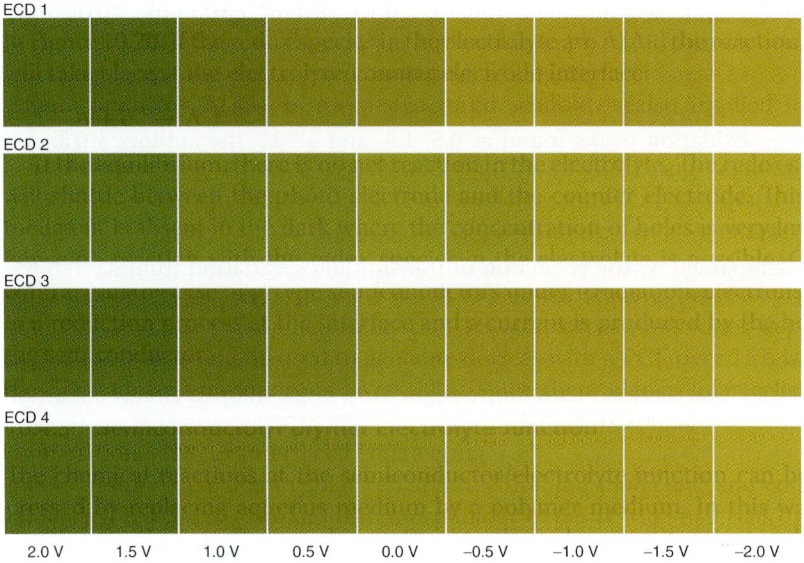

Figure 12.3 Variety of green and yellow colors in reflection mode as a function of the potential applied for ECDs in the configuration of glass|ITO|PB|PVB-based electrolyte|CeO_2–TiO_2|ITO|glass. Source: Assis et al. 2013 [8]. Reproduced with permission of Elsevier.

was attributed to protons from adsorbed water and/or the dissociation of LiI and I_2 into I^-, I_3^-, and I_5^- ions according to the Eq. (12.5) [31, 32]:

$$I^- + I_2 \leftrightarrow I_3^- + I_2 \leftrightarrow I_5^- \tag{12.5}$$

The presence of I_3^- and I_5^- ions was confirmed by Raman' analysis peaks at 113 and 140 cm^{-1}, which were attributed to asymmetric I_3^- [33] and asymmetric I_5^- stretchings [32]. The amount of I_5^- ions was lower than I_3^- as shown from the lower intensity of the former Raman peak.

Assis et al. [8] used the CIELAB colorimetric scale to define the color change from green to yellow of the ECD upon application of potential. Figure 12.3 depicts the variety of green and yellow colors that could be obtained with the PB|PVB electrolyte|CeO_2–TiO_2 ECD as a function of potential from −2.0 to 2.0 V. After 400 green/yellow cycles, a small reduction was observed in charge density by 0.6 mC cm^{-2} as well as reflectance values by 1.5% at both −1.5 V (yellow color) and +1.5 V (green color) color states. The ECD based on PVB electrolyte displayed memory effect for 120 minutes, whereby only small reflectance changes of 2.5% and 3.5% were observed, under open circuit condition, in the green-colored and yellow-colored states, respectively.

In the work of Alves et al. [20], plasticized electrolytes based on chitosan incorporated with varied contents of thulium triflate ($Tm(CF_3SO_3)_3$), salt, and glycerol as plasticizer were prepared. The authors employed the trivalent rare earth salt in order to produce electroluminescent device as thulium can emit blue light [34]. The authors obtained the highest conductivity of 3.5×10^{-7} S cm^{-1} at 30 °C at 20 wt% of $Tm(CF_3SO_3)_3$ in chitosan. Improvement

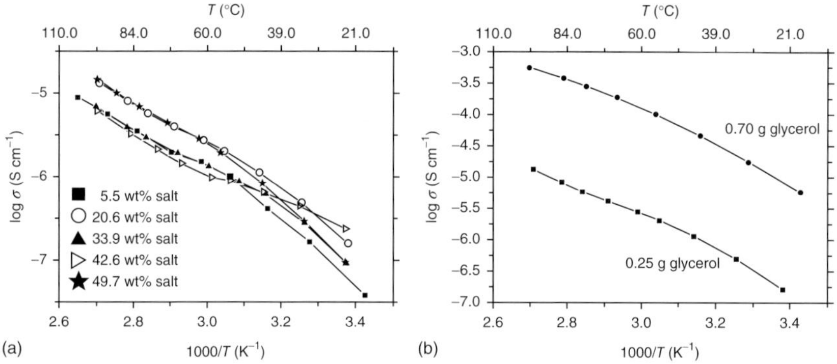

Figure 12.4 Conductivity plots of chitosan-based electrolyte containing varied contents of (a) $Tm(CF_3SO_3)_3$ and (b) glycerol. Source: Alves et al. 2017 [20]. Reproduced with permission of Elsevier.

in conductivity by two orders of magnitude to 1.6×10^{-5} S cm^{-1} was observed upon addition of glycerol, where weight ratio of chitosan:glycerol is 2 : 7. Figure 12.4 depicts the conductivity plots of chitosan-based electrolyte with variation in $Tm(CF_3SO_3)_3$ and glycerol. Glycerol facilitated the dissociation of ion aggregates and increased the amorphous content of the polymer. The increment of conductivity with temperature indicated that the ion transport mechanism of the chitosan-based electrolytes was governed by Arrhenius rule. Decomposition of the salt-containing electrolytes began between 130 and 180 °C due to loss of water since both chitosan and the salt are hydrophilic. The authors reported no glass transition in the electrolytes with the added salt, which indicated the high amorphousness of the samples. Atomic force microscopy (AFM) analysis correlated the roughness of the salt-containing electrolytes to the conductivity results. The highest roughness value of 20.7 nm was obtained for the best conducting sample containing both $Tm(CF_3SO_3)_3$ and glycerol. The configuration of the ECD employed was ITO|PB or WO_3|best conducting electrolyte|CeO_2TiO_2|ITO. CV analysis showed two cathodic peaks located at −0.75 and 1.13 V and an anodic peak at 1.75 V. The transmittance change reported at 550 nm between bleached and colored states was 6%.

Wu et al. [35] have assembled and tested all solid supercapacitors with polyaniline (PANI) as anode, tungsten trioxide (WO_3) as cathode, and poly(vinyl pyrrolidone) (PVP) doped with lithium perchlorate ($LiClO_4$) as solid polymer electrolyte (SPE). The authors observed that their supercapacitor possesses rapid and reversible color changes, during the charge and discharge processes, from transparent to blue (WO_3 in potential range from 0 to −1 V) and light yellow to green (PANI in potential range from 0 to 1 V) while working in −1 to 1 V potential interval. The energy densities of 24.8 and 14.2 Wh kg^{-1} at power densities of 206.6 and 2048 W kg^{-1}, respectively, confirmed the device's good electrochemical performance. Furthermore, this research has opened a new possibility for application of ECWs and electrochemical devices.

Thermally stable poly(ε-caprolactone) (PCL)/siloxane ormolytes doped with manganese perchlorate ($Mn(ClO_4)_2$) were synthesized and characterized by impedance, FTIR, and EPR spectroscopies [36]. In addition to the low ionic conductivity values of the samples, i.e. 4.8×10^{-8} and 2.0×10^{-6} S cm^{-1} at room and 100 °C, respectively, these studies revealed some microscopic phenomena that occur in these systems. From FTIR spectra, it was observed that $Mn(ClO_4)_2$ doping promoted a formation of more oxyethylene/urethane and urethane/urethane aggregates when compared to the non-doped d-PCL(530)/siloxane host hybrid matrix. The EPR experimental results were analyzed by numerical simulations of the spin Hamiltonian, assuming two different Mn^{2+} species characterized by different coordination symmetries. The obtained fits, considered satisfactory, showed that Mn^{2+} species I are coordinated by d-PCL(530)/siloxane matrix in an axially distorted coordination symmetry, while species II are not. The Mn^{2+} ion interactions occur not only with the "free" urethane and ester C=O groups of d-PCL(530)/siloxane matrix but also with ClO_4^- ions along a bidentate configuration, and these interactions are moderately ionic. There were also detected "free" ClO_4^- ions, whose concentration decreased with salt content. Finally, these hybrids thermally stable up to 200 °C were considered interesting materials for potential application in several ECDs.

PVB doped with either lithium iodide (LiI)/I_2 or $LiClO_4$ was used as PE to assemble WO_3/PVB electrolyte/CeO_2–TiO_2 ECDs [37]. The electrochemical analyzes showed that the device with PVB-LiI/I_2 revealed better charge density of 7.8 mC cm^{-2} than the device with PVB-$LiClO_4$, which showed 2.9 mC cm^{-2} under applied potential of −2.0 V for 15 seconds. Besides this, the redox reversibility was of 1 for the device with PVB doped with LiI/I_2 and for other one 0.9. The same happened with transmission difference between colored and bleached states, where the first device reached 30% and the second one only 21%, measured at 633 nm. However, in spite of good electrochemical performances, the ECD with PVB-LiI/I_2 supported only 1200 color bleaching cycles, while the ECD with PVB-$LiClO_4$ more than 48 000.

12.5.3.2 Gel Polymer Electrolytes (GPEs)

This category of PEs includes those that are in gel form and those incorporated with plasticizers and ionic liquids. The addition of these additives often increases the ionic conductivity of the PEs.

Polyacrylonitrile (PAN)-based electrolytes incorporated with varied lithium bis(trifluoromethanesulfonyl)imide (LiTFSI) contents and dimethylformamide (DMF) as plasticizer were reported by Sim et al. [22]. The best conducting sample was found at 40 wt% of LiTFSI reaching an ambient conductivity of 2.54×10^{-4} S cm^{-1}, and this was the most amorphous sample. IR studies revealed that interactions had occurred between the carbonyl (C=O) group of DMF and both the nitrile (C≡N) group of PAN and the Li^+ ions dissociated from LiTFSI. Results from computational studies by Gaussian software have explained the higher tendency of DMF to interact with Li^+ ion than with PAN due to the shorter distance between the O atom of DMF and Li^+ ion (Figure 12.5).

The ion conduction is found to obey the Vogel–Fulcher-Tammann behavior, whereby the ion transport occurs predominantly through the free volume of the polymer [38]. The ECD employing WO_3 as working electrode and CeO_2–TiO_2 as

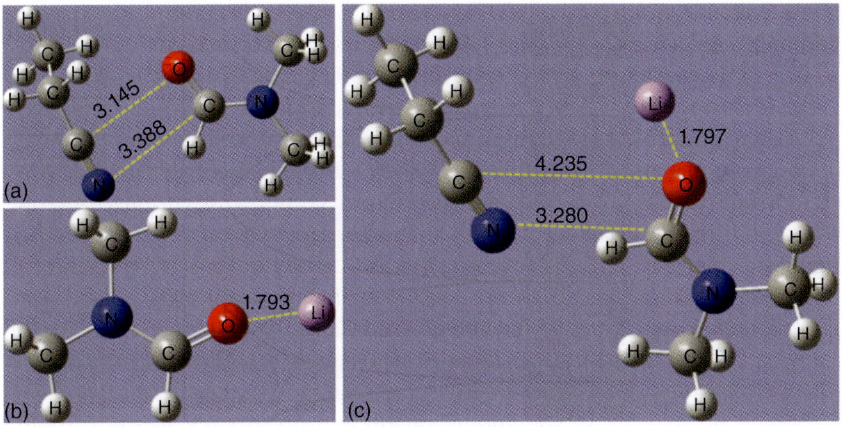

Figure 12.5 (a) Optimized structures obtained from Gaussian software of (a) PAN-DMF, (b) DMF-Li$^+$, and (c) PAN-DMF-Li$^+$ complexes. Source: Sim et al. 2017 [22]. Reproduced with permission of Elsevier.

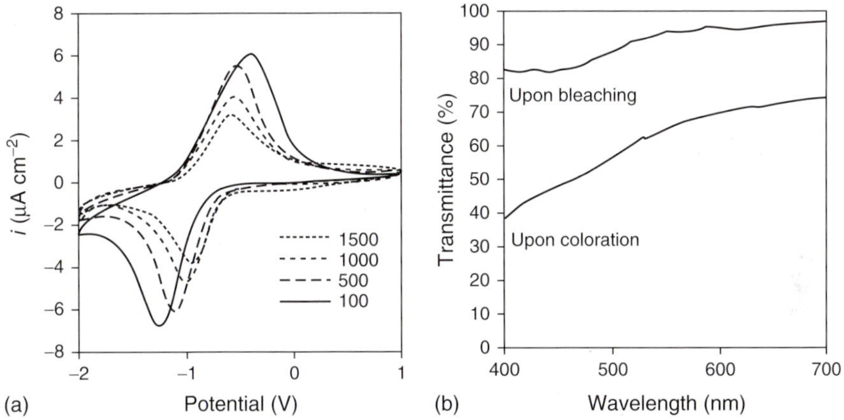

Figure 12.6 (a) CV curves of PAN electrolyte-based ECD at different cycles and (b) UV–Vis curves of the ECD upon coloration and bleaching. Source: Sim et al. 2017 [22]. Reproduced with permission of Elsevier.

the reference and counter electrodes produced light blue coloration at −1.25 V and bleaching at −0.40 V with transmittance variation of 29.1% at 550 nm. The ECD was colored and bleached up to 1500 times and showed a reduction in the current for both reduction and oxidation processes. Figure 12.6 illustrates the current density of the ECD as a function of potential at various cycles and the transmittance of the colored and bleached ECD for the first cycle.

A polymer blend comprising another polymethacrylate, i.e. PEMA and PVdF-HFP in the weight ratio of 70 : 30, was reported by Sim et al. [30]. Different plasticizers, namely, ethylene carbonate (EC) and PC, were added into the optimized PEMA/PVdF-HFP blend system containing 30 wt% of lithium trifluoromethanesulfonate or triflate ($LiCF_3SO_3$) salt. The best conductivity was obtained at 6 wt% for both EC- and PC-added systems with ambient conductivity of 1.05×10^{-4} and $1.46 \times 10^{-6}\,S\,cm^{-1}$, respectively. The better conductivity achieved for the EC-containing system was attributed to the higher number of free ions and higher amorphousness as compared to the PC system. An ECD employing WO_3 working and CeO_2–TiO_2 counter electrodes produced light blue coloration and bleaching at −1.3 and −0.8 V, respectively.

Plasticized PEs based on DNA obtained from salmon's sperm, glycerol, and varying erbium triflate ($Er(CF_3SO_3)_3$) contents from 10 to 50 wt% were studied by Leones et al. [29]. The best conductivity of $1.17 \times 10^{-5}\,S\,cm^{-1}$ was achieved with addition of 10 wt% of $Er(CF_3SO_3)_3$ at 30 °C. Linear relationship with Arrhenius equation was exhibited by the DNA-based electrolytes without and with the erbium salt. The E_a value decreased from 36.14 to 24.56 kJ mol^{-1} upon addition of 10 wt% of $Er(CF_3SO_3)_3$. The Er^{3+}-based electrolytes displayed electrochemical stability up to 3.5 V in the anodic region and up to 1.5 V in the cathodic region versus Li/Li^+. The transmittance variation of 7% was obtained between the colored state achieved at c. −1.5 V and bleached state at c. −0.3 V for the WO_3|electrolyte|CeO_2–TiO_2 configuration of ECD. The low transmittance variation was attributed to the slight opacity of the electrolyte.

Mîndroiu et al. [39] and Zgarian et al. [40] investigated the influence of varying amounts of Nile blue (NB) and Prussian blue (PB), respectively, as chromophores in their DNA-glycerol-based electrolytes. The DNA used was also extracted from waste produced by salmon processing industry [41]. Mîndroiu et al. [39] initially studied the effect of different amounts (9.1, 14.3, and 50 wt%) of glycerol in DNA and found that the weight ratio of DNA:glycerol of 1 : 1 improved the room temperature ionic conductivity of DNA from 5.7×10^{-9} to $9.0 \times 10^{-9}\,S\,cm^{-1}$. Moreover, the presence of glycerol in DNA matrix helped to produce transparent membranes. The authors incorporated varied wt% of NB (0.5, 1, 2, and 3) into the DNA–glycerol electrolytes to further enhance the ionic conductivity. The best conductivity of $3.8 \times 10^{-8}\,S\,cm^{-1}$ under ambient condition was obtained upon incorporation of 2 wt% of NB. The reason attributed for the enhanced conductivity was that NB intercalates with the double helix chains of DNA and causes structural transformation of the biopolymer [42]. For the PB-added DNA-based electrolytes, the best conductivity in the range of $10^{-6}\,S\,cm^{-1}$ was found at 2 wt% of the chromophore [40]. For both NB and PB, the addition above 2 wt% of respective dye into DNA–glycerol mixture caused a reduction in ionic conductivity, which was attributed to the aggregation of the chromophores, and thus lowered the mobility of ions.

Both Mîndroiu et al. [39] and Zgarian et al. [40] assembled and tested small ECWs with the configuration of glass|ITO|WO_3|electrolyte|CeO_2–TiO_2|ITO| glass. Charge insertion occurred at −2.8 V for both reports to induce coloration, while charge extraction occurred at +2 and +1.5 V in the work of Mîndroiu et al. [39] and Zgarian et al. [40], respectively. Although the electrolyte containing 2 wt% of PB exhibited higher conductivity by two orders of magnitude than that with 2 wt% of NB, the former produced lower inserted charge density of about 2.3 mC cm^{-2} as compared to about 2.5 mC cm^{-2} for NB-added sample when potentials were applied for 15 seconds/15 seconds. Bleaching process occurred faster than the coloration process for both cases. The amount of charge

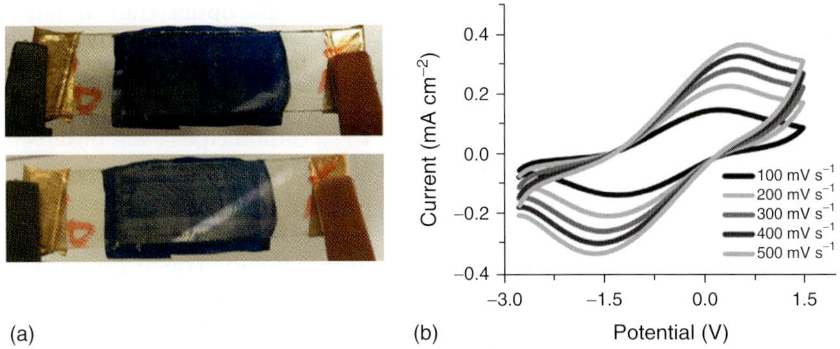

Figure 12.7 (a) Photographs of ECD employing DNA-based electrolyte during cathodic polarization (top) and anodic polarization (bottom) and (b) CV plots of ECD at different scan rates. Source: Zgarian et al. 2017 [40]. Reproduced with permission of Elsevier.

increased with increasing duration of applying potential. Finally, with increasing number of coloring and bleaching cycles, the charge density value decreased. Figure 12.7 illustrates the ECD device fabricated by Zgarian et al. [40], and the cyclic voltammograms of the ECD at varied scan rates.

PEs using chitosan as a matrix and cerium triflate as a salt were prepared by Alves and coworkers [6] and applied in small ECDs with WO_3/chitosan$_{33.32\%}$ $Ce(CF_3SO_3)_3/CeO_2{-}TiO_2$ configuration. The highest ionic conductivity of 1.46×10^{-6} and $8.74 \times 10^{-5}\ S\,cm^{-1}$ at 30 and 90 °C, respectively, was obtained for electrolyte with 33.32% of salt. Moreover, the authors observed that addition of glycerol promoted an increase of one order of magnitude of ionic conductivity of the samples reaching 1.67×10^{-5} and $4.93 \times 10^{-4}\ S\,cm^{-1}$ at 30 and 90 °C, respectively, for the sample with highest glycerol content, i.e. 0.7 g. The ECDs were tested between potentials of −2.0 to +1.8 V, revealing a decrease of charge density from −3.51 to $-2.82\ mC\,cm^{-2}$ after 200 color/bleaching cycles. The electrochromic contrast of $\Delta T_{VIS} = 5.0\%$ was obtained at 633 nm after switching the potentials between −2 and 1.8 V for 15 seconds.

Ramadan et al. [24] investigated the amount variation of glycerol as plasticizer, as well as acetic acid (CH_3COOH) or hydrochloric acid (HCl) as the source of proton in gelatin to form proton (H^+)-based electrolytes. Formaldehyde was used as the cross-linker. In the plasticized system containing a fixed amount of acetic acid at 0.95 ml, the highest ambient conductivity was achieved with addition of 32 wt% of glycerol at $2.56 \times 10^{-5}\ S\,cm^{-1}$. Higher contents of glycerol produced films with poor mechanical strength and thus would not be suitable for application in ECD. For the acid systems, the best conductivity of $1.28 \times 10^{-5}\ S\,cm^{-1}$ was obtained at 26 wt% CH_3COOH, which is an order of magnitude lower than that obtained in 2.78 wt% of HCl at $2.07 \times 10^{-4}\ S\,cm^{-1}$. The acid systems were prepared while maintaining the glycerol content at 0.99 and 2.39 ml in the CH_3COOH and HCl systems, respectively. The authors attributed the conductivity of all systems to the Arrhenius behavior, whereby the ionic transport occurred through hopping of the ions. All samples investigated exhibited an amorphous nature, as evidenced by the presence of two common broad peaks located at

$2\theta = 7°$ and 20° in the X-ray diffraction (XRD) analysis. The optical transmittance of the best conducting sample in the glycerol, acetic acid, and hydrochloric acid systems were very high, at least 96% at wavelength of 555 nm. The authors fabricated an ECD with the configuration of FTO|glass|WO_3|electrolyte|nickel oxide (NiO)|glass|FTO. At scan rate of 50 mV s^{-1}, a cathodic peak was observed at −1.1 V and accompanied by coloration of the ECD, whereas bleaching of the device occurred at 1.8 V. After 50 cycles, the current density of both cathodic and anodic peaks reduced, and the cathodic peak remained, while the anodic peak shifted to 1.4 V. The inserted charge density was found to be −41 mC cm^{-2}.

HPC, poly(ethylene glycol) (PEG300), tetrabutylammonium tetrafluoroborate (Bu_4NBF_4), and 4-dodecylbenzenesulfonic acid (DBSA) were used by Ledwon et al. [43] to obtain HPC-based solution in dichloromethane that was evaporated until thin membrane formation. The best PE composition sample displayed conductivities of 3.5×10^{-5} and 1.1×10^{-4} S cm^{-1} at 25 and 50 °C, respectively. Besides its good potential stability window of 3.5 and 3 V, measured between −2 and +1.5 V on ATO and ITO substrates, respectively, it also had good transparency and good adherence to different substrates. Therefore, this PE solution was deposited on either ITO|PB or ITO|PANI electrode and left to dry, resulting in gel polymer electrolyte (GPE) that was used to assemble small ECDs with glass/ITO/PB or PANI/HPC/PEDOT:PSS/ITO/glass configuration. The color change of these devices, measured by transmittance at 650 nm, was of 35% and 33% for ECD containing complementary electrochromic material of PB and PANI, respectively.

Zhang et al. [44] have prepared and characterized PVB-based GPE membranes doped with $LiClO_4$/PC. They used casting method to obtain the membranes that displayed the ionic conductivity of 4.0×10^{-5} and 6.2×10^{-5} S cm^{-1} at 25 and 50 °C. Then, they applied this GPE in an 5 cm × 5 cm ECD with glass|ITO|WO_3|PVB electrolyte|$Ni_{1-x}O$|ITO|glass configuration by using glass laminating method and high-pressure autoclave with high temperature to final assembly. The thickness of GPE was of about 300 μm. The electrochemical characterization of the device revealed the best optical modulation and coloration efficiency was of 65.8% and 175.34 cm^2 C^{-1} and that the device behaved almost constant during 300 color cycling tests.

Following some previous works, Costa et al. [45] have described a new dual function device by assembling electrochromic coating with dye-sensitized solar cells. The advantage of this device is that it does not depend on fixed power supply, so it can be used remotely. The authors called it DSC-EC and build some small prototypes with PEDOT:PSS as electrochromic material deposited on platinum counter electrode of DSC and two kinds of electrolytes: liquid (AN 50 from Solatronix) and semiliquid UV-cured PEO electrolyte. They obtained the best results with liquid electrolyte containing 8% of LiI that changed a color of $\Delta E = 30$ at applied c. 0.4 V and the highest energy conversion efficiency of 4.87% at 1 sun ($V_{oc} = 0.66$ V and $J_{sc} = 11.04$ mA cm^{-2}).

GPE of composite of PMMA and crystals of succinonitrile (SN) plasticized with polycarbonate (PC) was prepared by Wang et al. [23]. This GPE showed an ionic conductivity of 1.46 mS cm^{-1} for 4 : 1 of PC:SN composition. The electrolyte was then applied in a 5 cm × 5 cm ECD with glass or plastic|WO_3|PMMA-SN electrolyte|PB|glass or plastic configuration, which showed an optical contrast of 52.4% at 695 nm. The ECDs color change was reversibly modulated from 57.9% (in bleached state) to 5.5% (in colored state) by applying potentials of 1.8 and −2.0 V, respectively. The durability test was performed during 2250 color–bleaching cycles, and the ECDs transmittance change (ΔT) reached 44.5%, which was 85% of its original value or loss of 15%.

Other Li^+-conducting GPE obtained from solution of EC and PC with PMMA powder ($LiClO_4$:EC:PC:PMMA with 12 : 38 : 38 : 12 wt%) was studied by Solovyev et al. [46]. The authors applied this GPE to a solid-state ECD with WO_3 and Prussian blue ($Fe_4[Fe(CN)_6]_3$) thin films as electrodes and analyzed it as a function of PB thickness. They observed that a maximum coloration efficiency and transmittance modulation was achieved with 240 nm PB film. Its transmittance variation was of 59% and coloration efficiency of 43 $cm^2\,C^{-1}$ measured at 550 nm wavelength after applying ±2 V for 1 (coloration) and two to three minutes (bleaching). This ECD supported very well 1200 color/bleaching cycles with no significant changes in its electrochemical performance.

A new electrochromic compound of vinyl benzyl viologen (VBV) was synthesized and characterized for use in an ECD composed of PB, I-VBV, and ferrocene (Fc) was fabricated (PB/Fc/I-VBV) in which Fc acted as a redox mediator [47]. The PE was synthesized from methyl methacrylate (MMA) and ethoxylated trimethylolpropane triacrylate (ETPTA) with 0, 1, and 5 wt%; 0.1 wt% of 2-hydroxy-2-methyl-1-phenyl-1-propanone (HMPP), which was mixed with PC solution that contained 0.05 $mol\,l^{-1}$ of $VBV(BF_4)_2$; 0.5 $mol\,l^{-1}$ $LiBF_4$; and 0.05 $mol\,l^{-1}$ of ferrocene (Fc). This PE was then applied in between two transparent electrodes (ITO) at 60 μm from each other. The deposition of VBV at counter electrode was performed by applying 1.5 V for 30 seconds promoting a reduction of VBV^{2+} to $VBV^{+\bullet}$, and followed by a polymerization of MMA to PMMA by UV irradiation at 20, 40, and 60 seconds intervals. Among all the ECDs studied, the best one in long-term stability (remained 86.5% of ΔT after 10 000 cycles of 1.2 and −0.8 V for 5 seconds each) was PB/Fc/I-VBV, which electrolyte was UV irradiated for 40 seconds. Its transmittance change (ΔT) at 615 nm was of 60.6% under applied potentials of 1.2 and −0.8 V, and the bleaching and coloring times of 1.32 and 2.13 seconds.

Leones et al. [48] investigated the effect of incorporation of three different ionic liquids of fixed amount, namely, 1-ethyl-3-methylimidazolium ethylsulfate ($[C_2mim][C_2SO_4]$), 1-ethyl-3-methylimidazolium acetate ($[C_2mim][CH_3COO]$), and trimethyl-3-ethanolammonium acetate ($[Ch][CH_3COO]$), into agar-based electrolyte system containing glycerol as plasticizer. All agar-based electrolytes were translucent and displayed homogeneity with no phase separation between the polymer and IL. The electrolyte containing $[C_2mim][CH_3COO]$ achieved the best ambient conductivity of 2.35×10^{-5} $S\,cm^{-1}$ and the lowest glass transition temperature (T_g) of −60 °C. The conductivities of $[C_2mim][C_2SO_4]$ and $[Ch][CH_3COO]$ electrolytes did not differ much and were in the range of 10^{-5} and 10^{-6} $S\,cm^{-1}$, respectively. All electrolytes exhibited the Arrhenius behavior, where charge transport occurs by ion hopping mechanism. The activation energy (E_a) value of the agar-based electrolytes decreased upon addition of IL, and $[Ch][CH_3COO]$-containing sample exhibited an E_a of 24.3 $kJ\,mol^{-1}$. All the

IL-added electrolytes demonstrated good electrochemical stability up to 2.0 V. Among all the IL-added electrolytes, the sample containing $[C_2mim][C_2SO_4]$ showed the best transparency of 81% of transmittance, while the best conducting $[C_2mim][CH_3COO]$-added sample exhibited the lowest transparency with transmittance of 57%. Hence, an ECD configuration employing the $[C_2mim][C_2SO_4]$ electrolyte with WO_3 and CeO_2–TiO_2 as working and counter electrodes, respectively, was constructed. The coloration of WO_3 occurred at −1.3 V, and bleaching occurred at −0.3 V. This ECDs transmittance variation was of 13% at 550 nm.

Assis et al. [7] developed ECDs that had PB as electrochromic layer and CeO_2–TiO_2 as counter electrode [7]. These devices were tested with two natural macromolecules-based electrolytes that contained rare earth elements: DNA–$Er(CF_3SO_3)_3$ and agar–$Eu(CF_3SO_3)_3$ as ionic conductive membranes. These ECDs had configurations of glass–ITO/PB/DNA-$Er(CF_3SO_3)_3$–electrolyte /CeO_2–TiO_2/ITO–glass and glass–ITO/PB/agar–$Eu(CF_3SO_3)_3$–electrolyte/ CeO_2–TiO_2/ITO–glass. To obtain the electrochromic effect, the applied potentials were of −3 and 2 V, and the contrasts between colored and bleached states were of $\Delta T_{VIS} = 25 \pm 2\%$ and $35 \pm 2\%$, measured at 686 nm after 60 seconds of applied potential of −3 V, for ECD with DNA–Er^{3+} and Agar–Eu^{3+}-based electrolytes. These ECDs were tested for 1000 and almost 1500 color/bleaching cycles, revealing a decrease of both charge density and ΔT_{VIS}, which dropped to $-1.6\,mC\,cm^{-2}$ and $\Delta T_{VIS} = 5\%$ for ECD with DNA-based electrolyte and $-1.1\,mC\,cm^{-2}$ and $\Delta T_{VIS} = 4$–5% for ECD with agar-based electrolyte. These results show that natural macromolecules doped with rare earth triflate are promising candidates to be used as ionically conducting membranes in ECWs.

PMMA incorporated with 1-ethyl-3-methylimidazolium tetrafluoroborate ($EMIMBF_4$) as ionic liquid was employed to form a GPE in a report by Tang et al. [49]. Other components of the electrolyte system include $LiClO_4$ and PC. The composition of the ionic liquid-containing PMMA electrolyte is 1 M $LiClO_4$/PC:PMMA:$EMIMBF_4$ = 8.1 : 0.9 : 1. The electrolyte containing ionic liquid exhibit higher ionic conductivity of $2.9 \times 10^{-3}\,S\,cm^{-1}$ than PMMA-based electrolyte without the ionic liquid and is caused by increased ion concentration from the ionic liquid. ECDs made of WO_3 coated on FTO as the working electrode and counter electrodes, respectively, were fabricated. PMMA-$EMIMBF_4$ electrolyte displayed higher transmittance variation between colored and bleached states, by 10% than PMMA-based electrolyte. Besides the increased quantity of ions, reduced anodic potential from 2.0 to 1.5 V upon addition of the ionic liquid also led to greater modulation range of PMMA-$EMIMBF_4$ electrolyte. CV results showed the lower potential of anodic peak for PMMA-$EMIMBF_4$ as compared to PMMA-based electrolyte. The authors reported that the addition of cathodic species BF_4^- ions from the ionic liquid helped to reduce the anodic potential of the device. Two cathodic peaks were observed for both PMMA-$EMIMBF_4$ and PMMA-based electrolytes, which were related to different binding sites in the WO_3 film [50]. Bleaching of the ECD employing PMMA-$EMIMBF_4$ electrolyte occurred faster as compared to PMMA-based electrolyte. The coloration efficiency (CE) of the ECD with PMMA-$EMIMBF_4$ was found to be $55.3\,cm^2\,C^{-1}$ and was higher

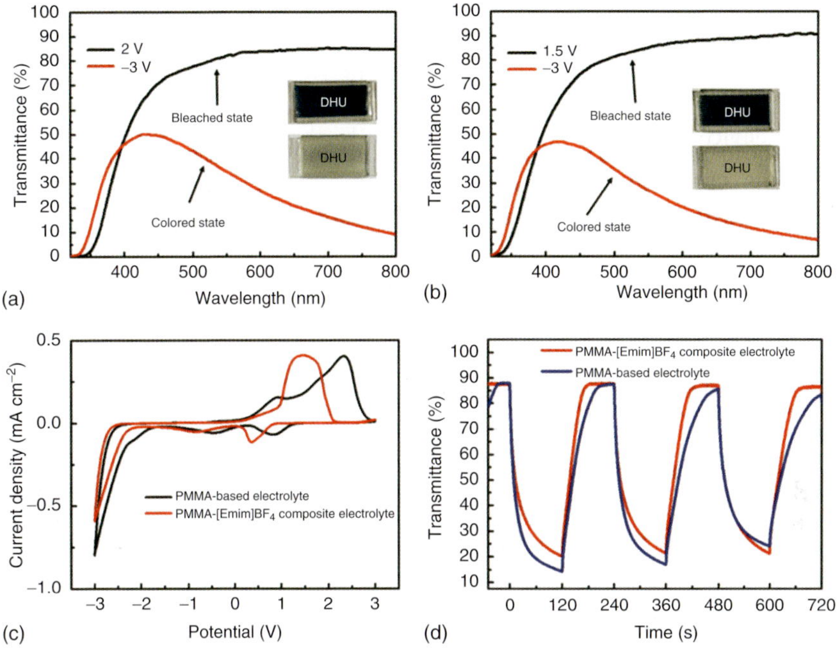

Figure 12.8 Transmittance of (a) PMMA-based electrolyte and (b) PMMA-EMIMBF$_4$ electrolyte; (c) cyclic voltammograms and (d) transmittance at 632.8 nm of ECDs employing PMMA-based electrolyte and PMMA-EMIMBF$_4$ electrolyte. Insets depict the photographs of the colored and bleached ECDs. Source: Tang et al. 2017 [49]. Reproduced with permission of Elsevier.

than that of PMMA-based electrolyte. This phenomenon was related to the higher ionic conductivity of the former, which led to faster ion intercalation in WO_3. Figure 12.8 illustrates the transmittance and cyclic voltammograms of PMMA-based electrolyte and PMMA-EMIMBF$_4$ electrolyte and the ECDs containing each electrolyte.

The electrochromic performances of ECDs using V_2O_5 as primary and WO_3 as complementary electrodes and solid-state membrane electrolyte of 0.3 mol l^{-1} of either LiTFSI or sodium bis-(trifluoromethanesulfonate)-imide (NaTFSI) in 1-butyl-3-methylimidazoliumbis-(trifluoromethanesulfonyl)-imide (BMITFSI) plasticized with PMMA were prepared and electrochemically tested by Mjejri et al. [51]. This ECD displayed good color reversibility between green and orange states upon applied potentials of −1.5 V (green state) and 1.5 V (orange state) for 60 seconds each. The reflectance changes were of 13–25% at 550 nm and changed to 17% and 27.5% after 200 chronoamperometric color/bleaching cycles. The authors concluded that this reflectance modulation is suitable for display application and that the polyol process used to obtain high quality V_2O_5 powder for thin film deposition by Doctor Blade technique.

Poly(4,4′,4″-tris[4-(2-bithienyl)phenyl]amine) (PTBTPA) films were polymerized from TBTPA and supporting electrolyte containing one of two ionic liquids: 1-butyl-3-methyl-imidazolium tetrafluoroborate (BMIMBF$_4$) or 1-butyl-3-methylimidazoliumhexafluorophosphate (BMIMPF$_6$) or tetrabutylammonium perchlorate (TBAP), on ITO conducting glass by Lv et al. [9]. Then these films were assembled in liquid prototype ECDs with electrolytes of either one of the same two ILs or TBAP in dichloromethane/ACN, a platinum (Pt) wire as counter electrode, and Ag/AgCl as the reference electrode. The authors observed that the device $BMIMBF_4$–PTBTPA–$BMIMBF_4$ shows better electrochemical stability when compared to other devices because of smaller BF_4^- doping ion when compared to PF_6^- or ClO_4^-. This smaller BF_4^- impacts less on the polymer film during repeated doping/dedoping processes. They also observed that the thickness of polymer film, i.e. the ion diffusion distance, plays a critical role for the EC switching speed.

Multicolored ECDs were made by Moon et al. [52] from free-standing EC gels. These gels were obtained by blending P(VdF-*co*-HFP), 1-butyl-3-methylimidazolium bis(trifluoromethylsulfonyl)imide [BMI][TFSI] or 1-butyl-3-methylimidazolium tetrafluoroborate [BMI][BF_4^-], heptyl viologen dication (HV^{2+}), and dimethyl ferrocene (dmFc) and assembled to ECDs by using "cut-and-stick" strategy, i.e. by obtaining gel membrane, cutting it in small piece, transferring to ITO conducting glass, and pressing other ITO glass against it. The multicolor properties of the ECDs were achieved by controlling the chemical equilibrium between viologen cation radical monomer (e.g. $HV^{+,\bullet}$) and the dimer ($(HV^{+,\bullet})_2$) by exchanging anions of the ionic liquids or by modifying the chemical structures of the viologen derivatives. The applied potentials were very small ranging from −0.8 to 0.0 V. The authors also proved that their EC gels are promising candidates to fabricate patterned, multicolored, flexible ECD on plastic.

Poly(propylene carbonate) (PPC), $LiClO_4$, and ionic liquid $BMIMBF_4$ were used to obtain electrolytes for ECDs with PANI/DBSA as electrochromic layer by Zhou et al. [13]. The electrochemical characterizations of these devices revealed that both PPC/$BMIM^+BF_4^-$ and PPC/$LiClO_4$/$BMIM^+BF_4^-$ electrolytes exhibit good electrochemical and spectroscopic performances. They showed high contrast ($\Delta A = 0.74$) switched between yellowish green (reduced state and low absorbance) to blue (oxidized state and high absorbance) in potential range of −2.5 to 2.5 V, depending on the electrolyte. They had short coloration and bleaching time and relatively low operation potentials, indicating that this kind of PE is a very promising for electrochemical devices applications.

Wang et al. [53] have designed and fabricated a flexible integrated energy storage smart window that have both an integrated supercapacitor and electrochromic functions. They prepared the electrode by deposition of PEDOT:PSS on PET film and then on the PEDOT:PSS top, an ordered PANI nanowire array. They used H_2SO_4-PVA (poly(vinyl alcohol)) as gel electrolyte layer, which was scrape-coated onto the electrode. To make this device power source independent, the authors proposed to integrate it with conventional solar cell. Therefore, these new devices have the properties of harvesting, storing, and using energy simultaneously.

Ngamaroonchote and Chotsuwan [10] describe a work on gel electrolyte based on cellulose acetate (CA) filled out with lithium perchlorate ($LiClO_4$) and PC as a solvent. After determining the best electrolyte composition, they applied it in ECD with poly(3-hexylthiophene-2,5-diyl) (P3HT) deposited on ITO conducting glass as electrochromic layer and bare ITO as an anodic layer. The ECD with glass/ITO/P3HTfilm/3.0 wt% of CA gel electrolyte/ITO/glass configuration was subjected to dc potentials of +1.0 and −1.0 V for three, five, and one seconds. The color change (ΔT%) of the device from transparent blue to dark purple was approximately 38–40%, measured at 580 nm, and did not change significantly during 1000 color/bleaching cycles.

12.5.3.3 Composite Polymer Electrolyte

PEs that contain high amounts of plasticizers or ionic liquids often suffer from poor dimensional stability. Therefore, ceramic inorganic particles known as fillers are usually added to produce composite PEs with improved mechanical strength.

Pehlivan et al. [54] investigated the incorporation of fumed 0.8 wt% SiO_2 and 0.7 wt% ITO nanoparticles into poly(ethylene imine) (PEI):LiTFSI:PC:EC (13 : 47 : 20 : 20) electrolyte system and studied their performances in ECD based on WO_3 and NiO as complementary electrode. The electrolyte added with SiO_2 showed the best ambient conductivity of 7.2×10^{-5} S cm^{-1}, while the ITO-added electrolyte exhibited a conductivity of 3.38×10^{-5} S cm^{-1}. SiO_2 particles can enhance ionic conductivity, while ITO can absorb near-infrared (IR) radiation to reduce solar energy transmittance while maintaining luminous transmittance. Both SiO_2- and ITO-added electrolytes displayed similar reduced average response time for coloration of the ECD from 74 to 31 seconds, and the average bleaching time remained at ≈146 seconds. The addition of SiO_2 and ITO into PEI-based electrolyte system slightly increased the transmittance variation between the colored and bleached states measured at 550 nm.

In the work of Chen et al. [55], PEO/PVdF blend in the weight ratio of 85 : 15 was employed to develop GPEs, which showed potential to be applied in ECD. The authors fixed the $LiClO_4$ salt content at 10 wt% and varied the contents of nanosized TiO_2. *N*-*N*-dimethylacetamide (DMAC) was used as the solvent in the PE system. The decomposition temperature of PEO/PVdF blend was found to be 390 °C, which is between that of PEO (375 °C) and PVdF (460 °C). XRD results showed that blending PEO with PVdF reduced the degree of crystallinity. The addition of $LiClO_4$ and increasing the doping of TiO_2 up to 2% into the PE system also continuously reduced the crystalline region. The absence of phase separation as observed in scanning electron microscopy (SEM) images of the polymer blend indicated good compatibility between PEO and PVdF. There is no correlation between the degree of crystallinity and the conductivity in this work, as the best conductivity of 6.98×10^{-6} S cm^{-1} was obtained in the polymer blend sample containing $LiClO_4$ with 0.5 wt% TiO_2. The conductivity decreased continuously when increasing TiO_2 contents higher than 0.5 wt% was incorporated. The authors attributed the drop-in conductivity above 0.5 wt% TiO_2 to increased ion–ion interaction, which blocked Li^+ ion transport. Blending of PVdF into PEO and addition of TiO_2 into the electrolytes improved the mechanical strength. Optical studies showed that the electrolytes exhibit high percentage transmittance between 89% and 96%, although increasing doping of TiO_2 lowered the transmittance. The authors did not test the electrolytes in ECD application.

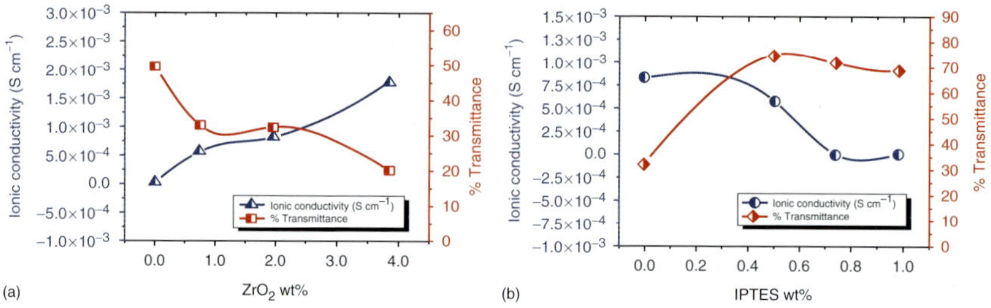

Figure 12.9 Conductivity and transmittance at 739 nm of PVdF-HFP-based electrolyte with respect to (a) ZrO_2 content and (b) IPTES content. Source: Puguan et al. 2016 [11]). Reproduced with permission of Elsevier.

Puguan et al. [11] investigated PVdF-HFP incorporated with $LiCF_3SO_3$, PC, DMF, and silane-functionalized zirconium oxide (ZrO_2) in PEs. Different amounts of unfunctionalized ZrO_2 filler was added with the aim to reduce the formation of crystallites in the polymer in order to increase the electrolyte's ionic conductivity. Addition of ZrO_2 increased the conductivity from 2.47×10^{-5} to 1.78×10^{-3} S cm^{-1} at 3.85 wt%. Reduced crystallinity of PVdF-HFP upon addition of ZrO_2 was evidenced by the disappearance of crystalline XRD peaks at $2\theta = 19.9°$ and 26.4° and reduced intensity of peak at $2\theta = 18.4°$. The addition of ZrO_2 into the electrolytes lowered the optical transmittance, which was attributed to light scattering due to aggregation of ZrO_2 nanoparticles. In order to enhance the optical properties of the electrolytes, the authors functionalized ZrO_2 using varied contents of 3-isocyanatopropyltriethoxysilane (IPTES) in order to reduce aggregation of the nanoparticles. Figure 12.9a,b illustrates the conductivity and transmittance as a function of ZrO_2 and IPTES, respectively. The transmittance of the electrolyte added with silane-functionalized ZrO_2 improved by around 40% at 739 nm upon addition of 0.5 wt% IPTES. This phenomenon was attributed to reduced aggregation of ZrO_2 caused by the covalent bonding of the trialkoxy group of IPTES ligand with the hydroxyl group of ZrO_2 [56]. Above 0.5 wt% IPTES, excess of IPTES caused the transmittance and the ionic conductivity to drop. ECD with the configuration glass|ITO|PANI:DBSA|electrolyte|ITO|glass switched from blue at −2.9 V to yellowish green at +2.9 V. Li^+ ions are inserted/extracted into/from the PANI:DBSA layer. The coloration time was 25 seconds, while the device took 36 seconds to bleach.

Faure et al. [4] developed a PE system employing LiTFSI salt as source of ion and varied the contents of submicron-sized TiO_2 particles as filler. Improvement of the ionic conductivity was achieved in TiO_2-doped electrolytes. The best conducting electrolyte was obtained at 2.5 wt% TiO_2 loading. In this work, ECD employing modified PEDOT known as PXDOT and copper foil (Cu) as the working electrode and counter electrode, respectively, was used. TiO_2 was utilized to mask the copper foil and to improve the visual color difference of PXDOT. The reflectance of the electrolytes increased from nearly zero to reach 58.4% at 25 wt% TiO_2 loading. The reflectance further increased upon addition of TiO_2 up to 10 wt% loading. The electrolyte containing 10 wt% of TiO_2 was chosen to be tested in ECDs as it helped to mask the defects produced during cycling of the device. The ionic conductivity of this electrolyte was found to be 1.0×10^{-5} S cm^{-1}. The authors studied the effect of three different potential ranges, namely, [−0.8, 1.3 V], [−0.9, 1.4 V] and [−1.1, 1.6 V], on the ECDs on the capacity and stability of the devices. Increasing the potential led to higher maximum capacity, while reducing the potential enhanced the stability of the devices by limiting capacity losses. The device cycled at the lowest potential range was able to withstand up to 296 000 cycles. This work demonstrated the importance of modulating the potential applied to ECDs in order to improve the performance and cycle lifetime.

12.6 Present ECDs Uses/Applications

Both electrochromic thin film and polymer or gel electrolyte are important for the development of modern and all-solid-state ECDs not only because of their aesthetic and technological value but also because of possibility of energy savings. These devices modulate visible light transmission, glare, solar radiation, and privacy while allowing views, so they can be installed as windows in architecture, vehicles, trains, and aircrafts glazing to dim the airplane windows [2], or sunglasses [57]. They can also work as displays that control the temperature of frozen food [58]. For instance, there are two applications that can be seen easily: the first one is as electrochromic rear-view mirrors that are already used in modern cars to glare attenuation during night driving [2] and the second one is as windows in Boeing 787 DreamLiner. There are also some exclusive buildings that have ECWs, and one of them is a Stadtsparkasse Bank in Dresden [5, 59]. These windows, installed in 1999, are based on WO_3 and FTO coatings and were supplied by Flaber Gmbh, owner of Pilkington Glass's ECD Technology. Other suppliers as Schott Glass Singapore and Asahi Glass in Japan also have such devices installed in some buildings over the world [5], and new companies such as Ynvisible [60] in Portugal have been established to compete in this new and promising market.

References

1 Deb, S.K. (1969). A novel electrophotographic system. *Appl. Opt. Suppl.* 3: 192–195.
2 Gentex (2013). The mirror as a module. http://www.gentex.com/automotive/mirror-module (accessed 11 June 2013).
3 Boening 787. (2007). DreamLiner. https://www.boeing.com/commercial/787/by-design/#/bigger-windows (accessed 9 December 2009).
4 Faure, C., Guerfi, A., Dontigny, M. et al. (2016). High cycling stability of electrochromic devices using a metallic counter electrode. *Electrochim. Acta* 214: 313–318.
5 Monk, P.M.S., Mortimer, R.J., and Rosseinsky, D.R. (2007). *Electrochromism and Electrochromic Devices.* New York, NY: Cambridge University Press.

6 Alves, R., Sentanin, F., Sabadini, R.C. et al. (2016). Influence of cerium triflate and glycerol on electrochemical performance of chitosan electrolytes for electrochemical devices. *Electrochim. Acta* 217: 108–116.
7 Assis, L.M.N., Leones, R., Kanicki et al. (2016). Prussian blue for electrochromic devices. *J. Electroanal. Chem.* 777: 33–39.
8 Assis, L.M.N., Ponez, L., Januszko, A. et al. (2013). A green-yellow reflective electrochromic device. *Electrochim. Acta* 111: 299–304.
9 Lv, X., Yan, S., Dai, Y. et al. (2015). Ion diffusion and electrochromic performance of poly(4,4′4″-tris[4-(2-bithienyl)phenyl]amine) based on ionic liquid as electrolyte. *Electrochim. Acta* 186: 85–94.
10 Ngamaroonchote, A. and Chotsuwan, C. (2016). Performance and reliability of cellulose acetate-based gel electrolyte for electrochromic devices. *J. Appl. Electrochem.* 46: 575–582.
11 Puguan, J.M.C., Chung, W.-J., and Kim, H. (2016). Ion-conductive and transparent PVdF-HFP/silane-functionalized ZrO_2 nanocomposite electrolyte for electrochromic applications. *Electrochim. Acta* 196: 236–244.
12 Saikia, D., Pan, Y.-C., Wu, C.-G. et al. (2014). Synthesis and characterization of a highly conductive organic-inorganic hybrid polymer electrolyte based on amine terminated triblock polyethers and its application in electrochromic devices. *J. Mater. Chem. C* 2: 331–343.
13 Zhou, D., Zhou, R., Chen, C. et al. (2013). Non-volatile polymer electrolyte based on poly(propylene carbonate), ionic liquid, and lithium perchlorate for electrochromic devices. *J. Phys. Chem. B* 117: 7783–7789.
14 Cholant, C., Westphal, T., Balboni, R. et al. (2017). Thin films of V_2O_5/MoO_3 and their applications in electrochromism. *J. Solid State Electrochem.* 21 (5): 1509–1515.
15 Dinh, N.N., Oanh, N.T.T., Long, P.D. et al. (2003). Electrochromic properties of TiO_2 anatase thin films prepared by a dipping sol–gel method. *Thin Solid Films* 423: 70–76.
16 Avellaneda, C.O. and Pawlicka, A. (1998). Preparation of transparent CeO_2-TiO_2 coatings for electrochromic devices. *Thin Solid Films* 335: 245–248.
17 Al-Kahlout, A., Pawlicka, A., and Aegerter, M. (2006). Brown coloring electrochromic devices based on NiO-TiO_2 layers. *Sol. Energy Mater. Sol. Cells* 90: 3583–3601. https://doi.org/10.1016/j.solmat.2006.06.053.
18 Schmitt, M., Heusing, S., Aegerter, M.A. et al. (1998). Electrochromic properties of Nb_2O_5 sol-gel coatings. *Sol. Energy Mater. Sol. Cells* 54: 9–17. https://doi.org/10.1016/S0927-0248(97)00218-3.
19 Assis, L.M.N., Sabadini, R.C., Santos, L.P. et al. (2015). Electrochromic device with Prussian blue and HPC-based electrolyte. *Electrochim. Acta* 182: 878–883.
20 Alves, R., Sentanin, F., Sabadini, R.C. et al. (2017). Innovative electrolytes based on chitosan and thulium for solid state applications: synthesis, structural and thermal characterization. *J. Electroanal. Chem.* 788: 156–164.
21 Pawlicka, A., Grote, J., Kajzar, F. et al. (2012). Agar and DNA bio-membranes for electrochromic devices applications. *Nonlinear Opt. Quantum Opt.* 45: 113–129.

22 Sim, L.N., Sentanin, F.C., Pawlicka, A. et al. (2017). Development of polyacrylonitrile-based polymer electrolytes incorporated with lithium bis(trifluoromethane)sulfonimide for application in electrochromic device. *Electrochim. Acta* 229: 22–30.
23 Wang, J.-Y., Wang, M.-C., and Jan, D.-J. (2017). Synthesis of poly(methyl methacrylate)-succinonitrile composite polymer electrolyte and its application for flexible electrochromic devices. *Sol. Energy Mater. Sol. Cells* 160: 476–483.
24 Ramadan, R., Elshorbagy, M.H., Kamal, H. et al. (2017). Preparation and characterizatioin of protonic solid electrolyte applied to a smart window device with high optical modulation. *Optik* 135: 85–97.
25 Przyluski, J. et al. (1993). Polymeric electrolytes - old problems and new approaches. *Mater. Sci. Forum* 122: 131–148.
26 Reiter, J., Krejza, O., and Sedlarikova, M. (2009). Electrochromic devices employing methacrylate-based polymer electrolytes. *Sol. Energy Mater. Sol. Cells* 93: 249–255.
27 Inganas, O., Carlberg, C., and Yohannes, T. (1998). Polymer electrolytes in optical devices. *Electrochim. Acta* 43: 1615–1621.
28 Neto, M.J., Sentanin, F., Esperança, J.M.S.S. et al. (2015). Gellan gum – ionic liquid membranes for electrochromic device application. *Solid State Ionics* 274: 64–70.
29 Leones, R., Fernandes, M., Sentanin, F. et al. (2014). Ionically conducting Er^{3+}-doped DNA-based biomembranes for electrochromic devices. *Electrochim. Acta* 120: 327–333.
30 Sim, L.N., Majid, S.R., and Arof, A.K. (2012). Characteristics of PEMA/PVdF-HFP blend polymeric gel films incorporated with lithium triflate salt in electrochromic device. *Solid State Ionics* 209–210: 15–23.
31 Tan, W.C., Alias, S.S., Ismail, A.B., and Mohamad, A.A. (2012). Effect of styrene-acrylonitrile content on 0.5 M NaI/0.05 M I_2 liquid electrolyte encapsulation for dye-sensitized solar cells. *J. Solid State Electrochem.* 16: 2103–2112.
32 Wang, H.X., Liu, X.Z., Wang, Z.X. et al. (2006). Effect of iodine addition on solid-state electrolyte LiI/3-hydroxypropionitrile (1:4) for dye-sensitized solar cells. *J. Phys. Chem. B* 110: 5970–5974.
33 De Freitas, I.N., Goncalves, A.D., de Paoli, M.A. et al. (2008). The role of gel electrolyte composition in the kinetics and performance of dye-sensitized solar cells. *Electrochim. Acta* 53: 7166–7172.
34 Bunzli, J.-C.G. and Piguet, C. (2005). Taking advantage of luminescent lanthanide ions. *Chem. Soc. Rev.* 34: 1048–1077.
35 Wu, X., Wang, Q., Zhang, W. et al. (2016). Preparation of all-solid-state supercapacitor integrated with energy level indicating functionality. *Synth. Met.* 220: 494–501.
36 Pereira, R.F.P., Donoso, J.P., Magon, C.J. et al. (2016). Ion conducting and paramagnetic d-PLC(530)/siloxane-based biohybrids doped with Mn^{2+} ions. *Electrochim. Acta* 211: 804–813.
37 Sentanin, F.C., Ponez, L., and Pawlicka, A. (2014). Electrochromic windows with PVB electrolytes. *Mol. Cryst. Liq. Cryst.* 604: 107–116.

38 Lin, Y., Lai, Y., Yuan, C. et al. (2013). A wider temperature range polymer electrolyte for all-solid-state lithium ion batteries. *RSC Adv.* 3: 10722–10730.

39 Mîndroiu, M., Zgârian, R.G., Kajzar, F. et al. (2015). DNA-based membranes for potential applications. *Ionics* 21: 1381–1390.

40 Zgarian, R.G., Tihan, G.T., Kajzar, F. et al. (2017). Chromophore doped DNA based solid polymer electrolyte for electrochromic devices. *Arabian J. Chem.* 10: 232–239.

41 Singh, T.B., Sariciftci, N.S., and Grote, J.G. (2010). Bio-organic optoelectronic devices using DNA. *Org. Electron.* 223: 189–212.

42 Lakowicz, J.R., Piszczek, G., and Kang, J.S. (2001). On the possibility of long-wavelength long-lifetime high quantum-yield luminophores. *Anal. Biochem.* 228: 62–75.

43 Ledwon, P., Andrade, J.R., Lapkowski, M., and Pawlicka, A. (2015). Hydroxypropyl cellulose-based gel electrolyte for electrochromic devices. *Electrochim. Acta* 159: 227–233.

44 Zhang, F., Dong, G., Liu, J. et al. (2017). Polyvinyl butyral-based gel polymer electrolyte films for solid-state laminated electrochromic devices. *Ionics*: 1–10.

45 Costa, C., Mesquita, I., Andrade, L., and Mendes, A. (2016). Photoelectrochromic devices: Influence of device architecture and electrolyte composition. *Electrochim. Acta* 219: 99–106.

46 Solovyev, A.A., Zakharov, A.N., Rabotkin, S.V., and Kovsharov, N.F. (2016). Electrochromic device with polymer electrolyte. *J. Electron. Mater.* 45 (8): 3866–3871.

47 Lu, H.-C., Kao, S.-Y., Chang, T.-H. et al. (2016). An electrochromic device based on Prussian blue, self-immobilized vinyl benzyl viologen, and ferrocene. *Sol. Energy Mater. Sol. Cells* 147: 75–84.

48 Leones, R., Sentanin, F., Rodrigues, L.C. et al. (2012). Investigation of polymer electrolytes based on agar and ionic liquids. *eXPRESS Polym. Lett.* 6 (12): 1007–1016.

49 Tang, Q., Li, H., Yue, Y. et al. (2017). 1-Ethyl-3-methylimidazolium tetrafluoroborate-doped high ionic conductivity gel electrolytes with reduced anodic reaction potentials for electrochromic devices. *Mater. Des.* 118: 279–285.

50 Surca Vuk, A., Jovanovski, V., Pollet-Villard, A. et al. (2008). Imidazolium-based ionic liquid derivatives for application in electrochromic devices. *Sol. Energy Mater. Sol. Cells* 92: 126–135.

51 Mjejri, I., Manceriu, L.M., Gaudon, M. et al. (2016). Nano-vanadium pentoxide films for electrochromic displays. *Solid State Ionics* 292: 8–14.

52 Moon, H.C., Kim, C.-H., Lodge, T.P., and Frisbie, C.D. (2016). Multicolored, low-power, flexible electrochromic devices based on ion gels. *Appl. Mater. Interfaces* 8: 6252–6260.

53 Wang, K., Wu, K., Meng, Y. et al. (2012). Integrated energy storage and electrochromic function in one flexible device: an energy storage smart window. *Energy Environ. Sci.* 5: 8384–8389.

54 Pehlivan, I.B., Marsal, R., Pehlivan, E. et al. (2014). Electrochromic devices with polymer electrolytes functionalized by SiO_2 and In_2O_3:Sn nanoparticles:

rapid coloring/bleaching dynamics and strong neara-infrared absorption. *Sol. Energy Mater. Sol. Cells* 126: 241–247.
55 Chen, P., Liang, X., Wang, J. et al. (2016). PEO/PVDF-based gel polymer electrolyte by incorporating nano-TiO_2 for electrochromic glass. *J. Sol-Gel Sci. Technol.* 81 (3): 850–858.
56 Luo, K., Zhou, S., Wu, L., and Gu, G. (2008). Dispersion and functionalization of nonaqueous synthesized zirconia nanocrystals via attachment of silane coupling agents. *Langmuir* 24: 11497–11505.
57 Murph, D. (2007). Electrochromic sunglasses change color on demand. http://www.engadget.com/2007/03/28/electrochromic-sunglasses-change-color-on-demand (accessed 28 July 2017).
58 Meulenkamp, E.A. (1997). Mechanism of WO_3 electrodeposition from peroxy-tungstate solution. *J. Electrochem. Soc.* 144 (5): 1664–1671.
59 Opto&Laser Europe (1999). Bank fits climate-control window, November 1999, p. 6.
60 Ynvisible (2017). Printed Electrochromic Displays. http://www.ynvisible.com (accessed 28 July 2017).

Index

Polymer Electrolytes: Characterization Techniques and Energy Applications,
First Edition. Edited by Tan Winie, Abdul K. Arof, and Sabu Thomas.

q

t